GENETICS

GENETICS

Critical Concepts in Social and Cultural Theory

Edited by Nanneke Redclift and Sahra Gibbon

Volume III

Genomic Applications: Consent, Ownership and Control

Routledge
Taylor & Francis Group
LONDON AND NEW YORK

First published 2006
by Routledge
2 & 4 Park Square, Milton Park, Abingdon, Oxon, OX14 4RN

Simultaneously published in the USA and Canada
by Routledge
270 Madison Avenue, New York, NY 10016, USA

*Routledge is an imprint of the Taylor & Francis Group,
an informa business*

Typeset in Times by RefineCatch Limited, Bungay, Suffolk
Printed and bound in Great Britain by TJI Digital, Padstow, Cornwall

British Library Cataloguing in Publication Data
A catalogue record for this book is available from the British Library

Library of Congress Cataloging in Publication Data
A catalog record for this book has been requested

ISBN10: 0–415–34230–9 (Set)
ISBN10: 0–415–34233–3 (Volume III)

ISBN13: 978–0–415–34230–8 (Set)
ISBN13: 978–0–415–34233–9 (Volume III)

Publisher's Note

References within each chapter are as they appear in the original complete work.

CONTENTS

ACKNOWLEDGEMENTS

The publishers would like to thank the following for permission to reprint their material:

The University of Chicago Press for permission to reprint John Maynard Smith, 'The concept of information in biology', *Philosophy of Science*, 67(2) (2000): 177–194. Copyright © 2000 University of Chicago Press.

The University of Chicago Press for permission to reprint Paul E. Griffiths, 'Genetic information: a metaphor in search of a theory', *Philosophy of Science*, 68(3) (2001): 394–412. Copyright © 2001 the University of Chicago Press.

The University of Chicago Press for permission to reprint Eva Jablonka, 'Information: its interpretation, its inheritance and its sharing', *Philosophy of Science*, 69(4) (2002): 578–605. Copyright © 2002 the University of Chicago Press.

Columbia University Press for permission to reprint B. Parry, 'Back to the future', from *Trading the Genome: Investigating the Commodification of Bio-information*, New York: Columbia University Press, pp. 349–265. Copyright © 2004 Columbia University Press. Reprinted with permission of the publisher.

Sage Publications for permission to reprint Adam Hedgecoe and Paul Martin, 'The drugs don't work: expectations and the shaping of pharmacogenetics', reproduced with permission from *Social Studies of Science*, 33(3) (2003): 327–364. Copyright © 2003 Sage Publications, by permission of Sage Publications Ltd.

The University of Chicago Press for permission to reprint Gísli Pálsson and Kristín E. Harðardóttir, 'For whom the cell tolls: debates about biomedicine', *Current Anthropology*, 43(2) (2002): 271–301. Copyright © 2002 the University of Chicago Press.

Taylor & Francis Books for permission to reprint Oonagh Corrigan, 'Informed consent: the contradictory ethical safeguards in pharmacognetics', in O. Corrigan and R. Tutton (eds), *Genetic Databases: Socio-ethical Issues in the Collection and Use of DNA*, London: Routledge, 2004, pp. 78–96.

Taylor & Francis Journals for permission to reprint Brian Rappert, 'Biological weapons, genetics and social analysis: emerging responses, emerging issues – 1', *New Genetics and Society*, 22(2) (2003): 201–213. www.tandf.-co.uk/journals

Blackwell Publishing for permission to reprint Dorothy Nelkin and Lori Andrews, 'DNA identification and surveillance creep', in P. Conrad and J. Gabe (eds), *Sociological Perspectives on the New Genetics*, Oxford: Blackwell, 1999, pp. 191–208.

Taylor & Francis Journals for permission to reprint Richard Tutton, 'Gift relationships in genetics research', *Science as Culture*, 11(4) (2002): 523–542. www.tandf.co.uk/journals

Annual Reviews for permission to reprint Henry T. Greely, 'Legal, ethical and social issues in human genome research'. Reprinted with permission from the *Annual Review of Anthropology*, 27. Copyright © 1998 by Annual Reviews, www.annualreviews.org

Robert Herdt for permission to reprint Robert W. Herdt, 'Enclosing the global plant genetic commons', paper prepared for delivery at the China Centre for Economic Research, 24 May 1999, pp. 1–19.

Blackwell Publishing for permission to reprint Alain Pottage, 'The inscription of life in law: genes, patents and biopolitics', in R. Brownsword, W. R. Cornish and M. Llewelyn (eds), *Law and Human Genetics: Regulating a Revolution*, Oxford: Hart, 1998, pp. 148–173.

Sage Publications Inc, for permission to reprint Kathryn C. Packer and Andrew Webster, 'Patenting culture in science: reinventing the scientific wheel of credibility', *Science, Technology and Human Values*, 21(4) (1996): 427–53. Copyright © 1996 Sage Publications Inc. Reprinted by permission of Sage Publications Inc.

Sage Publications for permission to reprint Sarah Franklin, 'Culturing biology: cell lines for the second millennium', reproduced with permission from *Health*, 5(3) (2001): 321–340. Copyright © 2001 Sage Publications. Reprinted with permission.

The University of Chicago Press for permission to reprint Shane Greene, 'Indigenous people incorporated? Culture as politics, culture as property in pharmaceutical bio prospecting', *Current Anthropology*, 45(2) (2004): 211–237. Copyright © 2004 the University of Chicago Press.

The University of California Press for permission to reprint Cori Hayden, 'From market to market: bioprospecting's idioms of inclusion', *American Ethnologist*, 30(3) (2003): 359–371.

Taylor & Francis Books for permission to reprint Donna Haraway, 'Genome', in *Modest_Witness@Second_Millennium. FemaleMan©_Meets_OncoMouse*™, London: Routledge, 1997, pp. 244–265.

Disclaimer

Part 1

THE INFORMATIONAL GENE AS A RESOURCE

38

THE CONCEPT OF INFORMATION IN BIOLOGY

John Maynard Smith

Source: *Philosophy of Science* 67(2) (2000): 177–194.

Abstract

The use of informational terms is widespread in molecular and developmental biology. The usage dates back to Weismann. In both protein synthesis and in later development, genes are symbols, in that there is no necessary connection between their form (sequence) and their effects. The sequence of a gene has been determined, by past natural selection, because of the effects it produces. In biology, the use of informational terms implies intentionality, in that both the form of the signal, and the response to it, have evolved by selection. Where an engineer sees design, a biologist sees natural selection.

A central idea in contemporary biology is that of information. Developmental biology can be seen as the study of how information in the genome is translated into adult structure, and evolutionary biology of how the information came to be there in the first place. Our excuse for writing an article concerning topics as diverse as the origins of genes, of cells, and of language is that all are concerned with the storage and transmission of information.

(Szathmáry and Maynard Smith 1995)

Let us begin with the notions involved in classical information theory. . . . These concepts do not apply to DNA because they presuppose a genuine information system, which is composed of a coder, a transmitter, a receiver, a decoder, and an information channel in between. No such components are apparent in a chemical system (Apter and Wolpert 1965). To describe chemical processes with the help of linguistic metaphors like 'transcription' and 'translation' does not alter

> the chemical nature of these processes. After all, a chemical process is not a signal that carries a message. Furthermore, even if there were such a thing as information transmission between molecules, transmission would be nearly noiseless (i.e., substantially nonrandom), so that the concept of probability, central to the theory of information, does not apply to this kind of alleged information transfer.
>
> (Mahner and Bunge 1997)

It is clear from these quotations that there is something to talk about. I shall be concerned only with the use of information concepts in genetics, evolution, and development, and not in neurobiology, which I am not competent to discuss.

1. The information analogy

The colloquial use of informational terms is all-pervasive in molecular biology. Transcription, translation, code, redundancy, synonymous, messenger, editing, proofreading, library—these are all technical terms in biology. I am not aware of any confusions arising because their meanings are not understood. In fact, the similarities between their meanings when referring to human communication and genetics are surprisingly close. One example must suffice. In "proofreading," the sequence of the four bases in a newly synthesized DNA strand is compared with the corresponding sequence of the old strand which acted as a template for its synthesis. If there is a "mismatch" (that is, if the base in the new strand is not complementary to that in the old strand according to the pairing rules, A-T and G-C), then it is removed and replaced by the correct base. The similarity of this process to that in which the letters in a copy are compared—in principle, one by one—with those in the original, and corrected if they differ, is obvious. It is also relevant that in describing molecular proofreading, I found it hard to avoid using the words "rule" and "correct."

Molecular biologists, then, do make use of the information analogy in their daily work. Analogies are used in science in two ways. Occasionally, there is a formal isomorphism between two different physical systems. Over fifty years ago, I worked as an aircraft engineer. One thing we wanted to know, in the design stage, was the mode of mechanical vibration of the future airplane. To find out, we built an electrical analogue, in which the masses of different parts of the structure were represented by the inductances of coils in the circuit, and elasticity by the capacitances of condensers. The vibrations of the circuit then predicted the vibrations of the aircraft. The justification for this procedure is that the equations describing the electrical and mechanical vibrations are identical. In effect, we had built a special-purpose analog computer. I remember being annoyed, later, to discover that I had been talking prose without knowing it.

Cases of exact isomorphism are rather rare. Much commoner is the recognition of a qualitative similarity, useful in giving insight into an unfamiliar system by comparison with a familiar one. A classic example is Harvey's recognition that the heart is a pump: it is unlikely that he would have had this insight had he not been familiar with the engineering use of pumps. A more controversial example is the fact that both Darwin and Wallace ascribe their idea of evolution by natural selection to a reading of Malthus's "An Essay on the Principle of Population." A third and more trivial example is that I was led to invent evolutionary game theory by analogy with classical game theory, which analyzes human behavior: as it happens, the main thing I got out of the analogy was a convenient mathematical notation. The point is that scientists need to get their ideas from somewhere. Most often, biologists get them by analogy with current technology, or sometimes with the social sciences. It is therefore natural that during the twentieth century, they should have drawn analogies from machines that transduce information. The first deliberate use of such an analogy, by August Weismann, occurred towards the end of the last century, and is described below. Of course, as I will demonstrate, if an analogy is only qualitative, it can mislead as well as illuminate.

But first I must address the criticisms by Mahner and Bunge quoted at the start of this article. First, is it true that there is no coder, transmitter, receiver, decoder, or information channel? This sentence does draw attention to some ways in which genetic transcription and translation differ from typical examples of human communication (Figure 1). In the human example, a message is first coded, and then decoded. In the genetic case, although we think of a message in coded form in the mRNA being translated at the ribosome into the amino acid sequence of a protein, it is perhaps odd to think of this 'de'-coding, since it was not 'coded' from protein to mRNA in the first place. I don't think this destroys the analogy between the genetic case and the second part of the human sequence. But it does raise a hard question. If there is 'information' in DNA, copied to RNA, how did it get there? Is there any analogy between the origins of the information in DNA and in Morse code? Perhaps there is. In human speech, the first 'coder' is the person who converts a meaning into a string of phonemes, later converted to Morse code. In biology, the coder is natural selection. This parallel may seem far-fetched, or even false, to a non-Darwinist. But it is natural selection which, in the past, produced the sequence of bases, out of many possible sequences, which, via the information channel just described, specifies a protein that has a "meaning," in the sense of functioning in a way that favors the survival of the organism. Where an engineer sees design, a biologist sees natural selection.

What of the claim that a chemical process is not a signal that carries a message? Why not? If a message can be carried by a sound wave, an electromagnetic wave, or a fluctuating current in a wire, why not by a set of chemical molecules? A major insight of information theory is that the same information

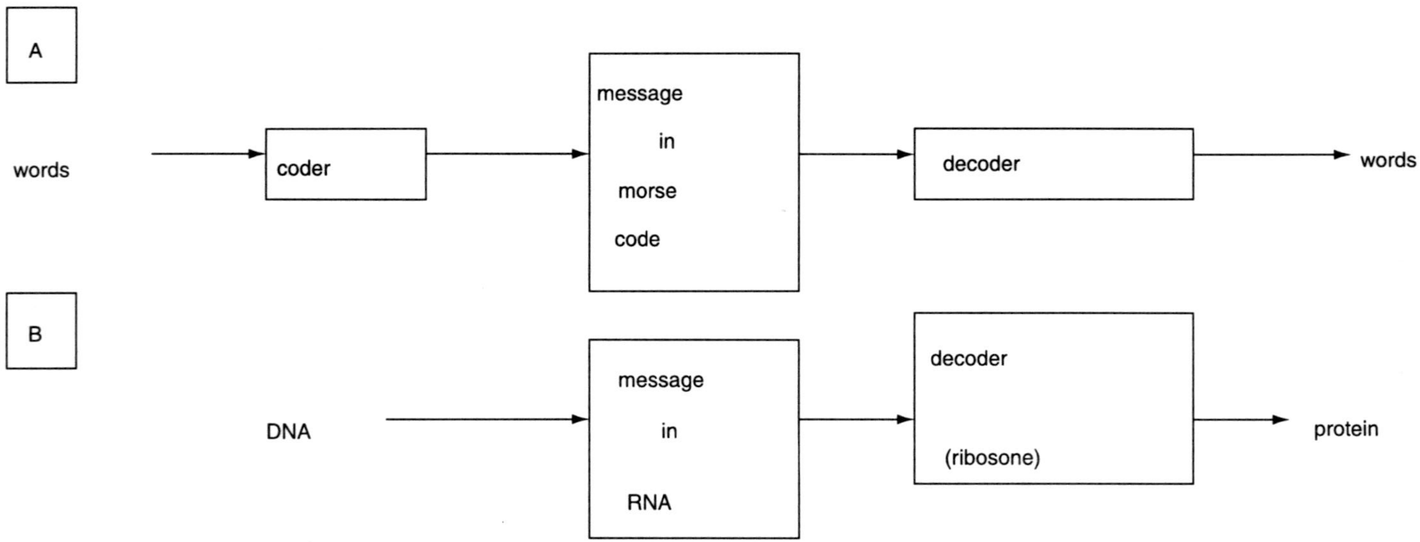

Figure 1 Comparison of A, transmission of a human message by Morse Code, and B, translation of a message coded in DNA into the amino acid sequence of a protein.

can be transmitted by different physical carriers. So far, engineers have not used chemical carriers, essentially because of the difficulty of getting information into and out of a chemical medium. The living world has solved this problem.

Finally, what of the objection that the concept of probability is central to information theory, but missing in biological applications? One could as well argue that information cannot be transmitted by the printed word, because typesetting is virtually noiseless. In information theory, Shannon's (1948) measure of quantity of information, $\Sigma p \log p$, is a measure of the *capacity* of a channel to transmit information, given by the number of different messages that could have been sent. The probabilistic aspects of Shannon's theory have been used in neurobiology, but rarely in genetics, because we can get most of what we need from an assumption of equiprobability. Given a string of n symbols, each of which can be any one of four equally likely alternatives, Shannon's measure gives $2n$ bits of information. In the genetic message, there are four alternative bases. If they were equally likely, and if each symbol was independent of its neighbors, the quantity of information would be two bits per base. In fact, the bases are not equally likely, and there are correlations between neighbors, so there is some reduction in quantity of information, but it is not very great, and is usually ignored: a greater reduction results from the redundancy of the code. In brief we do not bother with Shannon's measure, because 2 bits per base is near enough, but we could if we wanted to. As it happens, Gatlin (1972) wrote a whole book applying Shannon's measure to the genetic message. I'm not sure that much came from her approach, but at least it shows that the concept of probability does apply to the genetic code. There is a formal isomorphism, not merely a qualitative analogy.

There are difficulties in applying information theory in genetics. They arise principally, not in the transmission of information, but in its meaning. This difficulty is not peculiar to genetics. In the early days, it was customary to assert that the theory was not concerned with meaning, but only with quantity of information: as Weaver (in Shannon and Weaver 1949) put it, "this word 'information' in communication theory relates not so much to what you do say, as to what you could say." In biology, the question is, how does genetic information specify form and function?

I now describe five attempts, varyingly successful, to apply concepts of information in biology, ending with the problem of biological form. Then, in the concluding section, I use the analogy between evolution and engineering design by genetic algorithms to suggest how ideas drawn from information theory can be applied in biology.

2. Weismann and the non-inheritance of acquired characters

Weismann's assertion that acquired characters are not inherited is one of the decisive moments in the history of evolutionary biology. Darwin himself believed in "the effects of use and disuse." What led Weismann to such a counter-intuitive notion? Until I happened, rather by chance, to read *The Evolution Theory* (Weismann 1904), I thought that his reasons were, first, that the germ line is segregated early from the soma, and second, that if you cut the tails off mice, their offspring have normal tails. I thought these were poor reasons. There is no segregation of germ line and soma in plants, yet they are no more likely than animals to transmit acquired characters; and in any case all the material and energy for the growth of the germ cells comes via the soma, so what prevents the soma from affecting the germ cells? As to the mouse tails, this is not the kind of acquired character that one would expect to be transmitted.

I had, of course, done Weismann an injustice. There are two long chapters in *The Evolution Theory* devoted to the non-inheritance of acquired characters. The one argument not used in these chapters is the segregation of the germ line: this was important to Weismann for other reasons. His main argument is that there are many traits that are manifestly adaptive, but that could not have evolved by Lamarckian means, because they could not have arisen as individual adaptations in the first place: an example is the form of an insect's cuticle, which is hardened before it is used, and which therefore cannot adapt during an individual lifetime. It follows that adaptations can evolve without Lamarckian inheritance. But this does not prove that acquired characters are not inherited. His ultimate reason for thinking that they are not was that he could not conceive of a mechanism whereby it could happen. Suppose a blacksmith does develop big arm muscles. How could this influence the growth of his sperm cells, in such a way as to alter the development of an egg fertilized by the sperm, so that the blacksmith's son develops big muscles?

Explaining why he could not imagine such a mechanism, he wrote that the transmission of an acquired character "is very like supposing that an English telegram to China is there received in the Chinese language" (in fact, he uses the telegram analogy twice, in slightly different words). This is remarkable for several reasons. He recognizes that heredity is concerned with the transmission of information, not just of matter or energy. Second, he draws an analogy with a specific information-transducing channel, the telegram. Third, although his insight has been of profound importance for biology, his argument is in a sense fallacious. After all, if a sperm can affect the size of a muscle, why cannot a muscle affect a sperm? In fact, most of the information-transducing machines we use, such as telephones and tape-recorders, transmit both ways; they would not be much use if they did not. But some resemble the genetic system in that they transmit only one way. A CD player converts patterns on a disc into sound, but one cannot produce a new disc by singing at the player. I think that the non-inheritance of acquired characters is a contingent fact, usually but not always true, not a logical necessity. Insofar as it is true, it follows from the "central dogma" of molecular biology, which asserts that information travels from nucleic acids to proteins, but not from proteins to nucleic acids.

What, then, of the tails of the mice? Weismann tells us that, when he first spoke of his idea to a zoological meeting in Germany, people replied, "but this must be wrong: everyone knows that, if the tail of a bitch is docked, her puppies have distorted tails"—an interesting example of what Haldane once called Aunt Jobiska's theorem, "It is a fact the whole world knows." The mouse experiment was performed to refute this objection.

A failure to see that heredity is concerned with information, and that information transfer is often irreversible, has unfortunate consequences, as I know to my cost. As a young man, I was a Marxist and a member of the communist party. This is not something I am proud of, but it is relevant. Philosophically, Marxism is unsympathetic to the notion of a gene which influences development, but is itself unaffected: it is undialectical. I do not suggest that the only reason for Lysenko's views was his Marxism—he had less honorable motives—but I think Marxism must take some of the blame. Certainly, it made me uncomfortable with Weismann's views. I spent some six months carrying out an experiment to test them. The ability of an adult *Drosophila* to withstand high temperatures depends on the temperature at which the egg was incubated. Not surprisingly, I found that the adaptation is not inherited. For me, the exercise was perhaps not a total waste of time.

3. The genetic code

The analogy between the genetic code and human-designed codes such as Morse code or the ASCII code is too close to require justification. But there are some features that are worth noting:

(i) The correspondence between a particular triplet and the amino acid it codes for is arbitrary. Although decoding necessarily depends on chemistry, the decoding machinery (tRNAs, assignment enzymes) could be altered so as to alter the assignments. Indeed, mutations occur that are lethal because they after the assignments. In this sense the code is symbolic—a point I return to later.

(ii) The genetic code is unusual in that it codes for its own translating machinery.

(iii) The scientists who discovered the nature of the code, and of the translating machinery, had the coding analogy constantly in mind, as the vocabulary they used to describe their discoveries makes clear. Occasionally, they were misled by the analogy. An example is the belief that the code would be solved as linear B was deciphered—by discovering the Rosetta stone. What was needed was a protein of known amino acid sequence, specified by a gene of known base sequence. In fact, the code was not decoded that way. Instead, it was decoded using a "translating machine"—a piece of cell machinery which, provided with a piece of RNA of known sequence, would synthesize a peptide whose sequence could be determined. But despite such false trails, the information analogy did lead to the solution. If, instead, the problem had been treated as one of the chemistry of protein-RNA interactions, we might still be waiting for an answer.

In an article I came across only when this paper was almost completed, Sarkar (1996) describes in some detail the history of the idea of a 'comma-free code' (Crick et al. 1957). I agree with him that this proved to be a red herring, although I have suggested elsewhere (Maynard Smith 1999) that it was one of cleverest ideas in the history of science that turned out to be wrong. But it *was* wrong. It illustrates nicely the fact that analogies in science can be misleading as well as illuminating. But I think that Sarkar is over-eager to point to the failures of the information analogy and to play down its successes. For example, he does not explain that the discovery (Crick et al. 1961) of the relationship between DNA and protein—as a triplet code in which the correct 'reading frame' is maintained by accurately counting off in threes, and whose meaning can be destroyed by a 'frame shift' mutation—also arose from the coding analogy. It is intriguing that Francis Crick was one of the authors of both papers. As a second example, Sarkar's argument that the code does not enable one to predict amino acid sequences (because of complications such as introns, variations from the universal code, etc.) is seriously misleading; biologists do it all the time.

(iv) It is possible to imagine the evolution of complex, adapted organisms without a genetic code. Godfrey-Smith (1999) imagines a world in which proteins play the same central role that they play in our world, but in which their amino acid sequence is replicated without coding. In brief, he suggests that proteins could act as templates for themselves, using 20 'connector' molecules, each with two similar ends, one binding to an amino acid in the template, and another to a similar amino acid in a newly synthesized strand. In such a system, there would be no 'code' connecting one set of molecules to another set of chemically different molecules. I agree that such a world is conceivable, and that it lacks a code. I will argue below, however, that the notion of information, and the distinction between genetic and environmental causes in development, would be as relevant in Godfrey-Smith's world as it is in the real world.

4. Symbol and "gratuity"

Jacques Monod's (1971) *Chance and Necessity* did not get a good press from philosophers, particularly in the Anglo-Saxon world. But it contained at least one profound idea, that of gratuité (translated, not happily, as gratuity). Jacob and Monod (1959) had discovered how a gene can be regulated. In effect, a "repressor" protein, made by a second "regulatory" gene, binds to the gene and switches it off. The gene can then be switched on by an "inducer," usually a small molecule, lactose for this particular gene. What happens is that the inducer binds to the regulatory protein, and alters its shape, so that the protein no longer binds to the gene and represses it. The point Monod emphasizes is that the region of the regulatory protein to which the inducer binds is different from the region of the protein that binds to the gene; the inducer has its effect by altering the shape of the protein. The result is that, in principle, any "inducer" molecule could switch on, or off, any gene. Of course, all the reactions obey the laws of chemistry, as they must, but there is no chemical necessity about which inducers regulate which genes. It is this arbitrary nature of molecular biology that Monod calls gratuity.

I think it may be more illuminating to express Monod's insight by saying that, in molecular biology, inducers and repressors are "symbolic": in the terminology of semiotics, there is no necessary connection between their form (chemical composition) and meaning (genes switched on or off). Other features of molecular biology are symbolic in this sense: for example, CAC codes for histidine but there is no chemical reason why it should not code for glycine. (In passing, I have found the semiotic distinction between symbol, icon, and index illuminating also in animal communication (Maynard Smith and Harper 1995).)

Sarkar (1996) has an interesting discussion of Monod's notion of gratuity. He interprets Monod as arguing that 'the cybernetic account of gene regulation is of more explanatory value than a purely physicalist alternative', but says that this opinion is justified only if cases of gene regulation other than the lactose operon studied by Monod turn out to be of a similar nature. He concludes that 'attempts to generalise the operon model to eukaryotic gene regulation have so far shown no trace of success'. I think it would be hard to find a developmental geneticist who would agree with him. As I explain below, Monod's ideas are basic to research in the field.

Linguists would argue that only a symbolic language can convey an indefinitely large number of meanings. I think that it is the symbolic nature of molecular biology that makes possible an indefinitely large number of biological forms. I return to the problem of form later, but first I describe a story of how the information analogy led me up a blind alley, but at the same time prepared me for current discoveries in developmental genetics.

5. The quantification of evolution

Around 1960, I conceived the idea that, using information theory, one could quantify evolution simultaneously at three levels—genetic, selective, and morphological. The genetic aspect is easy: the channel capacity is, approximately, two bits per base. Things are complicated by the presence of large quantifies of repetitive DNA, but this can be allowed for. The selective level is tricky, but not hopeless. Suppose one asks, how much selection is needed to program an initially random sequence? If, reasonably, the selective removal of half the population is regarded as adding 1 bit of information, then 2 bits of selection are needed to program each base. The snag is that evolution does not start from a random sequence. Instead, an already programmed gene (or set of genes) is duplicated, and then one copy is altered by selection. However, one can still make a crude estimate of how much selection, measured in bits, is needed to program an existing genome. Kimura (1961), using Haldane's (1957) idea of the 'cost of selection', gave a more elegant account of how natural selection accumulates genetic information in the genome.

The hard step is to quantify morphology, but before tackling that question, I want to suggest that the quantification of genetic and selective information in the same units has one, perhaps trivial, use. Occasionally someone, often a mathematician, will announce that there has not been time, since the origin of the earth, for natural selection to produce the astonishing diversity and complexity we see. The odd thing about these assertions is that, although they sound quantitative, they never tell us by how much the time would have to be increased: twice as much, or a million times, or what? The only way I know to give a quantitative answer is to point out that, if one estimates, however roughly, the quantity of information in the genome, and the quantity that could have been programmed by selection in 5000 MY, there has been plenty of time. If, remembering that for most of the time our ancestors were microbes, we allow an average of 20 generations a year, there has been time for selection to program the genome ten times over. But this assumes that the genome contains enough information to spcify the form of the adult. This is a reasonable assumption, because it is hard to see where else the information is coming from.

How much information is needed to specify the form of the adult? Clearly, one does not have to specify the nature and position of every atom in the body, because not everything is specified. This suggested that one asks how much information is required to specify those features shared by two individuals of the same genotype—for example, monovular twins. For simplicity, imagine a pair of two-dimensional organisms (it is easy to extend the argument to three dimensions). Form an image of each as a matrix of black and white dots (in effect, pixels: again, one can extend the argument to more than two kinds of pixel). Start with minute pixels: then identical twins will differ. Gradually enlarge the pixels, until the images of identical twins are the same. Then the information required equals the number of pixels in the image.

It is only necessary to describe the method to see what is wrong with it. Imagine three black and-white pictures: the first a pattern of random dots, the second the Mona Lisa, and the third a black circle on a white ground.

The first would indeed require a quantity of information equal to the number of pixels. The Mona Lisa could be described in fewer bits, because of the correlations between neighboring dots, but would still require a lot of information. The circle could be specified by saying, if $(x - a)^2 + (y - b)^2 < r^2$, then black, else white (where ab is the center of the circle, and r its radius). One might argue that this is irrelevant, because genes don't know about coordinate geometry, but this would be a mistake. Most simple forms—and a circle is an example—can be generated by simple physical processes, so that all the genome need do is to specify a few physical parameters: for example, reaction rates can be fixed by specifying enzymes.

The fallacy of the "pixel" line of approach is that the genome is not a description of the adult form, but a set of instructions on how to make it: it is a recipe, not a blueprint.

6. Is the genome a developmental program?

There is, I think, no serious objection to speaking of a genetic code, or to asserting that a gene codes for the sequence of amino acids in a protein. Certainly, a gene requires the translating machinery of a cell—ribosomes, tRNA's, etc.—but this does not invalidate the analogy: a computer program needs a computer before it can do anything. For an evolutionary biologist, the point is that the translating machinery can remain constant in a lineage (although it needs an unchanging genetic program to specify it), yet changes in the genetic program can lead to changes in proteins.

It could be objected that a gene only specifies the amino acid sequence of a protein, but not its three-dimensional folded shape. In most cases, given appropriate physical and chemical conditions, the linear string of amino acids will fold itself up. Folding is a complex dynamic process: it is not yet possible to predict the three-dimensional structure from the sequence. But the laws of chemistry and physics do not have to be coded for by the genes: they are given and constant. In evolution, changes in genes can cause changes in proteins, while the laws of chemistry remain unchanged.

However, an organism is more than a bag of specific proteins. Development requires that different proteins be made at different times, in different places. A revolution is now taking place in our understanding of this process. The picture that is emerging is one of a complex hierarchy of genes regulating the activity of other genes. Today, the notion of genes sending signals to other genes is as central as the notion of a genetic code was forty years ago.

First, an experiment (Halder et al. 1995). There is a gene, *eyeless*, in the mouse. Mutations in this gene (in homozygotes) cause the mouse to develop without eyes, suggesting that the unmutated form of the gene plays some role in eye development. The normal mouse gene has been transferred to the fruitfly, *Drosophila*, and activated at various sites in the developing fly

(Halder et al. 1995). If it is activated in a developing leg, then an eye develops at the site: not, of course, a mouse eye, but a compound fly eye. This suggests that the gene is sending a signal, 'make an eye here'; more precisely, it is locally switching on other genes concerned with eye development.

Why should a mouse gene work in a fly? Presumably, the common ancestor of mouse and fly, some 500 million years ago, had the ancestor of the gene: this is confirmed by the presence in *Drosophila* of a gene with a base sequence very similar to the mouse *eyeless* gene. What was the gene doing in that remote ancestor? We do not know, but a plausible guess is that the ancestor had a pair of sense organs on its head—perhaps one or a small cluster of light-sensitive cells–and that the differentiation of these cells, from undifferentiated epidermal cells, was triggered by the ancestral gene.

This raises questions about the nature of the signals that are passing. I argued above that the inducers and repressors of gene activity are symbolic, in the sense that there is no necessary chemical connection between the nature of an inducer and its effects. In Jacob and Monod's original experiments, genes metabolizing the sugar lactose were switched on by the presence of lactose in the medium. This is obviously adaptive; there would be no point in switching on the genes if there was nothing for them to do. But if it was selectively advantageous for these genes to be switched on by a different sugar, say maltose, then changes in the regulatory genes that brought this about would no doubt have evolved.

Yet the experiment described above suggests that the gene responsible for initiating eye development has been conserved for 500 million years. If genes are symbolic, why should this be so? Words are symbols, and are not conserved. The words used to describe a given object change, so why has not the gene used to elicit an eye changed? The question is made more acute by the fact that signalling genes do sometimes acquire new meanings. In evolution, it often happens that a regulatory gene is duplicated: one copy retains its original function, and the other changes slightly, and acquires a new function. I think that the extreme conservatism of many signalling genes can be explained as follows. Regulatory genes are often arranged hierarchically: gene A controls genes B, C, D . . . and each of B, C and D control yet other genes. Adaptive evolutionary changes are likely to be gradual, and this rules out changes in the initial gene in a regulatory hierarchy. The gene *eyeless*, specifying where an eye is to develop, is likely to be such an initial gene, and so has been conserved. But the point I want to make here is that it is hard even to think about the problem if one does not think of genes sending signals, and if one does not recognize that the signals are symbolic.

To date, then, there is talk of genes 'signalling' to other genes, of the genome 'programming' development, and so on. Informational terminology is invading developmental biology, as it earlier invaded molecular biology. In the next section I try to justify this usage.

7. Evolution theory and the concept of information in biology

I start with a concept of information that has the virtue of clarity, but which would rule out the current usage of the concept in biology. Dretske (1981) argues as follows. If some variable, A, is correlated with a second variable, B, then we can say that B carries information about A; for example, if the occurrence of rain (A) is correlated with a particular type of cloud (B), then the type of cloud tells us whether it will rain. Such correlations depend on the laws of physics, and on local conditions, which Dretske calls 'channel conditions'.

With this definition, there is no difficulty in saying that a gene carries information about adult form; an individual with the gene for achondroplasia will have short arms and legs. But we can equally well say that a baby's environment carries information about its growth; if it is malnourished, it will be underweight. Colloqually, this is fine; a child's environment does indeed predict its future. But biologists draw a distinction between two types of causal chain, genetic and environmental, or 'nature' and 'nurture', for a number of reasons. Differences due to nature are likely to be inherited, whereas those due to nurture are not; evolutionary changes are changes in nature, not nurture; traits that adapt an organism to its environment are likely to be due to nature. For these reasons, the nature-nurture distinction has become fundamental in biology. Of course, the distinction could be drawn without using the concept of information, or applying it specifically to genetic causes. However, as the examples discussed above demonstrate, informational language has been used to characterize genetic as opposed to environmental causes. I want now to try to justify this usage.

I will argue that the distinction can be justified only if the concept of information is used in biology only for causes that have the property of intentionality (Dennett 1987). In biology, the statement that A carries information about B implies that A has the form it does because it carries that information. A DNA molecule has a particular sequence because it specifies a particular protein, but a cloud is not black because it predicts rain. This element of intentionality comes from natural selection.

I start with an engineering analogy. An engineer interested in genetic algorithms wants to devise a program to play a competitive game. For simplicity, he chooses Fox and Geese, a game played on a draughts board in which four 'geese' try to corner a 'fox'. (As it happens, I played with the 'evolution' of a program to play this game as long ago as the 1940s. Without a computer, I could not tackle more difficult games, but Fox and Geese proved easily soluble). He first invents a number of 'rules' for the geese (e.g., keep in line, don't leave gaps, keep opposite the fox). Each rule has one or more parameters (e.g., for the gap rule, specifying the position of any gaps). He then arranges for a bit string to specify these parameters, and the weightings to be given to the different rules when selecting the next move. He then does a typical genetic algorithm experiment, starting with a population of random strings, allowing each to play against an efficient fox, selecting the most successful, and generating a new population of strings, with random mutation. For a simple game like Fox and Geese, he will finish up with a program that wins against any Fox strategy; things are a bit harder for chess.

This procedure is illustrated in Figure 2A. If, instead of using a genetic algorithm approach, the engineer had simply written an appropriate program, no one, I think, would object to saying that the program carried information, or at least instructions, embodying his intentions. By analogy, I want to say that, in the process illustrated in Figure 2A there is information in the bit string, which has been programmed by selection, and not by the engineer. This usage is justified by the fact that, presented with a bit string and the moves that it generated, it would be impossible to tell whether it had been designed by the engineer directly, or by selection between genetic algorithms.

Biological evolution is illustrated in Figure 2B. It differs from 1A in two ways. First, a coding stage is present. Second, selection based on success in the game is replaced by survival and reproduction ('fitness') in a specific environment. I do not think the latter difference is important.

I think that the analogy between figures 2A and 2B justifies biologists in saying that DNA contains information that has been programmed by natural selection; that this information codes for the amino acid sequence of proteins; that, in a much less well understood sense, the DNA and proteins carry instructions, or a program, for the development of the organism; that natural selection of organisms alters the information in the genome; and finally, that genomic information is 'meaningful' in that it generates an organism able to survive in the environment in which selection has acted.

The weakness of these models, both engineering and biological, is that they do not tell us where the 'rules' come from. In the engineering case, the success of the procedure depends on the ingenuity with which the rules were chosen. In the biological case, the rules depend on the laws of physics and chemistry; organisms do not have to invent, or evolve, rules to tell a string of amino acids how to fold up. But there are higher-level rules, depending on the following facts: that cells divide repeatedly; that every cell contains a complete genome; that cells can signal to their neighbors; that genes can be switched on or off by other genes; and that states of gene switching can be transmitted through cell division to daughter cells. Research in developmental biology is concerned with identifying regulatory genes, and with identifying the higher-level rules whose parameters the genes control.

It should now be clear why biologists wish to distinguish between genetic and environmental causes. The environment is represented in Figure 2B by the 'channel conditions'. The laws of physics do not change, but the local environment may do. Fluctuations in the environment are a source of noise in the system, not of information. Sometimes, organisms do adapt to

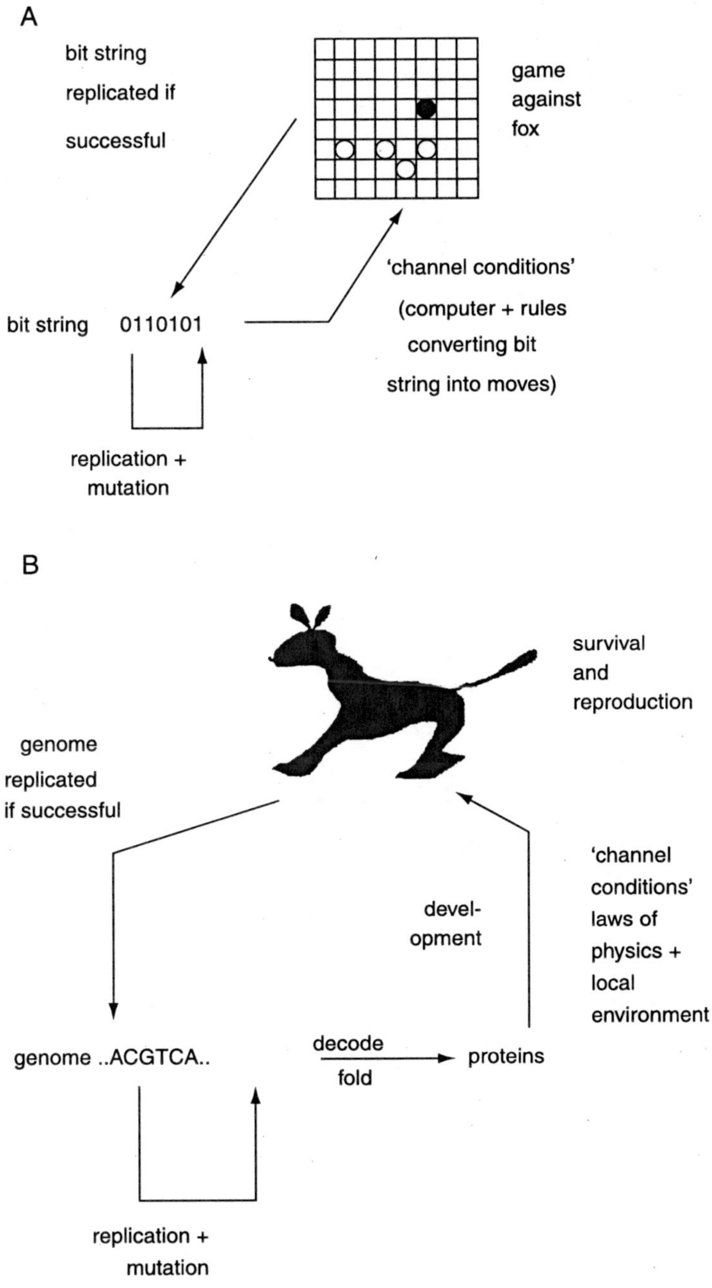

Figure 2 Comparison of **A**, selection of a 'genetic algorithm' to play a game of Fox and Geese, and **B**, biological evolution.

changes in the environment during their lifetime, without genetic evolution. For example, pigment develops in the skin of humans exposed to strong sunlight, protecting against UV. Such adaptive responses require that the genome has evolved under natural selection to cope with a varying environment. What is inherited is not the dark pigment itself, but the genetic mechanism causing it to appear in response to sunlight.

This has been a natural history of the concept of information in biology, rather than a philosophical analysis. The concept played a central role in the growth of molecular genetics. The image of development that is emerging is one of a complex hierarchy of regulatory genes, and of a signalling system that is essentially symbolic. Such a system depends on genetic information, but the way in which that information is responsible for biological form is so different from the way in which a computer program works that the analogy between them has not, I think, been particularly helpful, although it is a lot nearer the truth than the idea that complex dynamic systems will generate biological forms "for free". A less familiar idea that has been central both to molecular biology and to development is Monod's notion of "gratuity", which I think is most clearly expressed by saying that molecular signals in biology are symbolic.

Given the central role that ideas drawn from a study of human communication have played, and continue to play, in biology, it is strange that so little attention has been paid to them by philosophers of biology. I think it is a topic that would reward serious study.

8. Conclusions

In colloquial speech, the word 'information' is used in two different contexts. It may be used without semantic implications; for example, we may say that the form of a cloud provides information about whether it will rain. In such cases, no one would think that the cloud had the shape it did because it provided information. In contrast, a weather forecast contains information about whether it will rain, and it has the form it does because it conveys that information. The difference can be expressed by saying that the forecast has intentionality (Dennett 1987), whereas the cloud does not. The notion of information as it is used in biology is of the former kind; it implies intentionality. It is for this reason that we speak of genes carrying information during development, and of environmental fluctuations not doing so.

A gene can be said to carry information, but what of a protein coded for by that gene? I think one must distinguish between two cases. A protein may have a function directly determined by its structure—for example, it may be a specific enzyme, or a contractile fiber. Alternatively, it may have a regulatory function, switching on or off other genes. Such regulatory functions are arbitrary, or symbolic. They depend on specific receptor DNA sequences, which have themselves evolved by natural selection. The activity of an enzyme depends on the laws of chemistry and on the chemical environment (e.g., the presence of a suitable substrate), but there is no structure which can be thought of as an evolved "receiver" of a "message" from the enzyme. In contrast, the effect of a regulatory protein does depend on an evolved receiver of the information it carries: the *eyeless* gene signals "make an eye here," but only because the genes concerned with making an eye have an appropriate receptor sequence. In the same way, the effect of a gene depends on the cell's translating machinery—ribosomes, tRNAs, and assignment enzymes. For these reasons, I want to say that genes and regulatory proteins carry information, but enzymes do not.

A very similar conclusion about the concept of information in biology has been reached by Sterelny and Griffiths (1999). In particular, they write, "Intentional information seems like a better candidate for the sense in which genes carry developmental information and nothing else does." Justifying this view, they add, "A distinctive test of intentional or semantic information is that talk of error or misrepresentation makes sense." In biology, misrepresentation is possible because there is both an evolved structure carrying the information, and an evolved structure that receives it.

In human communication, the form of a message depends on an intelligent human agent; forecasts are written by humans (or by computers that were programmed by humans), and are intended to alter the behavior of people who read them. There are intelligent senders and receivers. How, then, can a genome be said to have intentionality? I have argued that the genome is as it is because of millions of years of selection, favouring those genomes that cause the development of organisms able to survive in a given environment. As a result, the genome has the base sequence it does because it generates an adapted organisms. It is in this sense that genomes have intentionality. Intelligent design and natural selection produce similar results. One justification for this view is that programs designed by humans to produce a result are similar to, and may be indistinguishable from, programs generated by mindless selection.

References

Apter, M. J. and L. Wolpert (1965), "Cybernetics and Development I. Information Theory", *Journal of Theoretical Biology* 8: 244–257.

Crick, F. H. C., J. S. Griffith and L. E. Orgel (1957), "Codes Without Commas", *Proc. Natl. Acad. Sci. USA* 43: 416.

Crick, F. H. C., L. Barnett, S. Brenner and R. S. Watts-Tobin (1961), "General Nature of the Genetic Code", *Nature* 192: 1227–1232.

Dennett, D. (1987), *The Intentional Stance*. Cambridge, MA: MIT Press.

Dretske, F. (1981), *Knowledge and the Flow of Information*. Cambridge, MA: MIT Press.

Gatlin, L. L. (1972), *Information Theory and the Living System*. New York: Columbia University Press.

Godfrey-Smith, P. (1999), "On the Theoretical Role of Genetic Coding", *Philosophy of Science* (in press).

Haldane, J. B. S. (1957), "The Cost of Natural Selection", *Journal of Genetics* 55: 511–524.

Halder, G., P. Callaerts and W. J. Gehring (1995), "Induction of Ectopic Eyes by Targeted Expression of the *Eyeless* Gene in *Drosophila*", *Science* 267: 1788–1791.

Jacob, F. and J. Monod (1961), "On the Regulation of Gene Activity", *C.S.H. Symp. Quant. Biol.* 26: 193–211.

Kimura, M. (1961), "Natural Selection as a Process of Accumulating Genetic Information in Adaptive Evolution", *Genetical Research, Cambridge* 2: 127–140.

Mahner, M. and M. Bunge (1997), *Foundations of Biophilosophy*. Berlin, Heidelberg, New York: Springer-Verlag.

Maynard Smith, J. (1999), "Too Good To Be True", *Nature* 400: 223.

Maynard Smith, J. and D. G. C. Harper, (1995), "Animal Signals: Models and Terminology", *Journal of Theoretical Biology* 177: 305–311.

Monod, J. (1971), *Chance and Necessity*. New York: Knopf.

Sarkar, S. (1996), "Biological Information: A Skeptical Look at Some Central Dogmas of Molecular Biology", in S. Sarkar (ed.), *The Philosophy and History of Molecular Biology*. Dordrecht: Kluwer Academic Publishers, 187–231.

Shannon, C. E. (1948), "A Mathematical Theory of Communication", *Bell Syst. Tech. J.* 27: 279–423, 623–656.

Shannon, C. E. and W. Weaver (1949), *The Mathematical Theory of Communication*. Urbana: University of Illinois Press.

Sterelny, K. and P. E. Griffiths (1999), *Sex and Death: An Introduction to the Philosophy of Biology*. Chicago: University of Chicago Press.

Szathmáry, E. and J. Maynard Smith (1995), "The Major Evolutionary Transitions", *Nature* 374: 227–232.

Weismann, A. (1904), *The Evolution Theory* (trans. J. A. and M. R. Thomson). London: Edward Arnold.

39

GENETIC INFORMATION

A metaphor in search of a theory

Paul E. Griffiths

Source: *Philosophy of Science* 68(3) (2001): 394–412.

Abstract

John Maynard Smith has defended against philosophical criticism the view that developmental biology is the study of the expression of information encoded in the genes by natural selection. However, like other naturalistic concepts of information, this "teleosemantic" concept is equally applicable to many non-genetic factors in development. Maynard Smith also fails to show that developmental biology is concerned with teleosemantic information. Some other ways to support Maynard Smith's conclusion are considered. It is argued that on any definition of information the view that development is the expression of genetic information is misleading. Some reasons for the popularity of that view are suggested.

1. Introduction: information talk in biology

> An organism's physiology and behaviour are dictated largely by its genes. And those genes are merely repositories of information written in a surprisingly similar manner to the one that computer scientists have devised for the storage and transmission of other information . . .
>
> ("Drowning in Data" 1999, 97)

The only controversial feature of this statement is the claim that behavior is largely genetic. It is conventional wisdom that insofar as the traits of an organism are subject to biological explanation, those traits express information coded in the organism's genes. Conventional wisdom has recently been defended in this journal by the eminent biologist John Maynard Smith

(2000a,b) I will argue, however, that the only truth reflected in the conventional view is that there is a genetic code by which the sequence of DNA bases in the coding regions of a gene corresponds to the sequence of amino acids in the primary structure of one or more proteins. The rest of "information talk" in biology, and the claim that biology "is, itself, an information technology" ("Drowning in Data" 1999, 97), is on a par with the claim that the planets compute their orbits around the sun or that the economy computes an efficient distribution of goods and resources. It a way to talk about correlation that, in some cases, allows a useful application of the mathematical theory of communication and in others plays no theoretical role but merely reflects the current cultural prominence of information technology. Taking "information talk" in biology too seriously is not merely a journalist's error. Many biologists, when asked to talk about their discipline in broad, philosophical terms, would represent it in the same light. Nevertheless, as Sahotra Sarkar has noted:

> there is no clear, technical notion of "information" in molecular biology. It is little more than a metaphor that masquerades as a theoretical concept and . . . leads to a misleading picture of possible explanations in molecular biology.
>
> (Sarkar 1996, 187)

Correcting the conventional wisdom is important in its own right, but it may also clear the ground for an important new approach to the debate over "genetic determinism." Expanding on a remark by Susan Oyama (2000a), Phillip Kitcher has joked that arguing against genetic determinism is like "battling the undead." (Kitcher 2001) The "interactionist consensus" that all traits depend on both genetic and environmental factors has been reaffirmed again and again, but the discovery that genes are involved in the development of a trait continues to be used to infer that the trait will be hard to change by manipulating the environment. This inference persists in the face of widespread agreement that there are always other causal factors involved in the development of "genetic" traits and many examples of "genetic" traits being modified via these factors. For example, genetic diseases can be treated by environmental (i.e., drug or dietary) interventions. "But why, then," Kitcher's frustrated interactionist asks, "do we always end up discussing whether genotypes are all-powerful in development?" (Kitcher 2001, 404) In my view, an important part of the answer has been given by Susan Oyama. (2000a,b) Genetic causation is interpreted deterministically because genes are thought to be a special kind of cause. Genes are instructions—they provide information—whilst other causal factors are merely material. The intuitive notion of information is a semantic notion, carrying the implication that genes, unlike other causal factors, are about, or directed at, the outcomes they help to produce. Little wonder, then, that the gene-trait relationship seems intuitively more context-independent than the relationship between traits and other causes. The temperature of the nest determines the sex of a crocodile, the eggs developing as male at intermediate temperature and female at extreme temperatures, but outside that very particular context the cause—a mound of rotting vegetation with a temperature between *n* and *m* degrees—retains no connection with masculinity. In contrast, even when the informational gene fails to produce its effect, it remains directed at that effect. A "gay gene" is an instruction to be gay even when the person carrying it is straight.

If Oyama is correct, then finding a new way to think about genetic causation is a necessary prerequisite to laying the unquiet ghost of the nature/nurture controversy and learning to think clearly about the interaction of genetic and other factors in development. But the deflation of information talk would be a clarification of the real nature of explanation in molecular biology even if Oyama's larger claim about the causes of genetic determinism is incorrect, as Kitcher has argued. (2001) Hence, the position I will defend here is a limited one that I have elsewhere called the "parity thesis." (Griffiths and Knight 1998) Any defensible definition of information in developmental biology is equally applicable to genetic and non-genetic causal factors in development. Definitions of information on which genes contain developmental information, but methylation patterns or incubation temperatures do not, are illegitimate because they cannot be "naturalized"—they ascribe properties to genes that cannot be grounded in physical and biological facts. The view that genes are distinguished from other factors by being the locus of developmental information is biologically illegitimate. Whether this illegitimate view is an important cause of the persistence of genetic determinism is a question for another day. (Griffiths, forthcoming)

2. Two concepts of information

Concepts of information can be divided into two rough categories: *causal* information concepts and *intentional* information concepts. Causal conceptions of information derive from the mathematical theory of communication. (Shannon and Weaver 1949) Mathematical information theory studies only the quantity of information in a physical system. The quantity of information in a system can be understood roughly as the amount of order in that system, or the inverse of the entropy (disorder) that all closed physical systems accumulate over time. This measure says nothing about the content of information. However, there is a related causal notion of information content. (Dretske 1981) Information flows over a channel connecting two systems, a receiver that contains the information and a sender, the system that the information is about. There is a channel between two systems when the state of one is systematically causally related to the other, so that the state of the sender can be discovered by observing the state of the receiver. The causal information content of a signal is simply the state of affairs with which it reliably correlates at the other end of the channel. Thus, smoke carries information about fire and disease phenotypes carry information about disease genes.

The second category of information concept deals with intentional information or "semantic information." (Godfrey-Smith 1999a,b) It is information in this sense that human thoughts and utterances are supposed to contain. If there is a relationship between intentional information and causal information it is a complex and distant one. We think about things with which we have only the most tenuous causal connection (e.g., undiscovered galaxies) and about things that do not exist (e.g., phlogiston or Pope Joan). The relation between thoughts and these various objects of thought is "intentionality" and the question of how physical systems like brains can exhibit intentionality is one of the most vexed issues in the philosophy of mind. One of the distinctive features of intentional information is that it can be false—it can misrepresent how things are. (Godfrey-Smith 1989) It is difficult to reproduce the phenomena of misrepresentation using a causal notion of information. A signal cannot both correlate with a source and not correlate with it, nor can a signal correlate with a source that does not exist. The most promising attempts to give a naturalistic account of intentional information are the so-called "teleosemantic" theories to be discussed below, according to which a sign represents whatever evolution designed it to represent. Genetic information is usually described as if it made sense to speak of a phenotype misinterpreting the message in the genes and hence appears to be intentional information. For instance, it is a common view amongst contemporary evolutionary psychologists that a genome contains a "disjunctive program" or "open program" for psychological development. The genetic message takes the form "develop like *this* under these circumstances, like *that* under other circumstances." Evolutionary psychologists seek to explain cultural differences between human minds as just this kind of disjunctive response of the human developmental program to various environmental factors. (Cosmides, Tooby, and Barkow 1992) In contrast, no one says that the human genome encodes the instruction "when exposed to the drug thalidomide grow only rudimentary limbs." But this would, indeed, be one branch of the disjunctive program if we were talking about the causal information in the human genome. When the relevant channel is contaminated by thalidomide, human genes send this causal information. The fact that the notion of a disjunctive program is not applied to outcomes that are thought to be pathological, accidental, or otherwise "unintended" suggests that the information in the program is being conceived as intentional information.

3. Maynard Smith on genetic information

The obvious way to defend information talk in biology is to argue that it is just like the many other, scientifically respectable, uses of causal information. This has been the traditional strategy. In an address to the Fifth International Congress of Evolutionary and Systematic Biology in 1996, John Maynard Smith made use of this conventional defense. Information talk in biology is to be interpreted "more or less in the spirit of information theory." (author's notes) The disadvantage of this defense is that it implies parity between genetic and non-genetic causes in development. Information in the sense legitimated by information theory is the systematic dependence of a signal on a source, a dependence that is created by a set of channel conditions. In the case of development, the genes can be taken to be the source, the life-cycle of the organism to be the signal and the channel conditions to be all the other resources needed for the life-cycle to unfold. But it is a fundamental feature of information theory that the role of source and channel condition can be reversed. The old television "test-card" did exactly that, holding the transmission constant so that the television engineer could read off the state of what were previously channel conditions. The source/channel distinction is imposed on a natural causal system by the observer. A source is simply one channel condition whose current state the signal is being used to investigate. If all other resources are held constant, a life cycle can give us information about the genes, but if the genes are held constant, a life cycle can give us information about whichever other resource we decided to let vary. So far as causal information goes, every resource whose state affects development is a source of developmental information. (Johnston 1987; Gray 1992; Griffiths and Gray 1994; Oyama 2000a)

The fact that causal information conforms to this "parity thesis" has been accepted by many of the participants in the current debate over the interpretation of genetic information. (Godfrey-Smith 1999a; Sterelny and Griffiths 1999; Maynard Smith 2000a,b; Kitcher 2001) In his most recent discussion of causal information, Maynard Smith notes that:

> With this definition, there is no difficulty in saying that a gene carries information about adult form; an individual with the gene for achondroplasia will have short arms and legs. But we can equally well say that a baby's environment carries information about growth; if it is malnourished, it will be underweight.
>
> (Maynard Smith 2000a, 189)

In response to the threat of parity, Maynard Smith has abandoned the idea that biological information is causal information. He continues:

> informational language has been used to characterize genetic as opposed to environmental causes. I want now to try to justify this usage.
>
> I will argue that the distinction can be justified only if the concept of information is used in biology only for causes that have the property of intentionality . . . A DNA molecule has a particular sequence because it specifies a particular protein, but a cloud is not black because it predicts rain. This element of intentionality comes from natural selection.
>
> (Maynard Smith 2000a, 189–190)

Maynard Smith proposes to analyze the intentionality of genetic information using *teleosemantics*, the philosophical program of reducing meaning to biological function (teleology) and then reducing teleology to natural selection. (Millikan 1984; Papineau 1987) Whilst there is considerable controversy about whether such reductions can be successfully carried out, teleosemantics remains one of the more popular programs for naturalizing intentionality. Maynard Smith begins by arguing that natural selection is strongly analogous to computer programming using the "genetic algorithm" technique. The genetic algorithm programmer randomly varies the code of a computer program and selects variants that best perform the desired task. In the same way, mutation randomly varies the genes of organisms and natural selection chooses the organisms that have the highest fitness. Just as the function of the selected computer program is to perform the task for which it was selected, the biological function of successful genes is to produce the developmental outcomes in virtue of which they were selected. Such genes are intentionally directed onto, or about, those effects. The defective hemoglobin gene in some human populations, which has been selected because it sometimes confers resistance to malaria, carries teleosemantic information about malaria resistance.

However, teleosemantic information is fundamentally unsuited to Maynard-Smith's aim of avoiding parity. The most developed version of the teleosemantic theory of genetic information is Sterelny, Dickison, and Smith's "extended replicator theory." (1996) Sterelny and his collaborators recognize from the outset that teleosemantic information exists in both genetic and in some non-genetic developmental causes. Sterelny (2000) reiterates this view in his reply to Maynard Smith. Russell Gray and I have argued that teleosemantic information exists in a very much wider range of developmental causes than Sterelny et al. suggest (Griffiths and Gray 1997), but my argument here follows even on the more conservative view. Teleosemantic information exists in any inheritance system that is a product of evolution, including epigenetic inheritance systems. The term "epigenetic inheritance system" is used to denote any biological mechanism which produces resemblances between parents and offspring and which works in parallel with the inheritance of nuclear and mitochondrial DNA. (Jablonka and Lamb 1995; Jablonka and Szathmáry 1995) Every organism inherits a great deal besides its DNA. To develop normally the egg cell must contain features such as basal bodies and microtubule organizing centers, correct cytoplasmic chemical gradients, DNA methylation patterns, membranes and organelles. Changes in these other resources can cause heritable variation that appears in all the cells descended from that egg cell.

One of the best understood of these mechanisms is the DNA methylation inheritance system. A methylation pattern is a series of additional chemical groups attached to a DNA sequence. Methylation patterns block transcription of any genes they cover and they are replicated by the methylation copying system in all the cells descended from a given cell. Differences in methylation are important in tissue differentiation during the lifetime of a single organism, but they can also pass between the generations. Methylation patterns are often applied to the DNA in a sperm or egg by the parent organism in what is usually interpreted as a mechanism for influencing development in the offspring. This form of methylation inheritance has excited a great deal of interest because of it is easy to see how it could play a role in conventional, micro-evolutionary change. A typical example comes from some controversial British research on behavioral differences in male and female children. The proposed mechanism for the transmission of these behavioral differences is that female humans methylate a sequence of the X-chromosome in their gametes, so that males, who get only one X-chromosome and get it from their mother, cannot transcribe the genes in that region. Hence certain gene products are denied to all males. Males demethylate that sequence in their sperm cells, so females get readable copies of these genes on the X-chromosomes they receive from their fathers. (Skuse et al. 1997) Naturally, when this research was reported in the media it was announced as the discovery that sex differences are in the genes: "Genes say boys will be boys, and girls will be sensitive." (Radford 1997)

Epigenetic inheritance research has focused on mechanisms within the cell. Developmental systems thinkers have emphasized a wider range of epigenetic mechanisms. (Gray 1992; Griffiths and Gray 1994, 2001) The characteristics of epigenetic inheritance systems within the cell are shared by many extra-cellular structures. Some castes of the aphid *Colophina arma* require a growth spurt as part of their life-cycle. These and only these castes inherit the micro-organisms that make the chemicals on which this growth spurt depends. (Moran and Baumann 1994) The morphology of queens and the colony structure of the fire ant *Solenposis invicta* differ radically between genetically similar lineages of the species because of stably replicated pheromonal nest "cultures." (Keller and Ross 1993) Any queen raised in a colony with a particular culture will found a colony with the same culture, as shown by moving eggs from one culture to the other. Many parasites, both vertebrate and invertebrate, maintain associations with particular host species over evolutionary time through "host-imprinting." Thus, some insects lay their eggs on the plant whose leaves they tasted as larvae. Some parasitic finches lay their eggs in the nest of the host species which they imprinted on as chicks. (Immelmann 1975) "Host switching" occurs on the rare occasions when this mechanism malfunctions and the mother lays her eggs on the wrong plant or in the wrong nest. On the still more rare occasions when these misplaced eggs flourish, the imprinting mechanism will ensure that this new, epigenetic mutant form reliably reproduces itself. Clearly, all the mechanisms discussed here are candidates for adaptive explanation. This means that the physical traces by which these mechanisms operate are as likely to have biological functions as any other putative adaptation and thus, on the teleosemantic approach recommended by Maynard Smith, that these traces contain information. Hence, just like causal information, teleosemantic information obeys the parity thesis and does not constitute the principled difference between genetic and other causes that Maynard Smith seeks.

The mechanism of epigenetic inheritance that takes us furthest from the nucleus is so-called "niche construction." (Odling-Smee 1988; Odling-Smee, Laland, and Feldman 1996; Laland, Odling-Smee, and Feldman 2001) Many features of an organism's niche exist only because of the effects of previous generations of that species on the local environment. One of the earliest examples of this phenomenon to be clearly recognized was the co-evolution of the eucalypts with the current pattern of bush fires in Australia. (Mount 1964) It is, however, unclear whether the *collectively* constructed features of a species' niche can be regarded as part of each *individual* organism, and hence whether they can be assigned biological functions and teleosemantic information content. More theoretical work is needed on the interpretation of this form of inheritance.

4. Stability and inheritance

Opponents of the parity thesis, such as Maynard-Smith, are well aware of the role of non-genetic factors in development and have a standard strategy for discounting them. This strategy is to question the intergenerational stability of non-genetic factors, and/or their potential to produce the kind of variation upon which natural selection can act. Thus:

> Differences due to nature are likely to be inherited, whereas those due to nurture are not; evolutionary changes are changes in nature, not nurture; traits that adapt an organism to its environment are likely to be due to nature.
>
> (Maynard Smith 2000a, 189)

> The special status of genetic factors is deserved for one reason only: genetic factors replicate themselves, blemishes and all, but non-genetic factors do not.
>
> (Dawkins 1982, 99)

Clearly, the phenomenon of epigenetic inheritance defeats the simplest version of this defense. Developmental systems theorists have argued that we should define "inheritance" so that something is inherited just if it passes from generation to generation in such a way that evolution can act on its variant forms. Hence, every element of the developmental matrix which is reliably replicated in each generation and which plays a role in the production of the evolved life cycle of the organism counts as something which is inherited. (Griffiths and Gray 1994, 2001) The more conservative "extended replicator" view goes some distance in the same direction. (Sterelny, Dickison, and Smith 1996)

It follows that genes cannot be singled out as the sources of developmental information on the grounds that they and they alone persist through lineages long enough for cumulative selection to act upon them. Lineages can be selected for having good methylation patterns or good symbionts or being imprinted on a good host. These features can persist for evolutionarily significant periods of time. (Gray 2001, Griffiths and Gray 2001) Some other factor needs to be added to the insistence on stable inheritance in order to defend the special, informational status of genes. One candidate is the idea that genetic causes are unique in being "symbolic" or "semiotic." In his most recent discussion, Maynard Smith claims that genes are symbolic and other developmental factors are not, because there is no intrinsic connection between the nature of the gene and the developmental outcome it produces. (Maynard Smith 2000a, 185) In his commentary on Maynard Smith, Peter Godfrey-Smith points out that the apparent difference is generated by looking at the *distal* effects of genetic causes, which depend on an array of other causal factors, while concentrating on the *proximal* effects of non-genetic causes, which, like the proximal effects of DNA, are uniquely determined by physical laws. (Godfrey-Smith 2000, 203) In response to this criticism, Maynard Smith introduces a second role for biological teleology in his account of information. A cause is semiotic or symbolic when it has its effect via an "evolved receptor" that has been selected to confer one of many possible causal "interpretations" on the signal. (Maynard Smith 2000b, 215) Whatever its virtues as an analysis of the nature of biological information, this new suggestion fails to generate a principled difference between genetic and epigenetic causes. For example, the mechanisms of habitat and host imprinting described above are clearly "evolved receptors" in Maynard Smith's sense. They have been selected to confer a highly specific "interpretation" on an otherwise meaningless chemical input, but in these cases the signal does not take the form of a DNA sequence.

There is something fundamentally puzzling about Maynard Smith's decision to advocate a teleosemantic interpretation of information talk in biology. If genetic information is teleosemantic information then it can have only a distant relationship to the actual genetic code. The genetic code clearly does not derive its semantic aspect from teleology. The codon CCC codes for proline even when it is part of a section of junk DNA with no selection history or when it has been inserted by an incompetent biotechnologist who intended it to mean leucine. This problem cannot be fixed by regarding the biological functions of these individual DNA codons as derived from the general functions of that type of codon, since not all codon/amino-acid pairings are adaptations. Although there is increasing evidence that some pairings are the result of selection, others are very probably the result of direct chemical affinities between ancestral RNAs and amino acids. (Knight, Freeland and Landweber, 1999) The exact conclusion of current debates over the meaning of the genetic code is not critical, however. No imaginable outcome could lead to the revision of the tables of the standard and variant codes to be found in the current literature. The codes are based on actual causal relationships, not on judgments about adaptive function.

More tentatively, I would suggest that insofar as information talk in developmental biology more generally is related to concepts of biological function, it is to the ahistorical, causal-contribution notion of function, rather than the evolutionary concept of adaptive function. In an important study of functional language in biology, Ronald Amundson and George Lauder have suggested that the causal-contribution notion of function predominates in sciences such as physiology and anatomy. (Amundson and Lauder 1994) Like physiologists and anatomists, developmental biologists are primarily concerned with proximal explanations (how mechanisms work at the current time) rather than ultimate explanations (why they evolved).

5. Evolvability and information

At this point I will leave Maynard Smith's most recent, teleosemantic analysis and turn to what I believe is a more promising approach found in his earlier work. Maynard Smith and Eörs Szathmáry have argued that the genetic inheritance system and cultural transmission in humans are the only two systems that display what they call "unlimited heredity." (Maynard Smith and Szathmáry 1995) Only these "unlimited" inheritance systems, they argue, should really be thought of as "coding" for outcomes. To make their distinction, Maynard Smith and Szathmáry argue that most inheritance systems can only mutate between a limited number of heritable states which can be specified in advance. Habitat imprinting, for example, can only lead organisms to choose different habitats and DNA methylation can only choose whether existing genes will be on or off. The genome and language, however, both have a recursive syntactic structure. Their basic constituents can be put together in many different combinations and these combinations can be of any length. Hence these inheritance systems have an unlimited number of possible heritable states. At first sight it may not clear why this provides a justification for viewing those systems and no others as transmitting information. One way to grasp the thought behind Maynard Smith and Szathmáry's proposal is to see them as providing an objective justification for fixing the source/channel distinction so that the genome or culture becomes the information source. The other causes of heritable variation are mere channel conditions because they have relatively few alternative settings. Only in genes (or culture) do we find a signal source with enough possible states to signal the vast range of possibilities that evolution needs.

The distinction between limited and unlimited heredity systems is an important one, and may provide insights into one of the key innovations that enabled the diversification of early life into the vast range of forms we see today. I will argue, however, that it is putting too much weight on the distinction to use it, as Maynard Smith and Szathmáry seem to wish, as a basis for the claim that developmental information resides in the genes and that other causal factors provide mere material support to decode this information. A developmental systems interpretation, according to which developmental information exists in the whole matrix of material resources that are need to reconstruct a developmental outcome, is equally consistent with the distinction between limited and unlimited heredity systems and with the insights into the evolutionary process that it provides.

The limited/unlimited distinction seems to support the idea that genes are the signal and the rest of the developmental system is a channel because it suggests that the genes have a vastly greater capacity to "signal" alternative outcomes. In effect, Maynard Smith and Szathmáry propose to partition the total number of developmental outcomes that can be generated by a developmental system between the various inheritance systems that make up that system. The number of outcomes allotted to an inheritance system measures its "limitedness" and the genetic inheritance system is allotted far more outcomes than any other system. But how is this partition of outcomes to be made? Maynard Smith and Szathmáry assume that the number of permutations of DNA codons, or perhaps of entire genes, is the relevant measure for the genetic inheritance system and that some corresponding measure of the number of permutations of physical parts is the appropriate measure for other inheritance systems.

But these are not appropriate measures. For any given inheritance system the range of physical changes that count as evolutionary changes is restricted to those that can be made use of by the rest of the existing system. The point is a familiar one with respect to the notion of "syntactic structure" for human languages. Not all physical differences between syntactic objects are *syntactic* differences. Differences in handwriting or in the accent with which one speaks, for example, are not syntactic differences.

The main lesson of the major evolutionary transitions that are the focus of Maynard Smith and Szathmáry's work is that evolution creates entirely new kinds of developmental systems that massively expand the possible interpretations of existing developmental resources, including genes. One way to see this is to consider how many evolutionary possibilities come into existence when a "limited" heredity system undergoes a change. Consider, for example, the evolutionary possibilities that can be "signaled" by a base pair substitution in the DNA of a eukaryote cell that cannot be signaled by those substitutions in a prokaryote cell. As Maynard Smith and Szathmáry themselves describe, that vast swathe of evolutionary possibilities came into existence through evolutionary change mediated by a limited heredity system. Membranes cannot be constructed without an existing membrane template into which to insert newly synthesized proteins. Hence, major changes to the partitioning of the cell require variation to arise through the membrane heredity system, not through mutations of the DNA.

In effect, the measure of "limitedness" that Maynard Smith and Szathmáry adopt allots to the genetic inheritance system all the outcomes that can be generated by making changes to that system across the full range of possibilities for the other systems while allocating to the other systems only the number of outcomes they could produce given one possible genome. That is why it is not an appropriate measure if the limited/unlimited distinction is to be used to judge the capacity of inheritance systems to "signal" (cause) evolutionary possibilities. It would be no less (and no more) meaningful to allocate to the genetic inheritance system only the range of outcomes it could generate given one state of the other inheritance systems. The other inheritance systems would receive all the outcomes that could be generated by permuting the genetic system in the presence of the remaining states of that inheritance system.

A closely related point is that the huge potential of differences in genes, language and perhaps pheromones in social insects to cause new evolutionary possibilities stems from the fact that these differences "mean something" to the rest of the developmental system. If the rest of the system surrounding the genes were such that indefinitely many base-pair combinations collapsed into only a few developmental outcomes, then the genetic inheritance system would not be unlimited. It is not hard to imagine cellular machinery with this result—the existing genetic code is substantially redundant in just this way, with several codons corresponding to the same amino acids. Hence, the unlimited nature of the genetic inheritance system is more accurately seen as a property of the developmental system as a whole and not of the genome in isolation. The language of separate "inheritance systems" can itself be highly misleading in some biological contexts. Methylation inheritance, for example, is a mechanism of gene regulation, making it odd to describe it as a separate "system" from the genetic "system." From a "selfish replicator" viewpoint a gene and the methyl groups attached to it are separate replicators, but no more so than any two genes. From a developmental systems viewpoint genes and methylation patterns are separate developmental resources, but they are elements of one developmental system. (Griffiths and Gray 2001)

It has never been part of the developmental systems tradition to deny that nucleic acids and natural languages are distinctive elements of developmental systems. The point of the "parity thesis" is to prevent these empirical differences turning into a kind of scientific metaphysics, as happens when genes are identified with information (or even "form") and everything else in development with mere matter. This distracts attention from the many ways in which non-genetic resources sometimes play biological roles more usually associated with genes. It also leads to the empirically inappropriate lumping together of very different non-genetic resources (the "environment"). From a developmental systems perspective, unlimited heredity is simply another important, empirical property of DNA, and perhaps of other resources, such as pheromones in their role in causing caste-differences in the morphology and behavior of social insects.[1]

6. Developmental switches and bioinformatics

It is clear that an analysis of "information" such that genes are the sole or main bearer of developmental information has still not been adequately defended. It is less clear why such an analysis is needed. Developmental systems theorists and other critics of "information talk" do not deny the unique biological role of genes as protein templates, nor the central role of gene products in development, nor do they deny "molecular Weismannism"—the prohibition on the inheritance of acquired characteristics via the genetic material (with the exception of a few retro-copying phenomena accepted on all sides). Conversely, advocates of the genetic program do not deny the existence, or even the evolutionary importance, of epigenetic inheritance. There are genuine disagreements among commentators on contemporary developmental biology over the relative promise of more program-like and more dynamicist models of gene regulation, but the lines of cleavage among the disputants do not align with those in the debate over genetic information, as Maynard Smith himself remarks. (Maynard Smith 2000b, 218) In any case, that dispute does not turn on whether all the information being processed in a developmental program comes from genetic factors.

If the idea that "biology is an information technology" is neither obviously true nor a "forced choice" on which biologists must take a stand if they are to do research, why is the idea so widely accepted? Partly, of course, because of the central role of information in the contemporary scientific world-view. (Keller 1995; Oyama 2000a) Another widely acknowledged factor is the misinterpretation of Weismann diagrams. (Griesemer and Wimsatt 1989; Sterelny and Griffiths 1999, 64; Sterelny 2000) The now standard diagrammatic representation of evolution shows a causal arrow from genes to phenotypes and a causal arrow from the genes of one generation to those of the next (e.g., Maynard Smith 1993, figure 8). There are no causal arrows that represent the influence of organisms on the reproduction of their genes, nor the many influences that organisms exert on their offspring in addition to reproducing their genes. Both the general cultural enthusiasm for information and the modern Weismann diagram are clearly important in giving the genetic program its air of common sense.

But there is another factor closely related to the themes of this paper that has been less widely acknowledged. This factor can be represented by a (very bad) argument:

(1) There is a genetic code.
(2) In molecular developmental biology there is talk of signals, switches, master control genes, and so forth.
(3) Therefore, the information flowing in (2) is information encoded in the sense of (1).

In this blunt form the argument sounds merely frivolous. But many discussions of molecular biology, especially those for a non-technical audience, insinuate something surprisingly close to it. In his recent paper, Maynard Smith presents something quite close to this argument in a section immediately preceding his analysis of biological information, entitled "Is the Genome a Developmental Program?" Here are the first and last sentences of the section and a representative passage from in between:

> There is, I think, no serious objection to speaking of a genetic code, or to asserting that the gene codes for the sequence of amino acids in a protein.
>
> However, an organism is more than a bag of specific proteins. Development requires that different proteins be made at different times, in different places. A revolution is now taking place in our understanding of this process. The picture that is emerging is one of a complex hierarchy of genes regulating the activity of other genes. Today, the notion of genes sending signals to other genes is as central as the notion of a genetic code was forty years ago.
>
> Informational terminology is invading developmental biology, as it earlier invaded molecular biology. In the next section I try to justify this usage.
>
> (Maynard Smith 2000a, 187–189)

While not quite the very bad argument given above, this series of points is clearly meant to suggest that since the genetic code is "real science" and not mere metaphor, it is only a matter of time before other information talk in biology becomes real science too. At best, this is a very weak inductive argument, at worst it is equivocation on the word "information." It is worth pointing out that a number of historians of science have suggested the exact opposite: that the history of information talk in molecular developmental biology is one of consistent retreat from literal to metaphoric in the light of increased understanding of molecular processes. (Sarkar 1996; Chadarevian 1998)

The very bad argument (hereafter "VBA") is one aspect of what Sarkar means when he says that loose information talk "leads to a misleading picture of possible explanations in molecular biology." (Sarkar 1996, 187) The information metaphor suggests a "bottom-up" explanatory strategy in molecular biology, rather than a "top-down" one. A bottom-up strategy tries to infer the developmental significance of a DNA sequence from the sequence itself, looking at the sequence to determine its product and looking at that product and other gene products to determine how they will interact in development. The alternative, top-down strategy begins by studying the developmental process, works out which gene products are involved and uses the sequence of these products to locate the DNA sequences from which they are made. A realistic picture of work in contemporary developmental biology is that it is at least as much top-down as bottom-up. Note also that the top-down strategy automatically locates and takes account of epigenetic causal factors in development, while the bottom-up strategy finds their involvement in developmental processes an obstacle to upward progress.

I recently came across a rather striking example of the use of the VBA to suggest that molecular developmental biology proceeds by a pure bottom-up strategy. Once again, sequential quotations can be fitted into the framework of the VBA:

> One great discovery was that a set of three rungs contains the information to make one amino acid. So the set of rungs AAA will make one amino acid, while the rungs ACG will make a different amino acid.
>
> After *[the human genome project]* we can look for the small section of DNA that holds the instructions to grow a new arm. . . . molecular biologists studying the common fruit fly found the section of DNA that controlled the growth of its eyes. They learnt to switch on the growth of eyes . . .
>
> But even when we have mapped the whole of the human DNA, there's another hurdle to jump before we can start growing arms.

> The DNA is huge! . . . That's why the science of bioinformatics has been invented.
>
> (Kruszelnicki 1998)

These quotes are from a leaflet issued to persuade students to enroll in a degree in bioinformatics. Bioinformatics is designed to produce biologically literate graduates trained in computational techniques that will allow them to handle the vast amounts of information *about* genes currently being generated by molecular biology. But deft use of the VBA suggests that bioinformatics deals with developmental information *encoded in* genes—the blueprint for life, and a much more exciting prospect for school-leavers than applied computing.

7. Conclusion: The code, the code and nothing but the code

The existence of the genetic code—the standard pairing of codons and amino acids and its seventy-odd minor variants—is beyond question. Beyond that, however, "information talk" in biology has nothing to do with the genetic code. Information talk in the various forms related to the mathematical theory of communication is equally applicable to non-genetic developmental factors. So too are the cybernetic locutions found in molecular developmental biology. Genes can be discussed in terms of their teleosemantic properties, but so can many non-genetic factors. It is also unlikely that this teleosemantic sense is what is intended by most information talk in molecular developmental biology, which is a science of proximate mechanisms, not of ultimate origins. There are numerous important differences between what DNA does in development and the roles played by other causal factors, but these differences do not map onto a distinction between informational and material causation. The present atmosphere, in which information talk is only applied to genes, makes that way of talking highly misleading. I have suggested here that it misleads people about the forms of explanation in molecular biology. I also believe that the asymmetrical use of information talk partly explains the persistence of genetic determinism, but that is an argument for another day.

Acknowledgement

A version of this paper was read at the International Society for the History, Philosophy, and Social Studies of Biology in Oaxaca, Mexico in June 1999. I am indebted to that audience and to Russell Gray and Robert Olby for comments on the manuscript.

Note

1 Another slightly different interpretation of the limited/unlimited heredity distinction would be to argue that only genes have the capacity to generate fine-grained response to selection and thus to give rise to adaptation. This point could be combined with the teleosemantic approach described above to yield the result that since only genes support genuine adaptation, only genes carry (teleo-) information. As Russell Gray has pointed out, this implicitly assumes that epigenetic inheritance systems must have evolutionary potential separately from one another and that only discrete, as opposed to continuous, variations count as alternative states of an inheritance system. The cards are being stacked against extra-genetic inheritance in numerous ways. However:

> Extragenetic changes can also be piecemeal and incremental. Just as natural selection can favor combinations of genes at different loci, so selection may favor combinations of endosymbionts. Quantitative variations in cytoplasmic factors, nest design, and habitat preferences could also all be passed on extragenetically. Thus, although combinations of these factors are not unlimited, they can be quite large enough to allow a fine-grained response to selection.
>
> (Gray 2001, 197)

References

Amundson, Ronald and George V. Lauder (1994), "Function without Purpose: The Uses of Causal Role Function in Evolutionary Biology", *Biology and Philosophy* 9(4): 443–470.

Chadarevian, Soraya de (1998), "Of Worms and Programs: *Caenorhabitis elegans* and the Study of Development", *Studies in History and Philosophy of the Biological and Biomedical Sciences* 29(1): 81–105.

Cosmides, Leda, John Tooby and Jerome H. Barkow (1992), "Introduction: Evolutionary Psychology and Conceptual Integration", in Jerome H. Barkow, Leda Cosmides, and John Tooby (eds.), *The Adapted Mind: Evolutionary Psychology and the Generation of Culture*. Oxford, New York: Oxford University Press, 3–15.

Dawkins, Richard (1982), *The Extended Phenotype*, New York: W. H. Freeman.

Dretske, Fred (1981), *Knowledge and the Flow of Information*. Oxford: Blackwells.

"Drowning in Data" (1999), *The Economist* (June 26): 97–98.

Godfrey-Smith, Peter (1989), "Misinformation", *Canadian Journal of Philosophy* 19 (5): 533–550.

—— (1999a), "Genes and Codes: Lessons from the Philosophy of Mind?", in Valerie G. Hardcastle (ed.), *Biology Meets Psychology: Constraints, Conjectures, Connections*. Cambridge, MA: MIT Press, 305–331.

—— (1999b), "The Semantic Gene", Manuscript.

—— (2000), "Information, Arbitrariness, and Selection: Comments on Maynard Smith", *Philosophy of Science* 67(2): 202–207.

Gray, Russell D. (1992), "Death of the Gene: Developmental Systems Strike Back", in Griffiths, Paul E. (ed.), *Trees of Life: Essays in the Philosophy of Biology*. Kluwer, Dordrecht: 165–210.

—— (2001), "Selfish Genes or Developmental Systems?", in Rama S. Singh, Costas B. Krimbas, Diane B. Paul and John Beatty (eds.), *Thinking about Evolution: Historical, Philosophical and Political Perspectives*. Cambridge: Cambridge University Press, 184–207.

Griesemer, James R. and William C. Wimsatt (1989), "Picturing Weismannism: A Case Study of Conceptual Evolution", in Michael Ruse (ed.), *What the Philosophy of Biology Is*. Dordecht: Kluwer Academic Publishers, 75–137.

Griffiths, Paul E. (forthcoming), "The Fearless Vampire Conservator: Philip Kitcher on Genetic Determinism", in Christoph Rehmann-Sutter and Eva M. Neumann-Held (eds.), *Genes in Development*. Durham, NC: Duke University Press.

Griffiths, Paul E. and Kussell D. Gray (1994), "Developmental Systems and Evolutionary Explanation." *Journal of Philosophy* XCI (6): 277–304.

—— and Russell D. Gray (1997), "Replicator II: Judgment Day", *Biology and Philosophy* 12(4): 471–492.

—— and Russell D. Gray (2001), "Darwinism and Developmental Systems", in Susan Oyama, Paul E. Griffiths and Russell D. Gray (eds.), *Cycles of Contingency: Developmental Systems and Evolution*. Cambridge, MA: MIT Press, 195–218.

Griffiths, Paul E. and Robin D. Knight (1998), "What is the Developmentalist Challenge?", *Philosophy of Science* 65(2): 253–258.

Immelmann, Klaus (1975), "Ecological Significance of Imprinting and Early Learning", *Annual Review of Ecology and Systematics* 6: 15–37.

Jablonka, Eva and M. J. Lamb (1995), *Epigenetic Inheritance and Evolution: The Lamarkian Dimension*. Oxford, New York, Tokyo: Oxford University Press.

Jablonka, Eva and Eörs Szathmáry (1995), "The Evolution of Information Storage and Heredity", *Trends in Ecology and Evolution* 10(5): 206–211.

Johnston, Timothy D. (1987), "The Persistance of Dichotomies in the Study of Behavioural Development", *Developmental Review* 7: 149–182.

Keller, Evelyn Fox (1995), *Refiguring Life: Metaphors of Twentieth Century Biology*. New York: Columbia University Press.

Keller, Laurent and K. G. Ross (1993), "Phenotypic Plasticity and 'Cultural Transmission' of Alternative Social Organisations in the Fire Ant *Solenopsis invicta*", *Behavioural Ecology and Sociobiology* 33: 121–129.

Kitcher, Phillip (2001), "Battling the Undead: How (and How Not) to Resist Genetic Determinism", in Rama S. Singh, Costas B. Krimbas, Diane B. Paul and John Beatty (eds.), *Thinking about Evolution: Historical, Philosophical and Political Perspectives*. Cambridge: Cambridge University Press, 396–414.

Knight, Robin D., Stephen J. Freeland and Laura F. Landweber (1999), "Selection, History and Chemistry: The Three Faces of the Genetic Code", *Trends in Biochemical Science* 24(6): 241–247.

Kruszelnicki, Karl (1998), "Why You Should Study Bioinformatics", University of Sydney Science Faculty publicity material.

Laland, Kevin, N., F. John Odling-Smee, and Marcus W. Feldman (2001), "Niche Construction, Ecological Inheritance, and Cycles of Contingency in Evolution", in Susan Oyama, Paul E. Griffiths and Russell D. Gray (eds.), *Cycles of Contingency: Developmental Systems and Evolution*. Cambridge, MA: MIT Press, 117–126.

Maynard Smith, John (1993), *The Theory of Evolution*. Cambridge: Cambridge University Press.

—— (2000a), "The Concept of Information in Biology", *Philosophy of Science* 67(2): 177–194.

—— (2000b), "Reply to Commentaries", *Philosophy of Science* 67(2): 214–218.

—— and Eörs Szathmáry (1995), *The Major Transitions in Evolution*. Oxford, New York, Heidelberg: W.H. Freeman.

Millikan, Ruth G. (1984), *Language, Thought and Other Biological Categories*, Cambridge, MA: MIT Press.

Moran, N. and P. Baumann (1994), "Phylogenetics of Cytoplasmically Inherited Microorganisms of Arthropods", *Trends in Ecology and Evolution* 9: 15–20.

Mount, A. B. (1964), "The Interdependence of the Eucalpyts and Forest Fires in Southern Australia", *Australian Forestry* 28: 166–172.

Odling-Smee, F. John (1988), "Niche-Constructing Phenotypes", in H.C. Plotkin (ed.), *The Role of Behavior in Evolution*. Cambridge, MA: MIT Press, 73–132.

——, Kevin N. Laland, and Marcus W. Feldman (1996), "Niche Construction", *American Naturalist* 147(4): 641–648.

Oyama, Susan (2000a), *The Ontogeny of Information*, 2nd ed. Durham, NC: Duke University Press.

—— (2000b), *Evolution's Eye: A Systems View of the Biology-Culture Divide*, Durham, NC: Duke University Press.

Papineau, David (1987), *Reality and Representation*. New York: Blackwells.

Radford, T. (1997), "Genes Say Boys Will Be Boys and Girls Will Be Sensitive", *The Guardian* (June 12): 1.

Sarkar, Sahotra (1996), "Biological Information: A Sceptical Look at Some Central Dogmas of Molecular Biology", in S. Sarkar (ed.), *The Philosophy and History of Molecular Biology: New Perspectives*. Dordrecht: Kluwer Academic Publishers, 187–232.

Shannon, Claude E. and W. Weaver (1949), *The Mathematical Theory of Communication* Urbana, IL: University of Illinois Press.

Skuse, D. H. et al. (1997), "Evidence from Turner's Syndrome of an Imprinted X-Linked Locus Affecting Cognitive Function", *Nature* 387(12 June): 705–708.

Sterelny, Kim (2000), "The 'Genetic Program' Program: A Commentary on Maynard Smith on Information in Biology", *Philosophy of Science* 67(2): 195–201.

Sterelny, Kim, Michael Dickison and Kelly C. Smith (1996), "The Extended Replicator", *Biology and Philosophy* 11(3): 377–403.

Sterelny, Kim and Paul E. Griffiths (1999), *Sex and Death: An Introduction to the Philosophy of Biology*. Chicago: University of Chicago Press.

40

INFORMATION

Its interpretation, its inheritance and its sharing

Eva Jablonka

Source: *Philosophy of Science* 69(4) (2002): 578–605.

Abstract

The semantic concept of information is one of the most important, and one of the most problematical concepts in biology. I suggest a broad definition of biological information: a source becomes an informational input when an interpreting receiver can react to the form of the source (and variations in this form) in a functional manner. The definition accommodates information stemming from environmental cues as well as from evolved signals, and calls for a comparison between information-transmission in different types of inheritance systems—the genetic, the epigenetic, the behavioral, and the cultural-symbolic. This comparative perspective highlights the different ways in which information is acquired and transmitted, and the role that such information plays in heredity and evolution. Focusing on the special properties of the transfer of information, which are very different from those associated with the transfer of materials or energy, also helps to uncover interesting evolutionary effects and suggests better explanations for some aspects of the evolution of communication.

1. Introduction

As many biologists and philosophers have noted, information was a central concept in twentieth-century biology, and is likely to remain central throughout the twenty-first century. However, the term information in its meaning-related sense is rarely defined, and as Oyama has shown, there are many problems in applying it to biological systems (Oyama 1985). Recently, in a target paper in *Philosophy of Science* (June 2000), John Maynard Smith has re-kindled interest in an analysis of the concept of information in biology, using an evolutionary approach. His article focused mainly on genetic information and the way it can be said to specify the form and function of living organisms. For Maynard Smith (hereafter, MS) the notion of biological information is intimately linked with the idea that the signal carrying the information (e.g., DNA) and the response to it are products of natural selection.

Before discussing how the concept of semantic biological information is to be defined and used, there is a need to justify the usefulness of "information talk." It can be argued that focusing on the diverse mechanisms underlying the processes of DNA replication, transcription, translation, signal transduction, and communication between organisms may be more useful than an attempt to generalize these processes under a common conceptual umbrella of "information" transmission, re-construction, processing, and so on. After all, genetic heredity, signal transduction within and among cells, and communication between individuals are processes that occur at different levels of organization and entail entirely different biochemical and physiological mechanisms, so relating them all to "information" may lead to distorted views of one or more or these biological processes. As Oyama (1985) has shown, this is a valid concern, particularly when the genetic system becomes the prototype for thinking about information. The problem is that the genetic system is highly specific and in many ways unusual, and an excessive focus on genes and their inheritance tends to lead to gene-based metaphors and analogies in other areas (for example, "memes" in cultural studies), which may be inappropriate or misleading. One can therefore choose to abandon general information talk, or alternatively suggest a general definition of semantic information in biology that recognizes that there are different types of information, and different ways of acquiring, re-constructing, and sharing it.

Suggesting a general definition of "information" requires finding a common denominator for the different types of things or processes that we intuitively recognize as "information" or as "information carriers"; moreover, this common denominator and the definition based on it should lead to useful generalizations or perspectives. Information theory provides such a common denominator, defining information as the measure of the probability of a particular message being selected from the set of all possible messages. It assumes the existence of sender, receiver, channel, encoding and decoding processes, and a set of messages which is compatible with this communication system. The theory calculates the probability of transmitting a particular message with specified accuracy (Shannon and Weaver 1949). However, this probabilistic, general measure of information is indifferent to function and meaning: for example, a DNA sequence coding for a fully functional enzyme contains as much information as a same-length sequence that codes for a completely non-functional enzyme; similarly, a string of nonsense words and a meaningful sentence of the same length are equivalent with respect to information content. Since what we seek is a meaning and function related concept of biological information, one that can be used in an evolutionary framework, the meaning-free concept of information theory is insufficient.

In everyday usage, as in biology, a train timetable, the molecules of distinctive food substances that are present in the milk received by a suckling mammal, a recipe for an apple pie, an alarm call, a DNA sequence, the concentration of a chemical in the nutrient broth of a bacterial colony, the appearance of the sky for an observing ape, the length of the day for a plant, and a piece of computer software are all called "information." We have here a motley collection made up of environmental cues (the appearance of the sky, day length), man-made instructions (train timetable, recipe, software), evolved biological signals (alarm call, chemical signal), and hereditary material (a very special type of evolved biological signal). What do they have in common?

A train timetable, a piece of software, a transmitted chemical, a recipe, an alarm call, an environmental cue, and a DNA sequence do share important general attributes. First, in all cases, there is a special type of reaction between a receiver organism (or organism-designed system) and what we may call the "source," and in all cases this reaction affects the actual or the potential actions of the receiver. Thus, a train timetable affects the potential actions of a reader; molecules of distinctive food substances in a youngster's milk may affect its future food preferences (probably by guiding the development of the nervous system and of specific receptors); a recipe for an apple pie is something that may affect the baking activities of a cook; an alarm call may affect the behavior of an animal that hears it; a DNA sequence may affect the phenotype of an offspring organism; the concentration of a particular nutrient may affect the aggregative response of bacteria; the appearance of the sky may affect the behavior of an observing ape; day length may affect the flowering time of a plant, software may affect the output of a computer.

Second, in all these cases, the reaction of the receiver to the source leads to a complex, regulated chain of events in the receiver, and in all cases this chain of events depends on the way the source is organized rather than on its energy content or its precise chemical constitution. Energy or material may be transferred from the source to the receiver, but it is not necessary for this type of reaction. In fact, the source may remain completely unchanged following the reaction, as happens with sources such as the appearance of the sky, day length, a recipe, a train timetable, and most types of software. More generally, there is no necessary correlation between the energy or material transferred from the source to the receiver and the magnitude of the receiver's functional response as measured in energy terms.

Third, in all cases the reaction to the source contributes to a type of response by the receiver that is beneficial over evolutionary time.

Fourth, in all cases variations in the form of the source often lead to corresponding variations in the form of the response: variations in DNA lead to variations in development; variations in train timetables lead to variations in the behavior of the traveler-reader; variations in the molecules in mother's milk lead to variations in the behaviour of the child; variations in the sky lead to variations in the behavior of the ape, etc. This contrasts with non-informational reactions involving material and energy transfer that change the state of the receiver. In the latter the effect on the receiver is proportional to the amount transferred, as, for example, when organisms consume food. With non-informational reactions, variations in organization are either unimportant or are important only to the extent that they contribute to the efficiency of energy or material transfer. Consider two identical pieces of meat, having exactly the same constituents and caloric content, one prepared as a steak the other minced and served as meatballs. Although when a starving human eats the two pieces of meat, one in the form of a steak, the other in the form of minced meatballs, there is a reaction of the eater with the meat, and the reaction may well change the state of the eater (she has gained calories and is not starving), there is nothing inherently different in the steak calories and meatball calories. Both steak and meatball will be digested and broken down to their constituent parts to be used as sources of energy and material. Neither steak nor meatball can be said to be interpreted in a steak-specific manner or meatball-specific manner by the individual. Of course, the digestive system is a complex evolved system and all the molecular reactions in this system are functional, specific, and dependent on the organization of the enzyme and the reactants. However, although the structure and function of enzymes depend on information, the enzymes and substrates do not have source/reciver relations. When food is eaten the receiver is the organism, and in this case the organism's adaptive response is *not* to variations in the organization of the meat, but to its caloric and material content. The two forms of meat—steak and meatball—are primarily sources of material and energy, not information (unless they are used as socially or culturally relevant signs about class or religion, for example).

On the basis of the common properties just listed, all of which are necessary for something to be referred to as "information," I shall suggest a semantic definition of information. I shall argue that when thinking about biologically relevant information, a functional-evolutionary perspective is necessary, but the focus should be neither on the evolution of the signal "carrying" the information, nor on the evolution of the final specific response. Rather, it should be on the evolution of the system mediating between the two—on the interpreting system of the receiver. The definition I suggest accommodates environmental cues as potential informational sources, and makes it easier to think about non-genetic information and its transmission through non-genetic inheritance systems.

2. A definition of biological information

A source—an entity or a process—can be said to have information when a receiver system reacts to this source in a special way. The reaction of the receiver to the source has to be such that the reaction can actually or potentially change the state of the receiver in a (usually) functional manner. Moreover, there must be a consistent relation between variations in the form of the source and the corresponding changes in the receiver.

What I mean by "form" is the *organization* of the features and/or the actions of the source, and specifically those aspects of organization with which the receiver reacts. The source eliciting the special type of reaction in the receiver will be referred to as "input" or "information cue," and the processes in the receiver that result in a regular and functional response will be called "interpretation." The term "signal" will be reserved only for evolved informational inputs, that is, evolved inputs produced by an evolved or otherwise designed source. "Functional" is here used to mean the consistent causal role that a part plays within an encompassing man-designed or natural-selection-designed system, a role that usually contributes to the goal-oriented behavior of this system. (I will discuss the term "functional" more fully below.)

The first thing to note about the proposed definition of information is that it gives a central role to the *interpretative system of the receiver of information*. Like MS's approach, my definition is based on evolutionary considerations, but it is different from the definition implied by MS (he does not offer a formal definition), which requires that the input, like the final response (output), must have evolved by natural selection. My definition requires instead that the processes mediating between input and output—the interpretation and evaluation processes of the receiver—are the products of natural selection. These processes, which during ontogeny develop in a context-sensitive manner, underlie the receiver's functional, and usually adaptive, response. The focus on interpretative processes is the central feature of the definition.

The second thing to note follows from this. My definition *does not* require that the form of the input must have evolved through natural selection. The definition allows for environmental cues such as a sky full of black clouds to be considered as informational sources, even though the cloudy sky has not evolved by natural selection. Most people will say that a sky full of black clouds provides information about rain to an intelligent vertebrate such as an ape. The clouds are not black as a warning for the observer, but their blackness is a reliable informational cue leading to the adaptive, learning-dependent, shelter-seeking response of the perceiving organism (and is often more reliable information than a TV weather forecast). We have to assume that there was past evolution of complex perceptual and cognitive processes that interpret this perceptual cue in an adaptive manner, leading to the anticipation of the rain, to the search of shelter, and to flexible responses to variations in the appearance of the sky. It is thus the receiver that actively constructs information: it is the receiver that makes the source, the black cloudy sky, into an informational input. Clearly, according to my definition, genes have no theoretically privileged informational status. They are just one type of informational source that contributes to the development of living organisms (Griffiths and Gray 2001, Sterelny and Griffiths 1999).

Of course, in order for external, non-evolved cues like a black cloudy sky to be interpreted adaptively, the interpretation system of the receiver must be able to respond to the cloudy sky, a recurrent environmental agent, by specifically altering its internal state. This usually involves a whole chain of causally related reactions. When we look at any signal transduction network, we see that the first response to a cue becomes a signal to the second element in the system, the second element's response becomes a signal to the third element, and so on. Most signals in a network are therefore both signals and responses (depending on their relative position as upstream and downstream elements in the chain of reactions). Only the primary input cue and the final output response may not have this dual function.

Communication is a special type of informational reaction, which occurs when both the source and the receiver have evolved (or been designed) to interact with each other on a regular basis. Therefore, when examining communication, the evolution of both the sender's and receiver's signaling and interpretation strategies require equal attention (Hauser 1997). MS's focus on signals means that he identifies the notion of information with communication, which is very restrictive.

A third point to note is that, according to the concept of information that I am suggesting, for a reaction to be considered as informational, the reaction of the receiver to the source must usually lead (in whatever circuitous manner) to an adaptive response. In other words, the reaction must have a *function* in the evolutionary sense. The notion of functionality as used and defined in evolutionary theory (Wright 1973, developed by Millikan 1984) is therefore central to the definition. However, my use of the term function is more general than the original usage of Wright, who defined a part's function as the part's effects which have evolved though natural selection. I combine Wright's notion with the sense of function used by Cummins (1975), who defined functions in terms of their causal role in the encompassing system's behavior, rather than with respect to evolutionary theory. I therefore employ "function" in a similar sense to that discussed by Kitcher (1993): as a general design-based notion. According to this view, the function of a part or a process (such as a reaction between an informational source and a receiver) has to be analyzed in terms of its causal role in the receiver system, which now or in the past contributed to the designed (by natural selection or by human intelligence) goal-oriented behavior of the encompassing whole (the receiver). Since, from an evolutionary perspective, in order to have a function a part (unit or process) must consistently contribute (or must have contributed in the past) to the most general goal-directed behavior of the encompassing system, i.e., the reproductive success of the organism, the notions of function and adaptive evolution through natural selection are tightly related. For a biological structure or process to have a non-incidental functional effect, it must have evolved by natural selection. However, focusing on the evolution of the receiver, which is the result of multiple and extremely complex past selective processes and various historical and developmental constraints, obviates the need to find a direct and straightforward relationship, brought about by natural selection, between the input and the response to it (which is what Wright's original definition of function required). This may be the way round Godfrey-Smith's (1993) and Amundson and Lauder's (1993) criticism of the evolutionary interpretation of function as too restricted to be of general applicability in biology. In my view, to be called "functional" the informational input has had to contribute to the evolution by natural or artificial selection of the receiver, but it may have done so in a very circuitous way.

My fourth point is related to the last one. It is that although the response has to have evolved, it is *not* necessary that it is a *directly* evolved specific adaptive response, rather than being the product of a more general evolved adaptive system. For example, an engineer can interpret a plan and build an airplane by using his highly evolved brain, but the ability to build airplanes has not evolved as a unit-trait through natural selection. Building airplanes is one of many adaptive responses of a human brain that has evolved for more general purposes. Or take the cloudy sky yet again: although the response to black clouds is adaptive, a specific "black-cloudy-sky-response" is unlikely to be an evolved unit-trait. The response is a manifestation of a more generally evolved cognitive system (the nervous system of the receiver), which is able to recognize and respond adaptively to environmental regularities. Hence, it is not always possible, or necessary, to find a simple, adaptive, directly-evolved relation between the input provided by the source and the response (output) of the receiver.

It can be argued that the information in an alarm call, a cloudy sky, or a chemical signal in a bacterial colony, has very little in common with what we call information in DNA. In the first category of cases it is easy to see what the response to the source adapts the receiver to (an alarm call warns the bird there are predators around; the cloudy sky alerts the ape to the coming storm; the chemical alerts the bacteria to imminent starvation). But it is not clear that the receiver system (the cell) "responds" to DNA, and that the response to DNA adapts the cell to anything in particular. Rather, DNA is simply "read" by the cell, so it is not information in the same sense. In other words, DNA is information "about" the cell or the organism, rather than "for" the cell or the organism. However, if instead of thinking about DNA in general we think about a particular locus with a particular allele, it is not difficult to think about the functional role of this particular allele in a particular set of environmental circumstances. Hence we can say that for all types of information, including alarm calls and pieces of DNA, a source S (allele, alarm call, cloudy sky, etc.) carries information about a state E for a receiver R (an organism or an organism-designed product), if the receiver has an interpretation system that reacts to S in a way that usually ends up adapting R (or its designer, if R is humanly designed) to E. (I am grateful to Peter Godfrey-Smith for suggesting this way of expressing the idea.) However, for S to be an information input rather than a source of energy or material, it is also necessary that it is its *form*, and variations in this form, rather than any other attribute, which affect R's specific response.

This leads to my fifth point: for there to be information, whether associated with evolved or non-evolved sources, it is necessary that there is a consistent relation between variation in some aspect of the *organization* of the source and the functional response of the receiver. This is true by definition for any evolved receiver system—there are always at least two alternative "interpretations" to a signal in an evolved system. At least one organizational state of the source leads to a process resulting in a functional response by the receiver, while all other organizational states do not. In most cases there are more than two alternatives (+ and –), of course. Thus, variations in the appearance of the sky lead to variations in the functional responses of the intelligent ape, variations in DNA lead to corresponding variations in the behavior of the cell, variations in an alarm call lead to variations in the response of group members, and so on.

3. Other properties of biological information

MS and his commentators referred to several other properties of genetic or genetically-coded biological information: informational signals are gratuitous (Monod 1971); they are relatively specific; they imply notions of misinterpretation or error; and they are intentional. In addition, MS noted that whether or not there is adaptive feedback from response to signal is contingent upon the particular properties of the information-processing system (see MS 2000, Sterelny 2000, Sarkar 2000, Godfrey-Smith 2000). All of these properties, which I discuss below, are compatible with my general definition of information, and are not limited to the specific case of genetic information.

When the informational inputs are signals produced by an organism as part of a communication or transmission system, rather than being abiotic cues (e.g., a cloudy sky or day length), they are gratuitous in that there is no necessary connection between their form and their function. This gratuity (also called semioticity, see Sarkar 2000) is an inevitable consequence of the *evolved* nature of any biological signaling system: if evolution had been different, the biologically produced signal associated with a given adaptive response could, and almost certainly would, have been different. It must be noted, however, that although *all* evolved signals are arbitrary or gratuitous in this general sense, for some signals their meaning is associated with their production cost. Take, for example, the meaning of the peacock's tail: as an honest indicator of its owner's fitness, the message of the tail to the peahen is associated with the cost of producing and maintaining it. With other types of signals, such as spoken words, there is no relation between the form of the signal and the cost of production. As Lachmann et al. (2001) have pointed out, when the receiver can immediately evaluate the honesty of signals, and can punish dishonesty, signals can be cheap to produce and their form is unrelated to the production cost. Thus, although all evolved signals are gratuitous, we need to qualify the way the form of a signal is related to its production and maintenance cost. Gratuity is also a feature of abiotic information cues: the fact that the cloudy sky is a reliable information cue for an ape is dependent on the peculiarities of the ape's perceptual system and brain, both products of evolution. A different evolutionary path would have led to another aspect of the environment being used by the organism as a cue alerting it to an imminent storm. There is nothing necessary in a particular aspect of the environment acting as information. "Information" is conferred by the receiver.

The way in which the status of the source as an informational input depends on the interpreting system is clear if we consider the following hypothetical case. Imagine, as James Griesemer (forthcoming) suggests, that in very ancient proto-cells, DNA was not used as an informational resource, but as a high-energy storage polymer. High-energy bonds were stored in linear, largely inert, DNA polymers that could be broken up when needed. The sequence of the DNA polymer was, of course, of no significance. Now add to the scenario the somewhat unlikely supposition that such a storage-DNA-polymer has, *by chance*, the precise sequence of the gene coding for the alpha chain of hemoglobin. This of course has no special consequences for the proto-cell, since there is no cellular system that can interpret this sequence in a specific way. As far as the proto-cell is concerned, any DNA sequence of the same length with the same AT/CG ratio is the same. Contrast this with a nonsense DNA sequence with an accidentally open reading frame in a modern cell: this sequence will be interpreted by a modern evolved cell (or cell-free extract) even if the sequence has been randomly generated by a DNA synthesizer machine. This latter case illustrates something obvious and important which has been noted by MS and others: once an input (even if randomly generated) conforms to a type that is recognized by an interpreting organism as an informational input, it will be interpreted—treated as information—even though this may well lead to a non-adaptive response. And of course, mistakes in the process of interpreting any input can also lead to non-adaptive responses. The notion of interpretation is always associated with that of occasional misinterpretation, or error.

Another point which MS noted is that the irreversibility of information transfer—in his terms, it moves from signal to response but not from response back to signal—is a contingent fact, *not* a necessary property of an information processing system. Often, but not always, there is no adaptive feedback from response to signal. Irreversibility is certainly true for some abiotic information sources: the response of the plant to day length does not affect the latter, and reading a recipe does not alter it. In the sense that (as the central dogma asserts) there is no back translation, the flow of information from DNA to protein is not reversible either. On the other hand, a cellular response such as a change in gene regulation (the culmination of signal transduction) often has feedback effects on signaling molecules earlier in the chain, which lead to the chemical modification of these molecules (Heldin and Purton 1996). And of course neural signals are often modified by the perceptual or motor response of the animal. Even with the genetic system, the idea that there may be an adaptive feedback between the response and the original source (DNA sequence) is not far-fetched, and does not depend on reverse-translation (Caporale 1999, Jablonka and Lamb in preparation). Clearly, when adaptive responses to the environment feed back into the interpreting system, they do so in a way that corresponds to its regulatory organization. In the case of the DNA system, what we know of gene regulation implies that such feedback must involve modifications in regulatory DNA sequences. Of course, the ability to make such an adaptive response must be the result of natural selection. In all signal transduction systems, environmental cues and their regular modifications become informational signals by virtue of the properties of the evolved receiver system.

MS has noted that the most important characteristic of biological information is that it is designed (by natural selection or by human intelligence), and in this sense is "intentional." Hence, MS's notion of intentionality is related to the notion of function (as is Millikan's (1984)). According to MS:

> In biology, the statement A carries information about B implies that A has the form it does because it carries that information. A DNA molecule has a particular sequence because it specifies a particular protein, but a cloud is not black because it predicts rain. This element of intentionality comes from natural selection."
>
> (MS 2000, 189–190)

In Maynard Smith's terms, A is seen as a signal that carries information about something (B, the protein, a trait), and we may presume that it is the organism that this information is for. According to my interpretation, A can be said to carry information for B and about the environment of B, because

B has been designed to so interpret A. A in this case is an informational input, which can, but need not be a signal, and B is the receiver of this input, which interprets it. Although signals, by definition, carry information for receivers and about senders, and hence can be said to be "intentional," a more general ascription of evolutionary "intentionality" is to the interpretation processes, because this applies to the transmission of information in general (about the environment and for the receiver), and not just to the special case of communication. In the case of the genetic system, DNA (like any intenal within-receiver signal, such as a hormone), is also part of the receiver itself. This is a special case of tightly evolved receiver-signal system. It is only when we think about the difference that *variations* in a particular allele make, that we can differentiate between the signaling aspect of DNA and its aspect as part of the receiver.

I have not addressed the question of the origin of information—how functional information first emerged (which is a different formulation of the origin of life question). The analysis of this question is beyond the scope of this article (for discussion see Maynard Smith and Szathmáry 1995). However, according to my definition, information is something that can exist only when there are living (or more generally, designed) systems. Only living systems make a source into an informational input. The more complex the organism, the more information it constructs.

4. The inheritance of information

My definition of information allows me to examine and compare different types of heritable information without a priori biasing the discussion towards the idiosyncrasies of genetic information and transmission. However, before I make such comparisons, bearing in mind Oyama's criticism of informational terms in biology (Oyama 1985, 2000, 2001), I need to stress that I use the term *transmission* not for the handing over of pre-existing entities, but to denote any process that results in an organization pattern from one entity being reproduced in another. Thus, when talking about heredity, an entity is related to others by special processes that lead to the reconstruction of its organization in those other entities. Examples of such special processes are DNA replication; the epigenetic mechanisms involved in reconstructing functional cellular states; socially mediated learning, etc. Inheritance implies that an entity is both a receiver and a potential source of information, and information is heritable when it leads to the reconstruction of some aspects of the internal organization of one entity in another. The process of transmitting information from one entity to another delimits what I call a "generation." This means that one can use "generation" in relation to both genetic (parent-offspring) and cultural (teacher-student) transmission. It also means that all the "offspring" of a given "parent" (whether genetic or cultural) belong to a single generation, even when produced at long time intervals, with some offspring being much older than others. Moreover, horizontal transmission—the transfer of information between individuals belonging to the same generation—can be treated in the same terms: whatever their age, individuals who possess information are the "older generation" with respect to those that acquire the information through interactions with them. Transmission from chronologically older to younger entities (which may sometimes be mediated through horizontal transmission) is, however, necessary for long-term evolution.

According to my definition of information, all processes of re-production and inheritance entail informational reactions. However, there are different types of inheritance systems, and different types of information (Jablonka, Lamb, and Avital 1998, Jablonka 2001). So although general concepts of information and inheritance are useful and necessary when considering the broader issues in evolutionary theory, these concepts have to be qualified if they are to be of more specific use. For this we need answers to a number of questions about information in the different inheritance systems. Is there heritable variation in information? If there is, is it 'limited' (with very few variants possible) or 'unlimited' (effectively with an infinite number of variants) (Maynard Smith and Szathmáry 1995)? How is heritable variation generated (are there evolved systems for generating variations, or is new variation the result of accidents and mistakes)? Is heritable information reproduced component by component (modular, as in DNA replication), or as a whole (holistic, as in an autocatalytic cycle) (see Maynard Smith and Szathmáry 1995, Jablonka and Szathmáry 1995, and Szathmáry 2000)? How is information transmitted (is there a dedicated system for the transmission of a given type of information)? Is transmission sensitive to the functional effects of the information and to the process of transmission itself? Can latent (potential, non-actualized) forms of information be transmitted? What is the dominant mode of transmission (vertical or non-vertical)? These questions imply that the genetic system is not the only system through which information is inherited, although it is the most fundamental one, and variations in the genetic system can affect all levels of biological organization and information.

I have summarized the properties of different evolved biological information systems that operate in inheritance in the form of two tables. Table 1 groups different information processing systems into general categories, and compares them with respect to the heritable variation associated with each category. Table 2 compares the way information is re-produced by different inheritance systems. Since some of the terms may be unfamiliar, I shall outline the salient properties of the each inheritance system. (For a more leisurely and detailed discussion see Jablonka 2001).

The genetic inheritance system (GIS) is based on the replication of DNA, a linear molecule made up of four different units. These units, the nucleotides containing the bases adenine (A), thymine (T), guanine (G), and cytosine

Table 1 Variations in information in different types of inheritance systems

Information system	*Types of heritable variation*	*Variations in information are transmitted by:*	*Evolved systems for generating adaptive variations (origin of variation)*	*Range of variation*
Genetic: DNA + DNA processing machinery	Variations in DNA sequences (point mutations, deletions, amplifications rearrangements, etc.)	Genetic inheritance system (GIS)	Present, but most variations are blind	Unlimited
Epigenetic cellular networks	Variable functional states of self-sustaining networks; variable architecture of molecular complexes (e.g., prions); variable transcripts; variable marks (e.g., methylation and protein marks)	Cellular epigenetic inheritance systems (EISs): self-sustaining regulatory loops; 3D templating of molecular structures; chromatin marking; RNA-mediated gene silencing	Specific adaptive feedbacks; also blind variation	Regulatory loops: limited at the level of the loop, unlimited at the cell level; 3D structures: limited at single complex level, unlimited at the cell level; RNA transcripts: limited at the transcript level; chromatin marks: possibly unlimited
Neuro-hormonal system (within-organism)	Variant physiological states	Reconstruction of developmental legacies	Specific adaptive feedbacks; also blind variation	Limited
Communication among organisms that is learning independent	Variant physiological states	Reconstruction of developmental legacies	Specific adaptive feedbacks; also blind variation	Limited
Communication among social non-human animals	Variant socially learned habits	Behavioral inheritance systems (BISs): socially mediated learning, sometimes leading to animal traditions	Present, and some blind variation	Limited at the level of a single habit, may be unlimited at level of lifestyle
Communication through symbols	Variations in symbols (e.g., linguistic)	Human symbolic culture (SIS)	Present and dominant	Unlimited at most levels

Table 2 Modes of information re-production through different inheritance systems

Type of inheritance system	*Alteration and re-production of variation*	*General dedicated copying system exists*	*Latent (non-expressed) information transmitted*	*Directions of transmission*
GIS	Modular	Yes	Yes	Mostly vertical
EISs				
Self-sustaining loops	Holistic	No	No	Mostly vertical
Structural templating	Holistic	No	No	Vertical and horizontal
RNA silencing	Holistic	Yes	Sometimes	Vertical and horizontal
Chromatin marks	Holistic and modular	Yes (for methylation)	Sometimes	Vertical and horizontal
Organism-level developmental legacies	Holistic	No	No	Mostly vertical
BISs	Holistic	No	No	Vertical and horizontal
SISs	Modular and holistic	Yes	Yes	Vertical and horizontal

(C), can be altered and replicated one by one. Maynard Smith and Szathmáry (1995) refer to this as "modular" replication. The replication machinery is insensitive to the sequence of the four units, and to the functional utility of the sequence. The modular nature of alteration and reproduction allows a large number of variations in sequence to be generated and transmitted. A DNA sequence of 100 nucleotides has more possible variations that the number of atoms in our galaxy, so, in the terms of used by Maynard Smith and Szathmáry (1995), heredity is unlimited. In sexually reproducing organisms, most genetic variation is generated through the sexual process—through the segregation and recombination of chromosomes during gamete formation, and the subsequent fusion of different gametes. New variations in DNA are often the results of chance mutations: they result from replication mistakes, unrepaired or misrepaired damage caused by physical and chemical insults, and the activity of mobile genetic elements. In addition, there are evolved cellular systems that can target mutations to specific regions in response to particular classes of stressful environmental circumstances (for reviews see Caporale 1999).

The genetic inheritance system is the only one of the types of inheritance that I shall describe in which the generation and the transmission of variations in information is largely independent of the developmental conditions of the organisms (even if we allow for some targeted mutations). But even with the GIS, there are internal filters and discriminatory criteria for what are acceptable DNA variations, so the organism is not a passive vehicle for all chance DNA changes. In addition to having an impressive repertoire of repair systems, foreign DNA that is introduced into a eukaryotic cell is met with some formidable intracellular "immune" responses: such foreign DNA, especially when it arrives in many copies, is usually silenced, and sometimes excised and mutated (see for example, Irelan and Selker 1997). Nevertheless, much genetic variation does seem to be transmitted without regard to its functional content.

Epigenetic inheritance systems (EISs) underlie cellular heredity—the inheritance, from one cell generation to the next, of functional and structural cellular variations that do not depend on differences in DNA sequence. A functional cell state may persist through many cell divisions, even when the stimulus that originally induced this state is no longer present. Cellular heredity is usually considered in the context of cell differentiation: in the same multicellular body, most cells have identical DNA, yet liver cells divide to give more liver cells, skin cells give more skin cells, and so on. The information underlying the specialized cell states is transmitted from mother cell to daughter cell in each lineage, even though the lineages have the same DNA. Epigenetic inheritance is not limited to within-organism inheritance, however. Variations in cellular states that are independent of DNA variations are transmitted between generations of unicellular organisms, and also between multicellular organisms such as plants, fungi, and mammals (for reviews see Jablonka and Lamb 1995, 1998, and in preparation).

There are several different mechanisms of epigenetic inheritance. Probably the most common EIS is based on self-sustaining regulatory loops, through which the activity-state of a metabolic network is maintained through positive feedback by one of the gene products in the network. The simplest case is when a gene positively regulates its own transcription In this case, a gene, which was originally turned on through an interaction with an external agent, produces a product that reacts with its own regulatory region to induce transcription, thus making the external agent unnecessary for its continued activity. Two genetically identical lineages of cells in an identical environment can thus have different heritable metabolic states if their inductive histories have been different.

A second type of EIS is based on templating: pre-existing three-dimensional cellular structures act as templates for the production of new daughter-structures. Such architectural templating is thought to underlie the multiplication and transmission of prions, the infectious protein particles that are associated with degenerative diseases of the nervous system such as BSE (bovine spongiform encephalopathy, or mad cow disease), scrapie in sheep, and CJD (Creutzfeldt-Jakob disease) in humans. In this EIS and the previous one, transmission is "holistic" rather than modular, because the whole structure, or the functional state of the loop as a whole, is transmitted from cell to cell.

The third EIS, chromatin marking, is based on the inheritance of the patterns of DNA-bound molecules such as proteins, RNAs, or small chemical groups. Collectively, these are referred to as "chromatin marks." Marks are established during development and as a result of environmental induction. The chromatin marks on a gene's DNA sequence affect the regulation of its transcription, and hence how a character develops. Crucially, some marks can be reproduced when DNA replicates. Cytosine methylation at CG sites is good example of cell-heritable chromatin marks. Cytosine, one of the four DNA nucleotides, can have a methyl group attached to it (C^{m}), or it can lack it (C). When cytosine has guanine next to it, the two strands of DNA have mirror image symmetry: CG on one strand is complementary to GC on the other; similarly $C^{m}G$ is complementary to GC^{m}. Patterns of DNA methylation can be reconstructed following semi-conservative DNA replication because when regions of DNA with $C^{m}G$ sites replicate, the old strands retain their methylated state. The complementary GC sites on the new strands have non-methylated cytosine, so, immediately after replication, the DNA duplex is hemi-methylated; it is $C^{m}G/GC$. However, enzyme-complexes that recognize and have a special affinity for these non-symmetrical, hemi-methylated sites preferentially methylate the new strand. Since the enzyme-complex does not recognize and methylate sites that are symmetrically non-methylated, the original methylation pattern is reconstructed: a methylated site in the old strand ends up having a methylated complementary site, while a non-methylated site in the old strand has a non-methylated complementary site.

In this way the pattern of methylated and non-methylated sites is reproduced in both daughter molecules. Other types of chromatin marks, for example DNA-bound protein marks, are also reproduced following DNA replication, although the details of the processes are not understood.

The fourth EIS involves a recently discovered epigenetic mechanism known as RNA-mediated gene silencing (Fire 1999). The system is based on the silencing effect of small RNA molecules (known as siRNAs) that originate from much larger RNA transcripts. What seems to happen is that RNA molecules that have certain topological peculiarities are recognized by an enzyme that chops them up into small siRNAs. Then, with the help of special RNA-replicating enzymes, these siRNAs are replicated many times, and copies are transmitted to daughter cells when the cell divides. The siRNAs can also move from cell to cell, behaving like infectious agents or chemical signals. In ways that are not yet fully understood, siRNAs destroy copies of the large transcript from which they were derived. They also associate with copies of the gene that mothered their transcript. This leads to the formation of a stable chromatin mark (often a methylation pattern), which suppresses the activity of the gene and can be transmitted to the next cell generation.

With all four EISs, for functional or structural variations to be transmitted to the next cell generation, they have to have special properties: for the first EIS there has to be a special type of regulatory network-organization; for the second, a special structural organization; for the third, a particular type of chromatin marking; for the fourth, a peculiar RNA topology. Cellular variations that are not associated with these special types of organization cannot be passed to the next generations. Thus, unlike DNA variations, which can always be transmitted whatever the sequence, only a subset of the functional and structural states of cells is potentially heritable. Moreover, at the level of the single heritable unit (a self-maintaining loop, a three-dimensional structure, a chromatin mark, an RNA-mediated state of gene silence), the number of variant heritable states is usually limited: a self-sustaining metabolic loop is usually either "on" or "off"; a particular prion has very few heritable conformations; a protein mark on a given DNA sequence has few heritable alternatives; an RNA transcript is either recognized as peculiar and its gene is silenced or it is not. The methylation system is somewhat exceptional, since a CG rich gene has many potential heritable patterns of methylation, although probably only a small set of them may make a functional difference to the gene's transcriptional activity. In general, however, at the level of individual units, heritable variation in the epigenetic system is very limited.

Developmental legacies, the molecular products and consequences of the parents' developmental history, can sometimes be re-produced and regenerated in the offspring generation (see Griesemer 2000). For example, environmentally induced maternal effects in plants can be inherited for several generations (see Miao, Bazzaz, and Primack 1991); multigenerational interactions between symbionts and their hosts that involve direct transmission of the symbionts, or that occur through the re-construction of parental developmental effects can be reproduced (Paracer and Ahmadjian 2000, Sterelny 2001); food preferences and host preferences may be reproduced within lineages of insects (Thompson and Pellmyr 1991); gender behavior and sex ratio can be inherited in lineages of Mongolian gerbils (Clark, Karpiuk, and Galef 1993). In all these cases and many others, developmental legacies are re-produced in ways that are specific to the particular case. There is no general machinery for transmitting all variant developmental legacies, whatever they are. On the contrary, only a small set of the many developmental variations that an organism can exhibit are reliably reproduced in its offspring, and like most cellular epigenetic variations, these too usually show limited heritable variation.

Behavioral inheritance systems (BISs) lead to the reproduction of information about behavior—about behavioral preferences and patterns of behavior. The activities of one individual, the source, change the state of another individual, the receiver. The transmission and interpretation of behavioral information are through processes of socially-mediated learning. In several different ways, naïve individuals can learn and develop patterns of behavior similar to those of experienced individuals (often, but not necessarily, their parents). This type of re-production depends on the social organization of the relevant social unit (the family, extended family or non-kin group), on social attentiveness, and on several types of social learning. The mechanisms of socially mediated learning are quite general, but whether or a not a particular pattern of behavior is reproduced in descendants depends on the animal's memory, its learning capacity, the social setting, the nature of the developing behavior, and so on. When habits are regularly transmitted between generations through socially mediated learning, animal traditions are established (Avital and Jablonka 2000).

The symbolic inheritance system (SIS) is a specifically human inheritance system. The transmission of information through symbols allows a great deal of cultural and social evolution, and this has been modelled and discussed extensively (for a recent summary of the different approaches, see Laland and Brown 2002). Symbol-based information transmission can occur both vertically and horizontally, and in several different ways. These have somewhat different properties. Language, for example, is modularly organized at the level of sentences, and an enormous number of linguistic variations can be generated and transmitted. But a symbolic system in which information is acquired and transmitted through pictures has a more holistic organization, with the parts being more interdependent. Unlike BISs, symbolic systems can transmit *latent* information about behavior (e.g., non-acted-upon ideas, transmitted through written or oral traditions). In this, SISs resemble the genetic system, but this does not mean that the reproduction of information through SISs is insensitive to content. Some mechanical copying may occur (through photocopiers for example), but usually a large element of symbol-based transmission is sensitive to the content of the information, and to the psychological, social, and cultural development of the individual or the group. Thus, unlike most new information in the GIS, new symbolic information is targeted because its acquisition and transmission are guided by basic cognitive coherence rules and categories. Symbolic information, like all behavioral information, is also constructed and edited—it is tested and adjusted to fit existing ideas and habits before it is passed on. A feature peculiar to the construction of symbolic information is that future goals and future plans are involved in the construction process.

In addition to the direct reproduction of information through the various inheritance systems just described, information can also be transmitted indirectly, through ecological niche construction (Lewontin 1983, Odling-Smee 1988, Odling-Smee, Laland, and Feldman 1996). Organisms often create an environment for their descendants, providing a niche that is usually quite similar to their own. Through their activities, they modify the environment (think, for example, how beavers alter the landscape), and their descendants react and respond (often adaptively) to cues in the environment that their ancestors constructed. When the constructed niche is the society and the culture, niche construction is an inseparable aspect of the BISs and the SISs. The source of information is the environment, which is being constructed by the behavioral (or symbolic) information of the organisms that are interpreting it and altering it.

This brief survey of the different systems and ways through which information is transmitted between generations highlights several points. First, as Tables 1 and 2 both show, there are a lot of different information processing systems and different types of information, and we need to study them in their own right, rather than as shadows of the GIS.

Second, as Table 1 shows, although heredity may be limited at one level of organization, it can be unlimited at a higher level. Hence, a single site in DNA has limited heredity, with only 4 variant transmissible possibilities, while a DNA sequence of 10 nucleotides strung together has over a million variations. Similarly a self-sustaining loop may have only two transmissible states, but a cell with 20 independent loops would have more than a million transmissible variant states.

Third, a lot of heritable variation is targeted, and some of it is also edited and modified by the organism before it is passed on. Since the internal mechanisms that target and construct variation filter out a lot of noise, the low fidelity of transmission of a variant does not lead to the disintegration of the information transmitted by the inheritance system. Most of the new targeted and filtered variations that are transmitted have a high chance of being either neutral or adaptive.

Fourth, the tables show that a lot of information is transferred horizontally, among entities that do not have a genealogical parent-offspring (vertical) relationship. In the genetic system, horizontal gene transfer is usually virus-mediated, and such transfer can occur not only within a population, but also among individuals belonging to very different taxa. Patterns of behavior and ideas are also often transmitted horizontally across genetic lineages. Of course, when reproduction is looked at from the virus's or the idea's point of view, it can be claimed that information transfer is always vertical—viruses give rise, vertically, to more viruses, and ideas spread by various processes of vertical, symbolically mediated, social learning. However, neither viruses nor ideas have an independent reproductive ability—the reproduction of viruses depends on a cellular system of interpretation and replication, while the spread of ideas depends on the structure of the cognitive system of the individuals and groups that construct, transmit, and acquire them. Moreover, with ideas, both their acquisition and transmission are part of the developmental-learning processes of senders and receivers, which often feed back on their very content and structure. Nevertheless, with viral infection and spread, and with the systems through which ideas are transmitted and reconstructed, two parties—hosts (receivers) and viruses (senders), and the senders and receivers of ideas—partake in the process and co-evolve. Thus viruses will co-evolve with their hosts, and the way of transmitting ideas will co-evolve with the way these ideas are acquired. One interesting result of this co-evolution will be that the form of the virus (its nucleic acid sequence and often its resultant morphological form), and the form of an idea (how well it fits the cognitive biases of cultural agents) will change through evolution.

It is difficult to evaluate the relative importance of the different inheritance systems and the different modes of information transmission in evolution. The first problem is that for many of the inheritance systems, the empirical and theoretical work has not yet been done. The second is that the theoretical criteria that have been used for high evolvability are commonly derived from the peculiar properties of gene-inspired replicators and the genetic inheritance system. Sterelny (2001), for example, defines sets of conditions that are important for the high evolvability of different types of entities with different ways of transmitting information. His criteria are based on the properties of the replicators of these entities, and his notion of "replicator" is inspired by the prototypical replicator, the gene. As a result, when Sterelny finds that human cultural transmission does not score high according to these criteria, he concludes that cultural transmission is not an inheritance mechanism. However, the criteria that Sterelny uses are not equally relevant for the evolvability of entities that have non-genetic inheritance systems. For example, the "high fidelity of replication" criterion which Sterelny adopts is not as relevant when there are internal filtering mechanisms that ensure that the transmitted variations are often adaptive. Sterelny's reliance on gene-based criteria also overlooks the fact that the accumulation of socially learned variations need not occur along a single axis (e.g., one habit becoming increasingly more complex), but can affect other correlated habits and thus alter a whole lifestyle in a cumulative manner (Avital and Jablonka 2000).

The evolution of new types of information and of interpretations of information is a subject that has been receiving a lot of attention recently (Jablonka 1994; Maynard Smith and Szathmáry 1995; Jablonka, Lamb, and Avital 1998). I shall not go into this question here, but I want to point out that the evolution of new types of information and interpretation systems (and new sources of functionality) has occurred several times during the evolution of life. Moreover, when a new type of information evolves, the new information system, and the evolution based on it, can be decoupled from the evolution based on the information at the primary DNA level (which has enabled the new information system to evolve). An obvious example is the extensive decoupling of information at the cultural level from information at the DNA level. However, as Tables 1 and 2 show, when new information is associated with a new system of heredity, *all* types of heritable information have a degree of autonomy, and it is important to see how this autonomy affects evolutionary change (Sterelny 2001).

5. Information sharing

Defining information in the way that I have suggested, and recognizing that information is inherited in a variety of ways, not just through the genetic system, changes the way that one approaches various evolutionary problems. One of these is the evolution of communication. Communication involves either receiving a signal from a sender or a reciprocal exchange of signals, and for present purposes both will be said to involve "information sharing."

A major problem for evolutionary biologists is to explain the very existence of evolutionarily stable communication. When organisms communicate, information is transmitted: a signal sent by a signaler influences the internal state of a receiver, and usually affects its behavioral responses. In evolutionary terms, we need to know how the signaler benefits from accurately representing the state of the world through its signaling (rather than misrepresenting it), and how the receiver benefits from heeding the signal rather than ignoring it. If communication has evolved, then both sender and receiver should gain (on average) from sending and receiving a given signal. Yet sender and receiver often do not have coincident interests, and sometimes it may benefit senders to misrepresent the world, and benefit receivers to ignore signals. So why is communication so often stable? What are the evolutionary incentives for not sending misleading signals?

One answer is that the individuals involved do have coincident interests, for example genetic interests, so selection has favored reliable signaling in kin groups. Another possibility is that since signalers and receivers often change roles, only the senders of reliable signals will receive reliable signals, so the benefit comes from reciprocation (Hauser 1997). It is also possible that communication enhances group productivity, and selection between groups is important in maintaining systems of intragroup communication (Avital and Jablonka 2000). Yet another possibility is that receivers evolved to respond only to costly signals, because when a signal is costly to produce, and the costs are proportional to the signal's veracity, only costly and hence reliable signals will not be ignored (Zahavi and Zahavi 1997). Related to this is the possibility that when the cost of misleading (rather than the cost of producing an honest signal) is high, signals will be reliable (Lachmann et al. 2001). In general, it is necessary to assume that there are always some special types of social interactions, population-dynamic effects, or balance of costs that make communication or any other form of cooperation stable.

Given that conditions promoting evolutionarily stable cooperation exist, how does giving and receiving information differ from giving and receiving non-informational resources (material, energy)? Avital and Jablonka (2000) and Lachmann, Sella, and Jablonka (2000) have stressed that the most important difference between the sharing of informational and non-informational resources is that information sharing is not subject to the intrinsic conservation principles that characterize the sharing of non-informational resources. When an organism shares food with another, it loses the amount that it gives away. But an organism that shares information with others retains its own information, and the cost of information sharing need not be proportional to the number of individuals with whom the organism shares it. For example, take a mother chukar partridge who is being followed by her chicks as she forages for edible seeds, and emits a food call when she encounters an edible food source. She can be said to be sharing her information with the chicks while "demonstrating" to them what is good to eat. In doing so, she does not lose information about what is edible; in fact, by acting upon her information—going through the processes of interpreting it again—she is probably often consolidating it, making it more reliable. Practice bolsters memory, so in a sense the more information you give away the more you have! Moreover, "demonstrating" to eight rather than to four chicks does not require a proportional increase in the mother's expenditure of energy; the extra cost to the mother, if it exists, is negligible (Smith 1977, Avital and Jablonka 2000). Of course, information sharing can have a significant cost, and sharing information about a limited energy resource (food) is just as costly as the direct sharing of energy. But in many cases the cost of mutually sharing information is only a fraction of the cost that a solitary individual would incur to gain the same amount of information.

Lachmann, Sella, and Jablonka (2000) modeled a simple case of information sharing. They assumed that all individuals in a group (such as bacteria in a colony) interact with the environment and make a response to a particular cue (e.g., the concentration of a nutrient). The environmental cue is therefore seen as carrying information about the environment, and the organisms respond to the cue by changing their behavior, for example by releasing a chemical factor into the medium. Each individual can make mistakes in its interpretation of the cue, and therefore responses are error-prone. The responses of each individual are, in turn, used as a signal, a source of information, to other individuals, about the interpretation of the environmental cue the individual has received—about its evaluation of the environment. An organism is assumed to have more information about the environment if it has more signals available, and its fitness is proportional to how much information it has. Information is said to be shared when individuals make their responses to the external cue available to all the other members of the group. The results of the simple models are straightforward: information sharing leads to fitness gains because it reduces the costs of responding correctly to the environmental cue. This is true even when we take into consideration errors in transmitting information and the costs of sharing it. For an individual in a group, the cost of acquiring information decreases with growing group size, and the amount of information that it can gather increases with the size of the group. In sharing conditions, an individual pays less for more! Therefore, in cooperation-promoting social conditions, information sharing is expected to be common, as indeed it is. For example, myxobacteria measure and respond to the state of nutrient in their local environment, and the response of each individual is made available to other neighboring cells. As a result of this signaling (in which as many as 100,000 cells can participate), each cell has more information, and this leads to a global response: the cells aggregate and form a fruiting body when the information indicates that the food supply is globally exhausted, or alternatively, continue their semi-unicellular existence if food is sufficient (Kaiser 1999).

These special properties of information sharing may have non-trivial evolutionary consequences. For example, while the cost of acquiring information decreases with group size, the benefits from any additional information eventually level off, so that big collectives may dedicate only a fraction of their individuals to collecting information. This might guide the evolution of specialization and a division of labor, and be an important factor in the transition to new levels of biological organization, such as the transition to multicellularity (Lachmann, Sella, and Jablonka 2000).

Thinking in terms of information also forces one to pay close attention to the context in which information is acquired, used, and shared. One of the examples that Avital and Jablonka (2000) discussed is the role of information sharing in the long-term monogamous relationships of birds. They suggested that the choice of mate and the consolidation of the relationship between the mates might involve acquiring and sharing information that is individual and pair-specific. This means that mate choice does not depend only on identifying characters in a potential mate that are indicators of good genes; it also depends on obtaining information about the potential mate's ecological and social background from things such as song dialect, or past familiarity. These indicate how compatible the potential partner is with the more idiosyncratic and personal demands and needs of the particular individual seeking a mate. Once a mate has been chosen, mutual learning may produce non-transferable information. For example, in some African shrikes and other songbirds, the two mates perform a mutual singing display, a well-coordinated "duet" that is learned over a period of weeks. The learned duet is specific to the pair, and should one of the mates desert, a new duet would have to be learned, at considerable cost, with a new mate. Although the information the duet carries can be repeatedly used and re-adjusted by pair members, this pair-specific information is not readily applied to relationships with other individuals. Obviously, when information is individual and pair-specific in this way, the costs of deserting a mate are high, and the relative benefits of cooperation are higher than they would be when information is transferable. Avital and Jablonka (2000) have argued that the non-transferable aspect of information sharing between long-term monogamous mates may be one of the reasons for the remarkable stability of such cooperative relationships.

The few examples described above show that focusing on the special features of information highlights some important aspects of communication, and allows us to offer new evolutionary explanations for the observed stability of some types of communication.

6. Conclusions

I have suggested that thinking about information in terms of the receiver's interpreting system makes it possible to expand MS's notion of biological function-related information and formulate a general definition. According to my definition, a source becomes an informational input when an interpreting receiver can react to the form of the source (and variations in this form) in a functional manner. The definition that I have suggested is therefore a broad one, appropriate for information stemming from environmental cues as well as evolved signals. The relationship between the input and the receiver's response can be complex, since the evolution of the interpretation system mediating between the two will often relate input and response only indirectly. Nevertheless, focusing on the interpreting receiver makes it possible to compare information from different types of sources.

Although it is generally recognized that there are different types of biological information and information processing systems, it is less generally appreciated that there are also several types of hereditary and evolutionary phenomena associated with these systems. Information is transmitted not only through genetic inheritance, but also through cellular and organism-level epigenetic processes, through non-symbolic social learning, and through symbolic learning. Each of these non-genetic inheritance systems has its own regularities and constraints, which are very different from the regularities and constraints of heredity associated with DNA replication. Consequently, heredity and evolution through these systems have to be studied on their own terms, rather than as somewhat deviant reflections of genetic inheritance and evolution. Moreover, since biological information has special properties that make its transmission very different from the transmission of materials and energy, sometimes focusing on these special properties can uncover interesting evolutionary effects. For example, the evolution of communication among co-operating members of a group becomes more understandable.

I share with Maynard Smith and Szathmáry (1995) the conviction that the evolution of ways of storing and transmitting information has been a major theme in the evolution of life. As I see it, one of the most important aspects of this evolution has been the selection of systems that can interpret additional types of information, and the consequent emergence of new inheritance systems. I believe that these new inheritance systems were the basis of some of the major transitions that led to the evolution of new levels of complexity and individuality. Therefore, understanding these evolutionary transitions will require thinking about all types of heritable information, not just that which is transmitted through the genetic system.

Acknowledgements

Many many thanks to Marion Lamb, Michael Lachmann, James Griesemer, Susan Oyama, Ayelet Shavit, and Eytan Avital for their constructive and critical comments. I am also grateful to Peter Godfrey-Smith and an anonymous referee for their thoughtful reviews of the previous version of this paper.

References

Amundson, Ron, and George V. Lauder (1993), "Function without Purpose: The Uses of Causal Role Function in Evolutionary Biology", *Biology and Philosophy* 9: 443–469.

Avital, Eytan, and Eva Jablonka (2000), *Animal Traditions: Behavioural Inheritance in Evolution*. Cambridge: Cambridge University Press.

Caporale, Lynn H. (ed.) (1999), *Molecular Strategies in Biological Evolution*. Annals of the New York Academy of Sciences, Volume 870. New York: New York Academy of Sciences.

Clark, Metrice M., Peter Karpiuk, and Bennett G. Galef Jr. (1993), "Hormonally Mediated Inheritance of Acquired Characteristics in Mongolian Gerbils", *Nature* 364: 712.

Cummins, Robert (1975), "Functional Analysis", *Journal of Philosophy* 72: 741–765.

Fire, Andrew (1999), "RNA-Triggered Gene Silencing", *Trends in Genetics* 15: 358–363.

Godfrey-Smith, Peter (1993), "Functions: Consensus without Unity", *Pacific Philosophical Quarterly* 74: 196–208.

—— (2000), "Information, Arbitrariness, and Selection: Comments on Maynard Smith", *Philosophy of Science* 67: 202–207.

Griesemer, James (2000), "Development, Culture, and the Units of Inheritance", *Philosophy of Science* 67 (Proceedings): S348–S368.

—— (forthcoming), *Reproduction in the Evolutionary Process*. Chicago: University of Chicago Press.

Griffiths, Paul E., and Russell D. Gray (2001), "Darwinism and Developmental Systems", in Susan Oyama, Paul E. Griffiths, and Russell D. Gray (eds.), *Cycles of Contingency*. Cambridge, MA: MIT Press, 195–218.

Hauser, Marc D. (1997), *The Evolution of Communication*. Cambridge, MA: MIT Press.

Heldin, Carl-Henrik, and Mary Purton (eds.) (1996), *Signal Transduction*. London: Chapman and Hall.

Irelan, Jeffrey T., and Eric U. Selker (1997), "Cytosine Methylation Associated with Repeat Induced Point Mutation Causes Epigenetic Gene Silencing in *Neurospora crassa*", *Genetics* 146: 509–523.

Jablonka, Eva (1994), "Inheritance Systems and the Evolution of New Levels of Individuality", *Journal of Theoretical Biology* 170: 301–309.

—— (2001), "The Systems of Inheritance", in Susan Oyama, Paul E. Griffiths, and Russell D. Gray (eds.), *Cycles of Contingency*. Cambridge, MA MIT Press, 99–116.

Jablonka, Eva, and Marion J. Lamb (1995), Epigenetic Inheritance and Evolution: The Lamarckian Dimension. Oxford: Oxford University Press.

—— and —— (1998), "Epigenetic Inheritance in Evolution", *Journal of Evolutionary Biology* 11: 159–183.

Jablonka, Eva, and Marion J. Lamb (in preparation) *Evolution in Four Dimensions*.

Jablonka, Eva, and Eörs Szathmáry (1995), "The Evolution of Information Storage and Heredity", *Trends in Ecology and Evolution* 10: 206–211.

Jablonka, Eva, Marion J. Lamb, and Eytan Avital (1998), " 'Lamarckian' Mechanisms in Darwinian Evolution", *Trends in Ecology and Evolution* 13: 206–210.

Kaiser, Dale (1999), "Intercellular Signalling for Multicellular Morphogenesis", in Reg R. England, Glyn Hobbs, Nigel J. Bainton, and Dave McL. Roberts (eds.), *Microbial Signalling and Communication*. Cambridge: Cambridge University Press, 139–160.

Kitcher, Philip S. (1993), "Function and Design", *Midwest Studies in Philosophy* 18: 379–397.

Lachmann, Michael, Guy Sella, and Eva Jablonka (2000), "On the Advantages of Information Sharing", *Proceedings of the Royal Society of London*, series B 267: 1287–1293.

Lachmann, Michael, Szablocs Számadó, and Carl T. Bergstrom (2001), "Cost and Conflict in Animal Signals and Human Language", *Proceedings of the National Academy of Sciences of the United States of America* 98: 13189–13194.

Laland, Kevin N., and Gillian R. Brown (2002), *Sense and Nonsense: Evolutionary Perspectives on Human Behaviour*. Oxford: Oxford University Press.

Lewontin, Richard C. (1983), "Gene, Organism, and Environment", in Derek S. Bendall (ed.), *Evolution: From Molecules to Men*. Cambridge: Cambridge University Press, 273–285.

Maynard Smith, John (2000), "The Concept of Information in Biology", *Philosophy of Science* 67: 177–194.

Maynard Smith, John, and Eörs Szathmáry (1995), *The Major Transitions in Evolution*. Oxford: W. H. Freeman.

Miao, Shili L., Fakhri A. Bazzaz, and Richard B. Primack (1991), "Persistence of Maternal Nutrient Effects in *Plantago major*: The Third Generation", *Ecology* 72: 1634–1642.

Millikan, Ruth G. (1984), *Language, Thought and Other Biological Categories*. Cambridge, MA: MIT Press.

Monod, Jaques (1971), *Chance and Necessity*. New York: Knopf.

Odling-Smee, John F. (1988), "Niche Constructing Phenotypes", in Henry C. Plotkin (ed.), *The Role of Behavior in Evolution*. Cambridge, Mass: MIT Press, 73–132.

Odling-Smee, John F., Kevin N. Laland, and Marcus W. Feldman (1996), "Niche Construction", *American Naturalist* 147: 641–648.

Oyama, Susan (1985), *The Ontogeny of Information: Developmental Systems and Evolution*, Cambridge: Cambridge University Press. 2nd revised edition 2000. Durham, NC: Duke University Press.

—— (2000), *Evolution's Eye: A System View of the Biology-Culture Divide*. Durham, NC: Duke University Press.

—— (2001), "Terms in Tension: What Do You Do When All the Good Words Are Taken?", in Susan Oyama, Paul E. Griffiths, and Russell D. Gray (eds.), *Cycles of Contingency*. Cambridge, MA: MIT Press, 177–193.

Paracer, Surindar, and Vernon Ahmadjian (2000), *Symbiosis: An Introduction to Biological Associations*. Oxford: Oxford University Press.

Sarkar, Sahotra (2000), "Information in Genetics and Developmental Biology: Comments on Maynard Smith", *Philosophy of Science* 67: 208–213.

Shannon, Claude E., and Warren Weaver (1949), *The Mathematical Theory of Communication*. Urbana: University of Illinois Press.

Smith, John W. (1977), *The Behavior of Communicating*. Cambridge, MA: Harvard University Press.

Sterelny, Kim (2000), "The 'Genetic Program' Program: A Commentary on Maynard Smith on Information in Biology", *Philosophy of Science* 67: 195–201.

—— (2001), "Niche Construction and the Extended Replicator", in Susan Oyama, Paul E. Griffiths, and Russell D. Gray (eds.), *Cycles of Contingency*. Cambridge, MA: MIT Press, 333–349.

Sterelny, Kim, and Paul E. Griffiths (1999), *Sex and Death*. Chicago: University of Chicago Press.

Szathmáry, Eörs (2000), "The Evolution of Replicators", *Philosophical Transactions of the Royal Society of London*, series B 355: 1669–1676.

Thompson, John N., and Olle Pellmyr (1991), "Evolution of Oviposition Behavior and Host Preference in Lepidoptera", *Annual Review of Entomology* 36: 65–89.

Wright, Larry (1973), "Functions", *Philosophical Review* 82: 139–168.

Zahavi, Amotz, and Avishag Zahavi (1997), *The Handicap Principle: A Missing Piece of Darwin's Puzzle*. New York: Oxford University Press.

41

BACK TO THE FUTURE

B. Parry

Source: *Trading the Genome: Investigating the Commodification of Bio-information*, New York: Columbia University Press, 2004, Ch. 7, pp. 349–365.

As I sit down to conclude this work, I have before me copies of the daily papers, each of which makes reference to the forthcoming Earth Summit on Sustainable Development that is about to commence in Johannesburg, South Africa. As exactly ten years have now elapsed since the Convention on Biological Diversity was first implemented, it seems an appropriate, and timely moment to reflect on how the introduction of this new regulatory regime has affected approaches to the utilization and stewardship of what has become, in the course of that same decade, one of the world's most valuable commodities: genetic and biochemical material and information. Interest in the collection and use of these resources began to escalate dramatically during the early 1990s as new technologies, new global environmental and economic policies, and restructurings in the corporate and scientific realms began to create new incentives to collect genetic and biochemical materials and new avenues for their exchange. The history of the collection and exchange of biological materials is extraordinarily long, stretching back hundreds of thousands of years. Within this context, it was easy to assume – and many have – that bioprospecting is simply a new descriptor for an age-old practice.

In this work I have tried to show that there are difficulties with this assumption. Although the collection and exchange of biological materials does indeed have a very long history, the application of sophisticated, indeed, entirely unprecedented biotechnologies has fundamentally altered the ways in which these resources are now collected and used – particularly within life-science industries such as the U.S. pharmaceutical industry. The change that has occurred in the way these materials are collected reflects a broader transformation that has occurred in drug development in recent years. The progressive introduction, and later dominance, of molecular approaches to the creation of pharmaceuticals and therapies has directed attention toward the components of living organisms. The genetic and biochemical materials and information embodied within them have proven to be extremely useful and valuable commodities, for both scientists and researchers, and interest has inevitably coalesced around the question of how to more effectively and efficiently collect and utilize these resources.

It soon became apparent to collectors that one means of improving the efficiency of the collecting process was to concentrate on collecting only those "key" or "essential" elements or components of biological organisms that were essential to processes of research and development, while divesting all other material that was considered to be extraneous or inessential. One way of achieving this was to produce a series of new engineered artifacts – such as cryogenically stored biochemical extracts and samples of tissue, cell lines, and sequence information – that could act as proxies for whole organisms. Of course, the genetic and biochemical material and information embedded in whole organisms has always been available for use; it has just not been as accessible as it is when embodied in these new forms. I have argued in this book that changing the way in which this genetic and biochemical material and information – what I have termed "bio-information" – is rendered, or presented, has had a profound effect on the practice and politics of biological resource exploitation.

In thinking about the way in which biotechnologies have transformed trade in biological materials, I have drawn a parallel between the actions of biotechnologies and informational technologies. Both, I have argued here, are employed to act on complex phenomena in order to produce from them proxies that provide consumers with, if nothing more, then at least those "key" or "essential" elements that they most desire. In both cases, producing these proxies involves a process of distillation – a de- and rematerialization. Much of the existing material or body of the original is divested, enabling the remaining information to be rendered in new, more lightweight, mobile, and transmissible forms. Just as new informational technologies have enabled particular resources to be structured in ways that have made them easier to circulate, store, and reprocess, so biotechnology has also enabled whole organisms to be rendered in ways that make the genetic and biochemical materials and information embodied within them much easier to transmit, store, reprocess, and recirculate.

These processes are of the utmost significance, as they allow the collectors of biological material to speed up the social and spatial dynamics of collecting from which power and profit derive. As I began this work by arguing, collecting is an inherently political activity, in that it involves annexing particular materials for exclusive use. The collector's power derives from his or her ability to acquire materials of interest, to concentrate them within particular locations where they can be ordered, controlled, and disciplined, and to then recirculate them (or not) within the marketplace to strategic advantage. Any factors that allow them to speed up the different phases of this process will be all to their advantage. I have argued that the ability to create new, highly transmissible bio-informational proxies that, unlike many of their historical counterparts, actually prove fungible for them in processes of research and development has allowed a select group of collectors to acquire, concentrate, and recirculate these commodities with far greater ease. This newfound ability, as well as the creation of a new regulatory paradigm that creates conditions within which these novel commodities may be exchanged on normative terms, have together enabled collectors to create a burgeoning and lucrative market economy for these bio-informational commodities.

Almost as soon as it became apparent that the introduction of new biotechnologies would inevitably create a demand and a market for genetic and biochemical material and information, concerns were raised as to how this market would be regulated and in whose interests it would operate. Much of the collecting that took place in earlier eras was undertaken in tropical developing countries, and these countries have remained the target of collecting expeditions. Many of these states were anxious to avoid being the object of a new wave of biocolonialism, and they began to lobby intensively for the creation of new policies and regulations that might govern the collection and use of genetic and biochemical resources, hoping to ensure that these activities were carried out in a more just and equitable fashion. This pressure was, in time, translated into the regulatory protocols that were introduced under the biodiversity convention and that have created the paradigm within which approaches to the governance of these resources have been shaped. During the late 1980s and early 1990s, an elite group of senior practitioners within the bioprospecting industry began to translate these protocols into a series of novel contracts that established the terms and conditions under which these resources could be exploited. A principal feature of these agreements, and one that played a central role in legitimating this new era of bioprospecting, was the inclusion of a formal benefit-sharing regime.

Some of the most interesting and significant characteristics of this emerging trade in bio-informational resources are the modes of transaction to which it has given rise. Bio-informational proxies, as I have noted here, privilege the genetic and biochemical information embodied in whole organisms at the expense of other attributes of the organism. This information (whether in a partially or wholly decorporealized form) may be used *successively* by a variety of consumers for a variety of different purposes. Collectors have quickly recognized that they may exploit their assets without relinquishing complete control of them. Rather than collecting and selling samples of material outright, many prefer to create an ongoing revenue stream by repeatedly selling *access*, on a short-term basis, to the genetic or biochemical information contained within them. This has given rise to forms of commodity exchange that are unprecedented in this domain, such as rental, licensing and pay-per-view. As I have illustrated here through empirical example, these resources are particularly valuable, as they can be used to form the basis of new products and processes.

If this trade is to be an equitable one, these benefit-sharing agreements must provide a mechanism that enables supplying countries to share in the profits that are generated from the many successive uses that are made of the bio-informational resources that have been collected within their borders. Agreements began with an awareness of this – the inclusion of a royalty mechanism reflected an acknowledged need to meet this aim. However, as I noted earlier, it is impossible to pay a royalty unless it is possible to determine to whom that royalty should be paid. This necessitates establishing a chain of consumption: ascertaining with certainty from whom the material was collected, by whom it has been used, for what purpose, and with what outcomes. It quickly became evident that this would require the introduction of a separate set of formal mechanisms for tracking and monitoring all of these successive uses.

This was realized through the development and implementation of a further set of contractual agreements that have become known as "Material Transfer Agreements." The use of MTAs was initially confined solely to transactions of bio-information that were understood to be commercial in nature. Almost immediately, however, it became clear that the distinction between commercial and noncommercial uses of such material and information (if it ever had existed) was being dramatically eroded by changes in the operating environment within the life sciences, particularly by the creation of a host of new dependencies and collaborations that established new links across traditional boundaries (e.g., corporate sponsorship of academic research, shared commercial and academic use of resources, and so on). Pressure thus mounted to extend the application of MTAs to *all* transactions that involve the transfer of collected biological materials. Although both corporate and scientific researchers recognize why, if MTAs are to be effective, they have to be applied uniformly, I have also detected their grave concern that extending such mechanisms to every possible exchange of material will be immensely burdensome, potentially unworkable, and, as I shall go on to suggest, morally corrosive.

This "compensatory" model is such that it also creates a need to address questions of informed consent. It has been argued that it would be unethical to change the use of collected genetic materials (for example, from a scientific to a corporate use) without inquiring whether the original provider agreed to that change of use. Every such change may yield an economic return, and as suppliers are potential recipients of a percentage of that return, it seems important that they be fully apprised of and in agreement with any proposed change of use. Although I have not addressed the issue in detail in this work, the task of securing such consents from suppliers is an immensely complicated one. Even in the most simple scenario – where a supplier provides material or information for an immediate use – it must first be possible for the collector to identify the particular individual or individuals from whom consent must be gained. This entails addressing complex and perhaps irresolvable questions about who "owns" these bio-informational resources (for example, individuals, communities, or states).

Acquiring "informed" consent also necessarily involves being able to adequately describe to such groups (once identified) how the material will be used in ways that are intelligible to them, by which I mean understandable within the context of their worldview and life experience. These difficulties multiply, first, as the numbers of transactions increase over time and, second, as the array of potential uses to which the material could be put exceeds that which any one person might reasonably be expected to comprehend or assess. At present, collectors have two choices. The first is to secure a *prior* informed consent for all potential uses; however, these consents are likely to be so broad in their compass as to lose all relevance. Such agreements are, in effect, a permission to do all things with the material even if those things cannot be specified, and in that case, to what have suppliers consented and how could their consent be said to be informed? Alternatively, collectors can insist that each subsequent user of the collected material return to the suppliers (assuming, of course, the suppliers are still there to be asked) to secure their further consent, on each and every occasion that the material is used in a fashion not previously anticipated or agreed upon. This will necessarily involve providing adequate explanations of what are, potentially, ever more complex technological applications (for example, gene splicing, nanotechnology, and so on).

Policymakers have responded to these complexities by introducing more and more layers of procedure designed to accommodate and regulate every possible eventuality, including those that cannot yet be known. Nowhere is this more evident than in the newly negotiated "Bonn Guidelines on Access to Genetic Resources and Fair and Equitable Sharing of the Benefits Arising out of Their Utilization." The agreement, which represents the culmination often years of regulatory development within the biodiversity convention, is immensely detailed; the proposed protocols for governing access to and use of these genetic and biochemical resources now take some twenty-five pages to articulate. They include recommendations that collectors introduce, among many other things, legally consistent mechanisms for securing prior informed consents; systems for protecting and encouraging customary uses of biological resources in accordance with traditional practices; mechanisms for extending and strengthening the application of MTAs; frameworks for improving local awareness and capacity to implement benefit-sharing agreements; and, indeed, ever more complex proposals for benefit sharing, including specific provisions on type, timing, mechanism, and distribution.[1] Together they constitute a system that is, it could be argued, ever more convoluted, onerous, and unwieldy to implement, and that may, in fact, serve only to deter prospective consumers.

Despite larding up these layers of contractual and legislative requirements (royalty agreements, MTAs, informed-consent agreements) in what is an increasingly baroque regulatory framework, it remains the fact that supplying countries and communities have yet to receive any *substantial* economic returns from the exploitation of their collected materials. It could be argued that this is simply because no products have yet been developed from the material. This seems surprising, however, given that this new era of collecting and highly sophisticated investigation and exploitation of collected materials has now been underway for nearly twenty years. It is an established fact that in recent decades some 25 percent of all prescription drugs have been derived from natural products, most of which have been collected in foreign countries. Are we now being asked to believe that despite the introduction of much more sophisticated technologies, no new drugs have been developed from such materials over the last twenty years, or at least not in the last fifteen years, since access- and benefit-sharing regimes were introduced? An alternate explanation would be that these collected biological materials have, of course, formed the basis of many commercial products during that period; it is just that this has not been established factually or acknowledged. What this suggests is that the existing regulatory system has failed to play its redistributive role, despite its elaborateness.

As I have argued here, this failure did not occur because such agreements have not been well intentioned but rather because changes in the way biological materials are rendered and utilized have combined to make the task of monitoring and compensating for their use extremely difficult. This problem is not, of course, confined to the biotechnology industry. As I have shown, the introduction of other new technologies – such as informational technologies – has created remarkably similar difficulties in other industries. The ability to transfer books, music, and films into a digital form has enabled these materials to be circulated, copied, and modified with extraordinary ease. The Associated Press reported on 21 August 2002 that it was estimated that 10 million people had tried to download the new and aptly named installment of the *Star Wars* saga, *Episode II: Attack of the Clones*, in the weekend after its release using Internet relay chat file-sharing systems, and that 4 million of them had succeeded.[2] Media executives and attorneys are so determined to stamp out these practices that they are now devoting a considerable percentage of their research and development budgets to devising new methods to prevent unlicensed replication of these products, such as encryption or hardware modification. Despite this, the industry is still finding the problem to be largely intractable. Immense amounts of time, energy, and resources are being dedicated to the task with little apparent success.

If those working in the bioprospecting industry hope to be able to trace all the uses that are made of collected genetic and biochemical materials and information over both space and time, it would seem that they too will need to employ some equally advanced technologies. Ever more elaborate and sophisticated mechanisms for monitoring and tagging the movement of genetic material and information are certainly being developed. DNA encryption, genetic fingerprinting, electronic identification (EID), molecular markers, and DNA profiling are all being promoted as the latest and most effective tools for establishing and maintaining the provenance of genetic components or sequence information as they are passed from one prospective user to another. All of which raises a question: How far down the road of tracing the successive uses of bio-information could, or should, we attempt to go?

There is no question that genetic and biochemical materials are used extensively in the biotechnology and life-sciences industries; however, they are primarily used when in these partially decorporealized or wholly informational forms. If benefit-sharing agreements are to work, they must be able to trace and take account of all the transactions that involve the use of genetic and biochemical resources, even when they are rendered in these elusive forms. The new tracing technologies that I outlined above could be applied to this task (they are not currently), although this would be an immensely complex, expensive, and time-consuming undertaking. Even systems such as the sustainable-timber certification scheme developed by the Forestry Stewardship Council, and the livestock passport scheme introduced to trace the movement of large whole organisms (such as trees and cattle) as they are transferred from one location to another through a succession of market transactions, have been derided for regularly failing to establish an identifiable "chain of custody."[3] It is important to note that these projects have faltered even though the material that they seek to trace is infinitely larger and more stable than anything that one might attempt to trace in the bioprospecting industry.

Despite this, a large proportion of the (relatively) small amount of money that has been generated through bioprospecting ventures has been channeled into creating a flourishing bureaucracy devoted to the task of attempting to track, monitor, and secure compensation for the use of tiny fragments of genetic and biochemical material and forms of extracted bioinformation to equally questionable effect. Many of the NGOs whose principal aim is the protection of biodiversity, and many institutions in developing countries, are now devoting a considerable proportion of their operating budgets to the task of developing, implementing, and overseeing what may well prove to be an ineffectual set of procedures. Developing-state governments are not immune from this process either. Many now feel compelled to develop detailed legislative and administrative structures that, they are told, will enable them to secure ongoing returns from bioprospecting operations. Given that this regulatory paradigm may fail to deliver major returns and that the opportunity costs of directing finite resources away from other crucially important areas, such as education and health provision, are so considerable, it becomes even more important to undertake an unflinching assessment of whether such investments can be justified or sustained.

Recognizing the limits of the existing regulatory paradigm is clearly of profound importance to all concerned. However, these matters also raise a larger question: Is there a risk that the existing regulatory system may not only fail to meet its goals but may, in fact, exacerbate the very problem of mistrust that it sought to remedy? As I have illustrated here, the new erea of collecting that began in the 1980s emerged in a climate of uncertainty – there were hopes that it might provide new opportunities and returns for developing countries but also concerns that the projects might reproduce the colonialist practices of an earlier era. The new regulatory protocols that were introduced under the biodiversity convention were explicitly designed to improve relations of trust – they sought to remedy past inequities by ushering in a new regime of just distribution of gains. This original impulse has since been translated into layer after layer of ever more detailed and complex forms of contractual regulation and practical requirements for transparency and accountability that have become sedimented over the past decade into a set of practices that have now been adopted globally.

The task of conforming to these new protocols is onerous and immensely consumptive of time, energy, and resources. Many scientific research institutes and, perhaps more importantly, many companies are so constrained or wearied by the need to meet these requirements that they have elected to either abandon in situ collecting altogether or to engage in these activities illicitly. Neither course of action brings any financial benefit to source countries. Those companies, organizations, and institutions that do decide to comply are burdened by the weight of administering contacts that attempt to pin down, through ever more finely calibrated mechanisms, the whereabouts and recent uses of every last bit and byte of genetic material and information, when it must be clear to most that this is a truly Sisyphean enterprise. The structural inability of this system to produce the promised returns, will, in time, I believe, give rise to a serious, profoundly corrosive, and depressing sense of frustration in all parties.

The notion that it is possible to secure accountability and trustworthiness through the imposition of more complex systems of behavioral auditing is certainly not confined to this realm – it has been widely embraced in many areas of medicine, science, academia, and public life in recent years. As the moral philosopher Baroness Onora O'Neill suggests, "a prominent feature of this widespread movement to improve accountability has been an increasing reliance on more formal procedures, including contracts, letters of agreement and financial memoranda that impose highly complex conditions."[4] Formalization of procedure appears to have a number of advantages that are, as she suggests, "constantly mentioned by its advocates: mutual clarity of expectations, clear performance targets, defined benchmarks of achievement, enhanced accountability."[5]

What her insightful analysis highlights, however, is that the *failure* of these formal mechanisms to deliver desired outcomes may ultimately engender, rather than remedy, feelings of mistrust. Ironically, although a strict auditing of these transactions promises to deliver accountability through transparency, in fact what it implies is that those involved in such exchanges are somehow inherently untrustworthy and must thus be subjected to continuous external monitoring. Although this may rankle, subjects might be prepared to tolerate it if they believe that the application of more detailed forms of auditing and monitoring will actually deliver equitable working relationships, standards, and outcomes. As Thompson notes, however, most eventually come to believe that such procedures in fact "only create further levels of bureaucracy and inefficiency . . . and set in motion a process that may exacerbate rather than alleviate the problems they were intended to address."[6] This may, as he concludes, contribute to deepening a culture of distrust.

This argument can, it would seem, be applied to an analysis of the development and application of regulatory paradigms in the bioprospecting industry over the past decade. These were introduced primarily to ameliorate concerns that source countries had about the inherent untrustworthiness of collectors – an attitude that had its genesis in the colonial era. In the interests of ensuring that the contemporary bioprospecting operations were carried out in a just and equitable fashion, stakeholders began to devise mechanisms that would regulate access and ensure an equitable distribution of benefits. These mechanisms and attendant procedures have become progressively more formal and more elaborate as the years have gone by, and this has, perhaps, ironically engendered further feelings of mistrust. These feelings are at risk of being exacerbated by that fact that *despite* the introduction of these increasingly complex procedures, remarkably little money has been returned to source countries from these bioprospecting operations.

Although we continue to receive estimates of the truly astonishing amounts of income that have been reaped from the sale of products developed from natural materials sourced, in most cases, from developing countries (estimated to be between $75 and $150 billion annually in the pharmaceutical sector alone), it remains the fact that none of these countries have received any significant proportion of this income – these increasingly convoluted regulatory and compensatory mechanisms have largely failed to produce the desired outcome.[7] An unfortunate consequence of this is that developing countries are being affirmed in their belief that this new wave of bioprospecting is unlikely to return the benefits it promised. Their confidence in the project has been so undermined that many are now threatening to close their borders to all collectors, including those that are undertaking collections for scientific purposes such as taxonomic identification. It would be truly unfortunate if this were to occur, as such activities are central to the project of preserving global biodiversity.

There is also clear evidence that overregulation is deterring companies from pursuing natural-products research. Rather than opening up new possibilities for the creative and sustainable use of biological materials and information, these new multitiered levels of regulation act to progressively choke off such opportunities. There may be those who are of the view that this is no bad thing – that biodiversity is best left unexploited. However, in an increasingly globalized world, the likelihood of biodiversity remaining so is remarkably low; it would be naïve to think otherwise. The question, it seems, is not whether these resources will be exploited but rather how they will be exploited and how this market might be organized such that it operates on a more equitable basis. Given all that we have ascertained from this research, it does not seem inappropriate to propose a radical rethinking of approaches to the regulation of this burgeoning global trade in genetic and biochemical material and information. I believe that by going back to the past, it may be possible to see a new way forward into the future.

Prior to the introduction of the biodiversity convention, access to genetic and biochemical materials was unrestricted. Countries, collectors, institutions, and individuals all held an equal right to access these resources, investigate them, and employ them in any number of creative ways. There seems to be a consensus of opinion that this was generally desirable – that the ability to access them freely stimulated innovative and inventive approaches to their use. It was also evident, however, that the power to utilize these resources had historically, and would in the immediate future, continue to be concentrated in the hands of those collectors with links to the most sophisticated laboratories in the developed world. This inequity created unequal power relations and a genuine probability that most of the profit that derived from the exploitation of these materials would remain in the hands of those individuals. A property-rights war broke out as a consequence: supplying countries began to claim rights over their genetic and biochemical resources, while corporate interests began to claim rights to the engineered artifacts that they created from them.[8] The outcome was the introduction of new benefit-sharing regimes designed to ensure that a proportion of the profits that accrued from the commercial exploitation of these collected materials be returned to those countries that had supplied them.

The principle of sharing benefits with supplying countries is not, I believe, disputed by any party, even those companies that would be expected to relinquish a portion of their profits to meet this requirement. In the course of my research, I did not find any executives who were opposed to paying a royalty of between 1 percent and 5 percent of net profits on products derived from natural materials that were not collected domestically. Most large pharmaceutical companies (those that generate large numbers of products and profits) now fully accept that they must pay for their raw materials. Ideally, they would prefer to only pay for those that prove to have some long-term utility for them, and this is understandable. They are unconcerned about the addition of this extra cost, in part because it is unlikely to be borne by them directly – it will simply be added to the sale price of the product as a production overhead, alongside costs such as advertising and packaging.

Companies are also now acutely aware of the adverse publicity that attaches to attempts to secure raw materials without paying some appropriate form of recompense to the supplier, and they have no desire to attract such opprobrium. Most are concerned to act as responsible and environmentally conscious corporate citizens. What companies emphatically *do not* wish to do, however, is to spend ever greater proportions of their operating budgets complying with an unnecessarily cumbersome and unpredictable regulatory system. Regulatory protocols (such as benefit-sharing agreements and MTAs) are being continually reworked and refined to take account of new developments. Understanding and implementing these changes is an enterprise that absorbs an increasing number of administrative and legal personnel in corporate, scientific, and academic organizations, whose services may be more usefully deployed on other tasks.

Source countries find themselves in a similar position. What they ideally want is to receive some significant and ongoing economic returns for the use of their resources, without having to commit an inordinate percentage of their extremely scarce resources to the task of developing, administering, and monitoring new policies and regulatory regimes. Although I have not had an opportunity to calculate it here, it would be interesting to conduct an assessment of how the costs that supplying countries have incurred in setting up and implementing bioprospecting legislation, policies, and agreements compare with the returns that have so far been realized from such ventures. Such an analysis would almost certainly reveal that the system is in negative equity. If the system as it currently exists does not appear to be satisfactorily meeting its aims, then other, alternative schemes must be proposed.

What I offer here is nothing more than one possible alternative derived from this detailed critique of current practice. Although radical in conception, it retains and respects the original impulses that informed the development of new regulatory approaches to the collection and use of genetic and biochemical resources, such as access and benefit-sharing agreements. I offer it not as a fully realized proposal, for it is clearly not, but rather in the hope that it might stimulate some further debate and discussion about the shape that future regulatory paradigms in this field might take. The prosposal that I intend to outline here is informed by ideas being developed in parallel domains such as the media and software industries. They have been devised in response to the need to evolve more sophisticated ways of dealing with the use, both licensed and unlicensed, of commodities and resources that have proven to be highly transmissible, readily replicable, and, as a consequence, extraordinarily slippery.

It seems to me that what matters most to collectors and suppliers in the final analysis is the question of what products are developed from collected natural materials. Samples of material and information may be used in many speculative endeavors; they may prove to be of use in some and not in others. All that really concerns people is whether those materials go on to form part of a patented or copyrighted product – such as a drug or sequence database – that is marketable and that generates profits. Suppliers have a right to expect that in such instances they would receive a small royalty for having provided the resources on which these products are based. Within the bioprospecting industry, royalties range between 1 percent for material collected at random to 15 percent for material that already has indications of proven efficacy against specific diseases. However, there is a general agreement, and evidence and probability would seem to suggest, that only a comparatively small proportion of collected samples would fall into this latter category. The majority of collected samples currently attract royalty payments of between 3 and 5 percent.

I propose that we abandon the task of attempting the trace all of the myriad uses that are made of collected genetic and biochemical materials and information and concentrate instead on working to secure a voluntary, global agreement from the pharmaceutical industry that they will add a sum of between 3 and 5 percent of their profit ratio to *all those products that they currently have in the marketplace that are based on collected natural materials.* This levy would remain in place for as long as those products are sold. In order to ameliorate some of the inequities that have characterized collection processes in the recent past, it does not seem inappropriate to suggest that this levy also apply to products, such as vincristine and vinblastine, that were derived from materials collected without recompense but which continue to generate substantial profits for drug companies.

I would suggest that a similar duty be levied on users of genetic-sequence databases and indigenous-knowledge databases. This could be easily added to the charges that institutions should, and do, routinely require to access these types of information. These levies, which would amount to not more than a few cents on most proprietary products or informational transactions, would be paid directly into a superfund that would ideally be administered by a global regulatory agency such as the Global Environment Facility.[9] Countries and communities from the developing world might then apply directly to the fund with proposals for development and conservation projects.

As I have suggested here, the task of attempting to establish the provenance of collected materials is highly problematic, and it would not be fruitful, in my view, to devote too much attention to trying to establish which countries had provided what material or information. There will undoubtedly be cries of indignation about the idea of abandoning a strict auditing of who gave what to whom and from where. Some countries will argue that they have given much more material than others, while others might argue that, while they have given less, the material that they have given has proven to be much more valuable. While such claims may well be true, it will be impossible to determine their veracity unless we are able to establish, with complete certainty, how those materials have in fact been used over time and space – a task that requires access to actuarial information that will always remain just out of reach. While we have awaited the successful completion of this potentially unworkable auditing process, years and years have slipped by without the production of any form of substantial compensation.

This scheme, might, I believe, yield a number of substantial and immediate returns for both suppliers and collectors. It is my fervent belief that what supplying countries most desire is to secure access to an ongoing income stream that they may utilize for a variety of conservation and development projects that can provide ecological, technological, cultural, economic, or social benefits. We all, collectively, live on a small and increasingly imperiled planet. As processes of globalization advance, we all become much more dependent on one another – environmental and economic collapses in one region of the world can have catastrophic consequences in another. There would undoubtedly be complexities associated with the disbursement of these funds, and I do not wish to downplay them here. The world's more biodiverse countries or those that have played host to a greater number of contemporary collecting projects may wish to argue for a greater proportion of the funds, and they may be justified in doing so. My abiding sentiment, however, is that most citizens of the world would feel that a considerable gain had been made if income from the superfund were used to reverse imminent environmental destruction, *wherever that may be occurring*, and I feel certain that consumers of pharmaceuticals would consider this an extremely appropriate and welcome use of any small surcharge that had been levied upon them.

Another considerable advantage of this scheme is that it would yield an immediate, substantial, and ongoing revenue stream. Instead of having to wait twenty years or more for income that may never be realized, supplying countries would be able to make immediate application for funds for applied conservation and development projects that need to be undertaken urgently. Second, the money could be accessed directly, through existing United Nations environmental aid and donor programs, for example, without the need to generate dedicated systems of disbursement. Third, the portion of funds that is now spent on monitoring contractual agreements and tracing successive uses could also be reduced, releasing resources for other social purposes. I would not propose completely abandoning the project of monitoring the ways in which materials are used, as questions of how such materials are collected and used clearly have important social, cultural, and political implications. These tasks should continue, but they need not be seen as a necessary prerequisite for the release of compensation, as they are now.

Collecting institutions, companies, and research institutes would also benefit greatly from being released from the burden of having to account for every successive use that they make of collected material and from having to implement and monitor highly nuanced and complex access and benefit-sharing agreements. This would result in a substantial reduction of their administrative costs. The savings that are made might be used, in the corporate world, to perhaps cross-subsidize further research and development of basic drugs for use in developing countries. Scientific, academic, and other research institutes might well apply their savings to the pursuit of further basic research for which federal funds have been cut, such as species-identification programs.

All of these parties would also benefit immeasurably from a return to what might be thought of a "moderated free-access regime." The collection of species for taxonomic and other purposes would no longer be hindered, and students would be free to pursue crucial field research. This may draw attention to areas or species at risk of further destruction. Collectors from both developed and developing countries would be equally free to fully explore all the potential applications of biological materials and to work experimentally with those materials to produce new inventions and innovations. These may, in time, become marketable products that will provide a future revenue stream. Although such a scheme may seem too revisionist, it is more likely to fulfill the shared aim of finding new, creative ways to justly and equitably develop genetic and biochemical resources and information for the broader benefit of humankind than those that we have available to us today. While the difficulties of administering such a fund may seem insurmountable, it is difficult to imagine that they would exceed those associated with the current system. As radical as it seems, this new superfund model may well act to unblock some of the obstructions that now threaten to fatally impede the operation of existing regulatory and compensatory frameworks.

Undertaking this research project has involved immersing myself in a shifting debate about how best to govern the use of a new commodity: variously embodied types of bio-information. Although we continue to imagine that biological material is the same type of resource that it always has been, this detailed research, I hope, has revealed that it is in fact a resource that is being construed, represented, and utilized in new and unexpected ways. I have argued here that we must consequently learn to evolve sophisticated and custom-made systems of regulation that might reflect the complexities that attend its changing construction and use. My motivation in undertaking this research has not been to deride or undermine the inventive approaches to the regulation of this new commodity that have been developed in the last ten years but rather to build upon and learn from such endeavors. I am indebted to all those practitioners in the bioprospecting industry whose willingness to be frank about the shortcomings of the existing system has enabled us to reflect in a much more informed way on where and how our collective approach to the uses and regulation of this new resource might be improved.

By drawing out the parallels that exist between biotechnologies and other informational technologies, I hope that I broadened understandings of how biological materials are constructed and used as commodities in the contemporary life-sciences industry and revealed the capacity that this has for creating a new and highly lucrative resource economy in bio-information. I also hope that this analysis will lead us to reflect on questions of social and economic justice and of our individual obligations to ensure that this new resource economy operates in a just and equitable fashion. While the collection of plant and animal samples for use in the pharmaceutical industry may be declining, the collection and use of human tissue and organs is on the rise. The issues that I raise in this work – questions of what the status of various biological derivatives and "works" might be and of who should have access to them and under what terms and conditions – will have growing implications for us as the flourishing life-sciences industry pushes forward with the creation of DNA and stem-cell banks, DNA-sequence databases, and, indeed, enters other realms that remain, as yet, unimagined and uncharted. As we move into the twenty-first century, the century, perhaps, in which the engineering of life will become our central preoccupation, it seems appropriate that we continue to focus our collective attention on the question of how we, as a global community, wish to manage the commodification of life.

Notes

1 The "Bonn Guidelines on Access to Genetic Resources and Fair and Equitable Sharing of the Benefits Arising out of Their Utilization" of the Convention on Biological Diversity is available at http://www.biodiv.org/doc/meetings/cop/cop-06/official/cop-06-06-en.doc.

2 Associated Press Release Online, "Theaters to Warn on Movie Piracy," 21 August 2002.

3 The livestock passport scheme is designed to regulate the transfer of individual cattle within the EU. See A. Kirby, "Form Filing Infuriates Farmers," BBC News Special, 14 September 1999, available at http://news.bbc.co.uk/l/hi/special_report/1999/09/99/farming_in_crisis/442052.stm.

4 O. O'Neill, *Autonomy and Trust in Bioethics* (Cambridge: Cambridge University Press, 2002), 130.

5 Ibid.

6 J. Thompson, *Political Scandal: Power and Visibility in the Media Age* (Cambridge: Polity, 2000), 254.

7 K. Ten Kate and S. Laird, *The Commercial Use of Biodiversity: Access to Genetic Resources and Benefit-Sharing* (London: Earthscan, 1999), 2.

8 For a fascinating discussion of the way in which property rights have been extended to encompass all manner of phenomena, including many forms of cultural property, see M. Brown, *Who Owns Culture? Native Rights and the Future of the Public Domain* (Cambridge, Mass.: Harvard University Press, 2003).

9 The Food and Agriculture Organization of the United Nations has considered introducing a superfund into which industrial countries and seed companies would pay a small percentage of profits from the sale of folk varieties to compensate farmers in developing countries. For a further discussion, see Rural Advancement Foundation International, *Conserving Indigenous Knowledge: Integrating Two Systems of Innovation* (New York: United Nations Development Program, 1994), 33; and D. Soleri et al., "Gifts from the Creator: Intellectual Property Rights and Folk Crop Varieties," in *Intellectual Property Rights for Indigenous Peoples: A Sourcebook*, ed. T. Greaves (Oklahoma City: The Society for Applied Anthropology, 1994), 25.

Part 2

EXPECTATION AND ACCOUNTABILITY

42

THE DRUGS DON'T WORK

Expectations and the shaping of pharmacogenetics

Adam Hedgecoe and Paul Martin

Source: *Social Studies of Science* 33(3) (2003): 327–364.

Abstract

This article examines one particular set of technologies arising from developments in human genetics, those aimed at improving the targeting, design and use of conventional small molecule drugs – pharmacogenetics. Much of the debate about the applications and consequences of pharmacogenetics has been highly speculative, since little or no working technology is yet on the market. This article provides a novel analysis of the development of pharmacogenetics, and the social and ethical issues it raises, based on the sociology of technological expectations. In particular, it outlines how two alternative visions for the development of the technology are being articulated and embedded in a range of heterogeneous discourses, artefacts, actor strategies and practices, including: competing scientific research agendas, experimental technologies, emerging industrial structures and new ethical discourses. Expectations of how pharmacogenetics might emerge in each of these arenas are actively shaping the trajectory of this nascent technology and its potential socio-economic consequences.

'Like it or not, we will soon have the ability to predict the variation in drug responsiveness of large numbers of individuals . . . A dramatic change in the practice of medicine over what we know today is certain. Each patient will have to be treated as the individual we already know he or she is.'

Charles Cantor (1999: 288)

Introduction

Much of the discussion surrounding the promise of the Human Genome Project revolves around its effect on clinical practice and the development of 'genomic medicine' (Bell, 1998). The biggest impact so far has been on drug discovery and development, and this article examines the rise of one particular set of technologies aimed at improving the targeting, design and use of conventional small molecule drugs – pharmacogenetics.

In the business press over the past five years, there has been increasing discussion of pharmacogenetics and 'designer drugs', their potential impact on the pharmaceutical industry and clinical practice, as well as the social and ethical issues they raise (Cookson, 1997; Carr, 1998; Shook, 2001). Pharmacogenetics may have a number of important consequences, most notably the development of new commercial strategies based on the targeting of particular drugs to patient sub-populations, patient stratification and the geneticization of common diseases, as well as raising the traditional problems associated with genetic testing (discrimination, confidentiality, etc.). However, much of the debate about the consequences of the introduction of pharmacogenetics has been highly speculative and has approached questions in a piecemeal manner, since little or no working technology is yet on the market. In contrast, we aim to outline a novel analysis of the development of pharmacogenetics, and the social and ethical issues it raises, based on what might be called the sociology of technological expectations (see Brown et al., 2000). This will emphasize the co-construction of particular scientific visions of the development of the technology, novel research agendas and experimental technologies, emerging industrial structures and new ethical discourses. Expectations of how pharmacogenetics might emerge in each of these arenas are actively shaping the trajectory of this nascent technology and its potential socio-economic consequences.

Conceptual approach

In trying to understand many of the developments in genetics and biotechnology social scientists, bioethicists and policymakers are faced with major problems as few, if any, technologies are widely used. In fact, one of the major criticisms levelled at the emerging field of genetic medicine is that it is surrounded by too much 'hype', speculation and unsubstantiated claims (e.g. Richards, 2001). As a consequence, the lack of any real substance to developments in this area makes it very difficult to accurately assess what might occur in the future and the ethical, social and legal consequences of different scenarios.

A novel approach will therefore be taken in this paper, based on a more critical understanding of how new technologies 'come into being'. Instead of seeing the speculative claims made about the future of pharmacogenetics as ephemeral and irrelevant, we argue that they are fundamental to the dynamic processes that create new socio-technical networks. Understanding the formation, mobilization and shape of these expectations or 'visions' is therefore central to the analysis of an emerging biotechnology.

A process of great uncertainty marks the early development of most new technologies and the search for formats, designs and applications which will both 'work' in technical terms and command support from potential developers and users. The analysis presented here will examine the key role of one kind of expectation, which we have called 'visions', in socially shaping emerging technological options and the design of physical artefacts and commercial products. Three groups of actors are particularly important in this respect: researchers whose visions guide the design of clinical experiments and prototype technologies; commercial firms who embed these visions in business strategies and potential products; and bioethical commentators who discuss the potential future social and ethical problems raised by different applications.

Previous work on the sociology of technological expectations has looked at the activities of researchers and technicians (van Lente, 1993), policymakers (de Laat, 2000) and, to a lesser extent, the activities of firms (Martin, 1999). However, almost no attention has been paid to the role of bioethicists in constructing future scenarios and possible social problems. This paper will argue that the creation of a bioethical discourse around a controversial technology is important as it both provides a negotiation space to explore the socially acceptable limits of the technology and acts as a means of enrolling support from key actors. In this particular case, leading advocates of the technology have attempted to construct options which distance pharmacogenetics from the serious ethical problems which surround other forms of genetic testing.

We therefore focus not just on technical and commercial issues, but also on the way in which ethics discourses are an integral part of the construction and shaping of technology. The contingency of technological development means that the impact and consequence of a new technology will depend on which set of technological options are adopted – some will be rejected as being too controversial. In the social negotiation that shapes emerging options, ethics may be a decisive factor. As a result, our focus will be on the way in which new technologies, new industries and new ethical, legal and social problems are co-constructed during the process of technical change through the creation of competing visions.

Conceptual framework and methodology

In the last 15 years a new sociology of technology has emerged, based on a critique of technological determinism (Bijker et al., 1987; MacKenzie & Wajcman, 1999). Instead of innovation and technological change being driven by an innate technical logic, the development of new technologies is seen as a fundamentally social process open to sociological analysis.

A number of different theoretical perspectives have been used to examine the creation of new technologies, including actor-network theory (ANT) (Callon, 1987), the social construction of technology (SCOT) (Bijker, 1995) and the analysis of large technical systems (Hughes, 1987). Although each takes a distinct approach they share several common features, notably the idea that the development of a new technology involves a range of heterogeneous social, technical, economic and political processes. In addition, these approaches argue that new knowledge is co-produced at the same time as new technologies and new socio-technical relations, through a process of mutual shaping. This paper will draw on the following concepts from within the sociology of technology:

The construction of socio-technical networks – Successfully introducing new technologies into routine use requires the alignment of a range of heterogeneous human and non-human actors and artefacts into stable socio-technical networks (Callon, 1987; Bijker, 1995). To achieve this, network builders might be involved in, for example, the creation of new social practices, new companies and new forms of state regulation, which emerge together during innovation. Network formation therefore requires the enrolment of various actors, the formation of alliances and the mobilizing of different social, technical and economic resources.

The social shaping of technology – As an integral part of the creation of stable socio-technical networks the emerging technology is socially shaped to reflect the activities and interests of the groups involved in the innovation process. This is mediated through the design, testing, selection and redesign of the various technological options and may result in the physical form of the technology changing dramatically over time. For example, as new groups of actors join the emerging network, they may favour particular options over others and shape the future direction of research and design (Bijker, 1995). Through an examination of the competing technological options, the changing designs and applications, and the role of the various groups involved, it thus becomes possible to analyse the physical development of a new technology in sociological terms.

The creation of visions and the enrolment of support – An important process in the formation of networks is the creation of particular promises, 'visions' or expectations for how the technology might be used in practice and sold as a commodity (van Lente, 1993; Martin, 1999; Brown et al., 2000). For example, in van Lente's study of the development of Tenax, a new paper for insulating high-voltage cables, the scientists involved in the project created the expectation that the technology was a way of solving technical questions of electricity transmission, coping with increased electricity demand and improving market share for their company. These expectations subsequently shaped the behaviour of key actors and their development of the technology.

During the early stages of the introduction of a radically new technology a number of competing expectations for how it might be used often coexist (Pinch & Bijker, 1984). These may be associated with the formation of different network configurations and the emergence of alternative designs or technological options. With respect to the formation of these networks, van Lente has asked 'What is the role of expectations in the alignment of heterogeneous elements that together shape technology?' (van Lente, 1993: 34). Using case studies of three 'promising technologies' he demonstrates that expectation statements are a resource for actors involved in innovation, in that they can help to:

- *legitimize*, justify, back their arguments, give reasons in general;
- *mobilize* funds, attention of other actors;
- allow decision-making and *reduce the uncertainty* inherent in technological developments (van Lente, 1993: 187, original emphasis).

Expectations therefore simultaneously act as a means of enrolling support and resources into the emerging socio-technical network, enabling interactions between actors and guiding the physical design of artefacts. They may also form part of a new set of cognitive structures that both enable and affect the development of the technology (Bijker, 1995).

In general, most of the work on the sociology of technological expectations has focused on what might be called the articulation of visions or promises. For van Lente these are statements uttered in public or written down. Other authors have analysed the representation of expectations (Grin & Grunwald, 2000; Michael, 2000), their narrative construction (Deuten & Rip, 2000) and metaphors of the future (Wyatt, 2000). In contrast, our emphasis will be on both the articulation of expectations in scientific and bioethical discourse in the form of specific visions, and their embodiment in the design of experiments and the formation of new biotechnology companies as a result of the decisions made by innovators. Visions therefore constitute a particular class of expectation which both project and anticipate how the future might emerge, and provide a strategic framework for actors as they attempt to construct particular sociotechnical networks. The deployment of this concept will help us to examine the way in which expectations are translated from discursive structures into a heterogeneous set of professional practices, social organizations and physical artefacts.

Much of the current discussion concerning both the science and economics of pharmacogenetics is speculative and differing options for its development can be clearly identified. This provides a good opportunity to chart the way in which an emerging discipline, set of technologies and industrial structures are being shaped by different visions during their early development (van Lente & Rip, 1998; Guice, 1999).

An important technique in the building of these visions is the formation of 'coalition[s] representing various elements of a prospective technical order' (Guice, 1999: 82). Such coalitions can clearly be seen in the alliances being formed between various industry actors and the collaborations between drug companies and academic researchers. Some commentators suggest that pharmacogenetics is the testing grounds for a new research paradigm which revolves around 'the collaboration between biotechnology companies, academic researchers and clinicians' (Pfost et al., 2000: 334). As a consequence, the development of this technology provides a good site for analysing the dynamic relationship between visions and network formation.

We use these theoretical tools to analyse the development of pharmacogenetics as a scientific concept, an experimental research practice, an emerging industrial sector and a bioethics discourse. The next section provides a brief history of pharmacogenetics from the 1950s until the present day and is based on a historical review of the scientific literature in this field. Then we describe the two main visions for pharmacogenetics being articulated by the scientific proponents of the technology. One is focused on the genetic basis of adverse drug reactions and the other revolves around the role of disease-related genes in determining drug response. Next we examine the industrial development of pharmacogenetics within the biotechnology and pharmaceutical industry, and the articulation of the two visions already identified. Finally we describe the bioethics discourse emerging around these two main technological options and outline the different ethical, legal and social problems associated with each. The conclusion draws together the different strands of the argument and offers some sociological reflections about the role of visions in shaping technologies, industries and ethical problems.

From pharmacogenetics to pharmacogenomics

Origins and early development

Traditionally pharmaceutical companies have adopted a 'one-size-fits-all' principle with regard to different people's reactions to drugs (e.g. Liggett, 2001). Although acknowledging variations in dosage (because of age, sex and body size) the general approach has been the development of mass-market therapies based on the assumption that everyone responds in the same way. Yet research scientists have long acknowledged genetic variation in drug response. Friedrich Vogel first introduced the term 'pharmacogenetics' in 1959, following groundbreaking work by Arno Motulsky two years earlier (Nebert, 1997; see Weber [2001] for a detailed historical review).

Around this time, a number of important discoveries were made, such as glucose-6-phosphate dehydrogenase (G6PD) deficiency in 1956 and the genetic variation in ethanol metabolism in 1964 (Kalow, 1990). Until the late 1980s, pharmacogenetics expanded slowly as a discipline, with specific drug-gene relationships becoming clear as and when investigators came across them, rather than there being any active search for such links. To a large extent, this was the result of relying on methods from classical genetics: individual studies followed by family and population analysis (Meyer, 1990).

Despite this long history, at an industry level there has been a profound reluctance to admit the extent of genetic variation and its effect on drug response (Corrigan, 2002; Glaser, 1998). This may be due to the fact that very little could be done, but another reason is the understandable unwillingness of the industry to admit to the huge scale and very real risks posed by genetically based adverse drug reactions (ADRs) (Abraham, 1997).

The molecular turn

This situation started to change with the emergence of molecular biology, which provided the means for improved investigation of the genetics of drug response. The 'molecular turn' in pharmacogenetics can be dated to 1988, when Frank Gonzalez and his colleagues isolated several alleles of the human CYP2D6 gene, part of the P450 cytochrome complex of genes responsible for drug metabolism in the liver (Gonzalez et al., 1988). This study opened the field up to the 'new genetics'; alleles of the P450 complex are associated with adverse drug reactions caused by several products, and have thus become a popular focus for contemporary research. Coming into the 1990s, a number of new technologies such as polymerase chain reaction and high-throughput sequencing gave scientists greater understanding of genetic variation and increased the interest in pharmacogenetic studies. In addition to these technical developments, there were also ideological changes which, in the wake of the Human Genome Project, started to reconstruct medicine in terms of genetics (Bell, 1998). Perhaps most importantly, pharmacogenetics finally aroused the interest of the pharmaceutical industry, perfectly combining the promise of the new genetic technologies with a focus on drug discovery and development. Around this time a new term began to be used to describe the discipline: pharmacogenomics. This concept has a variety of meanings, although most centre on the use of genome-wide screening for pharmacognetic reactions (Hedgecoe, 2003). In order to avoid confusion, this article tends to use the term pharmacogenetics, although many of the authors cited prefer pharmacogenomics.

Following the genome project and the development of powerful new gene sequencing and analysis technologies, a new form of research known as association genetics has started to transform the study of the relationship between genes and disease. Proponents claim that this new 'functional genomics' has enabled the unravelling of diseases at the genetic level and the identification of disease-associated genes. In addition, association studies allow the study of the relationship between particular genetic markers and specific reactions to drugs, without having to do traditional family-based linkage analysis.

In terms of testing, new technologies such as DNA arrays (also called 'gene chips') are being developed to allow the cheap, fast and large-scale screening of populations for genetic markers which are correlated to either positive or negative reactions to specific drugs (Evans & Relling, 1999).[1] Although still in its infancy, one such DNA array targeting the pharmacogenetically important cytochrome P450 complex is already commercially available (Sadée, 1999) and in the next few years more are expected to enter the market. For many in the field this will open the door to the routine use of pharmacogenetic tests in the clinic. However, before this can be done the validity and clinical utility of the technology will have to be demonstrated, and commercially attractive products and services will need to be developed. Although the general promise of pharmacogenetics became apparent in the mid-1990s, working applications have remained elusive. The following section will examine a number of specific visions for the scientific and clinical development of pharmacogenetics and explore how these are being used to design clinical experiments and construct future applications.

Scientific visions of pharmacogenetics

During the development of new medical technologies, researchers are involved in both designing and testing particular technical options and seeing if they can be incorporated into routine clinical practice (Martin, 1999). This process often takes a long time and involves heavy investment in human testing before even a prototype can be developed. The creation of visions for how a new medical technology might be used to treat patients can play an important part in this process. In particular, advocates of new technologies use visions to get access to funding, design research agendas and clinical experiments, and start the development of new drugs (Martin, 1999). An examination of the different visions surrounding the early stages of innovation can therefore shed light on these dynamic processes of enrolment and alignment. In terms of van Lente's ideas, these scientific visions 'legitimize' the concept of pharmacogenetics, justifying the research that has been carried out and providing a rational basis for further work. Whether it is because pharmacogenetics will reduce the harm caused by ADRs, or because it reveals the true mechanism of the underlying pathology, the visions painted in the scientific literature provide the starting point for any further developments.

Following a detailed review of the scientific and clinical research literature we have identified two main visions for the development of pharmacogenetics, with different authors trying to shape the discipline as a whole along different lines. In the first vision, pharmacogenetic variations are focused on drug metabolism and are independent of disease-causing genes. Here the emphasis is on the safety of a drug. In the second, the introduction of pharmacogenetics is centred on the genes associated with disease and will lead to changes in how common diseases are classified. Here the emphasis is on drug efficacy. The following sections examine each of these in more detail, describing how they are being articulated and embodied in practice with respect to both the use of established (already licensed) pharmaceutical products and the development of new medicines.

It might be tempting to view these visions in terms of 'controversy studies', which are such a reliable approach to the sociology of science and technology. We think this would be a mistake, since it would overstate the antagonism or competition between the two visions and their proponents. Both these visions rely on many of the same technologies (sequencing, DNA arrays, SNP analysis), and they overlap to a certain degree; there are many articles that discuss both approaches (e.g. Akhtar, 2002; Ensom et al., 2001; Evans & Relling, 1999; Ginsburg & McCarthy, 2001; Ingelman-Sundberg, 2001; Kleyn and Vesell, 1998; Mancinelli et al., 2000; Marshall, 1997; Norton, 2001; Persing & Cheek, 2000; Rusnak et al., 2001). However, while these visions can coexist, they are analytically and discursively discrete, are embodied in different research practices, shape firm strategies in distinct ways and have particular ethical features. Our point is to show that these differences exist, and suggest that they can best be explained in terms of constructing one of these two visions.

A vision of classical pharmacogenetics – ADRs and disease independence

The first vision is based on the notion that pharmacogenetics is mainly concerned with understanding how drugs are metabolized, rather than understanding the genetic basis of a particular disease (Housman & Ledly, 1998; Kalow, 2001; Meyer, 2000; Roses, 2000a, 2000b; Sadée, 1999; Schmitz et al., 2001; Steimer & Potter, 2002; Weber, 2001; Wolf et al., 2000). It therefore focuses on adverse reactions to compounds, the genetic variations in genes responsible for drug metabolism (pharmacokinetics and toxicology) and the use of this data as a means of improving the safety of medicines.

Articulating the vision

A vital actor in this vision is the set of genes that code for the cytochrome P450s, a family of enzymes that operate in the liver. These enzymes are responsible for the initial metabolism of many prescription drugs, and variations in the DNA (genotype) coding for them result in wide differences in how drugs are metabolized. For example, patients with an amplification of one of the P450 genes (CYP2D6) are classed as ultra-rapid metabolizers. This means that their body processes some drugs so quickly that no therapeutic effect is gained; in the case of some treatments, they might have to be prescribed 50 times the dose given to other patients (Wolf et al., 2000: 988). CYP2D6 alone is thought to be involved in the metabolism of about one-quarter of all prescription drugs, and alterations in this gene can lead to therapeutic failure, adverse effects and toxicity in particular subgroups of patient undergoing treatment. This is a major problem as over 30% of drugs that enter clinical trials are withdrawn from further development due to unfavourable drug metabolism. The P450s are widely cited by proponents of this vision (without exception; their well-researched nature makes them a reliable basis for the construction of this vision.

Another example used in articulating this vision are polymorphisms in the gene that codes for the enzyme thiopurine methyltransferase (TPMT), which is involved in the metabolism of a number of drugs commonly used to combat childhood leukaemia. Slow metabolizers of these drugs require up to one-fifteenth of the normal dose (to avoid a potentially fatal ADR), while fast metabolizers require much higher doses for effective treatment (Krynetski & Evans, 1999; Meyer, 2000).

What links the authors promoting this vision of pharmacogenetics is the focus on pharmacokinetics (drug metabolism) and in particular the P450 genes, as well as an emphasis on improving the safety of prescribing. For these authors: 'The aim of pharmacogenomics is to decrease adverse responses to therapy through determining new therapeutic targets and genetic polymorphisms that effect drug specificity and toxicity' (Wieczorek & Tsongalis, 2001: 1).

From a purely practical viewpoint, it is not hard to see how pharmaceutical companies could have an interest in developing a technology which, in a litigious environment like the US healthcare sector, may reduce the number of patients who have cause to sue drug manufacturers. Of course, the identification of pharmacogenetic reactions may also lead to a situation where companies that do not employ this technology are sued for endangering patients. This has already happened in the class action brought against SmithKline Beecham in late 1999, which alleges that the LYMErix vaccine for Lyme's disease reacts with the HLA-DR4+ polymorphism (present in up to 30% of the population) to produce autoimmune arthritis. The claimants suggest that the company knew about this pharmacogenetic reaction, but marketed the drug regardless (Wortman, 2001).

A key element of this vision is the separation of disease genetics from drug genetics; a genetic test in this case does not reveal anything about the prognosis or even the cause of the condition treated by the drug concerned. This serves a central purpose, as it roots the emerging discipline of modern pharmacogenetics in a well-tested and widely recognized biological effect: the metabolism of drugs by the P450 genes. Even sceptics about the role of genes in medicine would have difficulty casting doubt on the idea that many aspects of drug metabolism are genetically influenced.

How might this option be integrated into clinical practice?

In practice many large pharmaceutical companies already classify participants of their clinical trials as fast, slow or normal drug metabolizers using P450 genotyping, but in the clinical setting such testing is still very rare (Sadée, 1999). Clinical testing might involve the diagnostic genotyping of patients before they are prescribed drugs; those with the 'wrong' genotype would not be given that therapy, or would be prescribed an alternative dose. Genotyping of this sort has already become a practice in some (mainly cancer) centres, to reduce the risk of ADRs (Sadée, 1999).

Various proposals exist for the organization of testing, including third-party services, local hospital-based testing and 'desktop' devices in the physician's office. If such technology was available to identify those at risk of an ADR from a given drug, regulatory authorities might come under pressure to change the labelling of the drug to restrict its use to particular groups. In general, however, the introduction of this option for pharmacogenetics would not necessarily have a significant impact on established medical practices, but would simply bring in an element of pre-prescribing diagnosis. The UK Department of Health has recently carried out a study suggesting that P450 screening could be implemented in the near future, and would be cost effective if targeted at high-risk groups and key clinical areas (DoH, 2000).

The development of new drugs

Within this vision less attention has been given to how pharmacogenetics might be used to help develop new drugs. Instead the main focus here has been on the 'rescue' of drugs in development which have failed in late stage (phase II/III) clinical trials. Some commentators have suggested that if an effective drug fails to get marketing approval because it induces a genetically based ADR in a small percentage of patients, the trial could be redone using a pre-screening of the patient population to eliminate people at risk of ADRs (Regalado, 1999). If it was shown to be safe and effective in this group, the drug could then be licensed for use only with that particular genotype and with a clear warning on the label that people with the 'wrong' genotype could not be given the product. This might significantly increase the chance of drugs gaining marketing approval, but at the expense of restricting their use and a compulsory tie into diagnostic genotyping.

The way in which this vision is being articulated by its scientific advocates therefore involves the simultaneous construction of particular problems to be solved (i.e. drug safety), the solution of the ethical problems associated with genetic testing, and how the technology might be used to increase the profitability of the pharmaceutical industry through drug rescue. It therefore provides a rhetorical justification for the pursuit of a particular research strategy, but also offers a means of overcoming potential resistance and guides the development of business strategy.

A vision of disease dependent pharmacogenetics

The 'disease dependent' vision of pharmacogenetics aims to identify associations between genetic markers for drug response and those genes directly involved in the development of different forms of pathology (e.g. Anderson et al., 1999; Bell, 1997; Lichter & Kurth, 1997; Lindpaintner, 1999; Lindpaintner et al., 2001; Moyses, 1999; Persidis, 1998a, 1998b).

Articulating the vision

If a central theme of the ADR vision of pharmacogenetics is a concern with drug metabolism and safety, then the core of the disease dependent view is 'patient stratification' and improved efficacy. This is based on the idea that many common diseases are, despite their phenotypic or clinical homogeneity, caused by a number of different genetic changes:

> There is little doubt that many of our current disease definitions such as cancer, diabetes and heart disease are actually very different diseases at the molecular level that manifest themselves phenotypically in similar ways.
>
> (Persidis, 1998a: 210)

Conditions can therefore be reclassified in terms of specific genes or gene defects, linking variation in disease progression to variation in drug reaction, ensuring that one 'outcome from the pharmacognetics approach is a redefinition of what a disease is at the molecular level' (Persidis, 1998a: 210).

Since diseases can be split into a number of sub-conditions (on the basis of different genetic factors), patients can be separated into different groups according to this new classification, focusing on finding those people who can be classed as 'good responders'. In this vision, such division or patient stratification does not happen only at the level of the clinic, but further back in the drug development process, 'allowing more specific trials for single disease entities, eliminating some of the problems associated with enrolment of heterogeneous patients' (Kurth, 2000: 227). Stratification therefore offers the prospect of a reduction in the numbers of patients entered for clinical trials, and hence a financial saving for companies (Anderson et al., 1999).

How might this option be integrated into clinical practice?

In practical terms this particular technical vision might be integrated into the clinic by much greater use of gene-based diagnostics to help classify the disease suffered by a patient at a molecular level. In effect a 'molecular diagnosis' would be made, which would then form the basis for clinical management, including drug therapy, as a means of improving the efficacy of treatment. As with the first vision, genotyping might be organized in a number of ways, but the effect on diagnostic practice would be much more profound than simply screening for ADRs. In this case, genetic analysis would become a central and routine part of the treatment of many common diseases:

> A pharmacogenomic model of drug prescription . . . would start with the analysis of the patient's . . . DNA . . . A limited genome scan using preselected SNPs markers would be run . . . The patient's marker pattern would be compared with patterns of efficacy established for drugs that are generally prescribed for the disease being treated . . . [This] . . . would result in selection of a drug and dose which would be optimal for the patient.
>
> (Anderson et al., 1999: 266–67)

Painting such a scenario brings this vision into sharper focus, allowing clinicians and policymakers to see how pharmacogenetics might operate in the clinical encounter.

Proponents of this vision claim that because of such molecular diagnosis, doctors will be able to treat patients' 'real' illness, instead of simply dealing with a vague collection of symptoms that could correspond to a number of distinct conditions. This reclassifying role seems likely, especially since it has already occurred in a number of conditions, most notably diabetes (Bell, 1998; Hedgecoe, 2002).

Furthermore, there have already been a number of clinical research studies designed to emphasize the potential for pharmacogenetics to significantly redefine common diseases in relation to established drug therapies. In the same way as the previous vision heavily cited studies of the cytochrome P450s to root itself in solid research, so the disease-centred vision tends to cite some of the following studies, to 'prove' that pharmacogenetics can tell us something about underlying disease mechanisms.

A key point to note about some of these examples is not just that pharmacogenetics can lead to the stratification of disease along genetic grounds (e.g. asthma, schizophrenia), but that these stratifications can also map onto divisions in prognosis and clinical development of disease (e.g. Alzheimer's, heart disease and cancer). For example, in a widely cited study Kuivenhoven et al., 1998) showed that variations in the genetic locus responsible for the cholesteryl ester transfer protein (CETP) were related to variations in response to the drug prevastatin, commonly used to treat high cholesterol levels. However, the most important finding was that the mutation associated with the greatest risk of heart attack (TaqIB) is also associated with the most positive response to treatment. This means that a genetic test to decide on whether to prescribe prevastatin is also a prognostic test for the type and severity of heart disease a patient might develop. In this way some pharmacogenetic tests overlap with diagnostic genetic tests.

In the overall strategy associated with this vision of pharmacogenetics, research does not therefore have to chase after comprehensive explanations of the mechanisms underlying common diseases in order to redefine those diseases in genetic terms, provide a prognosis or predict the likely effect of therapy. If a strong correlation between a genetic marker and drug response or disease prognosis can be demonstrated, then this will have immediate clinical value, even if the biology of this link remains a mystery. This opens up the prospect of the rapid introduction of routine clinical genetic testing based on nothing more than a statistical association between a marker and the response to a particular drug.

The development of new drugs

In terms of new drug development, the stratification of diseases at the molecular level might enable the possibility of designing new drugs aimed at specific genotypes. In this way companies may be able to develop therapies for highly targeted patient groups, which may be unresponsive or react adversely to current treatments.

> In the context of drug development, the segmentation of patient populations according to genotype has the potential to define groups of individuals with the most beneficial therapeutic ratio. Since the size of clinical studies is affected by the magnitude of the therapeutic effect, increasing the proportion of responders should allow smaller and faster trials. Such small 'smart' studies will provide a counterbalance to the large 'dumb' studies used to detect small treatment effects in large populations.
>
> (Moyses, 1999: 199)

One consequence of such an approach would be the segmentation of drug markets and the creation of specific therapeutic niches. While this might improve therapy for some people, it could also lead to a situation where it becomes commercially unattractive to create new drugs for small numbers of people with rare 'unresponsive' genotypes.

Rather than being based on the construction of new problems to be solved (i.e. ADRs) the way in which this vision is being articulated by its scientific supporters focuses on the potential for creating a new type of medicine. Here the emphasis lies on new and powerful ways of seeing and diagnosing disease, the creation of a more scientific medicine, potential improvements in

Table 1 Clinical studies where disease-associated markers indicate drug response

Disease	*Gene marker*	*Drug*	*Comments*	*Reference*
Alzheimer's	APOE	Cholinesterase inhibitors	Carriers of APOE4 allele have reduced effect from drug	Poirier et al., 1995*
Atherosclerosis	Cholesteryl ester Prevastatin transfer protein (CETP) gene	Mutation TaqIB associated with greatest risk of heart attack also associated with most positive response to treatment	Kuivenhoven et al., 1998	
Ovarian cancer	P53	Common anti-cancer drugs	P53 makes tumours chemoresistant: more aggressive treatment required	Shelling, 1997
Asthma	ALOX5 promoter locus	ABT-761	Carriers of certain variations in gene showed improved response	Drazen et al., 1999
Breast cancer	HER2 overexpression	Herceptin	Herceptin, a monoclonal antibody, only effective in tumours overexpressing the HER2 protein, which have a worse prognosis	Cobleigh et al., 1999
Schizophrenia	Serotonin neurotransmitter receptor 2A	Clozapine	Certain alleles (carried by ~ 50% of pop.) lead to improved response to drug	Arranz et al., 2000

* We are aware that the link between Tacrine and APOE4 is regarded with suspicion by many psychiatric geneticists, and the results of this particular paper have been largely rejected. Yet this research is still cited in reviews of the literature, and thus plays an important role in the construction of this vision.

patient care and reducing the costs of drug development. As with the previous example, this vision provides a rhetorical justification for the pursuit of a particular research agenda, but also offers a guide to the way in which health services and professional practice should evolve, and strategies to improve the profitability of the pharmaceutical industry.

The creation of these two visions is not just a discursive or cognitive phenomenon, but is also embodied in research agendas and the design of clinical experiments. A commitment to a particular vision not only implies a judgement about the benefits of a certain scientific strategy, but also a commitment to: specific types of research questions (safety vs efficacy; prescribing practices (screening for ADRs vs a 'molecular diagnosis'; drug development strategies (drug rescue vs targeting new drugs at specific genotypes; and even definitions of disease (e.g. disease stratification).

The successful demonstration of the utility of these new ways of using and developing drugs in research settings gives weight to the initial claims made for the development of pharmacogenetics in general. However, these visions will only translate into widespread clinical practice if the pharmaceutical industry embraces them and if the important ethical, legal and social problems they raise can be successfully negotiated. The next section will examine the way in which pharmacogenetics has stimulated the creation of a new industrial sector and how these different visions are guiding the formation of company strategies for the commercialization of particular technical options.

The creation of a pharmacogenetics industry: the integration of genomics into the drug development process

The success or failure of pharmacogenetics will ultimately depend on how well it is integrated into the pharmaceutical industry. Drug companies drive the development and adoption of new medicines and constitute the most powerful group supporting the creation of this technology. They have already committed considerable investment to this area, with advocates predicting that the rise of pharmacogenetics will be beneficial to industry, clinicians and patients (Regalado, 1999). For the patient, it is claimed that pharmacogenetics will allow doctors to prescribe drugs more effectively and reduce the chances of adverse reaction. However, the greatest promise is for industry, where it is hoped that the technology will result in increased sales, cheaper and faster clinical trials and the 'rehabilitation' of drugs which have been removed from the market after being considered dangerous (Sadée, 1999).

Companies are therefore involved in developing commercial plans to exploit the promise of pharmacogenetics and translating the scientific visions described above into business and technology strategies for the creation of new products and services. We identify distinct commercial strategies based on each of the two main visions, which have been used to guide internal investment decisions, create inter-firm collaborations and attract finance from investors. Furthermore, these visions are also embodied in the physical design of the products and services being developed by firms. In this sense, visions serve as strategic resources to 'mobilize' funds and enrol the support of other actors to the technology. While the visions articulated in the scientific literature legitimize pharmacogenetics, the actions of industry bring those visions to life, both in terms of actual (or future products) and in the corporate literature and business plans based around these competing approaches.

The structure of the pharmacogenetics industry

As with many other biomedical technologies, such as diagnostic imaging or gene therapy, the development of a new technology has both resulted in, and depended on, the creation of a new set of industrial actors (Blume, 1992; Martin, 2001b). In this sense, new industries are co-constructed during the process of innovation and socio-technical network formation. The destiny of an emerging biotechnology depends on the existence of firms which have invested heavily or been established solely to exploit it, as only companies have the financial, managerial and commercial resources to bring new drugs to market. At the same time, the fate of these firms is intimately tied to the success of the technology. If a technology fails, so do many firms aligned with it.

Pharmacogenetics is no exception to this general pattern and a new industrial sector has been created in the last five years to exploit the commercial opportunities promised by the technology. The results of a comprehensive survey of the North American and European biotechnology industries[2] identified two distinct groups of companies involved in the development of pharmacogenetic technology: small biotechnology firms dedicated to creating 'platform technologies' for the analysis of the genetic basis of drug response, and large pharmaceutical companies involved in clinical drug development. Tables 2 and 3 summarize the results of this survey and give a list of both these groups of companies. This classification is, of course, well known within the pharmaceutical industry and is used by annual reviews of the industry as a useful way of comparing different kinds of firms. Our point is that these differences do not just represent firms' age, size or cash flow, but also their research strategies and attitude towards the two visions outlined above.

In total, some 48 biotechnology firms and 12 large pharmaceutical companies have started to develop pharmacogenetic technology. American firms dominate and the majority of the biotechnology start-ups have been founded in the last five years. Most firms are pursuing a range of different strategies and it is beyond the scope of this paper to analyse them all in detail. However, six of the leading biotechnology firms in the field of pharmacogenetics have been identified (three European, three US) and information about their strategies, products/services and collaborations is presented in Table 4; all are using large genetic databases and the techniques of association genetics in combination with SNPs to locate markers correlated with particular drug responses.

Table 2 Biotechnology firms developing pharmacogenetics technology

US (41 firms)	Acadia	InforMax
	Affymetrix	Interleukin
	Celera	Lynx
	Clingenix	MassTrace
	CuraGen/454 Corporation	Molecular Innovations
	DiaDexus	Molecular Simulations
	EmerGen	Myriad Genetics
	Exelixis	Nanogen
	First Genetic Trust	Orchid BioSciences
	Genome Therapeutics	Millennium Predictive Medicine
	Genomica	Pharmagene
	Genaissance Pharmaceuticals	PharmaSeq
	Genelabs Technologies	PPD Pharmaco
	Genelex	Quintiles Transnational
	Gene Logic	Rosetta Inpharmatics
	GeneticXchange	Sequenom
	Genome Pharmaceuticals	Signalgene
	Genometrix	Third Wave Technologies
	Illumina	Transgenomic
	Incyte	Variagenics
		WITA Proteomics
Europe (7 firms)	DeCODE Genetics	Genset
	ExonHit	Oxagen
	Gaifar	Oxford Glycosciences
	Gemini Genomics	

Table 3 Large pharmaceutical companies developing pharmacogenetics technology

US	*Europe*
Abbott	Amersham Pharmacia
Bristol Myers Squibb	AstraZeneca
Genzyme	Bayer
Janssen	Glaxo SmithKline
Pfizer	Hoffman-La Roche
Warner Lambert	Novartis

Visions of pharmacogenetics embedded in the strategies of leading firms

Closer examination of the leading firms' technology and business plans shows how the two visions described above are being embedded in firm strategies for the commercial development of pharmacogenetics. Analysing a

Table 4 The Strategies of Leading Biotechnology Firms Developing Pharmacogenetics

Name	*Location*	*No. of staff*	*Strategy*	*Products/services*	*Alliances related to pharmacogenetics*
CuraGen	US	300	Identification of genetic markers linked to drug response – emphasis on safety (ADRs)	Sell data to improve clinical development (trials) of new drugs – aim to reduce negative response/ toxicity. Develop diagnostics to detect ADRs	Bayer, Ono Pharmaceuticals
DeCODE Genetics (Encode subsidiary)	Iceland	500	Identification of genetic markers linked to drug response – emphasis on safety (ADRs)	Develop diagnostics to identify patients most at risk of ADRs to commonly prescribed medicines	Affymetrix, Hoffmann-La Roche
Gemini Genomics	UK	N/a	Identification of genetic markers linked to drug response – emphasis on safety (ADRs)	Sell data to improve clinical development (trials). Patents on CYP alleles and licensing use for development of diagnostics to detect ADRs	–
Genaissance Pharmaceuticals	US	120	Identifying genetic markers linked to response to commonly prescribed medicines – emphasis on efficacy (good responders)	Sell data to improve clinical development (trials) of new drugs aimed at good responders and develop diagnostics to improve marketing of licensed products to good responders	AstraZeneca, Gene Logic, Janssen, Pfizer, Seqenom
Genset	France	530	Identifying genetic markers linked to drug response – emphasis on safety (ADRs)	Sell data to identify people at risk of, and co-developing diagnostics to screen for, ADRs from licensed drugs. Selling data to target new drug development at people at low risk of ADRs	Abbott, Pharmacia and Upjohn, Sanofi-Synthelabo

continued overleaf

Table 4 Continued

Name	*Location*	*No. of staff*	*Strategy*	*Products/services*	*Alliances related to pharmacogenetics*
Variagenics*	US	N/a	Identifying genetic markers linked to drug response – emphasis on efficacy (good responders)	Sell data to improve clinical development (trials) of new drugs and develop diagnostics to target new medicines to responsive genotypes	Amgen, Boehringer Ingelheim, Covance, Isis, Quintiles

* As a sign of how quickly the pharmacogenetics industry is moving, and of the fragile nature of many of these companies, we note that at the end of January 2003 Variagenics merged with the 'biopharmaceutical' company Hyseq, to form a new firm: Nuvelo Inc. From the company website, Nuvelo seems focused on the development and marketing of new drugs, and thus, Variagenics' pharmacogenetic goals seem to have fallen by the wayside.

range of publicly available documents, including annual reports, websites, press releases, US Securities and Exchange Commission (SEC) filings and press stories, it was striking how important the articulation of a commercial strategy based on one or other of these visions was to these companies. Just as scientific investigators use the creation of expectation to enrol support and resources, firms appear to deploy visions of how their technology might be made profitable in order to attract investors, collaborators and customers. The creation and marketing of these visions is an important part of the work involved in starting a new biotech firm (Martin, 2001b).

Firm strategies focusing on finding genetic markers associated with ADRs

Four of the leading firms described in Table 4 are involved in developing technology based on the ADR-centred vision of pharmacogenomics with its emphasis on drug safety. The French firm Genset was the first biotechnology company to move into the field of pharmacogenomics (i.e. large-scale industrial pharmacogenetics) when it established a landmark US$23 million research collaboration with Abbott in 1997. Under the terms of the agreement, Genset aims to discover genes and markers associated with particular drug responses and Abbott will then produce tests using this information for diagnostic purposes. In particular, the collaboration focused on Abbott's asthma drug Zileuton, which causes liver damage in a small number of patients. In July 1999 Genset identified genes associated with Zileuton-induced liver toxicity and has subsequently filed patent applications covering these markers. The intention is to develop a genetic test to pre-screen asthma patients so that those with the 'wrong' genotype are not given Zileuton. The company hopes that this might increase the use of the drug.

In early 2001 CuraGen became one of biggest players in the field when it established a US$124 million collaboration with the large pharmaceutical company, Bayer. They aim to evaluate Bayer's preclinical 'pipeline' of new drugs early in development in order to reduce the cost of development, decrease the time to market and create 'safer and more efficacious products'. In particular, the firm intends to compile a database of gene-based markers and information that will enable scientists to predict potential drug toxicities early in their development. According to the company website:

> Through the application of this knowledge, CuraGen scientists can identify patients as being negative responders or as having a high susceptibility for a toxic reaction, prior to their enrolment into clinical trials. This knowledge can ultimately prevent toxic side effects and even catastrophic failures of potential blockbuster therapeutics. This same knowledge can also be applied in the development of diagnostics for use in personalized medicine. By determining in

> advance which drugs are safest and most efficacious for patients, CuraGen is helping to ensure that the most appropriate drugs are prescribed at the most opportune times, reducing side effects and limiting adverse reactions.[3]

Gemini Genomics' approach is also based on classical pharmacogenetics in that it involves studying the genetic basis of drug metabolism. In particular, the firm is involved in identifying polymorphisms in the cytochrome P450 (CYP) enzymes. In 2000 Gemini was granted a patent describing polymorphisms in the CYP2D6 and CYP2C19 genes, which can identify individuals with ultra-rapid drug metabolism. It plans to create diagnostic tests using this information. These polymorphisms' importance is highlighted by another small firm, Genelex, which has started marketing genetic testing for CYP2D6 variants direct to the public and medical practitioners. It claims that ADRs are the fourth leading cause of death and the single largest source of malpractice payouts in the US and urges doctors to 'be one of the first healthcare practitioners to enter the era of personalized medicine.'[4]

DeCODE Genetics, through its subsidiary Encode, has also taken a similar approach that emphasizes the benefits of this vision for improving the safety of prescription medicines. To this end, it plans to:

- use genetic information to determine proper dosage levels and establish which drugs patients should avoid because of possible side-effects;
- develop clear descriptions for individuals – perhaps in the form of a smartcard that can be carried in a wallet or purse – stating their most likely response to common drugs. By making such information easily and immediately available to individuals and physicians, the chances of drug treatments leading to dangerous reactions and side-effects should be significantly reduced.[5] As a means of trying to realize this vision of a 'smartcard', DeCODE has established a collaboration with Affymetrix to develop a 'gene chip' that will predict the responsiveness of individual patients to treatments for common diseases.

These examples suggest that the vision of pharmacogenetics based on the reduction of ADRs is shaping the commercial strategies of a number of leading firms in terms of the research they are undertaking, the intellectual property they are creating, their choice of collaborators, and the products and services they plan to sell. The focus is both on improving the safety of already licensed products and the development of new compounds.

Firm strategies focusing on the identification of 'good responders'

In contrast to the significant commercial interest in identifying ADRs, only two of the leading companies – Genaissance and Variagenics – are currently (January 2002) developing their commercial strategies around the disease dependent vision of pharmacogenetics as applied to both new and approved drugs. For new medicines, both Genaissance and Variagenics aim to examine the genetic variation among patients in clinical trials to create 'smarter' trials through the design of protocols that result in the inclusion of those patients most likely to benefit from the proposed therapeutic product. Here the emphasis is on finding good responders, rather than those at risk of ADRs, by identifying polymorphisms in disease-associated genes. This might reduce the size and, hence, the cost, of late-stage clinical trials.

In relation to drugs already on the market, Genaissance seeks to maximize pharmaceutical company product sales by first creating diagnostic tests that predict drug response and which might be used in tandem with the prescribing of a medicine. Second, the company is looking to integrate genetic variation information into drug marketing strategies to enhance a market leading position or to address problems such as poor market penetration, competitive pricing, risk of therapeutic substitution, and limited patent life. Finally, the firm hopes to use pharmacogenetic data to target new markets and obtain approval for new drug indications.

In particular, the firm hopes to find genetic markers that identify individuals who will respond better to a particular drug within a particular class of compounds, compared to other drugs in the same class, or who will respond better to one class of drugs compared to another class. It is attempting to demonstrate the potential of this approach with the statins, which are used to treat patients with high cholesterol and lipid levels, and who are at an increased risk of cardiovascular disease. According to the firm:

> This is a highly competitive market with multiple approved products seeking to gain increased market share. Currently, the market is approximately $11 billion worldwide and is forecast to at least double in size by 2005. Identification of genomic markers that would allow the right drug to reach the right patient would allow a company to further differentiate its product and may improve patient compliance, which are both particularly important factors when maximizing profit from drugs that are taken over the course of a lifetime.[6]

Genaissance is currently organizing clinical trials to test the predictive power of its genetic markers for a number of already licensed statin drugs. It then intends to auction this information to pharmaceutical firms involved in marketing these drugs to give the highest bidder a significant advantage over its competitors. Interestingly, this approach is at odds with the conventional wisdom about pharmacogenetics, which claims that the era of 'blockbuster' drugs (single compounds marketed to large populations) is ending. Instead, some have argued that the arrival of pharmacogenetically based 'minibusters' (McCarthy, 2000) might allow companies to 'segment and grow' particular markets (Anderson et al., 1999: 268).

As with the 'ADR vision', this idea of how pharmacogenetics might develop is embodied in the plans, strategies and activities of companies and is playing a major role in shaping the behaviour of these key actors. In particular, commercial strategies based on this vision aim to help improve the marketing of already licensed drugs to good responders and the development of new drugs aimed at highly profitable genotypes. However, while there are important differences between the commercial realization of each of these two visions, they share some common objectives, most notably, a reduction in the time and cost and improvement in the success rate of clinical trials, and the expansion of markets for established drugs through proprietary tests and an expansion of indications.

Integration of pharmacogenetics into the mainstream pharmaceutical sector

The success of these visions in creating interest in pharmacogenetics has fuelled mergers and close working relationships between large drugs companies and smaller firms specializing in diagnostics, platform technologies and the creation of genetic databases (Persidis, 1998b). Already over 150 pharmacogenetics alliances between large and small firms have been created since 1997.[7] Furthermore, prior to the widespread introduction of pharmacogenetics, most large drug companies are already preparing themselves for the impact of this technology on their core activities by collecting DNA samples from the participants of their clinical trials (Martin, 2001a). Even if a trial is not intended to investigate the pharmacogenetic effects of a particular drug, companies reason that such samples may be useful in the future if the drug has problems later in its development.

In terms of our explanatory framework, this review of the pharmacogenetics industry shows how the two visions articulated in the scientific literature have become embodied in the policies, structure and products of individual pharmaceutical companies and biotechnology firms. This has involved the translation of particular discourses into a heterogeneous range of artefacts, organization structures, work practices and material investments. In this way visions provide the strategic framework for actors to change the present in order to build the future.

Despite the large sums being invested in pharmacogenetics, there is still a high degree of contingency and uncertainty surrounding this technology, perhaps best represented by the tension between placing the emphasis on ADRs (which the industry is very worried about making public) and strategies aimed at identifying 'good responders'. Furthermore, very few pharmacogenetics products or services are actually on the market. Only with time will it be possible to assess if either of these visions translates into commercially viable tests and drugs.

Critically, the extent to which companies are able to realize the benefits offered by pharmacogenetics will depend on a range of social and technical factors, including: how well genetic testing is able to predict drug response in practice; how commercially attractive a particular application is; the ability to integrate the technology into routine clinical work; and the social, ethical and regulatory issues raised by each option. In particular, the resolution of the controversial ethical issues raised by pharmacogenetics will be a prerequisite for the successful creation of new markets. The way in which these bioethics debates are being constructed is explored in the next section.

The social problems raised by pharmacogenetics: bioethics as a form of expectation

As is the case with many other novel genetic technologies, bioethicists, the media and some scientists claim that there are pressing social and ethical problems associated with pharmacogenetics (Chadwick, 1999; Clarke et al., 2001; Issa, 2000; Issa & Keyserlingk, 2000; Kreeger, 2000; Rothstein & Griffin Epps, 2001). Issues raised in this context include the general concerns posed by genetic testing (data privacy, informed consent, discrimination by insurers and employers), access to healthcare and the safety of drugs developed for particular genotypes. However, the focus of this analysis will be on the way in which the two visions discussed above are reflected in the bioethics discourse surrounding pharmacogenetics, and the way in which that discourse is shaped to support these visions. To what extent does bioethical discourse play a role in creating expectations and constructing particular technical options?

Initially, bioethics might seem peripheral to the eventual outcome of technological developments, but we believe that this would ignore important changes that have taken place within bioethics as an academic discipline over the past few years and what one might call the creeping 'corporatization of bioethics'. There is growing concern among some bioethicists that: 'the pharmaceutical and biotechnological industries are funnelling more and more cash into the pockets of academics who teach and study ethics' (Elliott, 2001; see also Callahan, 2000). While Carl Elliott's concerns about the effect of this money on the erosion (both perceived and actual) of bioethics' reputation are beyond this paper, he lists several examples of the way in which firms have engaged with bioethics discourse, ranging from the withdrawal of funds[8] to the support for academic bioethics departments.[9]

The pharmaceutical industry is increasingly concerned about debates in professional bioethics. One of the clearest statements of this comes in Rahul Dhanda's recent book *Guiding Icarus: Merging Bioethics with Corporate*

Interests. Dhanda is in charge of bioethics at Interleukin Genetics, a pharmacogenetic biotechnology company. He suggests that as biotech start-ups become more successful, it will not be 'enough for a company to have faith in its technology anymore; the company must also be responsible for it . . . As biotechnology progresses, it must do so hand in hand with ethics' (Dhanda, 2002: 5–6). Dhanda contradicts worriers like Elliott and suggests that, rather than companies using bioethics as a form of PR, to legitimize decisions already made, biotechnology firms need bioethicists' impartial analysis, and that over time '. . . bioethics can become a seamless part of the corporate landscape. Standard operating procedures and corporate culture will each require reconception' (Dhanda, 2002: 7–8). He cites a report by PriceWaterhouse Coopers on the future of health care which recommends that biotech companies develop a bioethical framework within which to make decisions, and that they consult with professional bioethicists (PriceWaterhouse Coopers, 1999).

Added to this is the role that bioethics has played as a 'political broker' in debates over biotechnology regulation in the European Union (Salter & Jones, 2002) and the claim that pharmaceutical companies have, compared with academics, been reluctant to explore pharmacogenetics, largely because of the possible ethical issues involved (Jazwinska, 2001). When we claim that bioethics may be the decisive factor in deciding whether or not a particular vision becomes reified in scientific and company practice, we are simply building on a well-documented trend in the relationship between bioethics and the pharmaceutical industry.

To examine the bioethics discourse surrounding pharmacogenetics, we searched both the scientific and bioethics literature for articles dealing with the ethics of pharmacogenetics, using the definitive bibliographic database covering bioethics, *Bioethicsline*, which has now been incorporated into the National Library of Medicine's *PubMed* database (www.ncbi.nlm.nih.gov/entrez/query.fcgi). We found a total of 29 English language journal articles on *PubMed*, which included the terms 'ethics' and 'pharmacogenetics' or 'pharmacogenomics' in either their heading, abstract or keywords.[10] Strikingly, only one of these papers was in a non-scientific journal, and this was contained in a specialized law and science publication (*Law and Human Genome Review*). Fifteen papers focused on a discussion of the ethical issues specifically raised by pharmacogenetics. Of these, bioethicists (mainly lawyers) wrote six, academic scientists or clinicians four, with four papers coming from commercial sources and one from a journalist. The majority of the other 14 articles touched on this topic, but only as part of more general discussions of post-genomic medicine. Three more papers were found using the Science, Social Science and Arts and Humanities Citation Indices, one of which dealt with technical issues to do with clinical trials (Rioux, 2000), one with the organization of DNA databases (Lavori et al., 2002) and one which focused on the possible problems raised by pharmacogenetics and the use of ethnic groupings (Foster et al., 2001). All three were published in technical scientific journals rather than the bioethics literature.

In addition to the use of electronic databases, a manual search was also undertaken of the scientific literature described in the earlier sections of this paper to find articles which discussed the ethics of pharmacogenetics, but which were not indexed as covering this topic in *PubMed*. A further 20 articles were found which covered this area, although in almost all cases they were written by scientific researchers and the discussion of ethics was part of broader discussions of either the technical or commercial development of the technology.[11]

From this review of the literature a number of features can be clearly identified. First, this early discussion of the ethics of pharmacogenetics is mainly organized within the scientific community through publications in scientific journals. Very little debate has taken place within dedicated bioethics journals. Furthermore, it is also noteworthy that almost no social science work has been done on the ethical, legal and social issues raised by pharmacogenetics. Second, the discussion seems to be dominated by researchers involved in developing the technology, with independent professional bioethicists being in a minority. Third, with most of the articles not explicitly about just the ethics of pharmacogenetics, the debate is often closely tied to other discussions on the scientific, clinical and commercial prospects for the technology. It therefore appears that the construction of a bioethics discourse around pharmacogenetics is mainly being shaped by researchers advocating the development of the technology and is closely tied to the creation of scientific and commercial expectations.

The papers found in this search were then examined to see if they referred in any way to the visions identified above. While much of the discussion was of a generic nature and covered general problems raised by genetic testing, the two technical options were clearly visible in the discussion of the bioethics of pharmacogenetics. In van Lente's terms, explicitly constructing these visions in terms of ethical debates reduces the uncertainty associated with developing a potentially controversial technology. What we see here are authors using the discursive space provided by ethics debates to clarify the potential ethical issues associated with their particular vision, and in turn to influence possible future response on the part of regulators and the public.

Ethical expectations associated with the ADR vision of pharmacogenetics

Commentators who discussed the 'ADR approach' to pharmacogenetics often drew an ethical distinction between testing for genetic diseases and testing for drug reactions, and claimed a reduction in the ethical problems in this context. For example, Housman and Ledley note that:

> Pharmacogenomics is not aimed at achieving seminal insights into proximal causes of disease, but rather simple improvements in patient care. Pharmacogenomics should not be confused with predisposition or predictive testing.
>
> (Housman and Ledley, 1998: 492)

Allen Roses, a senior scientist at Glaxo SmithKline, takes this further by arguing that 'Pharmacogenetics is not really about disease diagnosis' (Roses, 2000b: 1358) and insists on the need to separate public perception of the ethics of pharmacogenetics from disease testing: 'Clear language and differentiation of respective ethical, legal and societal issues are required to prevent inaccurate vernacular usage creating a confused public perception of "genetic testing" ' (Roses, 2000a: 857–65; or see McCarthy, 2000: 142).

It therefore appears that supporters of this vision are making strenuous efforts to distance it from ethical controversies, claiming that this type of testing will reveal nothing about prognosis or disease susceptibility. But ethical issues are not simply passive; they reflect back on the visions and influence the practices being proposed. For example, Roses accepts that to avoid the possibility of ADR/disease SNP overlap, it would make sense to 'select polymorphisms with no known relationship to any disease diagnosis'. But even if there were some overlap, 'if the magnitude and severity of drug-induced [ADRs] is considered, the benefits far out-weigh the costs' (Roses, 2002: 547).

Alternatively, when the ethical issues surrounding the technology are mentioned, they are sometimes discussed in terms of the ethical duty to use pharmacogenetic testing, as a way of reducing harm to patients: 'One day it may be considered unethical not to carry out such tests routinely to avoid exposing individuals to doses of drugs that could be harmful to them' (Wolf et al., 2000: 989). Similarly Roses calls possible over-regulation of pharmacogenetic tests a 'tragedy' (Roses, 2000b: 1361).

While many in the pharmaceutical industry view this application of pharmacogenetics, which aims to reduce the numbers of ADRs, as being ethically responsible, there are some who seriously doubt the wisdom of applying the ADR vision to the rescue of drugs which fail late-stage trials. For example, Bill Haseltine of Human Genetic Sciences:

> sees a big problem in developing drugs that might work wonderfully in 75% of the population but be extremely harmful to the other 25% . . . [Pharmacogenetics] . . . means manufacturing and producing a drug that could either kill or severely harm a substantial number of people. And the only thing that stands between your drug and that harm is a diagnostic test that has to be interpreted correctly.
>
> (Quoted in Regalado, 1999: 47)

The ethical discourse around this particular vision creates a specific set of expectations about the problems associated with the future development of pharmacogenetics. In particular, advocates argue that testing for ADRs is quite distinct from genetic testing in general, and that the technology does not therefore raise significant ethical, social and legal problems. In this sense, it is being constructed as relatively problem-free. In contrast, the rescue of drugs which have failed late-stage trials because of serious ADRs is highly problematic.

Ethical expectations associated with the disease dependent vision of pharmacogenetics

While it may be true to say that the ADR-centred vision of pharmacogenetics does not imply particular ethical problems, this is not the case for many disease-centred pharmacogenetic tests. Some leading scientific advocates of this option acknowledge that the distinction between pharmacogenetics and genetic testing in this context cannot be maintained: 'there is little reason to assume that pharmacogenetic data for common complex disorders will differ significantly from primary disease susceptibility data in their potential to raise these issues' (Lindpaintner et al., 2001: 81; see also Anderson et al., 1999; Chamberlain & Joubert, 2001).

In the case of the disease-centred vision, a test for a pharmacogenetic reaction may also become a genetic test for the disease itself, with all the attendant ethical problems such as confidentiality, consent, use of test results, etc. For example, 'as soon as a patient with breast cancer submits her *trastuzamab* [Herceptin] prescription, the pharmacist and the cashier, as well as the data entry clerk, the claims adjuster, and any number of additional personnel involved in health care administration and reimbursement will, by inference, know that patient's HER-2-expressor status' (Lindpaintner et al., 2001: 81).[12]

As a consequence of this position, the ethical expectations raised as part of this vision are more problematic and might imply many of the same responses (e.g. genetic counselling, tight regulation) used to manage traditional genetic testing. This might make it harder to integrate this option into clinical practice. As in the previous vision, ethical issues are used to propose changes to practice, in this case to 'structures of confidentiality and individual choice regarding the use of all types of medical data' which may need to be 'redefined' in the light of the widespread use of pharmacogenetics (Lindpaintner, 1999: 488).

This brief discussion of the ethical debate surrounding these two visions highlights a number of key points. In particular, the way in which advocates are constructing the ADR-centred vision allows the insulation of pharmacogenetics from public disquiet about genetic testing and the misuse of information by third parties such as insurers and employers. If, as claimed, pharmacogenetics tells such people nothing about a person's current (or future) state of health, then it might be far simpler both in terms of the ethics of its use, winning public support and how it should be regulated. Yet this is in direct contrast with the alternative vision, which requires pharmacogenetic tests to be seen in the same light as more 'traditional' genetic testing. These ethical positions reinforce core aspects of the two visions: that the ADR-centred one focuses on drug metabolism and the disease-centred one 'unveils' the true causes of diseases. In this way these ethical positions help construct the visions, but also attempt to overcome likely barriers to the introduction of the technology and stress the potential benefits of pharmacogenetics.

The way in which scientific advocates of pharmacogenetics are playing a leading role in constructing particular bioethical discourses around the technology highlights another key point. Debates over the bioethics of controversial genetic technologies provide what might be called a 'negotiation space' which can be used to explore the socially acceptable limits of their use and how they might be governed. Through a process of creating a series of imagined scenarios or bioethical expectations for the application of a technology it is possible to examine the potential ethical, social and legal issues raised and gauge the extent to which a particular technical option will meet popular resistance. Given that public and professional support is essential for the adoption of biomedical technologies, these anticipatory debates serve a vital role in the construction of stable socio-technical networks and shape the creation of governance regimes.

The history of gene therapy well illustrates this point. The controversy surrounding proposals to clinically develop gene therapy in the US in the 1980s only subsided, and clinical trials were allowed to start, once a regulatory regime had been put in place which commanded widespread support (Martin, 1998). However, this was the final outcome of nearly a decade of bioethical debate and social negotiation, and initially placed severe restrictions on the technical options that could be developed (Martin, 1998). Some potential applications of gene therapy, most notably germ line therapy which involves the modification of future generations, were widely opposed and specifically excluded by the political settlement that underpinned the regulatory regime.

Bioethics debates therefore play a central role in creating visions which can form the basis for the enrolment of social support from key groups of actors, mobilizing resources and translation into practice. At the same time they influence the selection and shaping of technical options through a process of anticipatory negotiation over what is acceptable and how particular applications should be regulated. In this way the creation of particular bioethical discourses should be seen as an integral part of the socio-technical processes invovled in the shaping of artefacts, and the construction of applications and markets for emerging genetic technologies.

Conclusions

Controversial early stage genetic technologies are difficult to analyse, both in terms of the direction of their development and the social and ethical issues they raise. In this paper we have drawn on insights from the sociology of technological expectation to provide evidence about the different ways in which pharmacogenetics is currently moving, how its development might be shaped, and the ethical problems different options might raise. In particular, the concept of competing technological visions provides a powerful framework for organizing the analysis of early stage innovation, giving guidance about the potential trajectory of the technology, but without losing sight of the highly contingent nature of technical change.

Heterogeneous expectations and the formation of socio-technical networks

The preceding sections have described the construction of different types of expectation in three distinct social locations: within scientific practice through publications and the design of experiments; within the biotechnology industry through the creation of companies and the articulation of business plans; and within bioethics discourse. As we have shown, the creation of these two visions is not just a discursive or cognitive phenomenon, but is embodied in a range of heterogeneous artefacts, actor strategies and material practices. A commitment to a particular vision not only implies a judgement about the benefits of a certain scientific technique, but also a commitment to specific types of research questions (safety vs efficacy), prescribing practices (screening for ADRs vs a 'molecular diagnosis'), drug development strategies (drug rescue vs targeting new drugs at specific genotypes), commercial products (diagnostic tests vs selling data on efficacy), business relationships (collaborations on ADRs vs efficacy) and even definitions of disease (e.g. disease stratification). Visions provide a framework within which the future shape and application of a technology are constructed, as they act as both an aid for decision-making and a focus for the mobilization of actors and resources. In this way, new technologies, new industries and new ethical, legal and social problems are co-constructed and mutually shaped.

Instead of seeing the speculative claims made about the future of pharmacogenetics as ephemeral and irrelevant, this paper has argued that they are fundamental to the dynamic processes that create new socio-technical networks. The development of particular options for a new technology requires the construction of socio-technical network around them through a process of enrolment and alignment. Visions of these different options play a central role in organizing such networks. In addition to getting particular technologies to work in both experimental and clinical settings, four main groups of actors are required to create stable networks which will allow the introduction of a new medical technology into practice: industry, clinicians, patients and regulators. Visions provide industry with ideas about how profits might be made, clinicians with a guide to how a new technique can be integrated into practice, patients with the hope of improved care, and regulators with a framework for governing an emerging technology.

In the future, new groups of actors may become involved in the creation and use of pharmacogenetic technology and might be associated with the construction of new technical options, new visions and new ethical problems. This could take the development of the technology in new directions. However, we believe that the early formation of particular socio-technical networks around specific visions and applications will have a major impact in creating particular trajectories and could shape all subsequent activity. In extreme cases the formation of robust stable networks may lock a technology into a fixed niche and make some new visions and applications impossible to realize.

However, it should be stressed that this is not simply a one-way process, as the manner in which each group problematizes different technical options, and the extent to which their support can be enrolled, will ultimately decide the fate of a particular vision. If a key actor cannot be 'won over' then that vision will have great difficulty being successfully translated into material and social reality. In order to enrol support and develop stable socio-technical networks around particular options it will be essential to overcome any major difficulties that a group might have with a given application. This is especially important in the area of human biotechnology, which has proved to be highly controversial and where promising technologies have been killed off by public hostility.

Bioethics, the expectation of social problems and the shaping of technical options

It is in this context that the role of bioethics as a form of expectation and a means of social negotiation must be understood. The work of researchers and bioethicists in anticipating the likely social problems raised by pharmacogenetics might help ensure that either options are chosen for the development of the technology which are less controversial or safeguards that reassure the public are put in place before the technology is widely used. The creation of bioethical scenarios associated with particular visions which can act as a negotiation space is instrumental in the process of socio-technical alignment.

Such alignment enables the formation of new markets, as these require users, such as doctors and patients, to adopt new forms of practice. A central task in the enrolment of end users is the creation and maintenance of confidence about the risks and benefits of treatment. In the medical domain such risks are socially managed through both statutory regulation and professional governance processes, and bioethical debates play an important role in shaping these frameworks. Tight regulations controlling controversial technologies can be seen as embodying the outcome of social and political negotiations conducted through bioethical debates. They help ensure the acceptability of a specific set of options for a new biotechnology and underpin its diffusion into widespread use. Furthermore, a stable and widely supported regulatory framework also gives commercial developers the confidence to commit the heavy investment required to create new drugs and therapeutics. This is particularly important in biotechnology where product lead times can be as long as 10–12 years before a product reaches the market and investment has to be sustained throughout this period in anticipation of success.

Given the importance of ensuring social acceptability for new medical technologies, engagement with bioethics by researchers and companies may be seen as just as important to the creation of a new technology as getting particular techniques to 'work'. The increasing prominence of bioethics within biotechnology is clear. A similar reflection might also be made about scientific support for the funding of social science research on bioethical problems through the US National Institutes of Health ELSI programme and the UK Wellcome Trust's Medicine in Society programme.

By creating a more complete picture of the process of technical change in biotechnology, as well as carrying out empirical research on the ethical, legal and social issues raised by emerging genetic technologies, social scientists such as ourselves are also helping construct a set of future expectations. It is therefore impossible to maintain the long-established view that public policymakers and commentators such as bioethicists and social scientists are simply responding to the consequences of technical change. Through the anticipation of social and ethical problems and a critical engagement with the process of innovation, they are also helping construct and shape the future. The implications of this reflexive point have perhaps still to be digested by many who are working on the social studies of science and technology.

Acknowledgements

The authors would like to thank Brian Rappert, Graham Lewis, Andrew Smart and Andrew Webster for their input and feedback, as well as four anonymous *SSS* referees. Both authors are supported by grants from the Wellcome Trust: Paul Martin by grant #018381 ('The Clinical and Commercial Development of Pharmacogenetics') and Adam Hedgecoe by a Postdoctoral Training Fellowship #GR061491 MA ('Pharmacogenetics and Genetic Reclassification of Disease').

Notes

1 The basic principle behind genetic association studies is the statistical correlation between specific DNA sequences and particular diseases or drug responses.

Instead of trying to make disease-gene associations using raw sequence data, researchers are using single-nucleotide polymorphisms (SNPs) (Evans & Relling, 1999: 488). Human populations are to some extent genetically heterogeneous; i.e. the exact sequence of a particular gene varies within a population. The variation is generally limited to a relatively small number of such single base-pair changes, which are stable and inherited across generations. The hope is that specific SNPs will be found to closely correlate with the particular disease or drug reaction being investigated. Specific genes that react to particular drugs will tend to be inherited with certain SNPs. Thus, screening for those SNPs will indicate a person's reaction to a drug, regardless of whether the gene involved has been identified. This is a technically demanding procedure involving high-speed DNA screening on a huge scale, coupled to complex statistical analysis using massive data processing.

2 This survey was conducted using commercial databases (Genetic Engineering News Company Database, BioSpace) the biotechnology trade press (*GEN, Genome News, Nature Biotechnology*) and web searches. Company documents and news stories were then collected and used to classify and group firms. The most well-established companies in the field were analysed in greater detail.

3 CuraGen website, January 2001, <*http://www.curagen.com/technology/frameApps.htm*>

4 Genelex website, December 2001, <*http://www.healthanddna.com/drugreactiontest.html*>

5 Encode website, January 2002, <*http://www.encode.is/main_english.asp?group_id*=42&page_id=12>

6 Genaissance website, January 2002, <*http://www.genaissance.com/our_solutions/bus_models.html*>

7 Dr Andy Smart, personal communication, January 2002.

8 In 2000 Eli Lilly cancelled its support for the independent bioethics institute, the Hastings Center, because of a special issue of *The Hastings Center Report* (30[2] March/April 2000) which was critical of Prozac (a Lilly drug) and other antidepressants.

9 The University of Pennsylvania Center for Bioethics is supported by, among others, Monsanto, DeCode Genetics, Geron, Pfizer, AstraZeneca, and Human Genome Sciences. Elliott also lists: the Sun Life Chair in Bioethics at the University of Toronto; Stanford University's Center for Biomedical Ethics' $1 million gift from SmithKline Beecham; and the Midwest Bioethics Center's 'Research Integrity Project' funded by a $600,000 donation from Aventis Pharmaceuticals (Elliott, 2001).

10 Ameen et al. (2002); Broder et al. (2002); Issa (2002); Weinshilboum (2002); Anderlik & Rothstein (2001); Carroll & Coleman (2001); Davis & Long (2001); Francke (2001); Jones (2001); Kozma (2001); Nebert & Bingham (2001); Oestreicher (2001); Renegar et al. (2001a, 2001b); Robertson (2001); Rothstein & Griffin Epps (2001); Thomas (2001); Arledge et al. (2000); Fujiki (2000); Issa & Keyserlingk (2000); Lea (2000); van Ommen (2000); Kaplan & Junien (2000); Myers (2000); Hinde (2000); Emilien et al. (2000); Issa (2000); Motulsky (1978); Stern (1971).

11 Lindpaintner (1999); Chadwick (1999); Housman & Ledley (1998); Kreeger (2000); Roses (2000a, 2000b); McCarthy (2000); Regalado (1999); Wolf et al. (2000); Lindpaintner et al. (2001); Jazwinska (2001); Chamberlain & Joubert (2001); Ginsburg & McCarthy (2001); Anderson et al. (1999); Phillips et al. (2001); Sadée (1999); Clarke et al. (2001); Roses (2002); McLeod & Evans (2001); Akhtar (2002).

12 Not all commentators accept that Herceptin and HER2 testing is an example of pharmacogenetics: see Haseltine (1998); Rusnak et al. (2001).

References

Abraham, John (1997) 'The Science and Politics of Medicines Regulation', in Elston (1997): 151–82.

Akhtar, Saghir (2002) 'Pharmacogenomics: Are Pharmacists Ready for Genotyped Prescribing?', *The Pharmaceutical Journal* 268 (2 March): 296–99.

Ameen, M., Catherine H. Smith, and Jonathon N.W.N. Barker (2002) 'Pharmacogenetics in Clinical Dermatology', *British Journal of Dermatology* 146(1) (January): 2–6.

Anderlik, Mary and Mark Rothstein (2000) 'Privacy and Confidentiality of Genetic Information: What Rules for the New Science?', *Annual Review of Genomics and Human Genetics* 2: 401–33.

Anderson, Wayne, Craig Fitzgerald and Penelope Manasco (1999) 'Current and Future Applications of Pharmacogenomics', *New Horizons: Science and Practice of Acute Medicine* 7(2) (April): 262–69.

Arledge, Teresa, Andrew Freeman, Julian Arbuckle, Michael Mosteller and Penelope Manasco (2000) 'Applications of Pharmacogenetics to Drug Development: The Glaxo Wellcome Experience', *Drug Metabolism Reviews* 32(3–4): 387–94.

Arranz, Maria J., J. Munro, J. Birkett, A. Bolonna, D. Mancama, M. Sodhi, K. Lesch, J. Meyer, P. Sham, D. Collier, R. Murray and R. Kerwin (2000) 'Pharmacogenetic Prediction of Clozapine Response', *The Lancet* 355 (6 May): 1615–16.

Bell, John (1997) 'Genetics of Common Disease', *Philosophical Transactions of the Royal Society of London* 352: 1051–55.

Bell, John (1998) 'The New Genetics and Clinical Practice', *British Medical Journal* 316: 618–20.

Bijker, Wiebe (1995) *Of Bicycles, Bakelites and Bulbs: Towards a Theory of Sociotechnical Change* (Cambridge, MA: MIT Press).

Bijker, Wiebe, Thomas Hughes and Trevor Pinch (1987) *The Social Construction of Technological Systems* (Cambridge, MA: MIT Press).

Blume, Stuart (1992) *Insight and Industry: on the Dynamics of Technological Change in Medicine* (Cambridge, MA: MIT Press).

Broder, Sam, Arthur Caplan and William Evans (2002) 'Therapeutic Horizons – The Human Genome', *Journal of the American Pharmaceutical Association* 42(5 Suppl 1): S22–23.

Brown, Nik, Brian Rappert and Andrew Webster (eds) (2000) *Contested Futures: A Sociology of Prospective Techno-science* (Aldershot: Ashgate Press).

Callahan, Daniel (2001) 'Doing Good and Doing Well', *The Hastings Center Report* 31(2): 19–21.

Callon, Michel (1987) 'Society in the Making: The Study of Technology as a Tool for Sociological Analysis', in Bijker et al. (1987): 83–103.

Cantor, Charles (1999) 'Pharmacogenetics Become Pharmacogenomics: Wake Up and Get Ready', *Molecular Diagnosis* 4(4) (December): 287–88.

Cardon, Lou, Rammana Idury, Timothy Harris, John Witte and Robert Elston (2000) 'Testing Drug Response in the Presence of Genetic Information: Sampling Issues for Clinical Trials', *Pharmacogenetics* 10: 503–10.

Carr, G. (1998) 'Survey: The Pharmaceutical Industry', *The Economist* 345 (21 February): 1–16 of supplement.

Carroll, A. and C. Coleman (2001) 'Closing the Gaps in Genetics Legislation and Policy: A Report by the New York State Task Force on Life and the Law', *Genetic Testing* 5(4) (Winter): 275–80.

Chadwick, Ruth (1999) 'Criteria for Genetic Screening: The Impact of Pharmaceutical Research', *Monash Bioethics Review* 18(1) (January): 22–26.

Chamberlain, John and Pieter Joubert (2001) 'Opportunities and Strategies for Introducing Pharmacogenetics into Early Drug Development', *Drug Development Today* 6(11) (June): 569–74.

Clarke, Angus, Veronica English, Hilary Harris and Frank Wells (2001) 'Report on Ethical Considerations', *International Journal of Pharmaceutical Medicine* 15: 89–94.

Cobleigh, Melody, Charles Vogel, Debu Tripathy, Nicholas Robert, Susy Scholl, Louis Fehrenbacher, Janet Wolter, Virginia Paton, Steven Shak, Gracie Lieberman and Dennis Slamon (1999) 'Multinational Study of the Efficacy and Safety of Humanized Anti-HER2 Monoclonal Antibody in Women who have HER2-overexpressing Metastatic Breast Cancer that has Progressed after Chemotherapy for Metastatic Disease', *Journal of Clinical Oncology* 17: 2639–48.

Cookson, Clive (1997) 'Personalised Treatment', *Financial Times* (12 August 1997): 10.

Coombs, Rod, Ken Green, Albert Richards and Vivien Walsh (2001) *Technology and the Market: Demand, Users and Innovation* (Cheltenham: Edward Elgar).

Corrigan, Oonagh (2002) 'A Risky Business: The Detection of Adverse Drug Reactions in Clinical Trials and Post-Marketing Exercises', *Social Science and Medicine* 55: 497–507.

Davis, Alison and Rochelle Long (2001) 'Pharmacogenetics Research Network and Knowledge Base: 1st Annual Scientific Meeting', *Pharmacogenomics* 2(3) (August): 285–89.

De Laat, Bastiaan (2000) 'Scripts for the Future: Using Innovation Studies to Design Foresight Tools', in Brown et al. (2000): 175–208.

Department of Health (2000) *Harnessing the Potential of Gene Science: A Strategy for the DH* (London: Strategy and Planning, DoH).

Deuten, J. Jasper and Arie Rip (2000) 'The Narrative Shaping of a Product Creation Process' in Brown et al. (2000): 65–86.

Dhanda, Rahul (2002) *Guiding Icarus: Merging Bioethics with Corporate Interests* (New York: John Wiley & Sons).

Drazen, Jeffrey, Chandri Yandava, Louise Dube, Natalie Szczerback, Richard Hippensteel, Antonino Pillari, Elliot Israel, Nicholas Schork, Eric Silverman, David Katz and Jeffrey Drajesk (1999) 'Pharmacogenetic Association between ALOX5 Promoter Gnotype and the Response to Anti-Asthma Treatment', *Nature Genetics* 22: 168–70.

Elliott, Carl (2001) 'Pharma Buys a Conscience', *The American Prospect Online* 12(17), September 24(96) October 8, www.prospect.org/ Accessed 27 January 2003.

Elston, Mary (ed.) (1997) *The Sociology of Medical Science and Technology* (Oxford: Blackwell).

Emilien, G., M. Ponchon, C. Caldas, O. Isacson, and J.M. Maloteaux (2000) 'Impact of Genomics on Drug Discovery and Clinical Medicine', *Quarterly Journal of Medicine* 93(7): 391–423.

Ensom, Mary, Thomas Chang and Payal Patel (2001) 'Pharmacogenetics: The Therapeutic Drug Monitoring of the Future', *Clinical Pharmacokinetics* 40(11): 783–802.

Evans, William and Mary Relling (1999) 'Pharmacogenomics: Translating Functional Genomics into Rational Therapeutics', *Science* 286 (15 October): 487–91.

Foster, Morris W., R. Sharp and J. Mulvihill (2001) 'Pharmacogenetics, Race and Ethnicity: Social Identifiers and Individualized Medical Care', *Therapeutic Drug Monitoring* 23(3): 232–38.

Francke, Uta (2001) 'The Human Genome Project: Implications for the Endocrinologist', *Journal of Pediatric Endocrinology and Metabolism* 14 Suppl 6: 1395–408.

Fujiki, Norio (2000) 'A Biography and Bibliography: The Recent Trends in Bioethics and Medical Genetics in Japan (Part II)', *Law and Human Genome Review* (13): 183–210.

Ginsburg, Geoffry and Jeanette McCarthy (2001) 'Personalized Medicine: Revolutionizing Drug Discovery and Patient Care', *Trends in Biotechnology* 19(12) (12 December): 491–99.

Glaser, Vicki (1998) 'Pharmacogenomics: Laying the Foundation for Prescriptive Medicine', *Genetic Engineering News* 1 January 1998: 1, 9, 31 and 39.

Gonzalez, Frank J. et al. (1988) 'Characterization of the Common Genetic Defect in Humans Deficient in Debrisoquine Metabolism' *Nature* 331 (6155) (4 February): 442–46.

Grin, John and Armin Grunwald (2000) *Vision Assessment: Shaping Technology in 21st Century Society* (Springer: Berlin).

Guice, John (1999) 'Designing the Future: The Culture of New Trends in Science and Technology', *Research Policy* 28: 81–98.

Haseltine, William (1998) 'Not Quite Pharmacogenetics' letter, *Nature Biotechnology* 16: 1295.

Hedgecoe, Adam (2002) 'Reinventing Diabetes: Classification, Division and the Geneticization of Disease', *New Genetics and Society* 21(1) (April): 7–27.

Hedgecoe, Adam (2003) 'Terminology and the Construction of Scientific Disciplines: The Case of Pharmacogenomics', *Science, Technology and Human Values* 28(4) (in press).

Hinde, Annabel (2000) 'Evolution, not Revolution', *Trends in Biotechnology* 18(6): 230–31.

Housman, David and Fred Ledley (1998) 'Why Pharmacogenomics? Why Now?', *Nature Biotechnology* 16: 492–93.

Hughes, Thomas (1987) 'The evolution of Large Technical Systems', in Bijker et al. (1987): 51–82.

Ingelman-Sundberg, Magnus (2001) 'Pharmacogenetics: An Opportunity for a Safer and More Efficient Pharmacotherapy', *Journal of Internal Medicine* 250: 186–200.

Issa, Amalia (2000) 'Ethical Considerations in Clinical Pharmacogenomics Research', *Trends in Pharmacological Science* 21: 247–49.

Issa, Amalia (2002) 'Ethical Perspectives on Pharmacogenomic Profiling in the Drug Development Process', *Nature Reviews Drug Discovery* 1(4) (April): 300–08.

Issa, Amalia and Edward Keyserlingk (2000) 'Apolipoprotein E Genotyping for Pharmacogenetic Purposes in Alzheimer's Disease: Emerging Ethical Issues', *Canadian Journal of Psychiatry* 45: 917–22.

Jazwinska, Elizabeth (2001) 'Exploiting Human Genetic Variation in Drug Discovery and Development', *Drug Discovery Today* 6(4) (February): 198–205.

Jones, Judith K. (2001) 'Pharmacogenetics and Pharmacoepidemiology', *Pharmacoepidemiology and Drug Safety* 10(5) (August/September): 457–61.

Kalow, Werner (1990) 'Pharmacogenetics: Past and Future', *Life Sciences* 47: 1385–97.

Kalow, Werner (2001) 'Pharmacogenetics in Perspective', *Drug Metabolism and Disposition* 29(4) (April): 468–70.

Kaplan, Jean-Claude and Claudine Junien (2000) 'Genomics and Medicine: An Anticipation. From Boolean Mendelian Genetics to Multifactorial Molecular Medicine', *Comptes Rendus de l'Academie des Sciences (Paris) Series III* 323(12) (December): 1167–74.

Kleyn, Patrick and Elliot Vesell (1998) 'Drug Variation as a Guide to Drug Development', *Science* 281 (18 September): 1820–21.

Kozma, Chahira (2001) 'Genetic Testing and Disease State Management', *Managed Care Interface* 14(11): 54–57.

Kreeger, Karen Young (2001) 'Scientific, Ethical Questions Temper Pharmacogenetics', *The Scientist* 15(12) (11 June): 32.

Krynetski, Eugene and William Evans (1999) 'Pharmacogenetics as a Molecular Basis for Individualized Drug Therapy: The Thiopurine S-Methyltransferase Paradigm', *Pharmaceutical Research* 16: 342–49.

Kuivenhoven, Jan Albert, J. Wouter Jukema, Aeilko Zwinderman, Peter de Knijff, Ruth McPherson, Albert Bruschke, Kong Lie and John Kastelein (1998) 'The Role of a Common Variant of the CETP Gene in the Progression of Coronary Artherosclerosis', *New England Journal of Medicine* 338(2) (8 January): 86–93.

Kurth, Janice (2000) 'Pharmacogenomics: Future Promise of a Tool for Identifying Patients at Risk', *Drug Information Journal* 34: 223–27.

Lavori, P., H. Krause-Steinrauf, M. Brophy, J. Buxbaum, J. Cockroft, D.R. Cox, L. Fiore, H. Greely, H. Greenberg, E. Holmes, L. Nelson and J. Sugarman (2002) 'Principles, Organization, and Operation of a DNA Bank for Clinical Trials: A Department of Veterans Affairs Cooperative Study' *Controlled Clinical Trials* 23(3) (June): 222–39.

Lea, Dale (2000) 'A New World View of Genetics Service Models', *Online Journal of Issues in Nursing* 5(3) (September): 5.

Lichter, Jay and Janice Kurth (1997) 'The Impact of Pharmacogenetics on the Future of Healthcare', *Current Opinion in Biotechnology* 8: 692–95.

Liggett, Stephen (2001) 'Pharmacogenetic Applications of the Human Genome Project', *Nature Medicine* 7(3): 281–83.

Lindpaintner, Klaus (1999) 'Genetics in Drug Discovery and Development: Challenge and Promise of Individualizing Treatment in Common Complex Disease', *British Medical Bulletin* 55(2): 471–91.

Lindpaintner, Klaus, Elizabeth Foot, Mark Caulfield and Ian Hall (2001) 'Pharmacogenetics: Focus on Pharmacodynamics', *International Journal of Pharmaceutical Medicine* 15: 74–82.

McCarthy, Alun (2000) 'Pharmacogenetics: Implications for Drug Development, Patients and Society', *New Genetics and Society* 19(2) (August): 135–43.

MacKenzie, Donald and Judy Wajcman (1999) *The Social Shaping of Technology* (Buckingham: Open University Press. Second edition).

McLeod, Howard L. and William E. Evans (2001) 'Pharmacogenomics: Unlocking the Human Genome for Better Drug Therapy', *Annual Review of Pharmacology and Toxicology* 41: 101–21.

Mancinelli, Laviero, Maureen Cronin and Wolfgang Sadée (2000) 'Pharmacogenomics: The Promise of Personalized Medicine', *AAPS Pharmsci* 2(1) (7 March), Article 4.

Marshall, Andrew (1997) 'Getting the Right Drug into the Right Patient', *Nature Biotechnology* 15: 1249–52.

Martin, Paul (1998) 'From Eugenics to Therapeutics: The Impact of Opposition on the Development of Gene Therapy in the USA', in Wheale et al. (1998): 139–58.

Martin, Paul (1999) 'Genes as Drugs: The Social Shaping of Gene Therapy and the Reconstruction of Genetic Disease', *Sociology of Health and Illness* 21: 517–38.

Martin, Paul (2001a) 'Genetic Governance: The Risks, Oversight and Regulation of Genetic Databases in the UK', *New Genetics and Society* 20(2) (August): 157–83.

Martin, Paul (2001b) 'Great Expectations: The Construction of Markets, Products and User Needs During the Early Development of Gene Therapy in the USA', in Coombs et al. (2001): 38–67.

Meyer, Urs (1990) 'Molecular Genetics and the Future of Pharmacogenetics', *Pharmaceutical Therapies* 46: 349–55.

Meyer, Urs (2000) 'Pharmacogenetics and Adverse Drug Reactions', *The Lancet* 356(11) (11 November): 1667–71.

Michael, Mike (2000) 'Futures of the Present: From Performance to Prehension', in Brown et al. (2000): 21–39.

Motulsky, Arno (1978) 'Bioethical Problems in Pharmacogenetics and Ecogenetics', *Human Genetics* Supplement 1: 185–92.

Moyses, Chris (1999) 'Pharmacogenetics, Genomics, Proteomics: The New Frontiers in Drug Development', *International Journal of Pharmaceutical Medicine* 13: 197–202.

Myers, Martin (2000) 'Use of Polymorphism Analysis Required Ethical Guidelines', *British Medical Journal* 321 (12 August) 12: 453.

Nebert, Daniel (1997) 'Pharmacogenetics: 65 Candles on the Cake', *Pharmacogenetics* 7: 435–40.

Nebert, Daniel and Eula Bingham (2001) 'Pharmacogenomics: Out of the Lab and Into the Community', *Trends in Biotechnology* 19(12): 519–23.

Norton, Ronald (2001) 'Clinical Pharmacogenomics: Applications in Pharmaceutical R&D', *Drug Development Today* 6(4) (February): 180–85.

Oestreicher, Paul (2001) '4th Annual Pharmacogenomics and Medicine Lectures', *Pharmacogenomics* 2(3) (August): 291–96.

Persidis, Aris (1998a) 'The Business of Pharmacogenomics', *Nature Biotechnology* 16: 209–10.

Persidis, Aris (1998b) 'Pharmacogenomics and Diagnostics', *Nature Biotechnology* 16: 791–92.

Persing, Brady and Denis Cheek (2000) 'Pharmacogenomics', *Nursing Clinics of North America* 35(4): 975–80.

Pfost, Daniel, Michael Boyce-Jacino and Denis Grant (2000) 'A SNPshot: Pharmacogenetics and the Future of Drug Therapy', *Trends in Biotechnology* 18: 334–38.

Phillips, Kathryn, David Veenstra, Eyal Oren, Jane Lee and Wolfgang Sadée (2001) 'Potential Role of Pharmacogenomics in Reducing Adverse Drug Reactions: A Systematic Review', *Journal of the American Medical Association* 286(18) (14 November): 2270–79.

Pinch, Trevor and Wiebe Bijker (1984) 'The Social Construction of Facts and Artefacts: Or How the Sociology of Science and the Sociology of Technology Might Benefit Each Other', *Social Studies of Science* 14: 399–441.

Poirier, Judes, Marie-Claude Delisle, Isabelle Aubert, Martin Farlow, Debmoi Lahir, Siu Hui, Phillipe Bertrand, Josephine Nalbantoglu, Brian Gilfix and Serge Gauthier (1995) 'Apolipoprotein E4 Allele as a Predictor of Cholinergic Deficits and Treatment Outcome in Alzheimer Disease', *Proceedings of the National Academy of Sciences* 92: 12260–64.

PriceWaterhouse Coopers (1999) *Healthcast 2010: Smaller World, Bigger Expectation*, www.pwchealth.com. Accessed 27 January 2003.

Regalado, Antonio (1999) 'Inventing the Pharmacogenomics Business', *American Journal of Health System Pharmacy* 56: 40–50.

Renegar, Gaile, Patricia Rieser and Penelope Manasco (2001a) 'Family Consent and the Pursuit of Better Medicines Through Genetic Research', *Journal of Continuing Education in the Health Professions* 21(4) (Fall): 265–70.

Renegar, Gaile, Patricia Rieser and Penelope Manasco (2001b) 'Pharmacogenetics: The Rx Perspective', *Expert Reviews in Molecular Diagnosis* 1(3) (September): 255–63.

Richards, Tessa (2001) 'Three Views of Genetics: The Enthusiast, the Visionary and the Sceptic', *British Medical Journal* 322 (28 April): 1016–17.

Rioux, Patrice (2000) 'Clinical Trials in Pharmacogenetics and Pharmacogenomics: Methods and Applications', *American Journal of Health System Pharmacy* 57(9): 887–98.

Robertson, John (2001) 'Consent and Privacy in Pharmacogenetic Testing', *Nature Genetics* 28(3): 207–09.

Roses, Allen (2000a) 'Pharmacogenetics and the Practice of Medicine', *Nature* 405 (15 June): 857–65.

Roses, Allen (2000b) 'Pharmacogenetics and Future Drug Development and Delivery', *The Lancet* 355 (15 April): 1358–61.

Roses, Allen (2002) 'Genome-based Pharmacogenetics and the Pharmaceutical Industry', *Nature Reviews Drug Discovery* 1: 541–49.

Rothstein, Mark and Phyllis Griffin Epps (2001) 'Ethical and Legal Implications of Pharmacogenomics', *Nature Reviews Genetics* 2: 228–31.

Rusnak, James, Robert Kisabeth, David Herbert and Denis McNeil (2001) 'Pharmacogenomics: A Clinician's Primer on Emerging Technologies for Improved Patient Care', *Mayo Clinic Proceedings* 76: 299–309.

Sadée, Wolfgang (1999) 'Pharmacogenomics', *British Medical Journal* 319 (13 November): 1–4.

Salter, Brian and Mavis Jones (2002) 'Human Genetic Technologies, European Governance and the Politics of Bioethics', *Nature Reviews Genetics* 3(10): 808–14.

Schmitz, Gerd, Charalampos Aslanidis and Karl Lackner (2001) 'Pharmacogenomics: Implications for Laboratory Medicine', *Clinica Chimica Acta* 308: 43–53.

Shelling, Andrew (1997) 'Role of p53 in Drug Resistance in Ovarian Cancer', *The Lancet* 349 (15 March): 744–45.

Shook, David (2001) 'The Personalized Future of Medicine', *Business Week Online* 16 August 2001.

Steimer, Werner and Julia Potter (2002) 'Pharmacogenetic Screening and Therapeutic Drugs', *Clinica Chimica Acta* 315: 137–55.

Stern, Curt (1971) 'The Place of Genetics in Medicine', *Annals of Internal Medicine* 75(4): 623–29.
Thomas, Sandy (2001) 'Pharmacogenetics: The Ethical Context', *Pharmacogenomics Journal* 1(4): 239–42.
van Lente, Harro (1993) *Promising Technology: The Dynamics of Expectations in Technological Development* (Enschede: Department of Philosophy of Science and Technology University of Twente).
van Lente, Harro and Arie Rip (1998) 'The Rise of Membrane Technology: From Rhetorics to Social Reality', *Social Studies of Science* 28(2): 221–54.
van Ommen, G. (2000) 'Human Genetics in Health Care', *European Journal of Pediatrics* 159 Suppl 3: S170–72.
Weber, Wendell (2001) 'The Legacy of Pharmacogenetics and Potential Applications', *Mutation Research* 479: 1–18.
Weinshilboum, R. (2002) 'The Genomic Revolution and Medicine', *Mayo Clinic Proceedings* 77(8): 745–46.
Wheale, Peter, Rene von Schomberg and Peter Glasner (eds) (1998) *The Social Management of Genetic Engineering* (Ashgate Publishing, Ashgate).
Wieczorek, Stacey and Gregory Tsongalis (2001) 'Pharmacogenomics: Will It Change the Field of Medicine?', *Clinica Chimica Acta* 308: 1–8.
Wolf, C. Roland, Gillian Smith and Robert Smith (2000) 'Pharmacogenetics', *British Medical Journal* 320 (8 April): 987–90.
Wortman, Marc (2001) 'Medicine Gets Personal', *MIT Technology Review* January/February http://www.technologyreview.com/articles/wortman0101.asp.
Wyatt, Sally (2000) 'Talking About the Future: Metaphors of the Internet', in Brown et al. (2000): 109–26.

43

FOR WHOM THE CELL TOLLS

Debates about biomedicine

Gísli Pálsson and Kristín E. Harðardóttir

Source: *Current Anthropology* 43(2) (2002): 271–301.

Abstract

In this article we situate the recent debates over the genetic, genealogical, and medical databases that are being constructed for the population of Iceland in their domestic context, focusing on the contribution of anthropology to the understanding of central issues at the intersection of biomedicine and society. Modern biotechnology and bioinformatics have opened up an entirely new world in which a multitude of different kinds of human body components as well as genetic and medical information can be isolated and used for commercial, medical, and scientific purposes. While these developments are generally met with criticism and opposition, responses vary from one context to another. The perspective of moral landscapes recently developed in economic anthropology is helpful for understanding the topography of moral debates and determining the options available for informing public decisions on these contested issues.

The way in which people organize their relations to one another as a matter of control not just over things . . . but over aspects of life and body [that define "people"] will loom large on the world agenda over the next decade. We can expect an explosion of concern with ownership.

Marilyn Strathern, 1998

The debates generated by plans for medical, genetic, and genealogical databases on the entire population of Iceland focus on the ownership and marketing of biomedical material and information, patients' rights, informed consent, and the protection of privacy and personal autonomy. We suggest that these debates and similar debates elsewhere on the control of biomedical information and the commodification of body components are usefully explored from the perspective of moral landscapes. The data we use for charting the topography of local debates derive from three sources: newspapers, in-depth interviews, and participant observation. Focusing on both the scale and the ethnographic content of the Icelandic debate, we attempt to situate it in the domestic context, emphasizing the contribution of anthropology to the understanding of some of the central issues of the genome era.

The intersection of biomedicine and the market inevitably gives rise to disputes on the storing, exchange, control, and ownership of human bodily information and material, including blood, gametes, enzymes, tissues, and organs. While parts of the body (organs, hair, skin, and nails) have long been isolated, marketed, and exploited to build up knowledge of physiology (Foucault 1973), the politicization and commodification of the body have recently increased as a result of a complex series of political and technological developments. Not only has the "bare life" of the human body become the center of totalitarian politics (Agamben 1995) but ongoing developments in biotechnology and bioinformatics are opening up an entirely new world of "biosociality" (Rabinow 1996*a*). These developments are transnational and occurring at an exponential rate, and they have taken place in a social environment where partnership of science and the market is considered the most efficient means of advancing knowledge and human well-being (Fox and Swazey 1992, Komesaroff 1995, Titmuss 1997 [1970], Starr 1998). Genes, some people have argued, will be "the currency of the future" (see Nelkin and Andrews 1998). Modern biopolitics shifts the focus to bodily information, genes, enzymes, and DNA sequences; the critical issue in post-genomic research, after the sequencing of the human genome, is the use and control of the information that can be derived from body components rather than the components themselves. Body components and information—genealogies, medical records, and genetic data on individuals and entire populations—are quickly absorbed into the marketplace, where they are exchanged as commodities. As a result, "the boundaries of what used to be perceived as basic research on the one hand and its medical applications on the other have been blurred in an unprecedented fashion" (Rheinberger 1995: 255).

Inevitably, with the commodification of the human body and bodily information, property, ownership, and access become central issues. As Strathern (1999: 23) argues, "if the world is shrinking in terms of resources, it is expanding in terms of new candidates for ownership." In fact, the issue of ownership has shaped both contemporary debates on biotechnology, particularly in relation to postgenome projects (Greely 1998, Moore 2000), and recent anthropological and legal thought (Boyle 1996, Radin 1996, Hann 1998). Thus, for a year or two there was an intense tug-of-war between communitarian and commercial perspectives on human genome projects. On the one hand, the nonprofit Human Genome Project planned to share data from its decoding of the 3 billion chemical units of DNA that make up the genetic pattern of humans. Its aim was to publish a full genetic map on the Internet by 2003. A private company, the Celera Corporation, challenged that plan. Not only did it announce that it would complete mapping the human genome in a fraction of this period by different means (and at a tenth the cost of the $2 billion publicly funded project), it also planned to apply for numerous patents on genes it had discovered in the process. Negotiations between the two projects collapsed. On March 14, 2000, President Bill Clinton and Prime Minister Tony Blair issued a joint statement in an attempt to bring things under control, urging nations, scientists, and corporations to share their information on the human genome. A few weeks later, the Celera Corporation and the Human Genome Project jointly announced that they had completed a "working draft" of the human genome. These developments raised interesting questions about genome property which, no doubt, will be actively contested in both the academic community and the realm of law and politics over the next few years.

Many of the developments associated with modern biotechnology and bioinformatics are met with criticism and organized opposition. Not only are the issues complex and diverse—intrusions into the relatively "sacred" space of the body, the commodification of body components, and the redefinition of the conditions for scientific research, to mention a few—but also they vary from one context to another. Anthropologists have just begun to study some of these developments and the debates about them, partly as a result of recent developments in economic anthropology and partly as a result of growing interest in various aspects of the human body—including interest in bodily experience (Jackson 1989), disability (Ingstad and Whyte 1995), medical ethics and practice (Lindenbaum and Lock 1993, Kleinman 1995), human genome research (Rabinow 1996*b*), gender (Martin 1989), and concepts and technologies of reproduction (Franklin and Ragoné 1998, Rapp 1999).

The perspective of moral landscapes (Appadurai 1986, Kopytoff 1986, Bloch and Parry 1989, Helgason and Pálsson 1997) both offers a useful approach with regard to people's responses to developments in biotechnology and helps to determine the options available for informing public decisions on contested issues such as those surrounding modern biomedicine and biotechnology. We shall illustrate our general argument about moral landscapes and contests on the commodification of the body with reference to recent Icelandic debates. Iceland has been the site of complex biotechnological developments growing out of controversial legislation establishing a Health Sector Database (Gagnagrunnur á heilbrigðissviði) passed by the Icelandic Parliament in December 1998. Plans for the database specify how and under what conditions to assemble medical records and how they are to be combined with genetic information on modern Icelanders and genealogical data on all living Icelanders and a substantial proportion of those who have ever lived for the purpose of tracking the presumed genetic bases of diseases and making the National Health Service more efficient. While preparations are already being made for the database and the entry of data, important issues remain unresolved.

The public debate that we analyze often simply refers to "the database issue" (*gagngrunnsmálið*), subsuming the three databases and their combination. We have therefore decided to label the three databases—medical, genetic, and genealogical—collectively the Biogenetic Project. This larger, collective project, it needs to be emphasized, does not result in a permanent database; special permission is required from commissions on ethics and data protection to combine the information the three databases contain for particular projects with detailed research protocols. Preparations are being made for the Health Sector Database, and some medical centers have already consented to the transfer of data. An important deal was struck by deCode Genetics, the company contracted to operate the Health Sector Database, and the Icelandic Medical Association in late August 2001 according to which the former agreed to develop methods that allow for the deletion of personal information from the database for those who may decide to opt out at some later stage. It also agreed to consider the standards on ethical issues of the annual general meeting of the World Medical Association as guidelines.

Similar projects with somewhat different designs and property arrangements are under way in several other contexts, including Australia, Britain, Estonia, Newfoundland, Norway, Quebec, Singapore, Sweden, and Tonga. In Sweden, a database known as Medicinska Biobanken is under construction in the northern region of Västerbotten. Blood samples have been collected over many years as contributions to the Swedish welfare state, often known, significantly, by the term Folkhemmet, literally "the people's home" (Høyer 2001). By the end of the 1980s, with the growth of genomics, Swedish politicians and researchers had come to the conclusion that if the resource represented by the medical records and blood samples of Västerbotten was to be tapped, a private genomics company would have to be involved. As a result, the regional council signed a contract with the University of Umeå and the newly established biotechnological firm UmanGenomics on the use and further development of a biobank. According to the agreement, UmanGenomics has the exclusive right to exploit for commercial purposes the genetic information obtained from blood samples collected in Västerbotten. Plans for a similar database in Tonga have been debated both locally and internationally, following the announcement in November 2000 that the Australian company Autogen Ltd. had secured an agreement with Tonga's Minister of Health for exclusive rights to the country's entire gene pool. This debate has been partly informed by a discourse on bio-prospecting associated with the Human Genome Diversity Project (Reardon 2001).

In Britain the Medical Research Council and the Wellcome Trust plan to establish the UK Population Biomedical Collection for advancing genetic epidemiological research (Kaye and Martin 2000). The collection will contain DNA samples from up to 500,000 adults aged 45 to 60 years that will be linked to personal medical records and family histories. The aim of the project is to explore interactions between genes, environment, and lifestyle, focusing on cancer and cardiovascular conditions. Ongoing collection of data from research subjects is assumed, and genotyping of participants is to be done in centralized facilities. Important aspects of the design of the British collection, including the selection of participants, the form of consent, monitoring and control, and the nature and degree of commercial involvement, remain unspecified.

In Estonia, the Human Research Act was passed in December 2000 to facilitate the establishment of a gene bank database that will contain phenotype and genotype data on the entire adult population, around 1 million people. Again, the chief purpose is to explore the genetic causes of common diseases. The major anticipated clients of the bank are research institutions and companies in the fields of bioinformatics, biotechnology, and pharmaceuticals.

One of the central contributions of anthropology to recent developments in biomedicine is to explore how and why people moralize about them and why the responses may differ from one context to another. A detailed comparison of the biogenetics projects of Britain, Estonia, Iceland and Sweden, the ways in which they have been discussed domestically and internationally, and the "bioethical" issues that have emerged in the process is presented elsewhere (Pálsson and Rabinow n.d.). Drawing upon the work of Dezalay and Garth (1996) on humanitarian markets, emphasizing transnational networks of bioethics and biopolitics, the comparison explores a striking contrast. While the Icelandic Biogenetic Project has been the center of controversy and widespread discussion, in Britain, Estonia, and Sweden there has been virtually no public dialogue on similar projects. Moreover, the international press and the transnational scientific and bioethical communities have been heavily focused on the Icelandic Biogenetic Project, whereas the British, Estonian, and Swedish efforts have received scant attention.

We focus here primarily on developments within Iceland—the public debates surrounding the Biogenetic Project and the interest groups involved. First, we explore reports and op-ed articles on the databases in the major local newspaper, *Morgunblaðið*. Secondly, we have interviewed many of the major players in the ongoing debates on biomedicine, both proponents and opponents of the project. Finally, we draw on our participant observation over the past few years in deCode Genetics (Pálsson) and the Icelandic Blood Bank (Harðardóttir).

The notion of moral landscapes both offers a powerful tool for the comparative study of debates on the commodification of the human body and facilitates understanding of the topography and dynamics of local debates. Having discussed the idea of moral landscapes, we present the proposal for an Icelandic biogenetic project in more detail, emphasizing its three components—genetic, medical, and genealogical. We go on to analyze the distribution and content of newspaper articles on the project and to discuss bodily commodities and the parallels and differences between biogenetic databases and other "resources," in particular fisheries.

The moral landscapes of modern biopolitics

The advance of modern biotechnology and related developments in cybernetics and bioinformatics have profound social implications which are currently being explored by a number of anthropologists in different contexts. One clear consequence of recent developments is the radical reevaluation of existing notions of kinship connections, personhood, and relatedness (Strathern 1992, Edwards et al. 1999) and what it means to be human (Hayles 1999). This is exemplified by discussions of the social implications of the idea of human cloning, the practices of surrogate motherhood, organ transplantation, and genetic engineering, and the way in which these products of bioscience could and should relate to the world of commerce. Another example is the moral discourse in many contexts on the commodification and integrity of the human body. The appearance in prehistory of the practice of burying the dead underlined the notion of the integrity of the body and the "quasi-property" rights of relatives to the corpse of the deceased. Because for humans everywhere the body is both the locus and the agent of experience (Csordas 1994), it is often represented as a bounded space resistant to invasion, extraction, and commodification (Campbell 1992; Marshall, Thomasma, and Daar 1996; Scheper-Hughes 2000; Sharp 2000; Andrews and Nelkin 2001). While humans everywhere and at all times tend to resist the alienation and marketing of body parts, responses to the commodification of body components exhibit considerable cultural and historical diversity. There are noticeable differences across the globe—between, for instance, Europe, Japan, the United States, and the Islamic countries—in attitudes towards body components, the consumption of biotechnological products, and its moral values.

One point of difference in the discourse on the human body is the extent to which it is morally "right" or "permissible" to remove, reproduce, and market as commodities components of the body (including blood, cell lines, and clones). For some people, somatic commoditization is inhumane and degrading, an offense against personhood and dignity, but for others it represents a humanitarian effort in that it increases the supply of body parts and therefore saves lives and reduces human misery. Although in every case a fairly clear distinction is made between the permissible and the unethical, the line separating the two seems to vary from one context to another. These differences can be observed at different levels, ethnically and locally as well as nationally and internationally (Thompson and Chadwick 2000). The consequences, then, of the realignment of the market and the bodily domain vary widely with the social, religious, and political context (Clarke and Parsons 1997).

Clearly, a comprehensive understanding of debates about biotechnical commodities needs to be based on an empirical approach that confronts the ethnographic complexities involved—a point emphasized by Sharp (2000). The metaphor of landscape, we suggest, can be a handy tool in such an approach. Thus, the moral environment of bodily commodities can be likened to a topography intersected by pathways, spheres, and boundaries (Appadurai 1986, Kopytoff 1986, Hirsch and O'Hanlon 1995, Helgason and Pálsson 1997). Kopytoff (1986) has usefully outlined a general polarized system of classification on the basis of which all social things are ordered. At one end of this system lies the ideal type of the perfect commodity—a thing with an exchange value fully commensurate with that of any other exchangeable item according to a single standard of value. At the other end lies the ideal type of absolute singularity that precludes the possibility of value comparison and exchange. Between these two ideal types lie the majority of social things, imbued with varying degrees of commodity status. The domain of the human body is a case in point. In each context, some parts of the body are more likely candidates than others for commodification and exchange.

The approach of the moral landscape discredits the dichotomous view that a thing is either a commodity and consequently inside the economy or not a commodity and, therefore, external to economic life. Instead, the landscape is seen as partitioned into interrelated spheres of exchange, each permeated by a specific set of social and material relations (Helgason and Pálsson 1997). This topography obstructs or provides pathways for the exchange of particular things with other things. The perspective of spheres and boundaries demands an important qualification, however. The driving force behind the moral economy is ultimately derived from the intentionality of individuals, not from normative scripts or blueprints. The landscape of the moral economy is the shifting construct of those who live with it as commodification is initiated, accomplished, and perhaps reversed by real people acting in a social environment. Debates about biomedicine and the commodification of body components take place in a dynamic, living social environment. It is the protagonists and consumers of these debates, with their diverse range of perspectives, values, strategies, and interests, that weave the moral strands that result in legislation and the formation of policy and tradition. A critical step, therefore, for tackling "ethical" problems associated with modern biomedicine is to adopt an approach that focuses on the social and cultural contexts into which new forms of biomedicine are introduced and the actual practices, reactions, and concerns of the actors—scientists, interest groups, policy makers, and the general public. The social biographies of bodily commodities are defined and charted through biopolitics—through the activities and concerns of the actors involved.

While the use of spatial metaphors is by no means essential to the study of the exchange of bodily commodities, it presents a useful way of conceptualizing the contrasts, similarities and interactions of different forms of exchange. Comparative ethnographic research on people's responses to the development of biomedicine and the moral landscapes which they constitute provides a better understanding of the market and the economic system in question as well as the options available for informing decisions and regulating production and consumption. In this endeavor, anthropology—medical, economic, cognitive, and otherwise—can play a crucial role, given its routine attention to detail, context, and ethnography. One important task for anthropologists—following the recent works of Finkler (2000*a*) and some others on the medicalization of kinship, imagined genetic communities, and the phenomenology of relatedness—is to explore the way in which people experience the new genetics and how it may inform particular notions of identity, relatedness, and citizenship.

A potentially alternative approach for both illuminating and managing the commodification of the human body is represented by neoclassical economics, an approach that assumes the landscape of supply and demand. In this approach, however, moral issues are deliberately ignored (see Sen 1987). Proponents of property rights ideology even revel in the fact that their suggestions to apply the neoclassical approach to bodily commodities are frequently met with repugnance, seeing themselves as the vanguard of reason against "shibboleth and shamanism" (see Joralemon 1995: 345). By ignoring social context and the validity of hostility to the process of commodification, standard neoclassical theory is effectively ruled out from having much to contribute to a study of differing responses to biotechnological and market innovations. Significantly, the notion of human organic materials as commodities is more problematic for some people than for others. Yet another approach is that of the bioethics developed by moral philosophers and legal experts. In this case the ethical dimension is, indeed, included as a central one, but it is usually normative and uniform, much like the universal monetary standard of neoclassical economics, based as it is on some of the yardsticks of classical Western philosophy. Most philosophically trained bioethicists, as Kleinman (1995: 46) observes, draw upon "the orthodox sources of the self in Western philosophical tradition," emphasizing the autonomy of the person, partly in an attempt at balancing the interests of individual and society. The neoclassical theorizing of economics and the approach of bioethics may offer useful insights into some of the issues and concerns associated with the commodification of the human body and its parts; both have, indeed, played an important role in recent policy discussions. Neither of them, however, is "thick" enough anthropologically speaking to illuminate the topography, dynamics, and comparative context of moral debates.

DeCode genetics, Inc., and the Icelandic Biogenetic Project

After nine months of national debate, in December 1998 the Icelandic Parliament passed a bill authorizing the construction of a National Health Sector Database. The central idea of the database project and the source of much of the controversy around it was the assembly of medical records for the entire Icelandic population. The license to construct the database was open to competition; the licensee would finance it, and it would belong to the National Health Service, with the licensee retaining privileged rights to commercialize it for 12 years. DeCode Genetics, which outlined the original plans for the database, received the license to construct it in return for a fee. Founded in 1996, deCode Genetics operates in Iceland but is funded by venture capital funds coordinated in the United States (Pálsson and Rabinow 1999, 2000, 2001). A rival biotechnological company, UVS, was established in the heat of the debate on the database, partly to challenge deCode's "monopoly" of biomedical research in Iceland. DeCode Genetics soon strengthened its financial position through a business arrangement with the pharmaceutical giant Hoffmann La Roche, an agreement that focused on research on 12 common diseases. The Health Sector Database is part of the larger Icelandic Biogenetic Project (fig. 1), the aim of which is to combine, under specific conditions and for specific purposes (see Gulcher and Stefánsson 2000), medical records, genetic information, and family histories (the so-called Book of Icelanders).

What, if anything, makes the Icelandic gene pool a valuable commodity? For one thing, because of its demographic history the Icelandic population may have a certain appeal for investors in genetic research. Iceland was settled from Scandinavia and the British Isles in the 8th or 9th century (the "official" date is 874). The population increased to about 10,000–20,000 in the first 60 years and apparently reached 70,000 by 1300. A series of population declines occurred in the following centuries, the most devastating of them being due to plagues in 1402–4 and 1708 and a volcanic eruption in

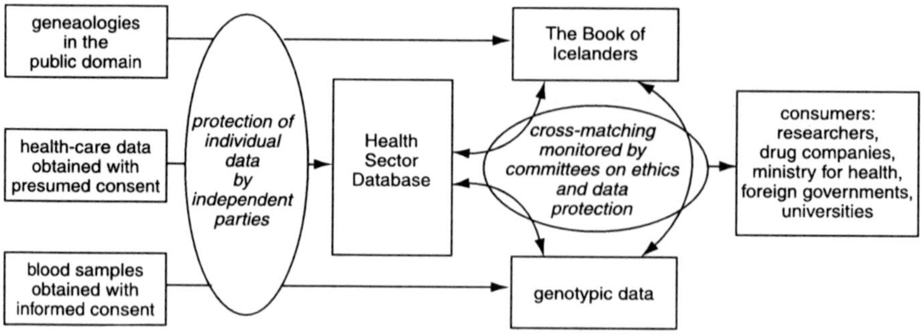

Figure 1 The Icelandic Biogenetic Project.

1784–85. By 1800 the population was only 40,000. As a result of the successive reduction of the population during times of plague and famine, the gene pool is fairly homogeneous; the branches of the family tree, so to speak, have repeatedly been cut off, with a limited number of lineages surviving to the present time (Helgason 2001). Geographic isolation has inhibited the replenishment of lost genetic diversity via immigration. Given these characteristics and the small size of the population (280,000 today), the Icelandic gene pool may have advantages for genetic and medical research.[1] No less important for the appeal of the Icelandic database project is the availability of detailed written records. The project is plausible and scientifically promising not only because the population has been small, relatively homogeneous, and comparatively isolated but also because Icelanders have a passion for keeping genealogical records. Finally, with a strong and centralized medical service Iceland offers extensive medical records that may be highly valuable for tracking the causes of common diseases.

Current work at deCode Genetics, independent of the planned Health Sector Database, typically begins with a contract with one or more physicians specializing in a particular disease with a potentially genetic basis. In their practice over the years, the physicians have constructed a list of patients with the particular symptoms in question. This list is passed on in an encrypted form to a research team within the company, which, in turn, runs the information provided through its computers, juxtaposing or comparing patients' lists and genealogical records by means of specialized software the company has developed. The aim is to trace the genes responsible for the apparent fact that the disease in question occurs in families. Such an analysis may show, for example, that of an original list of about 1,000 patients, 500 or so cluster in ten families. Knowing the genealogical relationships and using the standard procedures of genetics, the researchers are, theoretically, able to narrow down the search for the genes responsible for the disease. In the next step, the physicians affiliated with the research team collect blood samples from patients and some of their close relatives for DNA analysis. In the final stage of the research, specialists in statistics and bioinformatics evaluate the results of the genetic analysis, attempting to narrow down in probabilistic terms the genes responsible for the disease. In the words of the leader of one of the research teams, they "track the recombination of DNA through each generation, using cross-overs to further localize the location." In practice, this is a complex interactive process. Strategies of gene hunting are adopted and revised both intuitively in the laboratory or at the computer screen and in formal or informal meetings.

This procedure is not, of course, unique to deCode Genetics, not even in the Icelandic context. However, with the Biogenetic Project still on the drawing board, the power of genetic and epidemiological analyses may grow exponentially, with far larger samples and more families. The addition of medical records available since 1915 allows for the exploration of a set of new questions on the interaction among a number of variables apart from genetic makeup and genealogical connections, including variables pertaining to lifestyle, physical and social environments, the use of particular medicine, and degree and kind of hospitalization. The results may be useful, according to the designers of the project and the medical authorities, for pharmaceutical companies and for the medical service, yielding information about potential drugs, particular genes or proteins, and possible preventive measures in terms of consumption and lifestyle.

Two major biomedical collections in Iceland, the Icelandic Blood Bank and the National Tissue Bank, both of which belong to the national health-care system, remain independent of the Biogenetic Project. Established in 1953, the blood bank is the main collector of blood for medical purposes. It serves almost all of the hospitals in the country, and, as a result, it is a major source of biomedical data, with detailed information on blood donors and potential recipients of blood (mostly patients in hospitals). The blood bank collects blood within the framework of the gift. Although many Icelanders seem willing to donate blood for medical purposes, this does not necessarily mean that they agree to contribute information to research involving the Biogenetic Project. Those in charge of the blood bank in fact do not see "their" information as a part of that database. For them, linking the biomedical data of the blood bank with the Health Sector Database would mean risking the loss of blood donors. The other main independent collection, the National Tissue Bank, has an extensive collection of body tissues taken from patients in hospitals since 1917. While originally the chief purpose of the tissue collection was to document medical procedures and to sample pathological body components for both research and pedagogical purposes, given recent advances in biotechnology and genome research it may be extremely valuable for biomedical researchers interested in DNA material. The blood bank and the tissue bank, currently seen as separate collections, may eventually be connected to the Biogenetic Project. Among the key issues that are likely to be negotiated are those of payment, control, and property rights.

Even more important than the medical records and collections is the genealogical database now available in Iceland. Several kinds of records on families and genealogies are available—the Book of Settlement (written around 1125) and the Family Sagas, church registers, administrative records, and censuses. The national census of 1703 is arguably the first of its kind in the world. The modern Icelandic term for "anthropology," *mannfræði*, dates back to the medieval saga age. Literally, the term means "the study of humans," but traditionally it referred to knowledge about people, families, genealogies, and social relationships. The sheer wealth of records—"deep genealogy," in the rhetoric of deCode Genetics—is, indeed, an important asset for the database project; there are reliable continuous genealogical records as far back as 1650 and for some families all the way back to the time of settlement. As one of the deCode Genetics scientists suggested in a recent talk, "It's not just the gene pool that makes the Icelandic case interesting; it's the genealogy."

With the computer company Frisk Software, deCode Genetics is currently constructing a genealogical database of the entire Icelandic population, "The Book of Icelanders." The database is expected to cover half of those ever born in Iceland, approximately 600,000 names. Such a database will not only provide an invaluable historical dimension to the search for mutant genes and the causes of diseases but also is expected to be available on the World Wide Web for genealogy enthusiasts and the general public. It will allow its users to check in a split second the genealogical connections between any two Icelanders (for a discussion of the social history of family trees and the Book of Icelanders, see Pálsson n.d.). In recent years, the study of genealogies has become a popular pastime in Iceland. While the fascination with family trees is a common Western theme (Bouquet 1996; Carsten 2000), in Iceland it is somewhat extreme, associated to some extent with family reunions and urban nostalgia for the rural past.

Analysis of public discourse

The Icelandic Health Sector Database has been the center of controversy from the start. There have been several hundred articles about it in the Icelandic press, numerous television programs, a series of town meetings, and endless discussion and debate in Parliament, in shopping centers and cafes, and around dinner tables. Understandably, Icelanders have been quite attentive to the issues involved. There has been extensive coverage in the world press and on the major TV channels of deCode Genetics, its director, its scientific project, and the database under construction. Below we explore the domestic discourse on the Biogenetic Project, focusing on a single newspaper, *Morgunblaðið*. Public discourse as it appears in the mass media both informs and is informed by social context, the social relations in which it is embedded (Jaworski and Coupland 1999). This is usefully brought out in recent anthropological studies of environmental discourse (see, for example, Peace 1997). Moreover, as Gee (1999: 7) argues, public discourses "often influence each other in positive and negative ways, and . . . sometimes breed with each other to create new hybrids." The mutual effects of ongoing discourses on biomedicine and fisheries in Iceland are a case in point.

We have collected and analyzed all the articles mentioning the Biogenetic Project published in *Morgunblaðið* for 28 months after the introduction of the first bill on the Health Sector Database in the Icelandic Parliament. (Some of these pieces focus on the medical records—the Health Sector Database—while others discuss the larger idea, the combining of medical, genetic, and genealogical information in the form of the Biogenetic Project.) There are three main reasons for focusing on *Morgunblaðið*. First, with a circulation of some 54,770 copies every day, it is seen by almost every household in Iceland. According to surveys, about 64% of Icelanders and 71% of the inhabitants of the capital city Reykjavík read it every day. For all age-groups, *Morgunblaðið* is the most important newspaper. There are other newspapers, but they are far less significant. Secondly, *Morgunblaðið* prints almost every article sent to it, independent of its own policy on the issue in question, and therefore provides a fairly good idea of public discourse in the country. Finally, it is the only paper that can legitimately claim to be the newspaper of record, with a long and continuous history and an extensive on-line archive. *Morgunblaðið*, of course, not only reflects Icelandic discourse but actively *shapes* it as well through its near-monopoly of the newspaper market on the local scene, its current and historical links with the Independence party, and its strong stance on many public issues. This, however, does not reduce its usefulness for the present analysis. While the newspaper has generally supported the Biogenetic Project project in its editorials and its news coverage, most of the op-ed pieces it has published on the issue have been highly critical.

Our sample of articles is 569 items—news reports and op-ed pieces. Our method of analyzing the newspaper material was as follows: In the first round, we scanned all the newspapers, extracting items specifically focusing on the database. Then we searched the newspaper's web site (www.mbl.is) and downloaded all material mentioning the database. All the articles collected in this manner were then carefully read and classified into news and op-ed items. Next we divided the material in each category into "for," "against," and "neutral" in terms of its take on the issue. Finally, we explored in more detail the content of the newspaper material, focusing on ethnographic content. This is important to indicate the flavor of the Icelandic debates and to situate the actors, issues, and arguments.

Distribution of newspaper materials: for and against the biogenetic project

Figure 2 shows the distribution over time of articles (both news and op-ed pieces) on the Biogenetic Project over the period in question and the relative proportions of items for, against, and neutral. During this period, it should be kept in mind, four major events occurred: (1) In April 1998 a bill on a Health Sector Database was introduced in Parliament; (2) in July 1998 a revised bill was introduced; (3) in December 1998 Parliament ratified the second bill; and (4) in January 2000 the Ministry of Health granted deCode Genetics the license to construct the database. There are important shifts in both the volume of the discourse and the proportion of items for and against the Biogenetic Project depending on the formulation of concerns and issues in the course of the public debate. The large number of items during October 1998 represents an editorial move by the newspaper

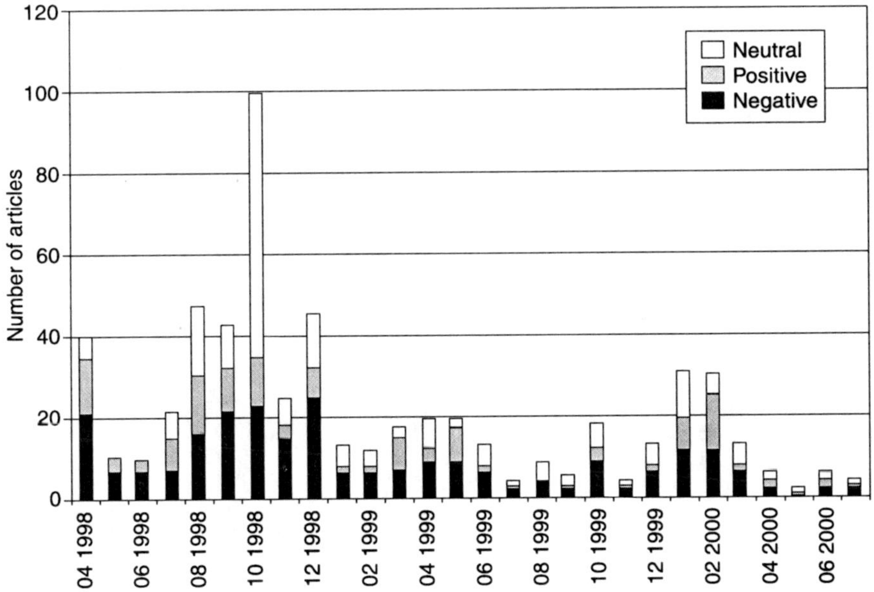

Figure 2 Newspaper material neutral, for, and against the Biogenetic Project (*Morgunblaðið*).

emphasizing the need to clarify important issues with detailed analysis and information.

The news items were typically neutral, in contrast to the op-ed pieces, which, by definition, usually had a rather strong and clear agenda. The op-ed items are perhaps more interesting in the sense that they represent the voice of at least some segments of the public. Figure 3 presents a breakdown of the op-ed items. Clearly, there are few neutral pieces, and in most months there are more negative items than positive ones. This conclusion is in striking contrast to the findings of opinion polls suggesting that the overwhelming majority of the public supports the Health Sector Database. In June 1998 a Gallup poll concluded that 58% of Icelanders supported the database, while 19% were opposed and 22% were neither for nor against. In April 2000 a Gallup poll concluded that 81% of Icelanders supported the database, while 9% were opposed and 10% were neither for nor against. Both surveys, it should be noted, were done at the request of deCode Genetics.

The authors of the op-ed pieces fall into five main categories: (1) physicians, (2) professional writers (creative writers and journalists), scholars, and experts on biology and informatics, (3) politicians, (4) spokespersons for particular companies, institutions, associations, or pressure groups, and (5) the general public or writers who identify themselves only by personal names. Figure 4 shows the numbers of items contributed by these categories of key players in

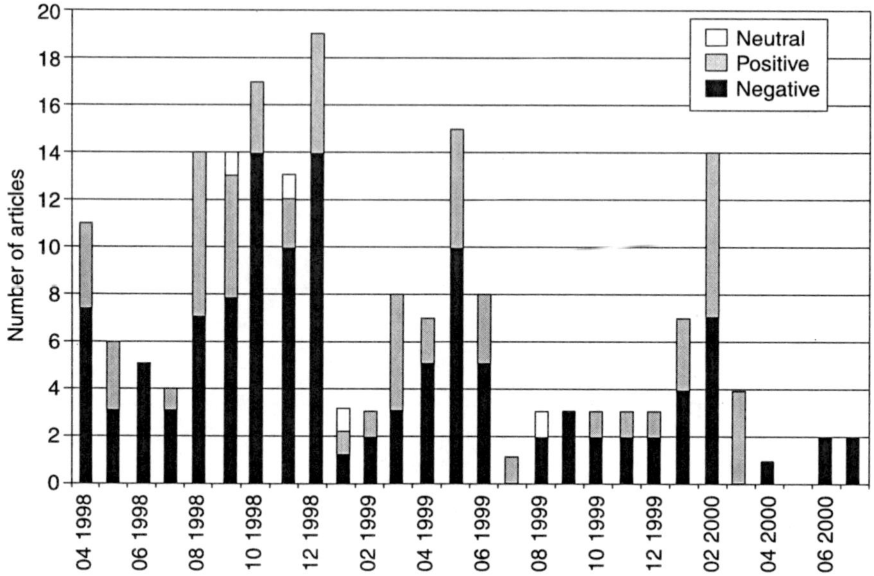

Figure 3 Op-ed articles neutral, for, and against the Biogenetic Project (*Morgunblaðið*).

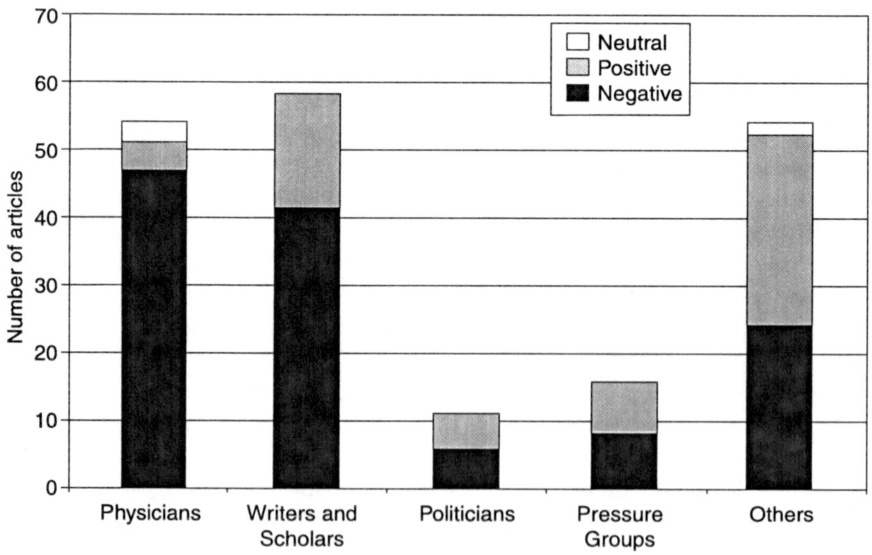

Figure 4 Key players in the public debate (*Morgunblaðið*).

the public debate, either for or against the Biogenetic Project. Physicians wrote no fewer than 28% of the items—understandably, perhaps, since they have compiled many of the medical records that are the subject of debate. Their contributions turned out to be overwhelmingly against the project. It should be noted, however, that only a few doctors were responsible for most of the negative items. One doctor was the author of 27.8% of the items, 15 of the 54 items in this category. Only in the fifth category, representing the general public, did positive items outnumber negative ones.

Ethnographic content

A strong ethical and political body—the Association of Icelanders for Ethics in Science and Medicine (Mannvernd)—was formed in direct response to the database project. The Association, at present the main platform for bioethical criticism in Iceland, is also well connected to the international scene of bioethics, providing the base from which many observers position themselves. Its main spokespersons are physicians, biologists, geneticists, and philosophers.

For one thing, the Biogenetic Project has raised important concerns about the protection of personal information. One member of the Association of Icelanders for Ethics in Science and Medicine, a university employee, wrote an op-ed piece entitled "I Don't Want to Be a Code Bar" (April 14, 1999): "I choose to trust neither the expected licensee nor the state, and to practice instead my own personal protection by requesting that no information about me will be transferred into the Health Sector Database." Another important concern for much of the opposition is that of informed consent. The Health Sector Database is to operate on the basis of the principle of *presumed* rather than informed consent; people can refuse to be included in the collective medical record, but if they don't, information on them will be automatically entered. One of the leading Icelandic geneticists and a board member of the Association of Icelanders for Ethics in Science and Medicine remarked in an op-ed piece entitled "The Bill for the Database is Sinking in the Harbor" (November 24, 1998): "To legalize presumed consent is to violate the positive attitude currently shown with respect to scientific research. The concept of presumed consent is an absolution for those who do not want to respect the individual right of self-determination."

Much of the discussion of the Biogenetic Project has focused on issues of property, ownership, and control. Thus, a fundamental debate has taken place concerning the ownership of and access to genetic information and medical records. One physician commented as follows (October 17, 1998): "I think it is a gross insult to my practice as a physician to appropriate documents that I record as memos and then to turn them into scientific data. This would be an assault if done by force. Therefore I protest, and so will others."

The property issue has often been discussed with reference or in comparison to ongoing debates about another thorny common property issue, namely, the allocation of individual transferable quotas (ITQs) to rights in fish (on the latter, see, for instance, Helgason and Pálsson 1997, 1998). The Icelandic fishing industry has been radically reorganized in the past 15 years or so with the application of market-based models to fisheries management. Critics argue that the current regime represents "the biggest theft in Icelandic history" (to use a frequent slogan from recent political campaigns), transferring wealth and property from the general public to a few privileged boat owners. Medical (and possibly genetic) information, it is argued, is a common-pool resource with some of the characteristics of the fishing stocks in Icelandic waters. Privileged access, permanent or temporary, should therefore be granted only in return for a fee to ensure equity and fairness. An editorial in *Morgunblaðið* made the following comparison (January 23, 2000):

> Don't we all agree that the industries should be treated equally? Why should boat operators not be requested to pay anything for their use of a resource belonging to the nation when deCode Genetics Inc. has to pay 6% of its profits from using the right to exploit medical data? Everyone must see that this doesn't make sense. Exactly the same arguments that are used to ensure returns for the use of the right to use medical data . . . apply to the main national resource, the fishing grounds.

For many of the critics, however, the commodification of biomedical data is problematic. Thus a physician commented (October 16, 1999): "The resources are no longer metal or spices but medical data and blood for entire nations. The returns will supposedly come later on in the form of free medicine. The rebirth of bartering!" Another physician emphasized the differences between fish and patients (December 4, 1998): "The authorities can maneouvre various things that they do not own in the legal understanding of the term, for instance, the fish in the sea. . . . Because of their lack of understanding of scientific ethics . . . the authorities haven't realized that patients have rights that fisheries products, . . . have not."

Perhaps the dominant focus in the debate is the fact that a private multinational company proposes to explore the genetic bases of common diseases in the entire population and to commercialize its results. A philosopher commented (November 21, 1998): "The nation has probably never before in Icelandic history come so close to being sold." Echoing the claims about "biopiracy" (Shiva 1997) popular in current debates on genetic research on indigenous groups and the Human Genome Diversity Project (Reardon 2001), such concerns have been raised internationally as well.[2]

Opposition to the Biogenetic Project, then, seems motivated by several concerns, including potential infringements of personal autonomy and privacy, the ethics of consent, and the threat of biopiracy. The debate about the database and deCode Genetics focuses above all, however, on the use and control of several kinds of information—derived from medical records, blood, and genealogies—which have not necessarily been controversial in the past. Much of the heat of the opposition is driven by an apparent sense among physicians of rather suddenly lacking authority in the biomedical domain, of losing dominion over information largely constructed and controlled by them in the past. The Icelandic opposition, thus, above all reveals concerns about potentially dramatic changes in the practices and structures of Icelandic bioscience and medicine. Some academics have alleged that the restrictions of access to information and resources implied in the privileged contract of deCode Genetics with the Icelandic state will inevitably result in the stagnation of bioscience. DeCode Genetics, they claim, will restrict access to medical data, although theoretically access will be free to independent researchers as long as they work only with data that will not violate the commercial interests of deCode Genetics. Many of the spokespersons for the Association of Icelanders for Ethics in Science and Medicine seem to see their scientific dominion as being threatened or devalued by deCode Genetics and the Biogenetic Project. Thus, the subtext of some of the debates centers on where cutting-edge research on the Icelandic human genome occurs today and where it should be located in the future.

In general, as already mentioned, the Icelandic public has supported deCode Genetics and the Biogenetic Project. One sign is the strong positive initial response of the Icelandic stock market, although the reaction of the market later on was rather mixed. Significantly, in both 2000 and 2001 the business magazine *Frjáls verslun* (Free Trade) declared deCode Genetics the most popular company in the country. Understandably, proponents of the project represent a more complex professional and social spectrum than the opponents, given the fact that they by far outnumber the critics. Those who have contributed op-ed pieces in support of the project emphasize concerns rather different from those of the opponents. DeCode Genetics and the Biogenetic Project, they argue, will make possible much more biomedical research in Iceland, creating many new positions for scientists and laboratory assistants both within the company and at the University of Iceland. DeCode Genetics, it is often pointed out, has attracted much investment from abroad and created numerous jobs for Icelandic scientists, many of whom in the absence of deCode Genetics would have been forced to seek employment abroad. The author of one op-ed piece commented (April 28, 1998): "When at last we envision a grand new plant in the economic flora, a project that draws upon science and human capital rather than pollutants, . . . many of us respond with fierce opposition. On its own, such a change in economic policy should improve public health."

Proponents of the project, some of whom represent particular patient groups, also emphasize medical advances. Thus, an op-ed piece suggested

(April 25, 1999): "Personally, I don't mind if someone makes a profit from using information about me, particularly if the person in question produces in the process new knowledge about inheritance and an untreatable disease that has colored my entire life and that of my family." Moreover, the supporters point out, there is a significant gap between current bioscientific and medical practices in Iceland (the practices of many of deCode Genetics's main opponents) and the types of protection of privacy and provisions for informed consent appropriately demanded for deCode Genetics (February 3, 2000): "People do not seem to realize that information from these medical records that apparently have to be guarded so carefully is, in fact, provided with far better protection in an encrypted database than in the archives of the cellars of medical centers throughout the country."

Some of the support for the Biogenetic Project reflects an underlying discontent with a vaguely defined medical establishment, in particular the control and expertise of physicians. A former teacher argued (April 7, 1999): "The outspoken encouragement for people to participate in group action in order to prevent the advance of knowledge is a shameful one, echoing perhaps the prejudices of the church in the early days of modern science. Such a campaign is not fitting in the case of people who belong to respected academic communities."

The Health Sector Database will become useful only if records from local clinics are passed on to the licensee for assembly. The Icelandic Medical Association has claimed from the beginning that the database would violate the relation of trust between physician and patient, since it operates on the basis of the principle of presumed consent. Now, as the database is under construction, some directors of local clinics, who are also physicians, are refusing to hand over "their" records, emphasizing their responsibility to their patients and pointing out that the latter have not consented to any such transfer of information about them. The main organized opposition to the project, the Association of Icelanders for Ethics in Science and Medicine, which is led by a physician, supports their refusal to hand over local records. Critics of the physicians, however, charge them with "patronizing" their patients and "medicalizing" the database issue. Some patient groups complain that the doctors are not consulting the patients and that they have no right to make decisions about medical records on their behalf. In effect, they argue, the doctors are claiming ownership and control of information that properly belongs to the community, the Icelandic state that has financed their recording and assembly. The association for patients with multiple sclerosis, which has cooperated with deCode Genetics from the beginning, has issued particularly strong statements in favor of the construction of the Health Sector Database.

The genealogical database has not been the focus of the debates in the op-ed pieces, although members of Parliament have at time drawn parallels between different sets of personal information. Nevertheless, family histories have been mentioned. One commentator offered the following remark, in an op-ed article entitled "Will the Interest among Icelanders in Family Histories Disappear with the Arrival of deCode Genetics?" (October 8, 1998): "Now a company in Iceland is recording in one place every piece of information documented in previously published works, including genealogies . . . and censuses. Unfortunately, the company is doing this without asking anybody for permission."

The actors in the domestic debate have sought to influence decision makers (in particular Parliament, political parties, Parliamentary committees, and the Ministry of Health) at various stages, through the media, public meetings, and informal lobbying. The outcome of the database issue—the laws themselves, the details on the recording of data, the monitoring of the construction and use of the database, and the encryption of data as well as the contract between the authorities and the licensee and the operating regulations on the license—is the collective result of a long and complex battle. While the two governing parties (Conservative and Progressive) have been committed to the database project throughout the political process, sometimes pushing their agenda through the political machinery with the brute force of their strong majority, the opposition has significantly shaped the course of events and the outcome. Both sides, then, have lost and won. This has not been a closed national debate. "Expert" witnesses from the transnational world of science and bioethics have often addressed the key issues in both the domestic and the international media (Pálsson and Rabinow n.d.).

Discussion

The arguments of the proponents of the Icelandic Biogenetic Project tend to emphasize the opportunities it provides in terms of medical advances, work, entrepreneurship, and private initiative, in the age of the challenging "new economy" and stagnant or declining fishing stocks. The opposition to the database project emphasizes ethical concerns, particularly those of consent and protection of privacy, as well as concerns with ownership and control. Both supporters and opponents seem, however, to be informed by "deeper" cultural and political considerations. While the general support of the database is partly rooted in the concept of the Nordic body developed during the nationalist era, the opposition is partly driven by both potential structural changes in Icelandic biomedicine and concerns with "capital"—concerns that parallel those of the recent politics of Icelandic fishing. With the construction of the database, physicians and academe are being removed from the discursive center of local biomedicine, making way for state officials, database staff, and shareholders in deCode Genetics. To some extent, this process parallels that of Icelandic fisheries; with the quota system, social honor and prestige were shifted from skippers and crews to quota holders and dealers on the stock exchange, leading to a radical reevaluation of the practical knowledge of fishers. Channeled through a discourse on ethics and human rights, the dissent has been directed, with some success, at resisting radical changes in the moral landscape of body components. While the discourse on fishing rights has informed discussions of the database, it has been applied somewhat differently to different aspects of the database: genetic data, medical records, genealogies, and the synergic combination of these sets of information. Some of the contours of the moral landscape are being altered in this process, but others remain intact.

The historical body: pedigrees, genetic and genealogical

Interestingly, the genealogical database, unlike medical records and genetic information, has not raised fundamental ethical questions of privacy and consent. Þorsteinn Jónsson, a competitor of Frisk Software, has already published extensive family records. One of his innovations was to publish photographs of Icelanders along with their family trees. Icelanders do not seem to be bothered by such exposure of identities, images, and relationships. On the contrary, they seem to welcome the opportunity to explore genealogies electronically and on a large scale. Significantly, the Recording and Treatment of Personal Information Law of 1989 (No. 121) specifically exempts "recording for the benefit of genealogical research and the documenting of family histories" (§ I.2). Restrictions on these along the lines of those applied to medical and genetic information would probably have caused a public outcry. The law, however, has been changed, as of 2000 (No. 77), after debates in Parliament on patients' rights and the protection of personal information partly triggered by the database issue. Thus, the recording of genealogical data no longer enjoys all the privileges it had earlier. It remains unclear, however, how the new laws will be interpreted and applied with respect to genealogical records and their publication.

On their own, the genealogical records have not been claimed as public property. Recently, however, a private company, Genealogia Islandorum, challenged deCode Genetics and Frisk Software in the courts for using previously published genealogical information to which it claimed the rights. Genealogia Islandorum argued that its property rights were being violated as others made a profit by marketing the products of its earlier work. DeCode Genetics responded to the challenge by announcing that genealogical information about Icelanders constituted a common heritage, the result of generations of scribes and collectors, and therefore it would be placing its genealogical program and database in the public domain. The critical issue, under Icelandic law, seems to be not one of marketing but rather one of the extent to which the new electronic database is the result of original, independent work or the "reproduction" of earlier texts. This is, indeed, a common issue in the information age, the core of what Boyle (1996: 177) refers to as "author talk."

Some element of "intertextuality" is unavoidable, given that the electronic database and earlier publications apply to the same population and that they are based on many of the same originals and earlier publications. DeCode Genetics and Frisk Software, however, have added extensively to earlier compilations with their independent research and, moreover, designed their own electronic format of storage and use. Since Genealogia Islandorum went bankrupt in the summer of 2001, its suit will probably be abandoned. There are two kinds of property issues, then, involving biomedical and genealogical records, both of which have been highlighted in recent developments following the granting to deCode Genetics of the license to construct the Health Sector Database. This latter issue, however, has been debated more in the courtroom than in the mass media.

Obviously, genealogies are culturally significant in the Icelandic context. This is not to say that Icelanders have been preoccupied with the structural relations of kinship. In fact, judging from the accounts of the Family Sagas (see Durrenberger and Pálsson 1999), friendship was a powerful relationship. Nor does the existence of "deep" genealogical records testify to a concern with "ethnic roots," in the modern understanding of the term. These records were originally constructed and maintained as part of claims about ownership of land. During the saga age and the following centuries, the notion of "Icelanders" was not an ethnic label but simply a reference to those inhabiting the island (Pálsson 1995, Hastrup 1998). With the "struggle" for independence from Denmark, however, during the 18th and 19th centuries, the notion of "Icelandicness" was invested with a heavy nationalistic load and forced upon the distant past. The nationalist idea of a "natural" relationship between the land, its people, and their language was born, an idea underlined by the local eugenics movement.

The general public support of the database plan must be seen in part in the context of a specific local history, the nationalist discourse of Icelanders, with its emphasis on the uniqueness of the Icelandic heritage, biological and cultural. Guðmundur Finnbogason, one of the most influential native intellectuals of the first half of the 20th century, invoked the "science of eugenics" to argue for the purification of the Icelandic race to preserve its spiritual and physical assets. He wrote in glowing terms about the eugenics movement in Germany, England, the United States, and Scandinavia, the ways in which it sought to improve the human race, and how Icelanders might benefit from it. While the "original race," he suggested, was "exceptionally good" and, fortunately, Icelanders had largely avoided "mixing with foreign blood" (1922: 202), there were some signs of "racial pollution" (*kynspell*). Finnbogason further suggested that some of the properties of modern Icelandic reflected an "Aryan" or "Viking" spirit. These ideas were largely informed by the Euro-American genetic discourse that foreshadowed the racial hygiene of the Nazi regime. An Icelandic physical anthropologist, Eiður Kvaran, trained in Germany between the two world wars, developed racial ideas along these lines to suggest that the literary heritage of Icelanders had a genetic basis. Kvaran (1934) examined the ways in which the people of the saga age attempted to avoid racial "degeneration"—infanticide, restrictions on marriage, etc. The Icelandic eugenics movement was a strong one, a significant part of the nationalist revival between the world wars (Finnbogason 1922, Hannesson 1924, Kvaran 1934).[3]

Current discussion of the Biogenetic Project clearly thrives on this nationalism. For some of the critics, the database signifies a return to the early eugenics. Many Icelanders seem proud of their "Nordic" roots and their genetic makeup (for a discussion of the Norwegian context, see Nielsen, Monsen, and Tennøe 2000), and they are eager to "offer" their "unique" blood samples and medical records, as the rhetoric goes, to science and the advancement of human well-being. Such a phenomenology of genetic citizenship is rather different from the discourse on "faulty" genes among some patient groups (Finkler 2000). The reference in current Icelandic rhetoric to the continuity with the Viking past may indirectly suggest a genetic notion of bounded citizenship. (The electronic version of this issue, on the journal's web page, contains images from the series "Birthmarks" by the Icelandic artist Katrín Sigurðoardóttir that explore these current concerns over the nature of individual and collective identity.) While all of those living in Iceland at present will be included in the Health Sector Database (except, of course, the approximately 20,000 people who have decided to opt out for ethical or ideological reasons) and the Book of Icelanders, in public perception some Icelanders are less equal than others, namely, "recent immigrants" (*nýbúar*, in current jargon) whose genetic and genealogical "roots" lie elsewhere. As we have seen, for biomedical and pharmaceutical companies the Icelandic gene pool is valuable capital precisely because of its presumed homogeneity and the direct genealogical link of "real" Icelanders, through a few lineages, with the medieval past.

Capital: biomedicine and fisheries

Concerns with capital, national sovereignty, and property rights reflect the tensions and factions in current fisheries debates between the political left and right—between neo-communitarians and advocates of common-pool arrangements, on the one hand, and, on the other, adherents of private property and the free market. The lines of political division around the database project, however, are neither clear nor simple. Significantly, while the Icelandic public has been fairly critical of the quota system in the fisheries and the closure and privatization it represents, most people are supportive of the database and deCode Genetics, and most of the political forces are split on the issue.

The Icelandic Parliament has emphasized that both biomedical data and fishing stocks are public property and that privileged access is only temporary, under some kind of arrangement of public trust. The common-pool nature of the fishing stocks was recently reinforced in a decision by the Icelandic Supreme Court after fierce debates on national fisheries policy. In December 1998, the same month that Parliament passed the law on the Health Sector Database, the court declared unconstitutional fisheries laws on individual transferable quotas which privileged those who derived their fishing rights from ownership of vessels during a specific period (during which their "fishing history" was established). This privilege, the court reasoned, violated both the constitutional rule against discrimination and the rule about the right to work. Later on, however, in another case, the court concluded with a somewhat contradictory declaration.

Although the Icelandic authorities have tended to emphasize the common-pool nature of both fishing stocks and biomedical data, fees have been applied only in the case of the Health Sector Database. It seems that the governing parties had learned from their political mistakes in the case of the fisheries. The text of the license granted to deCode Genetics for constructing the database specifies, partly in response to critics of the quota system in fishing, that the company must pay the Icelandic state a certain fee for the assembly and use of the records of the medical service. DeCode Genetics will pay for the cost of the agreement on the database and its construction and marketing. Also, it will pay 70 million Icelandic krónur ($704,650, 777,600 ECU) annually for the license to construct and use the database. Furthermore, the Icelandic state will receive 6% (up to 70 million krónur) of any annual profit that deCode Genetics may make from using the database. The annual revenue to the state will, thus, amount to something between 70 and 140 million Icelandic krónur.

As if to test the common-pool nature of biomedical data, a legal firm in Reykjavík recently offered its services to people who want to claim their share in the anticipated profit from the database. The firm advised people to exercise their legal right to withdraw from the Health Sector Database by writing to the Ministry of Health and then to offer reentry of information about them collectively through its service—in other words, to sell the information. Some of those who have objected to the database on ethical grounds, including the Association of Icelanders for Ethics in Science and Medicine, seem to support this initiative. The Ministry of Health, however, has declared that such marketing of medical information would be both unethical and illegal.

As we have seen, there are some parallels in terms of ownership between Icelandic fisheries and the domain of the Biogenetic Project. There are also, however, important differences. For one thing, there are different histories and contexts. The combination of medical, genetic, and genealogical information is not entirely new in the Icelandic context; many researchers have done this in the past under the aegis of the University Commission for Genetics (mostly funded by the U.S. Atomic Energy Commission from the 1960s onward), the Icelandic Cancer Society, and the Icelandic Heart Association. A database along the lines suggested by deCode Genetics, to be explicitly exploited for both medical and commercial gain, is a new concept. Icelandic medical records were fragmented and largely untapped commercially until deCode Genetics suggested combining them, linking them to genetic and genealogical information; indeed, at the time of writing they still do not constitute a "resource"—a single database. The fishing stocks in Icelandic waters, in contrast, have been used by the public for centuries, usually on the assumption of a common-pool regime.

Secondly, fishing stocks and medical information are somewhat different kinds of resource bases. Fishing stocks in most places are overused, sometimes by the logic of the so-called tragedy of the commons; the more we use them the less they yield. In the case of biomedical records, however, the escalation of "effort" does not necessarily threaten the "carrying capacity" of the resource base. Obviously, restriction of access to biomedical information limits the scope for prestige and financial gain somewhat as do enclosures in fisheries and other property regimes. Thus, the Health Sector Database may be quite valuable to those who first arrive on the scene (in this case deCode Genetics) but less so for latecomers. Once the records have been screened for, say, correlations regarding breast cancer, osteoarthritis, or Parkinson's disease, they may be of relatively little use to other researchers with an identical project. However, biomedical projects are rarely completely identical, as they explore different aspects of a highly complex interaction among a great number of variables. Moreover, the data continually change as a result of the addition of new cases and alterations in key variables relating to lifestyle, drug use, consumption, etc. Perhaps the exploration of a fishing territory with underexploited, unknown species by means of unspecified fishing gear is the proper maritime parallel to the exploitation of the biomedical data discussed here. Thus, databases may, up to a point, be renewable resources as long as the medical service operates.

Rights in bodies: who owns what and why?

Anglo-American law and, no doubt, many other legal traditions do not consistently recognize outright property rights in human bodies. Generally people are not allowed to sell their organs or those of their relatives, although some body parts, including hair, blood and, in some contexts, sperm and ova, may be sold (Marshall, Thomasma, and Daar 1996, Nelkin and Andrews 1998, Scheper-Hughes 2000, Sharp 2000). Organs have been typically exchanged within the framework of the gift (Gold 1996, Wilkinson and Garrard 1996). Biomedical information, however, does not fit easily into the legal tradition of the body (Greely 1998:488). The question whether cell lines derived from surgically removed tissue constitute private property and, if so, who is the legal owner is a contested one. Similarly, the issue of the human genome is legally complex and unsettled. While some policy-making bodies, including UNESCO, treat the human genome in a symbolic sense as the heritage of humanity, inviting the idea of international stewardship of the genome to avoid potential abuse, this "common heritage" concept has no legal foundation. Indeed, an alternative approach emphasizing the "familial" nature of human genetic material seems to be gaining acceptance in Western societies despite the growing social and legal fragmentation of the family in the current age (Knoppers 1999:23–24).

There is a rich anthropological, legal, and economic literature on the issue of property (see, for instance, McCay and Acheson 1987). Much of this literature underlines the distinctions between *tangible* property (land and livestock, for instance), and *intangible* (*intellectual*) property (including genetic and medical information). Many legal traditions emphasize one form or another of John Locke's notion of rights to the fruits of one's labor (Grubb 1998). For Locke, in "mixing" his or her labor with a particular resource base in the course of production the appropriator claims possession of the resource. While originally possession may be a weak form of entitlement or dispositional liberty in the context of so-called common-pool regimes, later on it often acquires the form of full-blown property. In the case of fisheries, the original formal allocation of rights to fish is often based on "fishing history," and sometimes such allocations later become permanent private holdings in the form of "quotas" or shares in the annual catch of a particular fishing stock (Helgason and Pálsson 1997).

The principle of rights to the fruits of one's labor is not, however, as straightforward as it may sound. What, for one thing, makes "fishing history"? What is the relative contribution of individual fishing skippers, crews, fleets, and communities? And how should the principle of rights to the fruits of one's labor be applied to the domain under discussion—the human body and biomedical information? Who are the actors mixing their labor with cell lines, genealogies, and medical information—the individuals from whom biomedical resources are derived, their families, the "experts" involved, the medical system, the community? An interesting case is the 1990 *Moore v. Regents of California*, which focused on the ownership of cell lines (see, for instance, Boyle 1996). In this case, a patient claimed that his physicians had taken tissue from him and created an immortal cell line of huge financial value. The California Supreme Court concluded that the extracted tissue and the cell line were not something to which the patient could claim property rights. By then the researchers had already patented the cell line they had invented. Did they establish their right to ownership and patenting of it by mixing their labor with the tissue in question? The issue is complicated by substantial cultural differences across the globe—the problem of comparing and translating more or less incommensurate notions of the boundary between individual and society, personhood, law, and ownership.

The U.S. National Heart, Lung, and Blood Institute refused to allow Boston University to consummate a deal with a private company for publicly acquired medical data from Framingham, Massachusetts. In some cases, however, national and international bodies are unlikely to be willing to take the financial risks involved in using genomic material and information, and this may apply to the construction of databases of the kind discussed in this article. Iceland, indeed, may be a case in point. Without private initiative, it has often been argued, the potentially powerful resource represented by the synergics of combining medical, genetic and genealogical information would probably be left unused. The main issue, in such cases, is how to prevent biopiracy, at the same time reconciling commercial and collective enterprises, and how to allocate differential access to the resource through some form of licensing, shared ownership, or public trust—in other words, maintaining public ownership of the resource and requesting reasonable compensation for the community. The solution in each case will depend upon the context—upon culture, politics, and history.

Conclusions

Biogenetic projects of the kind discussed in this article are not just innovative technical, scientific, and economic enterprises; they are radical experiments in biopolitics with potentially diverse social implications for individuals, families, and communities and the public at large. The broad responses by the international community to such experiments, however, vary from one case to another, depending on the moral landscapes of body components and biomedical information and the articulation of local and transnational concerns with civic virtue (Pálsson and Rabinow n.d.). While the construction of the British, Estonian, and Swedish databases has gone largely unnoticed at the international level, the Icelandic project has been the center of controversy. Moreover, the domestic responses are highly variable. Britons, Estonians, and Swedes have paid minimal attention to developments in their own contexts, whereas in Iceland there has been extensive public debate for over three years. Hundreds of newspaper articles have appeared on the issue in the local context, but the cultural debates offer minimal space for a true dialogue. No doubt, there are complex historical, cultural, and political reasons for these contextual differences, including differences in the organization of biomedicine, research, bioethics, and law. One important reason, we have argued, that the Icelandic debate has been so hot domestically has to do with the earlier and ongoing debates on neoliberal politics, in particular the reorganization of fisheries represented by the privatization of fishing rights.

A comparison of biogenetic databases and fisheries is pertinent for two reasons: First, in Iceland, debates on biomedicine have frequently referred to the "parallel" context of fishing, partly because of the importance of fishing for the national economy. Secondly, the history of property regimes in fishing may illustrate the options, dangers, and opportunities of alternative property regimes with respect to the human genome and other potential biomedical "resources." Modern fisheries management exhibits stunning complexity, combining commercial and communitarian perspectives in highly innovative ways, and this complexity may prove useful for both thinking about and managing the issue of the ownership of the human genome.

Although there are some parallels between the issue of fishing quotas and that of the Health Sector Database, there are important differences in terms of structure and history. This is underlined by the fact that while the public has been fairly critical of the privatization of fishing stocks it has generally supported the database project.

Modern biotechnologies, including biogenetic databases and human-genome projects, invite intriguing questions of scientific metaphors. What do people mean when they speak of human-genome research in terms of deciphering and editing the codes of the "book of life" (see Kay 2000)? What are the implications of such a metaphor for human self-understanding (Keller 2000)? Although it remains unclear exactly what a gene is (Rheinberger 2000), the successes of molecular biology have produced a discourse which presents the "blueprints" of life as being encoded in DNA molecules and genomes. Significantly, however, another scholarly discourse has recently emerged which underlines the unity of genes, organisms, and environment—the importance of mutual interactions: "a new way of talking is in the air, in keeping with the emergence of a new biology: molecular biologists, it appears have 'discovered the organism' " (Keller 1995:29). "Once it is realized," Ingold (2000:385) argues, "that capacities are constituted within developmental systems, rather than carried with the genes as a biological endowment, we can begin to see how the dichotomies between biology and culture, and between evolution and history, can be dispensed with." It remains to be seen how the theoretical contest between genetic determinists and contextualists will unfold. More important, perhaps, how these theoretical ideas relate to ideologies and moral landscapes is a matter of social context. Somehow each informs the other.

To what extent are the Icelandic Biogenetic Project and similar projects elsewhere rooted in nationalist thought? Are they locked up in the eugenics of the early 20th century, as some observers argue (Rose 2001), or are these completely separate projects? While all genetic disease studies have a eugenic dimension (Wikler 1999), since their aim is to eradicate or neutralize particular genetic disorders, there are important differences between eugenics and the new genetics. The difference (although perhaps not entirely clear-cut) is that the former was based on preventing procreation for those deemed to have "undesirable" qualities whereas the latter seeks to eliminate undesirable qualities (defined more restrictively—i.e., as "diseases") through different means, without inhibiting procreation. Also, while eugenics projects (see, for instance, Gallagher 1999), including that of the Nazi regime (Agamben 1995,

Hogle 1999), emphasized the virtues of bodily hygiene and gymnastics and the dangers of smoking, it would be silly to argue that modern fascination with aerobics and the resistance to smoking had anything to do with eugenics. The new modes of reasoning or "genetic imaginations" (Glasner and Rothman 1998, Simpson 2000) made possible by the mapping of the human genome do not necessarily suggest a eugenics movement; indeed, it is difficult to see why they should follow a predetermined, linear path.

The folk discourse on genetics and inheritance currently popular in Iceland and most other Western countries is complex and contradictory, much like the theoretical discourse of biology, anthropology, and genetics. One strand of folk discourse reduces many aspects of human behavior (as diverse aspects as alcoholism, gay culture, violence, and the ability to play chess) to the action of specific genes, while another emphasizes the role of context or the interaction of genes and environments (Pálsson n.d.). Given the social context—a context characterized by, say, unemployment and a prior history of ethnic conflict and racist thought—biological and genetic determinism can easily serve as the rationale for local eugenic regimes. The task remains to explore ongoing developments in biomedicine and the ways in which they inform, reflect, or interact with social context. This includes spelling out what may be termed the "epidemiology" of eugenics, outlining both the kinds of historical conditions that help to explain why eugenic ideas may become "contagious" (to continue the metaphor) and the social "antibodies" that may prevent their spread.

In the genome era, several kinds of maps (genetic and genealogical) are used to chart histories and relationships. Such maps are not only useful tools for research purposes; they are worthy in their own right, as are other maps (Orlove 1991), of anthropological attention, being the products of historical activities (Haraway 2000). The mapping and rhetoric associated with the Icelandic Biogenetic Project may implicitly reinforce a rather narrow notion of "Icelandicness," a notion underlined in the nationalist discourse of the last century. Metaphorically speaking, in reference to the notion of the Book of Life in Christian eschatology, "recent immigrants" in Iceland may be thrown into a lake of fire. While they live and work in Iceland they do not belong, from the point of view of the majority, to what Finkler (2000*a*:15) terms the circle of the "significant same." Iceland, however, in particular the capital city and rural fishing towns, is rapidly becoming a multicultural community, with a growing number of immigrants from different parts of the world. The genetic notion of citizenship implied by the Biogenetic Project is therefore likely to compete with other notions of citizenship that have increasingly been gaining force, emphasizing human rights and the empowerment of ethnic groups marginalized in the past. While an Icelandic human *diversity* project in the near future seems unlikely if not absurd, given the scale and composition of the Icelandic population, it remains to be seen how the competition between different notions of citizenship will be played out.

The occurrence of biotechnological innovation and commodification and the processes that take place as such changes are negotiated and debated in the social environment can, we have argued, drawing upon the works of Appadurai (1986), Kopytoff (1986), Bloch and Parry (1989), Helgason and Pálsson (1997), and some others, be viewed as periods of upheaval and formation as boundaries are erected or leveled and new pathways of exchange are created or blocked. It is the ideas, projects, and interests that motivate such changes in the topography of the moral economy that thus become the focus of attention. In the Icelandic context, as we have seen, the cultural biographies of genealogical records are different from those of biomedical information. Some of the legal clauses of the past specifically exempt the former from the restrictions generally applied to the recording and use of personal information. This underlines the ruggedness of the moral landscape of the commodities we have been discussing. At the same time, the case of the genealogies demonstrates how the debate about the Biogenetic Project has leveled out some of the contours in the moral landscape. Increasingly, genealogical records, which have traditionally been in the public domain without restrictions on recording and publishing, are being treated in the same legal fashion as other "sensitive" personal information.

Strathern (1998:216) envisages "an explosion of concern with ownership" as a result of increasing contests over what she calls "aspects of life and body." Indeed, human genetic material is progressively the object of a diversity of property claims by individuals, companies, ethnic groups, nation-states, and self-proclaimed spokes-persons for humanity. The apparent collapse of the grand scheme of the Human Genome Diversity Project was partly the result of property debates among "indigenous" groups through which genetic material was redefined as a cultural resource. Similarly, there have been heated debates about control over fragments of the national body in Iceland, France (Rabinow 1999), and elsewhere. Debates on the ownership of genetic material, however, represent more than territorial claims. What is at stake, as in most property claims (Hann 1998), is relations and identities. As Sharp (2000:309) argues, at the heart of debates on human genetic material is the atomization of the body, "raising questions about how increasingly minuscule human parts may still embody persons." For Euro-Americans in the age of genomics and biomedicine, trafficking in body parts, genetic material, and dispersed identities is increasingly a bare fact of life. Somewhat surprisingly, these developments provide illuminating insights into Melanesian processes and concepts emphasizing fragmented, dividual personhood (Strathern 1998, 1999). Likewise, the Melanesian world offers an unexpected commentary on the "West." Anthropology has much to offer on this score not only because its comparative perspective poses a challenge to universalist thinking but also because it has always, in one way or another, addressed humans as both natural and social beings. With the progressive dissolution of dualist thinking and the conflation of the natural and the social in the real world of biomedicine and bioinformatics, the classic philosophical anthropological project seems more important than ever.

Comments

JEFFREY H. BARKER
Department of Philosophy, Albright College, P.O. Box 15234, Reading, Pa. 19612–5234, U.S.A. (jeff@alb.edu). 26 XI 01

What does Kári Stefánsson, founder and CEO of deCode Genetics, have in common with a cod? Both are found in and around Iceland. As Pálsson and Harðardóttir have noted, both Stefánsson and the cod have played or are playing essential roles in the economic well-being of Iceland and Icelanders. Most important, both have been the subject of intense debate in and beyond Iceland concerning issues of the control and ownership of economically valuable societal resources.

The "fisheries" question in Iceland has at its heart equity and public trust. In the case of Stefánsson and deCode Genetics, the resource in question is what some see as one of the most valuable economic resources Iceland has to offer, its unique combination of a relatively homogeneous and historically isolated population and a long-standing tradition of genealogical and public-health record keeping. The latter resource is *bioinformation*—both existing information in the medical and genealogical records of the country and the potential information to be gained from a systematic correlation of the existing information with genetic information derived from living Icelanders. This is especially true with regard to the creation and exploitation of the Icelandic Health Sector Database. Like that of the cod, the exploitation of Icelandic medical and genetic information has created important issues of trust and public trust, though now the questions focus on politicians and policy makers, on university intellectuals, and especially on the practices and practitioners of biomedicine.

Pálsson and Harðardóttir's use of the perspective of moral landscapes helps illuminate the contours of the public debate both within Iceland and in the international bioethical and legal communities. Two features of the topography and dynamics of this landscape are important for understanding the complex and sensitive issues of trust and public trust as seen in the deCode controversy, contours that have received relatively little attention in international debates.

The first feature is one discussed by Pálsson and Harðardóttir in comparisons between the Icelandic Biogenetic Project and the development of individually transferable quotas in the Icelandic fisheries in the 1980s and 1990s. This is not a new feature. Icelandic history is dotted with examples of what Icelanders see as attempts by foreigners and some Icelanders to exploit the common resources of Iceland and the Icelandic people with little regard to the interests or will of Icelanders themselves. Post-World War II Icelandic history has been marked by the experience of being situated at a key strategic location in the North Atlantic, which led to occupation during World War II and to the building of an extensive NATO air base after the war. The decline of the fisheries in the postwar period, linked to overfishing by Icelandic and foreign boats, was punctuated by the series of "Cod Wars" between Iceland and Great Britain from the 1950s to the mid-1970s. This decline, seen as the product of internal and external exploitation of a common-pool resource, is one more part of the dynamics of the moral landscape that has emerged in the deCode controversy.

The second feature, discussed by Pálsson and Harðardóttir but receiving relatively little attention in the international response to the deCode controversy, is the question of Icelandicness. As Pálsson and Harðardóttir note, there is a relationship between the development of 19th- and 20th-century Icelandic nationalism and a sense of Icelandic cultural and "natural"—that is, genetic or proto-genetic—distinctiveness and identity. In its more humorous forms, including that promoted by the Icelandic tourist authorities, Icelandic exoticism takes the form of marketing Iceland as the home of Nordic isolates, of handsome and rugged "Viking" men and blonde and uniformly beautiful Icelandic women. In some more extreme forms, this sense of identity and even exoticism includes a pride in the culturally and biologically "Nordic" origins of many Icelanders. It is possible to exaggerate the influence of the eugenics movements on attitudes toward contemporary genomics in Iceland and even to exaggerate the connection between eugenics and deCode's projects, but there does seem to be an identifiable connection between the legacy of this sense of Icelandic nationalism and Icelandicness and support for those projects.

Finally, Pálsson and Harðardóttir are critical of the approaches and analyses of Icelandic and other bioethicists involved in the deCode controversy. Some of this criticism is appropriate, as bioethics has more than once been subordinated to a sometimes uncritical biopolitics when dealing with deCode. While the moral-landscapes perspective may provide an anthropologically "thicker" approach, bioethics itself has "thicker" critical interdisciplinary traditions, including narrative ethics, with broader and yet more sensitive methodologies than those of conventional ethical perspectives. Adopting these approaches can provide a different and yet often complementary perspective to that employed by Pálsson and Harðardóttir.

KAJA FINKLER

Department of Anthropology, University of North Carolina, Chapel Hill, CB3115 Alumni Building, Chapel Hill, N.C. 27514, U.S.A. (*kxf*9438@*email.unc.edu*). 21 XI 01

Because of its genetic homogeneity and its extensive genealogical records, Iceland has become a gold mine for the biotechnology industry. As a result, there is vigorous debate over whether the country ought to permit this fount of genetic data to be "mined" by commercial interests. The Icelandic biogenetic saga calls attention to concerns that have been addressed in various publications in bioethics (e.g., informed consent, privacy, and autonomy) and economics (e.g., the boundary between private and public property), to note but two areas of concern.

A gut response to Pálsson and Harðardóttir's account is to follow Lewontin and say that Iceland is selling its DNA—converting private into public property for the benefit of the international commercial sector. If we accept the current notion that genes are the essence of personhood, we could also add that Iceland is selling its very being. Indeed, some Icelanders seem to have made this connection in arguing in support of the Biogenetic Project that their genealogies delimit nationhood and "Icelandicness." Emphasis on genetic inheritance can easily slip into a mindless nationalism based on biological determinism that is reminiscent of mid-20th-century German history.

However, the paper calls attention to other intriguing issues. The arguments against the Biogenetic Project presented by Icelandic physicians and others in the health professions revolve around ethical concerns and physicians' fears of loss of control of their profession, but they may also imply recognition that specialists in molecular medicine will soon be replacing them and the traditional doctor-patient relationship. Elsewhere (Finkler 2000*a*, 2001) I suggest that with the new genetics all human beings may become *perpetual patients*, and the medicine of the future may consist of testing and treating asymptomatic patients with medicines tailored to the individual patient for which advertisements are already appearing on television. Patients may not need to encounter the physician face to face to be diagnosed and treated: they may only need to supply their DNA or report a family history of a disease, and doctor-patient interaction will become redundant.

The content of the debate suggests that the discourse about biogenetics is shaped by globalized themes rather than cultural concerns. We learn little about Icelandic culture from this debate but a great deal about bioethical, economic, and legal issues that are being raised, for example, in the United States. This raises the question whether the new genetics and its commodification transcend cultural shaping. While economists and political scientists tend to regard globalization as a homogenizing process, social scientists have claimed correctly that it is a dialectic process that is reshaped and hybridized in a local setting (see Finkler 2000*b*). Does the commercialization of the new genetics dictate a universal response? The irony is, of course, that while the Biogenetic Project in Iceland is produced by global forces and argued for and against in Anglo-American bioethical and economic terms, the unintended consequence is that it becomes a cultural force by creating for some a new cultural identity of "Icelandicness."

The discussion of the commodification of body parts leads me to wonder to what degree this type of commodification differs from the commodification of labor. A capitalist economy is usually defined by its treatment of not only goods but also labor and land as commodities. In capitalist societies people are required to "sell" themselves to obtain work. If we wish to question the ethics of the commodification of biogenetic materials of Icelanders or any other population, we must distinguish between the commodification of labor that is intrinsic to the market exchange economy and commodification of biogenetic materials.

The authors seem to confound micro and macro levels of analysis in their reference to my discussion of "faulty" genes (Finkler 2000*a*). My work focuses on individual people and their beliefs concerning genetic inheritance, whereas their macro analysis deals with a project that involves an entire population, presumably both healthy and ill. To make a valid comparison between the two instances it would be important to hear from individual sick Icelanders who believe their disease has been genetically transmitted. Interestingly, one of the various persons I discuss in my book regarded herself and her family as heroic precisely because they had survived an onslaught of "faulty genes," which could be comparable to the Icelanders' taking pride in their heritage of "healthy" genes.

MICHAEL GROSS
Department of Crystallography, Birkbeck College, Mlaet St., London WC1E 7HX, U.K. (*m.gross@mail.cryst.bbk.ac.uk*). 20 XI 01

In their interesting analysis of the public debate surrounding the Icelandic genome database projects, Pálsson and Harðardóttir describe how historical and cultural factors determine the amount of interest and the overall opinion which a population adopts towards modern biomedical projects involving the use of materials and/or information derived from human subjects and their genes.

In the case of the Icelandic project, a traditional interest in genealogy, along with a self-image of ethnic separateness and the traditional experience of sharing fishery rights as a national resource, has apparently fostered both intense interest in the project (hence the large amount of public debate) and its ultimate approval by large majorities of both the parliament and the population. (Maybe the metaphor "fishing for genes" struck a chord, too?) In striking contrast, a larger project in Estonia was approved without creating much of a stir. It is quite likely that in other European countries, including Germany, the concerns of eugenics and invasion of privacy would create a debate at least as intense as the Icelandic one but ultimately with a negative response.

Similar contrasts are seen in attitudes to other biological techniques such as the use of human embryonic stem cells. While a raging public debate in

Germany boiled over when a bishop accused researchers of cannibalism, both the public and religious leaders in Israel take a positive view of this technique. Contrasting views rooted in social, historical, and/or religious attitudes make it difficult to judge the new biomedical technologies "objectively." In the moral landscapes described by the authors, each population has to find its own way, preferably by an open and informed debate followed by democratic decision making.

Considering the length at which Icelanders discussed their database projects (implying that, as a nation, they have devoted some thought to the issue), one cannot find fault with them for generously sharing their genetic, medical, and genealogical information with science. In contrast, the lack of debate in cases like the Estonian project should raise the concern that the subjects of such silently approved projects have consented in a less than informed way. Thus, Iceland stands out as a model for both proponents and opponents of such projects.

STEFAN HELMREICH

Draper Program in Humanities and Social Thought, Graduate School of Arts and Science, 14 University Place, New York University, New York, N.Y. 10003, U.S.A. (*sgh2@nyu.edu*). 21 XI 01

In their cartography of "moral landscapes," Pálsson and Harðardóttir maintain that "the moral environment of bodily commodities can be likened to a topography intersected by pathways, spheres, and boundaries." They mean here to plant their discussion of controversies about biomedical databases in Iceland in the bedrock of substantivist economic anthropology, and, indeed, their language echoes that of Paul and Laura Bohannan (1968), who, in their work on the subsistence, prestige, and kinship traffics of the Tiv, famously wrote of "spheres of exchange" (a term Pálsson and Harðardóttir also use). What Pálsson and Harðardóttir seek to add to the orbit of this culturalist vision of economy is, of course, the arc of biopolitics, those trajectories of governmentality arrayed around management of the material and symbolic components of human "life," scripted these days in a genomic idiom. This makes good sense, especially since genomic biopolitics increasingly operates within the gravitational field of commerce.

Pálsson and Harðardóttir argue that cultural anthropology has much to contribute to the mapping of debates about the ownership of genetic material and information. I agree, but I would say that their contribution is more specific still. I think they show us what *maritime anthropology*—in dense dialogue with economic anthropology at least since Barth (1966)—can offer to these analyses. Maritime anthropology has long investigated the cultural constitution of ideas about property in media characterized as fluid, shifting hosts to common-pool resources (McCay and Acheson 1987, Pálsson 1991, Johnson 1996). Pálsson and Harðardóttir reveal that much of the public debate over deCode Genetics's contract to operate Iceland's Health Sector Database has made strategic reference to controversies familiar among Icelanders about the market-based remodeling of the Icelandic fishing industry. Talk of the national human gene pool—rendered in the genre of the database—has been funneled through analogies to the "common-pool" resource of fish in Icelandic waters—drawn in the abstract as possible fish to which various fishing agents might have rights of access. What has enabled this comparison? What are the similarities and differences in the "property" and "nature" at stake?

Pálsson and Harðardóttir argue that in both cases a resource that has been considered at once natural and national is being brought within a market logic. I would emphasize that part of what enables the comparison is the way the material bases of the resources in question have been transmogrified into abstract entities: genes, considered as hereditary substance and information, have been delaminated to bequeath their informational dimension to a database, while fishing stocks have been transubstantiated into imagined populations of possible fish to which access might be bought and sold. Borrowing phrasing from Pálsson and Harðardóttir, the *tangible* property aspects of genes and of fish have both been transformed into *intangible* properties as they have been redescribed as data, and this transformation is facilitated by the extension of the territory of the terrestrial nation to include the floating realm of such notional domains as Iceland's gene pool and fish stock. The resulting fetishization and commodification have had a similar effect on the fish in the sea and the genes in the pool; they have taken their material anchors away, allowing for the reinvention of both as property based on sheer potentiality. As Pálsson and Harðardóttir point out, however, there are important differences between genetic information in a database and fish in the ocean, not least of which is that actually existing fish populations offer limits to their exploitation. Pálsson and Harðardóttir's careful attention to popular Icelandic maritime comparisons allows us to think freshly about how "properties" of "national" genomes are sited in particular moral landscapes; the maritime anthropology they employ tunes us in to how to read across the relevant and locally adjacent cultural domains (see Yanagisako and Delaney 1995) of the Icelandic context.

Given the fluid disconnect between materiality and informatics in discussions of Icelandic genes and fish, though, I wonder whether "moral *landscapes*" is always the right metaphor. There is a collusion between Pálsson and Harðardóttir's landscape language and the mapping projects of the genome database workers and, presumably, of those in charge of allocating rights to fish in Iceland's waters. The trouble with mapping genes and mapping fish, of course, is that in both cases the objects leak beyond the borders of the map—genes because they point to the heterogeneous organismic and lab ecologies that make them legible (Keller 2000) and fish because they travel without regard for the boundaries of national waters. What if Pálsson and Harðardóttir's "topography intersected by pathways, spheres, and boundaries" were replaced by the shifting currents and confluences of a leakier frame, "moral seascapes"? The chart of the Biogenetic Project available through such repositioning might afford another way into Pálsson and Harðardóttir's compelling and consequential analysis of flows between fishly biopolitics and human biosociality.

ERIC HIRSCH
Department of Human Sciences, Brunel University, Uxbridge, Middlesex UB8 3PH, U.K. (*eric-hirsch@brunel.ac.uk*). 22 XI 01

Science as a global enterprise arises out of political and economic accomplishments realized in local settings. This has been one of the key findings of ethnographic studies devoted to scientific procedures. Certainly since the 1990s these local achievements have been increasingly dictated by issues of property and ownership, perhaps nowhere more than in the field of biomedicine. Pálsson and Harðardóttir's analysis is a timely and significant contribution to the controversies that attend the workings of science and the specific form assumed by contentions about the ownership of bodily information. The authors draw to our attention the diverse debates articulated in the Icelandic media regarding the use of local "bodily resources" for biomedical research. The important circumstance they emphasize is that the current controversy is informed by similar debates about fishing stocks which were previously transformed into new kinds of resources.

These Icelandic debates parallel in interesting ways those concerned with studies of authorship in biomedical research itself (see Biagoli 1998). As authorship takes on more the form of a "corporation," mirroring the market character of "big science," debates have arisen in biomedical publishing about how authorship should be deployed. This reveals a tension about what defines science—public, accessible knowledge—and the increasing tendency to limit access to this knowledge, treating it as if it were (intellectual) property. The debates described and interpreted by Pálsson and Harðardóttir were initiated by the claims made by the Biogenetic Project: what came to be perceived as a "common wealth" of records had been transformed into a commercially exploitable database.

Their analysis can also be usefully compared with Anderson's (2000) study of "possessing kuru." Here the focus is on the controversial Nobel Laureate figure of C. Gajdusek and his diverse transactions with the New Guinea Highlands Fore and medical persons and establishments in Australia and the United States. As Anderson demonstrates, Gajdusek had to negotiate a complex "moral landscape" in order to secure Fore blood and brains for his research and for his scientific reputation among senior colleagues on different continents. In particular contexts these bodily entities had to be alienated as if commodities (from the Fore), whereas in others they were transacted as if gifts (among scientific colleagues). Although Pálsson and Harðardóttir do not mention the founder and CEO of deCode, K. Stefánsson, he too has had to negotiate a complex network of transaction in order to maintain his viability and status. As Anderson (2000:73, quoting A. Strathern) observes, "Whether among Melanesians or among scientists, 'for big men it is important both to have large networks and to manage them well.' "

Such comparisons are illuminating, but there are also limits, and this is where the authors' crucial point about local concepts and morality is most pertinent. This is brought into sharper relief when one considers the importance of genealogies to the "resource" created by the Biogenetic Project. The genealogical records were seen as more important than the medical and genetic materials. They provided historical depth to the other, more recent information on Icelanders, and they have not been subject to the same controversy. Although they were used to validate claims to land ownership in the past, their contemporary use and dissemination is connected with different concerns. Perhaps most significant, genealogical knowledge is sought because of the need to obviate any form of local inequality. In effect, the continuing Icelandic interest in genealogical knowledge has everything to do with the relative positions of persons in the modern, egalitarian Iceland nation (cf. Pálsson and Helgason 1996). The role played by the major Iceland newspaper examined by the authors is a further indication of how this intense public scrutiny operates.

The authors end their discussion by noting that the "Melanesian world offers an unexpected commentary on the 'West.' " In fact, Papua New Guinea scholars are currently negotiating the new world of ownership and property promulgated by organizations such as the World Trade Organization and the World Intellectual Property Organization, which also assert their global claims. For example, Kalinoe (2001:3) has recently highlighted the relevance of genealogy to processes of everyday social life based on relations with the land in Papua New Guinea: "The 'root title' to this land is genealogy, followed by various relationships, such as one's relation to a mythical ancestor that perhaps first walked or worked the land." In contrast to the Icelandic case, the genealogical knowledge is *not* publicly displayed and distributed. Although Papua New Guinea people are explicitly concerned with what Westerners might recognize as "equality," the manner in which this is articulated is their own. Genealogical knowledge is not just about individual persons and their connections but about what Pálsson and Harðardóttir refer to as "dividual" persons implicated with others and their surroundings in multifarious ways. Routinely revealing such knowledge would be disclosing one's claims to act and thus the ability to appear equally powerful among others. As Kalinoe (2001:5) notes, extreme caution must attend the disclosure of this knowledge, even for a cosmopolitan such as himself. Although people "possess" this knowledge, it is not their "property"; rather, they are its custodians. Thus, as Pálsson and Harðardóttir emphasize, distinctive moral landscapes and their inherent knowledge conventions prominently influence the ways in which issues of ownership and property are perceived, debated, and ultimately codified, whether in law or as part of scientific procedures. And it is in this regard that anthropology has much to contribute to questions of elucidation.

ALF HORNBORG
Human Ecology Division, Lund University, Finngatan 16, 223 62 Lund, Sweden (*alf. hornborg@humecol.lu.se*). 19 XI 01

Pálsson and Harðardóttir invite us to reconsider popular debates on biomedicine in the light of recent perspectives from economic anthropology. They touch on several intriguing analytical vantage points for exploring these issues, leaving the reader convinced that this is indeed a fertile domain for anthropology. Attempts to commodify components of the human body undoubtedly elicit various responses in different cultural, social, and political contexts, evoking distinct notions of morality, personhood, kinship, ownership, and exchange.

I would strongly encourage further comparison between Iceland and Tonga. Here, too, is an ancient island nation proud of its genetic homogeneity, obsessed with genealogy, and distrustful of external contamination, debating whether to "sell" its gene pool to a biotech company (Australian-based Autogen Ltd.). But rather than the object of local, popular debate, the alleged "sale" of the Tongan gene pool by Minister of Health Dr. Viliami Tangi—announced on the Internet by Autogen on November 17, 2000, and confirmed by its German partowner Merck on June 28, 2001, but denied by Tangi himself (personal communication, August 1, 2001)—seems to have been largely shrouded in secrecy. Criticizing this lack of public discussion, Lopeti Senituli, director of the Tonga Human Rights and Democracy Movement, revealed some of the fabric out of which a Tongan debate would be fashioned (*European Pacific Solidarity Bulletin* 8[4–5]):

> The Tongan people are not, and have never been and never will be in a situation where they would need to even consider selling parts of their body in order to survive. The very thought is repugnant. . . . What is involved is the sanctified blood of human beings and not the genetic make-up of our pigs (with all due respect). . . . In any case we hope that these medical researchers will come with their own reserve of compatible blood from Australia because they might end up needing an emergency blood transfusion themselves. . . . The very thought that one human being (or company or government) can own the intellectual property rights over the parts of another human being's body conflicts with the sanctity of human life and the basic tenet of the Tongan's Christian faith that we were created in God's image.

Significantly, Autogen's chairman, Joe Gutnick, similarly referred to divine principles, arguing that "God wants that people should be able to enhance their life-style and live longer" (www.smh.com.au/news/0011/23). Yet, both Autogen and Merck have felt prompted to reject allegations of "biopiracy" voiced, for instance, by Greenpeace and the Indigenous Peoples' Council on Biocolonialism. Autogen's director of research, Greg Collier, blames the "misinformation" largely on "the mistakes made in Iceland," but a Tongan newspaper editor says that people in the islands were immediately distrustful of the project, adding that "there is no Tongan word for genes" (*Time Pacific*, August 20–27, 2001). A Health Ethics and Research Committee has been set up to investigate the project (www.rsnz.govt.nz/news/talks/sci-soc), in late August described as "on hold." The *Time Pacific* article concludes that, "for Tongans, heredity carried a moral value as well as a monetary one." The Oxford geneticist Bryan Sykes reflects: "[The region's DNA] is very homogeneous. It's the simplest, most straightforward population I've ever dealt with. . . . It's DNA that's been handed down to people by their ancestors. So it's a precious thing for individuals and for communities—certainly not something that should be bought and sold" (*Time Pacific*, August 20–27, 2001).

Sykes illustrates how Europeans project images of pre-modern purity onto these islands while in the process of converting them into merchandise. Although it would be difficult to extricate an essentially Tongan stance from the global moral discourses whose proponents have always been as eager as the "gene hunters" to colonize this virgin terrain, the emerging Tongan debate may offer an instructive variant of the moral topography evoked by Pálsson and Harðardóttir. Perhaps it could inspire a further elaboration of the posited relevance of frameworks from economic anthropology, emphasizing the moral incommensurability of capitalist enterprise and genetic continuity in terms of "things" versus "persons" (Kopytoff 1986) or "the short-term" versus "the long-term" (Bloch and Parry 1989). It might also suggest that we explore at greater depth the extent to which biotechnology and ethnonationalism may be embedded in similar ideologies of genetic essentialism and microdeterminism, predisposing both projects to underestimate the role of interaction with specific contexts or environments. It seems ironic that Autogen's gene hunt should have been prompted by the increasing frequency of "Western diseases" due to exposure to "the Western culture and diet" (http://dailynews.yahoo.com/h/nm/20001122), that is, environmental factors. The fact that national resources such as Tongan land and economy (and, in effect, Tongan genes) have been safeguarded historically (by a constitution of British derivation) probably accounts for some of the resentment currently directed at the small group of Chinese shopkeepers recently admitted into Nuku'alofa, but whether such ethnonationalism is significantly connected to folk models of genetic essence (as has obviously been the case in Iceland) has yet to be determined.

A few methodological issues raised by Pálsson and Harðardóttir's contribution would seem to have implications for the practice of both biomedicine and the anthropology of biomedicine. How reliable, for instance, are genealogies as sources of information on actual, biological kinship? What are the implications of the fact that genealogical knowledge is primarily cultivated by the elite? What are the advantages and limitations of newspaper material as ethnography?

The central issue here is biotechnology's challenge to Cartesian dualism. Far from maintaining a clear-cut distinction between mind and body, even "Westerners" identify in profound ways with their organisms. The increasingly conspicuous moral conundrum of biomedicine stems from the underestimated extent to which we not only *have* but *are* our bodies.

ANDRES METSPALU
Department of Biotechnology, University of Tartu, 23 Riia St., Tartu, Estonia (andres@ebc.ee). 27 XI 01

The human genome sequence, extensive genetic marker databases, and rapidly advancing information technology are the newly established bridgeheads of contemporary biomedical research. From "one gene–one Ph.D." science we are moving into systems biology, where entire genomes, transcriptomes, metabolomes, etc., can be analyzed in a single experiment or, as a cross-section, in an individual. The Icelandic Biogenetic Project, brave vision of Kári Stefánsson, has served the world as an ice-breaker—leading the way but also taking most of the pressure, internationally as well as internally. "For Whom the Cell Tolls" is an extensive and in-depth review of the Icelandic approach that dissects the moral, ethical, and public-perception aspects of this undertaking. The front line of the debate in Iceland (and also in Estonia) has been ethics (consent and privacy), but the real motive force behind the whole debate is ownership and control. Why should any private company make a profit using *my* health and genetic data?

Why not? Society is organized this way: governments build roads, but drug development is in the hands of private companies, and we are free to change this. People are not against the health database (just 7% in Iceland and 6% in Estonia disapprove). Rather, the opposition is coming from a small number of professionals. If one is sick, the privacy issue is not at the top of one's list; one usually cares much more about getting well again. Another attitude is also common in both countries and is reflected in the statement of the university employee from Iceland who does not trust either the company or the state and will not allow his health data to be included in the database. However, at the same time he expects the state to provide the best possible medical care. Things do not work this way. Altruism (donation of one's health data and blood sample to the database) is what a society can expect of its responsible members anywhere. We all have to take steps to improve the health of our children and to reduce human suffering because of disease. Medicine is entering a new phase of development in which genetic information will be taken into account in diagnostics and treatment much more than before. This is not a novel conception; it is just that new tools will be used.

The problem is one of education. Science and technology are developing so fast that it is increasingly difficult for the general public and even for some professionals to keep up with it. Lack of comprehension of genetics (and science in general) is a breeding ground for suspicion and confusion. Hundreds of articles in the press, TV shows, and radio broadcasts have done an enormous job in this field, educating people in both countries mentioned here and in the world in general. If we scientists do not do it, who will? Usually it is Hollywood, and this is not a good source of scientific information.

Finally, it is true that we are navigating uncharted waters, and in addition to novel scientific and technological solutions we are extending traditionally small-scale genetic-epidemiological studies into global enterprises affecting almost everyone in the society. Therefore it is important to have articles like this one in which this new situation is analyzed objectively in its historical, economic, and philosophical context.

LYNN M. MORGAN
Department of Sociology and Anthropology, Mt. Holyoke College, South Hadley, Mass. 01975–1426, U.S.A. (lmmorgan@mtholyoke.edu). 30 XI 01

Pálsson and Harðardóttir's analysis of the Icelandic debates is commendable not least for its sensitivity to the historical circumstances that foster or impede the commodification of bodies and body parts. The authors resist the temptation to portray commodification as the ugly and inevitable result of a savage and unfettered capitalism. They rightly insist that the commodification of body parts is ethnographically complex and must be understood within the context of frequently reconfigured social, political, and commercial arrangements. What, they ask, are the multiple factors that determine the "moral landscapes of body components"? Under what conditions does the intersection of the market, biomedicine, politics, and body parts lead to commodification?

Their temperance and ethnographic grounding are admirable at a time when some have been inclined to treat the commercial exploitation of body parts as the predictable teleological outcome of bio-capitalist expansion. The reactionary argument goes like this: Biotechnology firms funded by promissory capital and exempt from government restraint in an era of neoliberal reform are "enterprising up" the human body. As biotech firms and "big pharma" advertise and market their latest laboratory efforts to clone humans, sequence the genome, and develop artificial organs, they inevitably commodify human body parts and thus cheapen and degrade the moral dignity once bestowed upon the body. While there is certainly cause for alarm over state complicity in the corporate exploitation of body parts, Pálsson and Harðardóttir's article exemplifies the kind of attention that anthropologists and historians should devote to understanding the movement of bodies, body parts, and biodata across the boundaries of commodified exchange. It might be useful at this juncture to remember that the reverence attached to bodies today is itself a historical artifact and that body parts have frequently moved back and forth from commodified to noncommodified realms of value, circulation, and exchange. A historical example will illustrate my point.

A hundred years ago, anatomists and civil authorities in Baltimore formed an alliance to remove cadavers from the unsavory realm of grave robbing. In the process, they set in motion the historcial process through which cadavers and body parts would come to be regarded as revered and priceless objects, gifts, and prestige items. When Franklin P. Mall arrived to teach anatomy at the newly opened Johns Hopkins medical school in October 1893, he had his work cut out for him. He was scheduled to begin teaching gross anatomy on November 15, but there were no cadavers on hand. Much to his dismay, he was forced to rely on grave robbers to acquire bodies. Buying dead bodies was illegal, but in the absence of legal means of obtaining cadavers for dissection it was practiced more regularly than we might like to admit. In 1878, just two years before Mall enrolled in medical school at the University of Michigan, three "resurrectionists" (body snatchers) had been arrested in Toledo and "confessed to a contract for seventy bodies for Ann Arbor." The illicitly acquired cadavers—unearthed from graveyards across the Midwest—arrived at the anatomy laboratory in barrels marked "pickles," giving rise to a campus quip that "a medic is never happier than when he finds a fellow man in a pickle" (Holtz 1967:95). The same was true in Baltimore, where Henry Louis Mencken reported that in the 1880s an elderly "colored" woman "had been murdered by two thugs, and her remains sold to the janitor of the University of Maryland Medical School for fifteen dollars" (Mencken 1940: 294–95, cited in Gindhart 1989). For at least his first five years at Johns Hopkins, approximately 60% of the cadavers Mall received came from unspecified (and probably illegal) outside sources. His own description of anatomical acquisitions was deliberately vague, leaving the impression that the bodies made it to the anatomy lab by magic: "Late in the evening, a subject was left in the basement. The next day one came from the State, and a few days later another appeared in the basement." Mall and his colleagues were well aware that the medical schools were tainted by their association with grave robbing, so they embarked on a campaign to move the business of cadaver acquisition from the criminal realm to the safe haven of scientific and state control.

Together with the Baltimore Health Commissioner and the Anatomical Board, Mall and his colleagues worked to insure that the bodies of paupers and "unclaimed dead" would be donated to medical schools rather than buried. After several years of patient work, they put grave robbers out of work, saved the state the expense of burying the pauper dead, moved cadavers out of the realm of monetarized exchange, eliminated inter-state commerce in cadavers, improved the reputation of the profession of anatomy, and exerted strict proprietary control over the dead. No longer would bodies be exchanged for money (except to reimburse transit costs and such), but the commodity value associated with cadavers would arguably be enhanced when they were redefined as scientific specimens and gifts. This historical shift required intense social negotiations and the exercise of power to reconfigure the meanings attached to the dead and to realign the social relationships that governed access to cadavers. The moral landscape of bodies is still full of landmarks erected during that time, but we should not imagine that they were—or would remain—permanent fixtures of our moral geography. Fortunately, if we follow the lead of Pálsson and Harðardóttir, we need not make that mistake.

DOROTHY NELKIN
Department of Sociology, New York University, New York, N.Y. 10003, U.S.A. (*nelkind@juris.law.nyu.edu*). 21 XI 01

Pálsson and Harðardóttir employ the metaphor of the "moral landscape" to describe the debates over biotechnology databases. This metaphor is entirely appropriate and, indeed, brings out an endless stream of related ones. This landscape is bulldozed by market interests, beset by turf disputes over personal autonomy, besieged by territorial claims over the control of information, and embroiled in property disputes. Who owns the medical, genetic, and genealogical information stored in databases? This question has shaped the Icelandic debate and, more broadly, the troubled terrain at the intersection of biotechnology and society.

The growth of the biotechnology business with its patent practices has turned DNA into a commodity. The gene is not just the essence of a person but a kind of property, an item to be bought and sold. Describing the value of genes, scientists are using a commercial language of supply and demand, contracts, investment, and compensation. In the quest for valuable genes, body tissue is extracted like a mineral, mined like a resource, procured like land or goods, and banked like money. In a well-known legal dispute over the patenting of cell lines, *Moore v. University of California*, a lawyer proposed taking Moore's cell lines under the laws of eminent domain. Pathologists seeking access to archived human tissue in government or hospital repositories have called these tissue banks "national resources" or "treasure troves." The molecular biologist Leroy Hood has predicted that "your entire medical history will be on a credit card and physicians will look at humans as complex systems and medicine will be managed by mathematicians." And the director of the deCode project, Kári Stefánsson, has suggested that, like the network of centralized databases that allows us to withdraw money from automatic teller machines, gene banks are centralized repositories of personal information on health that enable the withdrawal of useful and valuable currency.

The commercial thrust of genetics these days places the human body in an ambiguous space in our moral landscape. Are genes the essence of the person or a commercial product? If genetic material is both the essence of the person and valuable property, what legal claims would a person have over his or her genes? Who in fact owns the human body?

The deCode saga has spawned a cottage industry of commentary on the moral landscape of population-based genomics—in particular, on the issues raised by the development of giant databases. These databases must be located in the context of commercial interests, and this is a key aspect of the deCode debate. But the Iceland situation is hardly unique. Hospitals and research centers have developed collaborative agreements with companies that, in exchange, have access to tissue samples and their data. Public health departments contract with private companies to store the blood spots collected from newborn testing. The University of Utah provides private firms with genealogical data from its extensive Mormon records. Boston University, which maintains the extensive records collected since 1948 from the famous Framingham Heart Study, has tried to form a private company to mine the data for genetic research.

The deluge of petitions and position papers in opposition to the deCode project reveals the range of issues shaping the topography of the databanking business. Objections revolved around the practice of presumed consent. The United States requires explicit consent for the banking or use of human tissue, but similar debates take place over the use of stored tissue samples. Can these samples, collected for one purpose, be used for another without renewed consent?

Related questions have to do with maintaining genetic privacy; maintaining anonymity is difficult in a small country. Many of Iceland's clinicians object to deCode because medical records are in their territory and they do not want them transferred to a central data base. They worry about destroying the fiduciary relationship between doctors and patients, and they believe that if patients know that their records will be centralized they may be less likely to divulge personal information. Mannvernd, a society of professionals opposed to the project, believes that it infringes on human rights, personal privacy, and accepted medical and scientific standards, and the Icelandic Medical Association has advised its members to refuse to submit patient records.

A further issue has to do with patient control over the use of the body. Some people want to control what is done with their DNA even if samples are kept anonymous, for the body has social meaning.

DeCode's critics have also raised questions of fairness. Given the potential profits to be derived from the database, who will share in the benefits—the government, patients, researchers, or the people whose records are involved? Is the collection of DNA data a new form of bio-colonialism?

Such objections have to do with property rights over the body. Do individuals own their DNA, or is it a common heritage that cannot be owned at all? Should valuable and vital information about a nation and its people be traded as "just another commodity on Wall Street"?

From the perspective of many scientists, the moral landscape of biotechnology is increasingly cluttered with impediments to research progress. But opposition persists, and one of the contributions of deCode may lie in generating public discussion and debate about the medical, genealogical, and genetic databases that are being stored in biobanks throughout the world.

ROBERT N. PROCTOR
Department of History, Pennsylvania State University, University Park, Pa. 16802, U.S.A. (rnp5@psu.edu). 16 XI 01

Pálsson and Harðarđóttir make it clear that our efforts to understand and evaluate Iceland's deCode debate must take into account local circumstances. Iceland in the 21st century is not America in the 1920s or Germany in the 1940s, and ideas or actions that have led to discrimination or even murder in one environment may prove harmless in another.

It is not irrelevant, for example, that Iceland is a democratic social-welfare state, with free medical care and support for the infirm and aged. This means that the discrimination that is routine in other countries is not as common or consequential here. You cannot lose your job or be denied health insurance in Iceland because you have gene X, Y, or Z. Privacy is not such a big issue (it is said) because everyone already knows everyone's business. It's an island.

It is also true, though, that the commercialization of Iceland's genetic heritage raises new questions that are by no means trivial. Prominent among these are questions about rights to ownership of genetic information, rights of access to medical records, and rights to privacy. Many people still seem to feel uncomfortable thinking of gene sequences as "text" (subject to copyright) and tissues as "inventions" (subject to patent). The situation is complicated by the fact that we are talking about our *bodies* (or, more often, other people's bodies), while also abstracting from—or brushing aside—wildly divergent notions of disease and well-being, not to mention purity, disgust, and the sacred.

The idea of "presumptive consent"—that people are your willing partners unless they go out of their way to object—raises ethical issues far beyond the realm of biotechnology. (Think of the junk e-mail you are presumed to want.) Presumptive consent is arguably no consent at all, an oxymoron. The fact that directors of some local clinics have been refusing to hand over patient records—claiming medical confidentiality—shows how deeply the deCode deal threatens traditional medical authority, as the authors nicely point out. The fact that lawyers have encouraged noncompliance in order to have people *sell* their compliance back to the state shows that commercialization is not a one-way street.

The question can be asked: Who really stands to benefit most from the enclosure of Iceland's genetic/genealogic commons? DeCode has already invested prodigious sums of money into Iceland in the form of jobs and buildings, helping to stanch the brain-drain that the country has suffered in recent years. Pálsson and Harðardóttir point out that the licensing agreement guarantees an annual payment of 70 million krónur to the Icelandic state for deCode's exclusive access to the database, plus a royalty of 6% of all profits. The Icelandic state is clearly getting its kickback, but what does the company expect from its investment?

The big prize, of course, would be to be able to identify genes that promote or prevent certain diseases—say, cancer, stroke, or heart failure. The idea is that detection of such diseases is easier in a place like Iceland, where you have reliable genealogies and a fairly homogeneous population. There are lots of reasons to question such a strategy, however. It is not clear that Icelandic genes are as homogeneous as is often supposed, and one wonders whether Iceland is really so different from other European countries, where bastardy has been routinely hidden to save face, grace, or inheritance. It is also not at all clear that we will ever find simple genes for big-ticket diseases like heart attack or cancer—nor is it clear whether useful drugs can be made from such genes, if found. The authors aptly compare the Biogenetic Project to "exploration of a fishing territory . . . by means of unspecified fishing gear": we don't even know if there are fish in these waters, or what kind of bait to use, or if any of the fish we might catch are worth eating.

This brings us to the matter of *hype*. Genes may or may not be "the currency of the future," but public health scholars certainly do despair when they hear how much money is spent on long-shot magic-bullet projects such as this, given how little is done to stop what we already know to be well-defined causes of disease—like tobacco, which kills millions of people worldwide every year, or faulty sewerage, for which genes are not likely to be found. There is an old Viking saying: "A man's fate should be firmly hidden to preserve his peace of mind." Genetic medicine has long suffered from the fact that, while it is not always hard to find out what is wrong with us, it is often hard to do anything useful about it (short of horrors). The day may be not far off when a person's genome may be sequenced and put on a computer chip; that says very little, though, about whether any kind of benefit is likely to come of it. There is something fishy in Iceland.

LESLEY A. SHARP
Department of Anthropology, Barnard College, New York, N.Y. 10027–6598, U.S.A. (lsharp@barnard.edu). 29 XI 01

New biotechnologies and associated forms of bioinformation frequently (or, even, inevitably) transform the human body and its parts into targets for commodification. In turn, commodification must be recognized as a highly complex, multifaceted, and particularized process, one driven, enhanced, or impeded by technical, sociopolitical, and economic desires and constraints. Finally, as Pálsson and Harðardóttir show, such forces mitigate the commercialization of the body. The developments that drive Iceland's Biogenetic Project further refine understandings of body commodification, a process that, clearly, must be investigated in reference to local, national, and global debates.

Human bodies and their parts have long been commodified. An especially troubling contemporary development, however, is the body's hyper-fragmentation into increasingly minuscule parts (Sharp 2000), and nowhere is this more evident than in genetics. The current obsession with mapping the human genome requires an atomization of the body that radically alters definitions of self, personhood, and, ultimately, humanity. Pálsson and Harðardóttir illustrate this concisely and clearly. Most intriguing is the manner in which the Icelandic Biogenetic Project now insists that the human body—whether corporal or corporate—be *reimagined*: as property, as a moral landscape, and as a historical artifact.

Pálsson and Harðardóttir's work confirms Strathern's (1998) prediction that the new biotechnologies would generate an "explosion of concern with ownership." When body fragments are commercialized (that is, excised, altered, and packaged and then traded or sold on the open market), individuals or corporations scramble to lay claim to them and they undergo rapid metaphorical transformation. Our corporal fragments are now harvested, extracted, and mined, reemerging as inanimate, soulless objects or even as precious ores (Andrews and Nelkin 2001, Nelkin and Andrews 1998). Pálsson and Harðardóttir thus compel us to consider who, in fact, owns the Icelandic body. This critical question nevertheless obscures more profound dilemmas that extend beyond issues of property. More important, the Icelandic case exposes how a host of social meanings have become entangled in debates over individual versus other sorts of corporate rights.

For example, the value of human genetic material hinges on its relationship to specific gene pools. How, then, is Iceland reconfigured and thus reimagined as a specialized terrain of knowledge, and how is this highly specialized "pool" rendered a precious commodity? Can the essence of Icelandicness be appropriated or alienated from these original "owners"? As Pálsson and Harðardóttir show, questions such as these parallel other national debates concerning alienated fishing rights. They are, in multiple ways, the same kettle of fish. Furthermore, an underlying premise is that the paired forces of commodification and commercialization generate new or intensify existing anxieties where individual and collective "fears and longings" (Brodwin 2000:1) are spawned by the arrival of entrepreneurs, venture capitalists, and multinational corporations. Within such domains, the human body is reimagined as an altogether different kind of cultural artifact, and social property acquires radically different meanings.

For Pálsson and Harðardóttir, the moral landscape is a manifestation of the moral economy, and this transfiguration is hardly mere metaphor. Current debates must be bracketed in reference to specific cultural and historical trajectories, but they are also rooted in very particular geographies. As science privileges the gene pool, isolated, bounded populations suddenly emerge as sites of intense interest and concern. Thus, although genomic projects exist in England and the United States, of greater interest to me here is the pairing of Iceland and Tonga. Islands have long fascinated naturalists and conservationists; in the genomic context they are of scientific worth of yet another sort, with the result that distant, isolated, and exotic locales now occupy center stage in 21st-century biopolitics. A myth of uniqueness dominates the scientific imagination, a construct that essentializes local populations assumed to harbor unusual genetic histories and therefore commercial potential. We should most certainly heed warnings of biopiracy (Hayden 1998, Shiva 1997): we are indeed witnesses to the advent of new forms of commercial appropriation of islanders' bodies and body fragments. Anthropology should serve as a watchdog (Scheper-Hughes 2000) in this emerging realm of scientifically driven neocolonial encounters.

The privileging of genetic isolation also exposes exoticization in another guise, one based on assumptions of genetic homogeneity and, ultimately, of purity. Pálsson and Harðardóttir link this to a history of Icelandic eugenics. This theme of assumed purity bleeds into other realms of social and historical importance. For example, what precisely is the relationship between biogenetic projects and the indigenous Icelandic obsession with genealogies? How might one ultimately reshape the other? Anthropologists have long understood that genealogies are only partial truths at best, easily refashioned to improve the social standing of some individuals while excising others. Is there a dialectic in the making here involving genetics and genealogies? Might this new relationship bear down on Icelandic history, too, altering, perhaps, the sagas themselves, which stand as precious emblems of the Iceland of long ago (and yet another myth of uniqueness)? Thus, the moral economy could, in the end, even affect historical morality, generating new notions of what defines Icelandic social facts.

BOB SIMPSON

Department of Anthropology, University of Durham, 43 Old Elvet, Durham DH13HN, U.K. (*robert.simpson@ durham.ac.uk*). 20 xi 01

Research and commentary on Iceland's Health Sector Database and its links to the private sector are beginning to emerge as an important case study of the complex issues that are thrown up around the new biotechnologies. Pálsson and Harðardóttir have made an important contribution to this corpus by drawing attention to the particular features of Icelandic society, culture, and history which give the deCode episode a distinctive local flavour. They provide some useful pointers as to how the issues that are invariably raised when the biotech industries roll into town can be put in an anthropological context. They highlight not just the form that the conflicts take but, more important, the ways in which what I have called ethical "hot-spots" (Simpson 2001) emerge. For example, we can see that issues such as privacy, consent, and ownership do not take the same form as they might in, say, Tonga, Estonia, or the United Kingdom.

One argument that features throughout the piece is that the bringing together of databases and biotechnology should not automatically invoke a grim eugenic future. An article of my own appears to be cited as an instance of this association's being made. The gist of that article is that whilst one should not automatically conjure up eugenics as an outcome, neither should one discount it. The linkages I propose are not necessarily with the "hard" eugenics epitomized by the ideology of the Third Reich but with the myriad "soft" forms that arise in the routine construction of identity, community, ethnicity, and nation. I put forward the notion of "imagined genetic communities" (à la Benedict) and raise questions similar to those associated with Pálsson and Harðardóttir's "Viking notions of bounded citizenship" or the possibility of genetic determinism's serving as the "rationale for local eugenic regimes." Put simply, as ideas about genes and DNA enter popular discourses, new entities become the means to express some very old ideas about boundaries, essences, and identities. As Pálsson and Harðardóttir suggest, we should explore where these linkages fit into the bigger picture that human genomics opens up.

The image of the moral landscape is used to good effect by Pálsson and Harðardóttir, although it is not always clear whether they are assuming the role of cartographers or of official guides. The idea is used to illuminate how the body and its constituents and products mean different things not only in different settings but also as they move between sectors within a single landscape. However, the metaphor is essentially spatial, and as such it obscures a critical temporal dimension to the debates. It is not simply that commodities are contested as they move, for example, from a blood bank to a pharmaceutical company but that their meanings may change over time and in ways that are difficult to predict or control. The biotechnology activity that is under scrutiny here is one that concerns substances that betoken growth, generation, and potentiality. Public anxieties are heightened by concerns about consequences that the regulation of traffic in these substances fails to allay. As the John Moore case demonstrated, consent freely given for one kind of treatment may be transmuted into other forms for which today's controls and assumptions of ownership are, in legal terms, redundant. Indeed, there is a futurological imperative running through the discourse of the new genomics which remains elusive in this analysis (cf. Rabinow 1999, Fortun 2001).

An intriguing feature of the piece is that it makes no mention of the company's chief executive officer, Kári Stefánsson, a central figure in the origins, development, and future of the whole enterprise. His part in the dispute surrounding deCode and the Health Sector Database expresses elements of cultural specificity of the kind that Pálsson and Harðardóttir are trying to foreground. For example, Stefánsson is often described as "charismatic," "the embodiment of the tall Viking" (Rose 2000:12) and something of a conquering hero bringing back wealth and prosperity to the homeland. The support group for patients with multiple sclerosis is identified as among the earliest supporters of the Health Sector Database, but no mention is made of the fact that Stefánsson had worked as a physician and a researcher with multiple sclerosis patients (Sigurdsson 2001:114). Whether consciously engineered or a ready idiom which simply unfolded around Stefánsson, his role has taken on saga-like dimensions, and it would be surprising if this were not a significant part of the appeal of the enterprise for the Icelandic public. But, if I might extend the play on Donne's Meditation 17 that figures in the title, it may be that for Pálsson and Harðardóttir "No man is an Iceland entire of itself."

Reply

GÍSLI PÁLSSON AND KRISTÍN E. HARDARDÓTTIR *Reykjavik, Iceland.* 17 XII 01

We appreciate the thoughtful comments upon our article by a group of scholars representing a diversity of both academic fields and empirical contexts. The authors, in our view, have seriously engaged with our main arguments, at times challenging and expanding on some of the points we make. All seem to agree about the importance of a consistent interdisciplinary effort to understand the broader implications of recent developments in biomedicine, in particular the articulation of science and the market and the moral landscapes in which biogenetic projects and bodily commodities are necessarily embedded. Below we discuss the key themes addressed by the commentators, some of which, of course, are overlapping. By necessity, a brief response such as this cannot cover everything.

The key notion of moral landscapes is taken up by some of the commentators. While Nelkin draws attention to a stream of related metaphors, Helmreich points out that it draws upon an early substantivist notion of spheres of exchange developed by several scholars, including Barth and the Bohannans (see Gudeman 2001) and Hornborg suggests further refinement in the genomic domain. Simpson claims that the spatial metaphor obscures the temporal dimension of the public debates we discuss. Yet, we have emphasized their dynamics, in particular with respect to the changing status of genealogies; not only do the cultural uses and meanings of genealogies vary from one ethnographic context to another (Gross, Hirsch) but their significance changes over time. While Polanyi (1944) posited a rather permanent distinction between natural and "fictitious" commodities, we suggest a highly unstable and shifting categorization.

The human body is a case in point. As Morgan shows with her historical example, body parts have frequently moved from commodified to noncommodified realms and vice versa. In the course of this history, some of the landmarks in the moral economy of the body have been erected while others have been demolished. Helmreich's notion of moral *seascapes*, we think, adequately captures the "leaky" nature of commodities in the genome domain. As some of the commentators point out, the moral landscape is not composed of bounded units but subjected to the globalizing forces of culture, science, and capital. One of the intriguing aspects of biotechnology is the fact that it represents, to continue the aquatic metaphor, "navigating uncharted waters" (Metspalu).

Some of the commentators discuss possible ways of situating oneself in the moral landscapes of biomedicine. Indeed, the tension among the commentaries partly reflects the contours of the moral landscape represented by our newspaper material. Thus, to simplify, some of the authors speak of Icelanders' generously sharing their biogenetic information in an altruistic spirit for the benefit of health and science, others seem to avoid the issue of positioning altogether, implying that the jury is still out and that it is important to resist the temptation of gut reactions, and still others emphasize the fishiness of the Icelandic case, the dangers of corporate power and biocolonial encounters, and the possibility of soft, if not hard, eugenics.

More generally, when does anthropology cease to describe the real world and begin to construct it? If, as Shore and Wright (1997:6) argue, policy is a "charter for action" much like Malinowski's notion of "myth," what happens to the anthropological enterprise if anthropologists are part of the myth-making narrative, as policy makers? Simpson identifies the diametrically opposed roles of the cartographer external to the system described and the official guide presumably firmly embedded in the enterprise. Neither of these is satisfactory for the anthropological enterprise. Anthropologists, ethnographers, and other explorers of the genome era are not external to the moral landscapes they outline; the impartial observer assumed by the modernist perspective is a mythical construct, since descriptions, whether we like it or not, are speech-acts with a momentum of their own. Nor should ethnographers be reduced to mere guides or servants; their primary role is to understand, to illuminate, and to probe beyond surface appearances.

Sharp and Simpson both seem to recommend the alternative role of the watchdog. We agree that a sustained, critical engagement with our field sites is essential for both intellectual and ethical reasons. There are several species of watchdogs, however, and much depends on which of them is adopted. When unleashed, the watch-dog can easily slip into the mode of the random aggressor unaccountable to anyone, assuming what might be called the role of the academic paparazzo. While the anthropologist is "always on the verge of activism" (Marcus 1998:122) and modern, multisited fieldwork is necessarily highly complex, activism is tricky business, as it can simply become an excuse for conditioned reflexes and theoretical stagnation—and, as a result, loss of credibility.

Barker makes the point that bioethics has "thicker" and more critical dimensions than we have suggested. While this is no doubt correct, the institutionalization of normative ethics remains one of the key developments of the genome era. Bioethicists in Iceland as elsewhere are engaged in a kind of commodification not unlike the one that is often the subject of their criticism. In the process of deconstructing one fictitious commodity, namely, body components and biomedical information, they busily manufacture another, the symbolic capital of their profession. That capital is no less contested than the capital of the biotechnology industry. Just as the moral landscape surrounding the use of cadavers has been partly informed by material interests (Morgan), the discourse of current bioethics on the exploitation of DNA and medical records is partly informed by professional concerns. Bioethics produces its own moral landscapes—its own contours and topographies—which warrant ethnographic study (Pálsson and Rabinow n.d.).

As some of the commentators (Helmreich, Proctor) point out, there is plenty of hype and virtualism in the genome domain. The rhetoric of genetics, we think, needs further exploration. The comparison with the "hard" social science of economics may be useful. Just as economists tend to emphasize what may be called the "alchemy of capturing rent," not only fetishizing an abstract economic construct but also making the world conform to their virtual vision (Carrier 1998:2), geneticists have reified the gene and elevated it to supreme status. In his portrayal of economics and its preoccupation with the objective, the harsh, and the mathematical, McCloskey (1993:76) describes neoclassicists as a motorcycle gang. Perhaps the practitioners of biotechnology are aptly described as the motorcycle gang of biology, preoccupied with "shares" in the gene pool rather than fishing stocks or other natural "resources" (Helmreich). But if modern biotechnology is all about imaginary futures and trivial hype, what could possibly be the point in social scientists' adopting the role of the watchdog?

The extracting and marketing of genetic information, however, raises important questions about ownership and control, as several of the commentators point out. Given the fact that genetic information is sometimes considered the essence of personhood (Finkler, Sharp), its marketing, patenting, and legal protection in the form of copyrights may represent the ultimate form of alienation in the Marxian sense of the term. The extent to which the extraction of DNA represents a new form of biocolonialism (Nelkin, Proctor) is an issue that demands theoretical and empirical scrutiny.

Hirsch raises an interesting question about the similarities between the key figures of modern bioscience and the big-men of Melanesia. Indeed, strong leaders who are dependent upon their skills in managing and negotiating extensive social networks, establishing the kind of trust discussed by Barker and Finkler, characterize accounts of both contexts. Thus, in the world of biotechnology Francis Collins, Kari Mullis, and Kári Stefánsson have become household names. No doubt it would be fruitful to explore some of the parallels between the leaders of the two contexts. The theoretical discourse on Melanesian big-men (and their relatives "great men" and "chiefs") is, however, ambiguous. Authorship, creativity, gender, and the personification of power remain contested issues in the ethnography of Melanesia, and the big-man concept does not easily, as Hirsch himself indicates, lend itself to immediate application in the world of biopolitics.

Some of the commentators (Barker, Metspalu, Nelkin, Simpson) bring the big-man of Icelandic bioscience to the fore. Simpson argues that "an intriguing feature" of our article is that we make "no mention" of Stefánsson, pointing out that his role has taken on "saga-like dimensions" and that this is probably one of the reasons for the "appeal" of deCode Genetics. We would point out that while Stefánsson was at the center of the storm around the company and the biogenetic project in the beginning, later on his person moved to the margin of the op-ed pieces that we analyzed. While his "saga-like" dimensions have become clichés in the international media coverage of biogenetic projects, we saw no need to reproduce them in our piece. Many of the contributors to the international media coverage and some bioethical, legal, and social science commentators have been seduced by the Viking rhetoric of deCode Genetics, which has prevented an adequate understanding of the context and issues involved. The heroic notion of the "skipper effect" (Pálsson 1994) seems just as problematic in the domain of biotechnology as in the case of fisheries. In our view, skill, authorship, and creativity are properties of "dividual" persons and communities of practice, not autonomous individuals.

Hornborg and Sharp urge us to consider the comparison with the case of Tonga. We agree that much of the fascination with the biogenetic projects and genealogies of Iceland, Tonga, and other islands stems from the fact they represent relatively small, and isolated populations, apparently appropriate for genome mapping and research because of their presumed genetic homogeneity. The very notion of a gene "pool" seems to engender a cosy image of boundedness and unity, not a complex and fleeting *ocean* of genes, to draw upon the imagery introduced by Helmreich. The island imagery does, indeed, entail certain risks (Proctor), including those of essentialist representation, eugenic ideology, and biopiracy. The perception of risk, however, depends on the context. Thus, German fears of the reemergence of Nazi eugenics manifest themselves in an exceptionally passionate and critical press (Gross).

Human blood often has a particular place in the moral landscape of bodily commodities; not only does it frequently occupy a central symbolic role but it is usually the source of the DNA used in genetic analyses. Thus, the extraction of blood is frequently subjected to the constraints of altruism (Metspalu). For some Tongan people (Hornborg) and in fact many others, blood is a sanctified substance that can be exchanged only within the framework of the gift. The purity of the gift, however, remains contested, as many anthropologists have emphasized. In the Icelandic case, some donors are more equal than others (Harðoardóttir 2002); for example, blood donation requires at least two years of residence in the country.

In the genome era, as several of the commentators emphasize, the human body has been progressively reimagined. Not only have people increasingly identified with their bodies, appreciating the fact that they both have and *are* bodies and thereby challenging the dualism of nature and society (Hornborg) but the drafting and mapping of the human genome represent unprecedented fragmenting and commodification of the body (Sharp). In the process, kinship has been medicalized in the sense that each of us is, to some extent, reduced to a perpetual patient, a carrier of potentially faulty genes (Finkler). The nature and implications of these developments need to be carefully explored by social scientists faithful to both human needs and the realities of the genome era, including people's accounts and experiences (Jackson n.d.). We agree with Finkler that it is important to explore the micro perspectives of patient groups and control groups.

Here, anthropology has a crucial role to play, sensitive as it is to particular sites, everyday experience, and holistic perspectives. What each of us makes of the "facts" of genomics and biomedicine is an empirical question that needs careful and systematic attention. In the long run, those of us who happen to enjoy the luxury of following the exotic developments of biomedicine have to take some kind of stance regarding what the key issues are and how they should be tackled; academic freedom entails both the obligation and the right to speak out on public concerns, bearing responsibility as both citizens and scholars. As Donne might have it, "Ask not for whom the cell tolls, it tolls for you!"

Acknowledgements

An early draft of this article was presented to the session "Jurisdictions of Authority and Expertise in Science and Medicine" at the annual meeting of the American Anthropological Association in San Francisco in 2000. We thank the organizer, Linda F. Hogle, as well as the discussants, Laura Nader and Lesley A. Sharp, for their comments. Moreover, we thank the editor, Benjamin S. Orlove, and five anonymous reviewers for their detailed reading of the manuscript and excellent suggestions for revisions. In writing the paper we have also benefited from discussions with a number of colleagues, including Anna Birna Almarsdóttir (Reykjavík), Eeva Berglund (London), Eric Hirsch (London), Stefan Helmreich (New York), Klaus Høyer (Copenhagen), Jane Kaye (Oxford), Susanne Lundin (Lund), Paul Rabinow (Berkeley), Sidsel Roalkvam (Oslo), Halldór Stefánsson (Heidelberg), Esa Väliverronen (Helsinki), and Lynn Åkesson (Lund). The research on which the paper is based is part of a larger Nordic project, "Bodily Commodities: Concerns about Biotechnology and Market Exchange," coordinated through the University of Iceland. The project has been supported by several funds, including the Nordic Committee for Social Science Research, the Icelandic Science Fund, and the University of Iceland.

Notes

1 Icelanders, however, are no genetic Robinson Crusoes. Throughout its written history, Iceland has regularly been visited by slaves, pirates, fishermen, and travelers. Recent research in biological anthropology, based on DNA analysis and demographic studies, indicates that while Icelanders are more homogeneous than most other European populations they nevertheless contain a surprising amount of genetic diversity for a small island population (see Helgason et al. 2000).

2 Thus, Richard Lewontin argued in an opinion piece in the *New York Times* (1999) that the Icelandic project amounted to the "selling of Icelandic DNA."

3 One commentator, Hannesson (1937: 225), a well-known physician, added in an exceptionally favorable review of Kvaran's work: "Now we *know* for sure that the way of thinking of the ancient past is in most respects correct and beneficial for a nation. . . . Now this issue receives great attention for most civilized nations, and it is even the core of the political thought of the Germans. Genetics has now provided full proof for most of the issues involved, but our ancient scripts have also provided guidelines in several respects."

References

ANDERSON, W. 2000. The possession of kuru: Medical science and biocolonial exchange. *Comparative Studies in Society and History* 42:713–44. [EH]

AGAMBEN, G. 1995. *Homo sacer: Sovereign power and bare life*. Standford: Standford University Press.

ANDREWS, L., AND D. NELKIN. 2001. *Body bazaar: The market for human tissue in the biotechnology age*. New York: Crown

APPADURAI, A. 1986. *The social life of things*. Cambridge: Cambridge University Press.

BARTH, FREDERICK. 1966. *Models of social organization*. Royal Anthropological Institute Occasional Paper 23. [SH]

BIAGIOLI, M. 1998. The instability of authorship: Credit and responsibility in contemporary biomedicine. *FASEB Journal* 12: 3–16. [EH]

BLOCH, M., AND J. PARRY. Editors. 1989. *Money and the morality of exchange*. Cambridge: Cambridge University Press.

BOHANNAN, PAUL, AND LAURA BOHANNAN. 1968. *Tiv economy*. London: Longmans, Green. [SH]

BOUQUET, M. 1996. Family trees and their affinity: The visual imperative of the genealogical diagram. *Journal of the Royal Anthropological Institute* 2:43–66.

BOYLE, J. 1996. *Shamans, software, and spleens: Law and the construction of the information society*. Cambridge: Harvard University Press.

BRODWIN, P. E. 2000. "Introduction," in *Biotechnology and culture: Bodies, anxieties, ethics*. Edited by P. E. Brodwin, pp. 1–23. Bloomington: Indiana University Press. [LAS]

CAMPBELL, C. S. 1992. Body, self, and the property paradigm. *Hastings Center Report*, September–October, pp. 34–42.

CARRIER, J. G. 1998. "Introduction," in *Virtualism: A new political economy*. Edited by J. G. Carrier and D. Miller. Oxford: Berg.

CARSTEN, J. Editor. 2000. *Culture and relatedness: New approaches to the study of kinship*. Cambridge: Cambridge University Press.

CLARKE, A., AND E. PARSONS. Editors. 1997. *Culture, kinship, and genes: Towards cross-cultural genetics*. London: Macmillan.

CSORDAS, T. Editor. 1994. *Embodiment and experience: The existential ground of culture and self*. Cambridge: Cambridge University Press.

DEZALAY, Y., AND B. GARTH. 1996. *Dealing in virtue: International commercial arbitration and the construction of a transnational legal order*. Chicago: University of Chicago Press.

DURRENBERGER, E. P., AND G. PÁLSSON. 1999. "The importance of friendship in the absence of states, according to the Icelandic sagas," in *The anthropology of friendship: Beyond the community of kinship*. Edited by S. Bell and S. Coleman, pp. 59–97. Oxford: Berg.

EDWARDS, J., S. FRANKLIN, E. HIRSCH, F. PRICE, AND M. STRATHERN. Editors. 1999. *Technologies of procreation: Kinship in the age of assisted conception*. London: Routledge.

FINNBOGASON, G. 1922. Mannkynbætur (Racial improvement). *Andvari* 47:184–204.

FINKLER, K. 2000*a*. *Experiencing the new genetics: Family and kinship on the new medical frontier*. Philadelphia: University of Pennsylvania Press.

——. 2000*b*. Diffusion reconsidered: Variation and transformation in biomedical practice, a case study from Mexico. *Medical Anthropology* 19:1–39. [KF]

——. 2001. The kin in the gene: The medicalization of family and kinship in American society. CURRENT ANTHROPOLOGY 42: 235–63. [KF]

FORTUN, M. 2001. Mediated speculations in the genomics futures markets. *New Genetics and Society* 20:139–56. [BS]

FOUCAULT, M. 1973. *The birth of the clinic: An archaeology of medical perception*. Translated by A. M. Sheridan. London: Tavistock.

FOX, R. C., AND J. P. SWAZEY. 1992. *Spare parts: Organ replacement in American society*. Oxford: Oxford University Press.

FRANKLIN, S., AND H. RAGONÉ. Editors. 1998. *Reproducing reproduction: Kinship, power, and technological innovation*. Philadelphia: University of Pennsylvania Press.

GALLAGHER, N. L. 1999. *Breeding better Vermonters: The Eugenics Project in the Green Mountain State*. Hanover: University Press of New England.

GEE, J. 1999. *Introduction to discourse analysis: Theory and method*. London: Routledge.

GINDHART, PATRICIA. 1989. An early twentieth-century skeleton collection. *Journal of Forensic Sciences* 34:887–93. [LMM]

GLASNER, P., AND H. ROTHMAN. Editors. 1998. *Genetic imaginations: Ethical, legal, and social issues in human genome research.* Aldershot: Ashgate.

GOLD, E. R. 1996. *Body parts: Property rights and the ownership of human biological materials.* Washington, D.C.: Georgetown University Press.

GREELY, H. T. 1998. Legal, ethical, and social issues in human genome research. *Annual Reviews in Anthropology* 27: 473–502.

GRUBB, A. 1998. "I, me, mine": Bodies, parts, and property. *Medical Law International* 3:299–17.

GUDEMAN, S. 2001. *The anthropology of economy.* Oxford: Blackwell.

GULCHER, J. R., AND K. STEFÁNSSON. 2000. The Icelandic Healthcare Database and informed consent. *New England Journal of Medicine* 342:1827–30.

HANN, C. M. 1998. *Property relations: Renewing the anthropological tradition.* Cambridge: Cambridge University Press.

HANNESSON, G. 1924. Norræna kynið (The Nordic race). *Andvari* 49:140–63.

——. 1937. Ritfregnir ("News"). *Skírnir* 111:222–25.

HARAWAY, D. 2000. "Deanimations: Maps and portraits of life itself," in *Hybridity and its discontents: Politics, science, culture.* Edited by A. Brah and A. E. Coombes. London: Routledge.

HARDARDÓTTIR, K. E. 2002. "Þeim rennur blóðið til skyldunnar": Blóðgjafir Íslendinga í mannfræðyilegu ljósi (Blood and the "obligation" to give: Anthropological perspectives on the exchange of blood in Iceland). M.A. thesis, Department of Anthropology, University of Iceland, Reykjavik, Iceland.

HASTRUP, K. 1998. *A place apart: An anthropological study of the Icelandic world.* Oxford: Clarendon Press.

HAYDEN, C. P. 1998. "A biodiversity sampler for the millennium," in *Reproducing reproduction: Kinship, power, and technological innovation.* Edited by S. Franklin and H. Ragoné, pp. 173–206. Philadelphia: University of Pennsylvania Press. [LAS]

HAYLES, N. K. 1999. *How we became posthuman.* Chicago: University of Chicago Press.

HELGASON, A. 2001. The ancestry and genetic history of Icelanders: An analysis of mtDNA sequences, Y chromosome haplotypes, and genealogy. Ph.D. diss., Institute of Biological Anthropology, University of Oxford, Oxford, U.K.

HELGASON, A., AND G. PÁLSSON. 1997. Contested commodities: The moral landscape of modernist regimes. *Journal of the Royal Anthropological Institute* 3:451–71.

——. 1998. "Cash for quotas: Disputes over the legitimacy of an economic model of fishing in Iceland," in *Virtualism: The new political economy.* Edited by J. Carrier and D. Miller. Oxford: Berg.

HELGASON, A., S. SIGURDARDARDÓTTIR, J. R. GULCHER, R. WARD, AND K. STEFÁNSSON. 2000. MtDNA and the origin of the Icelanders: Deciphering signals of recent population history. *American Journal of Human Genetics* 66:999–1016.

HIRSCH, E., AND M. O'HANLON. Editors. 1995. *The anthropology of landscape: Perspectives on place and space.* Oxford: Clarendon Press.

HOGLE, L. F. 1999. *Recovering the nation's body: Cultural memory, medicine, and the politics of redemption.* New Brunswick: Rutgers University Press.

HOLTZ, WILLIAM. 1967. Bankrobbers, burkers, and bodysnatchers. *Michigan Quarterly Review* 6(2):90–98. [LMM]

HØYER, K. 2001. Blod, aktier, og "etik." M.A. diss., Institute of Anthropology, University of Copenhagen, Copenhagen, Denmark.

INGOLD, T. 2000. *The perception of the environment: Essays in livelihood and dwelling and skill*. London: Routledge.

INGSTAD, B., AND S. REYNOLDS WHYTE. Editors. 1995. *Disability and culture*. London: Routledge.

JACKSON, M. 1989. *Paths toward a clearing: Radical empiricism and ethnographic inquiry*. Bloomington: Indiana University Press.

——. n.d. Familiar and foreign bodies: A phenomenological exploration of the human-technology interface. *Journal of the Royal Anthropological Institute* (incorporating *Man*). In press.

JAWORSKI, A., AND N. COUPLAND. Editors. 1999. *The discourse reader*. London: Routledge.

JOHNSON, JEFFREY. 1996. "Maritime anthropology," in *Encyclopedia of cultural anthropology*. Edited by David Levinson and Melvin Ember, pp. 726–28. New York: Henry Holt. [SH]

JORALEMON, D. 1995. Organ wars: The battle for body parts. *Medical Anthropological Quarterly* 9:335–56.

KALINOE, L. 2001. Expressions of culture: A cultural perspective from Papua New Guinea. Paper presented to the subregional workshop on intellectual property, genetic resources, and traditional knowledge, organized by the World Intellectual Property Organization in cooperation with the Government of Australia, Brisbane, Australia, June 25–27. [EH]

KAY, L. E. 2000. *Who wrote the book of life? A history of the genetic code*. Stanford: Stanford University Press.

KAYE, J., AND P. MARTIN. 2000. Safeguards for research using large-scale DNA collections. *British Medical Journal* 321: 1146–49.

KELLER, E. F. 1995. *Refiguring life: Metaphors of twentieth-century biology*. New York: Columbia University Press.

——. 2000. *The century of the gene*. Cambridge: Harvard University Press.

KLEINMAN, A. 1995. *Writing at the margin: Discourse between anthropology and medicine*. Berkeley: University of California Press.

KNOPPERS, B. M. 1999. Status, sale, and patenting of human genetic material: An international survey. *Nature Genetics* 22: 23–26.

KOMESAROFF, P. A. Editor. 1995. *Troubled bodies: Critical perspectives on postmodernism, medical ethics, and the body*. Durham: Duke University Press.

KOPYTOFF, I. 1986. "The cultural biography of things: Commoditization as process," in *The social life of things*. Edited by A. Appadurai. Cambridge: Cambridge University Press.

KVARAN, E. 1934. Um mannfræðilegt gildi forníslenskra mannlýsinga (On the anthropological significance of bodily description in medieval Iceland). *Skírnir* 108:83–101.

LEWONTIN, RICHARD. 1999. People are not commodities. *New York Times*, January 23.

LINDEBAUM, S., AND M. LOCK. Editors. 1993. *Knowledge, power, and practice: The anthropology of medicine and everyday life*. Berkeley: University of California Press.

MCCAY, B. M., AND J. M. ACHESON. Editors. 1987. *The question of the commons: The culture and ecology of communal resources*. Tucson: University of Arizona Press.

MCCLOSKEY, D. N. 1993. "Some consequences of a conjective economics," in *Beyond economic man: Feminist theory and economics*. Edited by M. Ferber and J. A. Nelson, pp. 69–91. Chicago and London: University of Chicago Press.

MARCUS, G. 1998. *Ethnography through thick and thin*. Princeton: Princeton University Press.

MARSHALL, P. A., D. C. THOMASMA, AND A. S. DAAR. 1996. Marketing human organs: The autonomy paradox. *Theoretical Medicine* 17:1–18.

MARTIN, E. 1989. *The woman in the body: A cultural analysis of reproduction*. Boston: Beacon Press.

MENCKEN, H. L. 1940. *Happy days, 1880–1892*. New York: Alfred A. Knopf. [LMM]

MOORE, A. D. 2000. Owning genetic information and gene enhancement techniques: Why privacy and property rights may undermine social control of the human genome. *Bioethics* 14: 97–119.

NELKIN, D., AND L. ANDREWS. 1998. Homo economicus: Commercialization of body tissue in the age of biotechnology. *Hastings Report* 20(5):30–39.

NIELSEN, T. H., A. MONSEN, AND T. TENNØE. 2000. *Livets tre og kodens kode: Fra genetikk til bioteknologi, Norge 1900–2000* (The tree of life and code of codes: From genetics to biotechnology, Norway 1900–2000). Oslo: Gyldendal Akademisk.

ORLOVE, B. S. 1991. Mapping reeds and reading maps: Representation in Lake Titicaca. *American Anthropologist* 18:3–40.

PÁLSSON, G. 1991. *Coastal economies, cultural accounts: Human ecology and Icelandic discourse*. Manchester: Manchester University Press. [SH]

——. 1994. Enskilment at sea. *Man* 29(4):901–27.

——. 1995. *The textual life of savants: Ethnography, Iceland, and the linguistic turn*. Chur: Harwood Academic Publishers.

——. n.d. The life of family trees and the "Book of Icelanders." MS.

PÁLSSON. G, AND A. HELGASON. 1996. "The politics of production: Equity, efficiency, and enclosure," in *Images of contemporary Iceland: Everyday lives and global contexts*. Edited by G. Pálsson and E. Durrenberger, 60–86. Iowa City: University of Iowa Press. [EH]

PÁLSSON, G., AND P. RABINOW. 1999. Iceland: The case of a National Human Genome Project. *Anthropology Today* 15(5): 14–18.

. 2000. Islande: Le cas deCode (Iceland: The case of de-Code). *Biofutur: Le mensuel Européen de biotechnologie* 206 (December):108–11.

——. 2001. The Icelandic genome debate. *Trends in Biotechnology* 19:166–71.

——. n.d. The Iceland controversy: Reflections of the trans-national market of civic virtue. MS.

PEACE, A. 1997. *A time of reckoning: The politics of discourse in rural Ireland*. Institute for Social and Economic Research Social and Economic Studies 59.

POLANYI, K. 1944. *The great transformation*. Boston: Beacon Press.

RABINOW, P. 1996*a*. *Essays on the anthropology of reason*. Princeton: Princeton University Press.

——. 1996*b*. *Making PCR: A story of biotechnology*. Chicago: University of Chicago Press.

——. 1999. *French DNA: Trouble in purgatory*. Chicago: University of Chicago Press.

RADIN, M. J. 1996. *Contested commodities*. Cambridge: Harvard University Press.

RAPP, R. 1999. *Testing women, testing the fetus: The social impact of amniocentesis in America*. London and New York: Routledge.

REARDON, J. 2001. The Human Genome Diversity Project: A case study in coproduction. *Social Studies of Science* 31: 357–88.

RHEINBERGER, H.-J. 1995. Beyond nature and culture: A note on medicine in the age of molecular biology. *Science in Context* 8(1):249–63.

——. 2000. "Gene concepts," in *The concept of the gene in development and evolution*. Edited by P. Beurton, R. Falk, and H.-J. Rheinberger. Cambridge: Cambridge University Press.

ROSE, H. 2001. *The commodification of bioinformation: The Icelandic Health Sector Database*. London: The Wellcome Trust.

SCHEPER-HUGHES, N. 2000. The global traffic in human organs. CURRENT ANTHROPOLOGY 41:191–224.

SEN, A. 1987. *On ethics and economics*. Oxford: Basil Blackwell.

SHARP, L. 2000. The commodification of the body and its parts. *Annual Review of Anthropology* 29:287–328.

SHIVA, V. 1997. *Biopiracy: The plunder of nature and knowledge*. Boston: South End Press.

SHORE, C., AND S. WRIGHT. 1997. "Introduction," in *Anthropology and policy: Critical perspectives on governance and power*. Edited by C. Shore and S. Wright. London: Routledge.

SIGURDSSON, S. 2001. Yin-yang genetics, or the HSD deCode controversy. *New Genetics and Society* 20:103–17. [BS]

SIMPSON, B. 2000. Imagined genetic communities. *Anthropology Today* 16(3):3–6.

——. 2001. Making connections: Reproductive choice and the new technologies in Sri Lanka. Paper presented at the Association of Social Anthropologists Conference "Rights, Claims, and Entitlements," April. [BS]

STARR, D. 1998. *Blood: An epic history of medicine and commerce*. London: Little, Brown.

STRATHERN, M. 1992. *After nature: English kinship in the late twentieth century*. Cambridge: Cambridge University Press.

——. 1998. "Divisions of interest and languages of ownership," in *Property relations: Renewing the anthropological tradition*. Edited by C. M. Hann, pp. 214–32. Cambridge: Cambridge University Press.

——. 1999. *Property, substance, and effect: Anthropological essays on persons and things*. London: Athlone Press.

THOMPSON, A. K., AND R. F. CHADWICK. Editors. 2000. *Genetic information: Acquisition, access, and control*. New York: Kluwer Academic/Plenum Publishers.

TITMUSS, R. M. 1997(1970). *The gift relationship: From human blood to social policy*. London: LSE Books.

WIKLER, D. 1999. "Can we learn from eugenics?" in *Genetic information: Acquisition, access, and control*. Edited by A. K. Thompson and R. F. Chadwick, pp. 1–16. New York: Kluwer Academic/Plenum Publishers.

WILKINSON, S., AND E. GARRARD. 1996. Bodily integrity and the sale of human organs. *Journal of Medical Ethics* 22: 334–39.

YANAGISAKO, SYLVIA, AND CAROL DELANEY. 1995. "Naturalizing power," in *Naturalizing power: Essays in feminist cultural analysis*. Edited by Sylvia Yanagisako and Carol Delaney, pp. 1–22. New York: Routledge. [SH]

44

INFORMED CONSENT

The contradictory ethical safeguards in pharmacognetics

Oonagh Corrigan

Source: O. Corrigan and R. Tutton (eds), *Genetic Databases: Socio-Ethical Issues in the Collection and Use of DNA*, London: Routledge, 2004, pp. 78–96.

Introduction

Since the late 1990s, an increasing number of DNA samples are being collected from patients and healthy volunteers participating in clinical drug trials, to be used for genetics research. This type of research is being carried out globally, usually at the instigation of large multinational pharmaceutical companies, many of which are building extensive genetic databases. While there has been a great deal of attention to and debate surrounding the collection of DNA samples for use in public genetic databases, there has been virtually no discussion of the formation of such databases being compiled on behalf of or by the pharmaceutical industry (see Lewis this volume). This is somewhat surprising given that much of the debate surrounding public databases concerns their potential commercial exploitation (Marks and Steinberg 2002). The systematic collection of samples by industry is linked to what is known broadly as pharmacogenomic research and first began to appear in the form of 'add on' studies to the main drug trial protocol in the late 1990s. 'Pharmacogenomics' is the name given to a broad-based pharmaceutical industry-led initiative, based on developments brought about by the Human Genome Project which aims to capitalize on this knowledge base to discover new therapeutic targets and interventions and to elucidate the constellation of genes that determine the efficacy and toxicity of specific medications. 'Pharmacogenetics' is a more specific term used to define the narrower spectrum of inherited differences in drug metabolism and disposition linked to individual genetic variations. The pharmaceutical industry is currently focusing their research efforts on the latter (using DNA collected during clinical trials), while also storing samples longer term for as yet unspecified pharmacogenomics research. Many large pharmaceutical companies are now engaged in genomic-based research and making substantial financial investment in this area.[1] Industry-sponsored clinical drug trials take place on a global scale, with perhaps several thousand patients worldwide simultaneously taking part in any single clinical trial.[2] Drug development is an extremely time consuming and costly operation, taking 10–15 years and costing an estimated US$300–600 million with a success rate of about 1 in 10, only a fraction of which will be 'blockbusters'. Development of new blockbuster drugs is becoming increasingly difficult (Horrobin 2000) and pharmacogenomics promises revolutionary forms of drug development with substantial financial gains for the industry:

> Industry analysts predict that, by improving medical outcomes by the use of pharmacogenomics-enhanced drugs and diagnostics, pharmaceutical companies could benefit to the order of US$200 million to US$500 million in extra revenue for each drug. . . . For these reasons, pharmaceutical companies have begun to integrate pharmacogenomics into drug development programs.
>
> (Bogdanovic and Langlands 1999: 182)

The collection of blood samples for pharmacogenomics-related studies first began to be added to clinical drug trials during the late 1980s. While it is difficult to get an accurate assessment of the scale of such research (see Lewis this volume) a company director of one of Europe's main laboratories conducting pharmacogenetic tests informed me that in the year 2000 his company had handled 12,000 samples, half of which were from UK drug companies and that there are at least two other such laboratories operating in Europe. Furthermore, he hinted at the global scale of these tests:

> [W]e [the Clinical Research Company that owns this laboratory] don't just do it for Europe, we do it on a global basis. So we've got a lab in the States, we've got a lab in Australia in Sydney, another lab in Cape Town South Africa, and a lab in Singapore.

Some pharmaceutical companies are taking quite a targeted approach and are engaged in collecting samples from patients in specific clinical trials. For example, one of the world's largest pharmaceutical companies is targeting selected conditions such as metabolic disorders and during an interview with one of the clinical genetic scientists from this company I was informed of genetic tests being carried out on a particular multinational trial for a drug to treat type II diabetes involving 11 trials in Europe and the US and 4000 patients. What distinguishes this type of genetic data collection from other forms of regional or national databases is its global dimension and its lack of public knowledge of or accountability for research carried out on commercial databases. The numbers of patients enrolled in such trials worldwide are increasing and a recent article published in the *New England Journal of Medicine* urges the collection of 'genomic DNA from all patients enrolled in clinical drug trials, along with appropriate consent to pharmacogenetic studies' (Evans and McLeod 2003: 547).

The collection of blood samples for this form of research has been facilitated by the pre-existing ethical regulatory mechanisms that govern the conduct of conventional clinical drug trials. A range of governance mechanisms are in place, in the form of national and international ethical codes and systems of 'independent' ethical review, and are based on the ethical requirement to obtain the informed consent of research subjects, and weigh the risks and benefits of participation in research. Informed consent in particular is an ideal that has gained increasing salience within Western healthcare policy, having become progressively something of an ethical panacea since it first emerged over fifty years ago in international guidelines governing biomedical research (Corrigan 2003). In pharmacogenomics-related clinical drug trials, consent is first sought from subjects to take part in the main clinical drug trial and, if they agree, they are subsequently asked to consent to the additional genetic aspect of the trial. In addition, subjects are informed that the sponsoring pharmaceutical company has sole property rights over the sample and the research results. This chapter, based on research supported by the UK's Wellcome Trust, examines the social and ethical implications of the development of pharmacogenetics in clinical trials and focuses on the issue of consent to these trials and the problems and contradictions surrounding this process.

The origins and compulsion of informed consent

> Informed consent is based on the principle that competent individuals are entitled to choose freely whether to participate in research. Informed consent protects the individual's freedom of choice and respects the individual's autonomy.
>
> (Council for International Organizations of Medical Sciences 1993)

During the past fifty years, the principle of informed consent, which has its roots in Western medicine and liberal theory (Beauchamp and Childress 1989, Burgess 2001) has become a cornerstone of contemporary biomedical research ethics, and been enshrined in a multitude of national and international ethics guidelines supported by mechanisms to ensure its adherence. Bioethicists writing on the historical emergence and development of clinical research ethics (Gray 1975, Faden and Beauchamp 1986, Beauchamp and Childress 1989, Annas and Grodin 1992, Katz 1992) point to a number of significant events occurring between the years preceding the Second World

War and the end of the 1970s, which they claim have raised ehtical concerns among those within the medical field and beyond. Furthermore, they claim that such concern has prompted the implementation of new policies, guidelines, and the general proliferation of a bioethics literature as now exists, to ensure ethical practice in clinical research involving human subjects. The field of research ethics in this context is constructed as progressive, concerned with checking the excesses of medical science and protecting the individual research subject from exploitation and harm.

The first critical event often cited is the emergence of evidence about horrific scientific and medical experiments conducted by Nazi physicians and scientists on prisoners during the Second World War. The *Nuremberg Code* (1947), constituted in direct response to the exposure of these details during the trials of the Nazi doctors, at the end of the war, was the first major international ethical statement to stipulate principles of biomedical research. It consists of ten principles, the first of which stipulates that 'the voluntary consent of the human subject is absolutely essential'; this principle has become the primary ethical consideration in all biomedical research (Annas and Grodin 1992). A year after the *Nuremberg Code*, a more general response to the inhumanities of Nazism was established in the *Universal Declaration of Human Rights*. The requirement of informed consent was declared a universal human right, grounded in the fundamental dignity and worth of every individual and supported by respect for the liberty and security of the person. In 1954, the World Medical Association (WMA) adopted a set of ethical principles that went beyond the *Nuremberg Code*, and further extended the principle of informed consent to include surrogate consent for incapacitated subjects. The WMA followed this in 1964 with the *Declaration of Helsinki*, a more comprehensive set of guidelines, which, while further emphasizing the principle of consent, also included a distinction between therapeutic and non-therapeutic research. Nevertheless, despite the production of these and other national professional guidelines, in practice, informed consent procedures were not initially generally implemented or adhered to consistently.

The problem of non-compliance was highlighted by two whistle-blowers, Henry Beecher in the US and Maurice Pappworth in the UK, who produced documented evidence of cases where patients had been subject to medical research without their knowledge or consent (Beecher 1966, Pappworth 1967). Beecher described the now renowned Tuskegee case, a study lasting over forty years (1932–1972) and involving nearly four hundred black men from Tuskegee, Alabama, suffering from syphilis. These men were told they were being treated for 'bad blood' and enrolled in the study to observe the manifestations of syphilis. However, the study team went to great lengths to ensure that these men were not informed of the true nature of their condition or given penicillin, which by 1947 had been shown to cure the disease. The publication of this and other cases revealed by Beecher and

Pappworth prompted implementation of further protective mechanisms to ensure adherence to informed consent procedures.[3]

During the 1970s, research ethics committees in the UK and 'institutional review boards' (IRBs) in the US were established to oversee research programmes and enforce compliance with international and professional ethics guidelines. The principal mandate of these committees was and continues to be the review of proposals to carry out research on patients or healthy volunteer subjects within the medical environment. These bodies act as gatekeepers to safeguard the welfare of subjects in clinical trials, ensure that prior written informed consent is obtained from patients, and that risk to subjects involved in biomedical research is minimized. Research ethics guidelines have continued to be updated over the last 30 years, and new ones have emerged to meet the changing terrain of medical research. The extent to which research ethics has been internalized by the commercial side of science and medicine is evidenced in a pharmaceutical industry-led initiative which has produced the most recent set of international policy guidelines promoting ethical standards in the context of clinical drug research (ICH 1998).[4] These guidelines promote an international ethical and scientific quality standard for designing, conducting, recording and reporting trials that involve the participation of human subjects, and claim consistency with the principles that have their origin in the *Declaration of Helsinki*. Great emphasis is again given to the concept of written 'fully informed consent', as well as specific stipulations about the nature of information that is mandatory for the researcher to disclose to the subject.

Interestingly, while ethics guidelines draw a distinction between therapeutic and non-therapeutic research, no distinction is drawn between commercial and non-commercial research. Such guidelines have emerged in almost complete absence of any real debate about the particular ethical issues that arise from pharmaceutical or other commercially sponsored clinical trials. Despite a number of recent articles published in the medical and bioethics literature highlighting the potential conflicts of interests of pharmaceutical-sponsored research (Steflox *et al.* 1998, Glass and Lemmens 1999), with the exception of guidance recommending that healthy subjects in taking part and physicians conducting such research are not unduly financially rewarded, there is at present no formal guidance for research ethics committees regarding such research.

Calculations of risk and benefit

The signficance awarded to the principle of informed consent in biomedical research cannot be overstated. However, as I have already mentioned, the governance mechanisms in biomedical research also draw on a risk–benefit calculation. Research ethics committees are required both to scrutinize informed consent procedures and to ensure there is a correct balance between the potential risks and benefits. The following extract from the newly updated *Declaration of Helsinki* illustrates the kinds of risk–benefit considerations that those involved in conducting clinical research are required to adhere to:

> Every medical research project involving human subjects should be preceded by careful assessment of predictable risks and burdens in comparison with foreseeable benefits to the subject or to others. . . . Physicians should abstain from engaging in research projects involving human subjects unless they are confident that the risks involved have been adequately assessed and can be satisfactorily managed. . . . Medical research involving human subjects should only be conducted if the importance of the objective outweighs the inherent risks and burdens to the subject. . . . Medical research is only justified if there is a reasonable likelihood that the populations in which the research is carried out stand to benefit from the results of the research.
>
> (World Medical Association 2000)

Furthermore, most guidelines also draw a distinction between therapeutic and non-therapeutic research, highlighting the necessity to ensure that in non-therapeutic research, risks to subjects are minimized. This is premised on the notion that an individual taking part in a therapeutic study at least has the chance of benefiting by receiving treatment, and thus it may be deemed reasonable for the individual to be subjected to a higher degree of risk than a subject taking part in a non-therapeutic study where there is no such potential benefit. For example, a subject's involvement in pharmacogenetics research is non-therapeutic and therefore, as I will subsequently discuss, has implications for the ways in which risk is assessed. Current European legislation requires research ethics committees to ensure that written information given to the prospective research subject includes explanations of twenty different aspects relating to the subject's involvement in the trial (ICH 1998). These include explanations of the 'purpose of the trial', the trial procedures to be followed, the 'reasonably foreseeable risks' or inconvenience to the subject, and responsibilities the subject undertakes as a result of involvement in the trial. Nevertheless, while physicians and research ethics committees are required to ensure that risks and benefits are carefully assessed, ultimately it is the prospective research subjects themselves who are obliged to assess the risks and benefits.

> When a patient gives consent to participate in a piece of research he should have been given enough information to make his own choice. The concept of risk/benefit analysis should be conveyed to patients as part of the procedure of seeking consent.
>
> (Royal College of Physicians 1990:11)

Although research ethics committees have the dual task of scrutinizing informed consent procedures and ensuring that there is a correct balance between the potential risks and benefits, in practice most research ethics committees in the UK and Institutional Review Boards (IRBs) in the US give more priority and attention to informed consent procedures rather than the analysis of risks and potential benefit.[5] In effect, research ethics committees manage risk–benefit considerations by facilitating the process of informed consent and ensuring that sufficient written information is given to the patient or healthy volunteer subject, and includes details of the anticipated risks and benefits so that the subject may assess the risk and choose whether or not to participate.

The establishment of international and national ethical codes of practice such as the major ones I have highlighted, and the policies and practices that have arisen to ensure their implementation, are most often understood as a progressive response to the growing awareness of the need for regulation to protect subjects in clinical research. The public revelation of the atrocities committed by the Nazis, the cases exposed by Beecher and Pappworth, and the growing attention placed on human rights issues during the 1960s have all been cited as major contributory factors in the emergence of regulatory practices and policies on clinical research. Couched within the discourse of human rights and liberal principles based on respect for autonomy, the concept of informed consent is presented in regulatory guidelines and has been adopted in health research policy as the optimal ethical solution to mitigate the dangers of exploitation and coercion of the research subject. Patient advocacy groups have also welcomed the growing recognition of the need for informed consent. For example, the UK charity patient group Consumers for Ethics in Research (CERES) actively promotes the right to informed consent and inclusion of research subjects as 'partners' or 'participants' in clinical research (CERES 1999).

We can see then that the various techniques and procedures encompassed within the bioethical framework governing research rely on the constitution of an autonomous free thinking, rational subject who ultimately bears the responsibility for assessing risks and benefits. The mechanisms designed to protect subjects from harm and exploitation do so by 'empowering' subjects and giving them the right to make informed decisions about research participation. However, the right to autonomy comes with the concomitant obligation to make a rational choice about whether to consent, having first considered the information and made an assessment of the potential risks and benefits. And there is a parallel obligation on the part of researchers to provide sufficient information for subjects to make an informed decision. I suggest that the informed consent process can be understood as a process that in effect constructs subjects as 'biological citizens' who have rights to be informed about biomedical research, and who simultaneously have obligations to make informed, reflective choices.

The notion of active citizenship has become increasingly prominent in political discussions and policy practices in Britain in the past 15 years (Marinetto 2003). This is seen as a significant development insofar as conventional practices of modern liberal democracies have been based on a representative mode of government in which the wider citizenry has a more passive role. However, we need to acknowledge that such participatory forms of active citizenship place an onus upon subjects to engage with the informed consent process. Although citizens can opt-out of this obligation – either by refusing to engage in this process and decline trial participation, or by accepting study participation based on 'blind trust' or in deference to medical authority – in the context of participatory citizenship such an action is seen as morally 'invalid'. Biological, scientific and technological citizenship requires that valid public acceptance must be deep and informed, subjective, voluntary, rigorous, and rich. Invalid acceptance involves thoughtless deference and blind trust' (Frankenfeld 1992: 460). Underpinning this model of active citizenship is a notion of an autonomous, rational, free thinking moral being.

Governed through freedom

That 'medical progress is based on research which ultimately must rest in part on experimentation involving human subjects' (World Medical Association 2000) is an a priori and, as such, medical research involving human subjects continues to proliferate. Furthermore, the current ethical system premised on informed consent is one that must be understood as most readily fulfilling the prior requirement of neo-liberal society to conduct research for the sake of 'medical progress', while at the same time appealing to an individualistic notion of 'free choice' and the protection of autonomy. The concept of 'freedom' here is understood in conventional liberal political and moral philosophical terms to be a terrain that exists outside the field of power. 'Free will' as understood by Immanuel Kant is the domain of every rational, moral autonomous agent:

> If a rational subject supposed his judgements to be determined, not by rational principles, but by external impulsion, he could not regard these judgements as his own. This must equally be true of practical reason: a rational agent must regard himself as capable of acting on his own rational principles and only so he can regard his will as his own. That is to say, from a practical point of view every rational agent must presuppose his will to be free. Freedom is a necessary prsupposition of all action as well as of thinking.
>
> (Kant 1995 [1948]: 40)

Libertarian philosophies too, such as those of John Stuart Mill, are concerned with the individuality of action and thought, and claim that people should be able to make their own choices and decisions, free from governmental intervention and control. However, I would argue that ideas such as 'freedom' and autonomy should not be understood as politically innocent premises.

Michael Foucault's genealogical approach reveals that power and knowledge are central components in the historical transformation of various 'regimes of truth'. His approach to the study of power rejects the traditional concept as one of an external dominating force possessed by one group over another. He offers instead a radically new concept of power, linking it inexorably to knowledge. In *The Birth of the Clinic*, Foucault (Foucault 1991 [1963]) charts the emergence at the end of the eighteenth century of what he calls the political anatomy of the body. Medicine, he argues, becomes an increasingly important vehicle for the exercise of surveillance and control, concerned with both 'anatomo-politics of the human body' and 'the biopolitics of the population':[6] By this, Foucault means that the technologies of this disciplinary apparatus serve to fabricate discrete individuals and, at the same time, extend the constantly normalizing medical gaze over the whole population. Furthermore, Foucault's historical analysis highlights the ways in which various rationalities of modern government are not mere mechanisms of social control and subordination of the human will but rather operate through the promotion of certain kinds of subjectivities. Bodies become the objects through which these disciplinary mechanisms are exercised, not through overt violence and coercion, but through a micro-politics of discipline (Foucault 1991). Biomedical research involving human subjects is now subjected to regulatory mechanisms to ensure research is no longer based on direct coercion, but through an individual's freely given informed consent. Nikolas Rose (1996) and others who have followed Foucault's line of analysis demonstrate the extent to which governmentality forges the modern embodied individual subject.[7] Rose highlights the way that, in advanced liberal society, the subject is situated in relation to a host of technologies and strategies aimed at governing through the regulated and accountable choices of autonomous agents:

> Advanced liberal rule . . . does not seek to govern through 'society', but through the regulated choices of individual citizens, now constructed as subjects of choices and aspirations to self-actualization and self-fulfilment. Individuals are to be governed through their freedom.
>
> (Rose 1996: 41)

I suggest that the individual human subject of biomedical research has been variously constructed. First, in the period prior to the adoption and implementation of ethical codes of practice, subjects were frequently constructed as passive, irrelevant or even as expendable objects of research who could be coerced or manipulated to take part. Second, the implementation of various protectionist measures and risk–benefit calculations constructs vulnerable subjects, subjects who need protecting from exploitation and harm. Finally, the increasing significance awarded to informed consent constructs subjects as empowered citizens who, given sufficient information, are able to make free, informed, rational and thus moral choices with respect to their participation. I suggest that such constructs obscure the extent to which subjects are treated and experience trial participation as all three. Furthermore, as I will discuss later in this chapter, none of these constructs encompasses the ways in which research subjects provide valuable labour and bodily material for pharmaceutical research, making them co-producers of drug products.

Clinical drug trials and sample collections: a multi-layered consent practice

As I have already indicated, currently blood samples for pharmacogenomics research purposes are being collected during otherwise conventional clinical drug research. Pharmaceutical companies' objectives towards the development of pharmacogenetics-based drug production are currently being pursued through studies of genetic variability in drug response. Pharmacogenetic tests are carried out to find correlations between an individual's drug-metabolizing enzymes and their response to the drug being taken. Pharmaceutical companies are also engaged with a wider range of research using collected DNA samples for the production of potential therapies based on an understanding of the genetics involved in individual susceptibility to disease.

New drugs developed by pharmaceutical companies are required to go through lengthy testing procedures before being granted a licence and made available for the general patient population. Following a period of lab testing including standard animal and *in vitro* tests, the new drug is tested in human subjects during several phases of clinical trials. At this stage, physicians are recruited by the sponsoring pharmaceutical company to act as 'trial investigators' and oversee the study and the administration of the drug according to the predetermined trial protocols. During phase I trials, the drug is tested on human subjects for the first time, and most of theses studies are conducted on healthy volunteer subjects in specially designated clinical trial suites (much like a private hospital ward).[8] Phase I studies are conducted either directly by the sponsoring pharmaceutical company in corporate premises, or the trials are subcontracted to specially designated clinical research organizations (CROs). Healthy volunteer subjects are financially rewarded for allowing their bodies to be used as experimental sites and most individuals taking part in such research do so primarily for economic gain (Corrigan 2000). Further phases of clinical drug trials (phases II to IV) involve greater numbers of subjects and are normally conducted on patients who suffer from the medical condition for which the test drug is an intended treatment. These studies are conducted in regular clinical settings such as hospital outpatient clinics, community clinics or GP surgeries and hospital wards. During these latter phases, patients are generally asked to participate in research by physicians during the course of routine clinical encounters.

Generally, drug trial subjects are asked to consent to three separate aspects of research. They are asked to consent to the main clinical drug trial, and then also to 'add-on' pharmacogenetic research in which they consent to an 'identified' genetic test for research related to drug effect, and a 'non-identified' test for unspecified future research. These three aspects are progressive insofar as giving consent to the second and third aspects is dependent on giving prior consent to the first and second. Patients being asked to take part in clinical drug trials are often given no prior notice about potential research but rather the possibility of participating in research is introduced during clinical consultation. This initial aspect of informed consent involves considerations relevant to participation in the drug trial itself. Subjects are given written and verbal information relating to the known risks of the drugs as well as details about procedures, such as the length of the study, the number of hospital visits required, and blood tests or X-rays associated with the study. In consenting to this first stage, patients agree to take part in a therapeutic drug trial and it is expected, as part of the consent process, that they will consider issues relating to the potential benefits and risks of the trial, including harms such as adverse drug reactions. Research examining the experiences of patients who have given their consent to take part in such trials demonstrates that despite the ethical rhetoric of informed consent based on adequate understanding of the information given to prospective subjects, in practice the reality is often very different. Despite the fact that patients are given information under headings such as 'possible risks' (usually relating to drug-induced side-effects), patients are often unable subsequently to recall the mention of any side-effects (Hassar and Weintraub 1976, Bergler *et al.* 1980, Cassileth *et al.* 1980, Estey *et al.* 1994).

Clinical drug trials are generally randomized so that neither the patient nor the medical team know whether a particular trial subject is taking the study drug, an alternative drug, or in some cases a placebo (a dummy pill). Studies show that trial subjects have poor understanding of the process of randomization in clinical drug trials (Cassileth *et al.* 1980, Jan and DeMets 1981, Snowdon *et al.* 1997), and as a consequence are sometimes under the misapprehension that they will be receiving an active treatment (Corrigan 2003). Furthermore, given the clinical context of these studies, from the perspective of the patient the boundaries between research and treatment are often quite fuzzy. In such circumstances, a request to consent is often interpreted as *guidance* to consent, insofar as patients have pre-existing expectations and norms regarding the role of the physician as a professional offering advice to patients about their best interests. Research in this area demonstrates that this so-called 'therapeutic misconception' is common (Bamberg and Budwig 1992, Dresser 2001). There is also evidence that some patients who consent to take part in clinical drug trials experience such decisions as 'burdensome' (Taylor 1988, Corrigan 2003); my own research indicates this is more likely to be the case for patients who have an acute condition or who are in pain or suffering great anxiety. Nevertheless, while research ethics guidelines recognize that for some patients giving 'adequate' consent might be difficult and such subjects are to be considered as 'vulnerable', these guidelines apply mainly to 'incompetent' subjects such as those who have some form of cognitive impairment (Council for International Organizations of Medical Sciences 1993, World Medical Association 2000). Such guidelines do not therefore address the varying needs of sick but otherwise competent patients. Although there is a growing awareness of problems relating to the practice of informed consent in clinical trials, solutions to such problems focus on improving written information given to patients, making sure that such information is comprehensive, detailing trial procedures and risks likely to be encountered by subjects (ICH 1998). In relation to consent to DNA tissue sample collections, here too problems and their solutions frequently centre on adequacy and comprehensiveness of information with solutions being proffered in terms of suggested comprehensively designed 'model' consent forms (Knoppers 1998, Merz and Sankar 1998). There is scant regard in both conventional bioethics guidelines and more socially nuanced policy perspectives of the need to appreciate more fully the social context in which consent is acquired. In the case of pharmacogenetics add-on studies to clinical drug trials, not only are patients being asked to consider complex issues related to the treatment of their condition and the potential risks and benefits of participating in the trial process, but they are further required to consider issues relating to the pharmacogenetics aspect of the trial.

In consenting to the second part of the study patients give the sponsoring pharmaceutical company permission to carry out genetic research on their blood sample, including linking it to personal medical information, such as details of their medical condition and family history. The sample at this stage is 'identified', meaning that the patient's name, medical history and DNA sample data all remain linked. According to an information leaflet produced by a pharmaceutical company to inform patients about such research: 'we will use the Identified Sample to look at genes that might be involved in the condition you have, and how those genes might affect the study medication you are receiving' (patient information leaflet 1998). Subjects are informed that samples will remain 'identified' for about six months after completion of the main drug trial study, and that such trials may take several years to complete. Indeed, patients may even be informed that their identified sample could be kept for up to a maximum of ten years.

The third part of the consent process involves a request to store the sample after completion of the trial for an unlimited period in order to carry out future, as yet unspecified, research. Subjects are informed that at this stage the sample will be 'non-identified'. By non-identified the pharmaceutical companies mean that the DNA sample, data derived from it and medical data concerning the subject will be stored but that it will not be possible to link this information to the name of the subject and as such anonymity will be secured. If subjects agree to consent to the third stage then identification markers linking the patient to the sample will be destroyed and the sample will be stored anonymously:

> [The pharmaceutical company], or people working with [the company] will store your blood sample (or genetic sample prepared from your blood). They will also store information about your medical history. After [the company] takes your identification off, they will store the sample, all medical information, and any genetics results for an unlimited amount of time.
>
> (patient consent form 1998)

The information given to patients and healthy volunteers is expected to be sufficient for making informed choices. However, unlike conventional clinical drug trials where information regarding potential risk relates mostly to the risk of bodily harm, apart from the slight possibility of bruising when a small blood sample is extracted from the subject, no other direct physical risks are involved in the pharmacogenetics aspect of the trial. Sponsoring pharmaceutical companies indicate that the risk of harm from participation in the genetic study is 'small' (patient information leaflet 1998). Risks are identified rather as relating to potential problems concerning confidentiality and anonymity. A patient information leaflet given to subjects gives assurances about sufficient measures in place to ensure that confidentiality and anonymity of this data are maintained. However, subjects are also informed that the pharmaceutical company 'cannot be certain that your genetic test results could never be linked to you' (patient information leaflet 1998). Although the risk of such harm is described as 'small', genetic information arising from such studies is potentially dangerous insofar as it may lead to discrimination in the workplace or difficulty in obtaining insurance if individuals are thought to have greater than average risk of ill-health. Indeed, an information video produced by the pharmaceutical company GlaxoSmithKline (GSK) aimed at informing clinical investigators about pharmacogenetic trials, emphasizes the need for doctors to ensure that the patient's privacy is protected. GSK representatives state that access to the data is to be restricted and such data must not be given to patients or insurers.

Insofar as consideration of potential benefits is concerned, unlike the main drug trial where patients may directly benefit as part of their involvement in the study, there is no likely direct benefit to the patient in taking part in the collection of samples for pharmacogenetics purposes. Whether or not patients in a clinical drug trial achieve the 'ideal' of autonomous decision-making, or they find the experience of consent empowering, there generally remains at least a potential therapeutic benefit to the patient in trial participation. Indeed, the hope of therapeutic effect is a reason often cited by patients for consenting to drug trials (Corrigan 2003). When clinical drug trials are non-therapeutic, as in the case of phase I trials conducted on healthy volunteers, subjects are financially rewarded for their participation. Other forms of non-therapeutic trials sometimes take place in the clinical setting, as for example in epidemiology or observational studies, but to my knowledge, these are not sponsored directly by pharmaceutical companies or other commercial organizations. Patients participating in pharmacogenetics research are not informed of the results of their genetic test even when such information may reveal that the patient is potentially a poor metabolizer for the study drug (or indeed for other drugs that rely on the same metabolizing enzyme) and thus are potentially at greater risk of experiencing adverse drug reactions. Instead, patients are informed that they are unlikely to benefit directly and an appeal is made to altruistic reasons for donation:

> [The pharmaceutical company] expects no immediate benefit to you although your participation in this research may eventually help people with migraine. . . . We may also learn which patients may suffer side-effects from certain medicines.
>
> (patient information leaflet 1998)

Risks to be considered for this aspect of the trial are deemed to be smaller insofar as at this stage it would not be possible for those involved in the trial to link the patient formally to the sample and the data. However, there remains a small risk that the patient could still be identified from the data relating their medical information with sample information, both of which remain linked. Furthermore, patients at this stage are consenting to as yet unspecified commercial research and have no future control over the kinds of research to which their samples may be used.

Should individuals not wish to consent to this third part of the study, they are informed that their sample will be removed and destroyed. However, as Graham Lewis (this volume) indicates, given the complex arrangements concerning storage of these samples by third party companies, many of whom are currently undergoing mergers, this endeavour may prove difficult and without effective oversight it may be hard to ascertain whether these promises are being adhered to.

Conclusion

In this chapter I have demonstrated that informed consent procedures have been introduced as a mechanism to protect the research subject's right to autonomy and appear to fit well with other strategies that encourage active forms of citizenship. Nevertheless, representing consenting patients as 'active participants' fails to acknowledge the extent to which they remain objects in the research process and are a potential vulnerable population. In the case of clinical drug trials where patients are not only being asked to consent to the main trial but are additionally being asked to consent to further pharmacogenetics-related studies, I suggest that the limits of consent are being overreached. Given evidence of the inability of patients who have decided to take part in clinical trials to comprehend information fully on trial procedures, along with their inability to recall information about potential risks, it is very unlikely that patients entering into pharmacogenetics-related trials will conform to the informed consent requirements of having given due careful consideration of all potential risks and benefits. As I have already indicated, consent is obtained in the clinical context, usually during consultations with a physician in the hospital or at an outpatient clinic. I suggest that asking patients to consent to further pharmacogenetics studies in this context is a form of exploitation, because they frequently already feel overburdened with information and anxiety, and in a weakened position to make independent decisions. Although in conventional clinical drug trials patients are subjected to additional risks over and above conventional therapy, insofar as there is only partial information on the efficacy and safety of the drug being tested in general, patients have at least the possibility of benefiting either from the comparative drug being tested or from the trial drug. However, as I have argued, there is no direct benefit to the patient in the pharmacogenetics study and thus compromises in the consent process cannot be offset against potential therapeutic benefit. This issue is of sufficient concern that in France, for example, non-therapeutic research is reviewed by different ethics review bodies from those reviewing therapeutic research, with the result that the conduct of pharmacogenetics add-on trials has been prevented.

More importantly, pharmaceutical companies, who have huge commercial incentives for developing genetic-based drugs, are sponsoring pharmacogenetics trials. While physicians and medical professionals are involved in conducting the actual trials, pharmaceutical companies pay physicians for this service. Following the John Moore case (1998–1990), in which a cell line was patented and sold to the drug company Sandoz for $15 million, issues of property rights are increasingly being raised in relation to tissue samples (Boyle 1992, Landecker 1999). As such, blood samples that are donated for genetic testing purposes during the pharmacogenetic part of a clinical trial may potentially be extremely valuable commodities in genetic-based drug development. That said, human subjects have been experimental vessels for the production of drugs and subsequent commercial profit for well over half a century. In pharmacogenetic clinical trials, it is not simply the issue of DNA tissue that is at stake, but also human bodies and their labour, insofar as taking part in the trial may mean that the patient is required to attend the hospital clinic for tests related to the trial at regular intervals. Not only does the collection of DNA samples provide the 'raw material for industry' (see Lewis this volume) but trial subjects are also worked upon and transformed into data that are essential drug product components.

Such a conclusion raises profound questions about the meaning of consent for these trials. Should patients be financially rewarded for their involvement, as is the case for healthy subjects in phase I trials, as well as the physicians who recruit, conduct and oversee the clinical trials? Or should the fact that patients in the main trial are likely to receive benefit in the form of therapy be sufficient reward for taking part in all or some of the three aspects? And to what extent should altruistic reasons on the part of patients be taken into consideration? While pharmaceutical companies are foremost profit-making organizations and economic generators, they are also the vehicle by which successive governments have chosen to develop and test new drugs. We are thus left with the challenge of determining how to regulate clinical drug trials and pharmacogenetics research in a way that better protects patients and research subjects, but does not overly restrict potential medical breakthroughs. Of one thing we can be certain informed consent is not a sufficiently robust concept (even were its implementation to be improved) to protect research subjects adequately – better protection is required if the risks that patients take as subjects in clinical trials are to be justified.

Acknowledgements

I would like to thank Jose Lopez, Martin Richards, Bryn Williams-Jones and Richard Tutton for their comments on earlier drafts of this chapter.

Notes

1 In 1998 the estimated pharmaceutical industry investment in genomics approached $2bn (Bogdanovic and Langlands 1999: 141).
2 For a more detailed description and information about the processes of each trial phase see Corrigan (2002).
3 Beecher (1966) compiled a list of more than twenty examples of unethical experiments, the results of which had been published in prestigious medical journals during a single year. Both Pappworth in the UK and Beecher in the US suggested that unethical research was widespread.
4 International Conference on Harmonisation (ICH 1998) is a pharmaceutical industry-led initiative aimed at standardizing good clinical practice (GCP) to provide a unified standard for the EU, Japan and the US.
5 A survey of 58 members belonging to six local research ethics committees in the UK area reveals that members rated their duty 'to ensure prospective subjects understand the implication of taking part in the study as the most important aspect of their work' (Kent 1997). This was given priority over and above the duty to protect subjects from harm. In the US IRBs have also been shown to prioritize informed consent over risk–benefit analysis (Weijer 2000).
6 These are concepts expanded more fully by Foucault (1979).

7 Governmentality studies focus on the technical character of liberal and neo-liberal government. This approach analyses the multi-form tactics of government, where the term 'government' is not to be understood simply in terms of the exercise of state power, but as a complex array of strategies of rule where techniques of power and domination interact with techniques of the self. For a comprehensive range of discussions on the concept of 'governmentality' by its various exponents see especially Barry *et al.* (1996).

8 Phase I trials are also conducted on patients with life-threatening conditions such as advanced or aggressive cancer.

References

Annas, G. and Grodin, M. (eds) (1992). *The Nazi Doctors and the Nuremberg Code*, New York: Oxford University Press.

Bamberg, M. and Budwig, N. (1992). 'Therapeutic misconceptions: when the voices of caring and research are misconstrued as the voice of curing,' *Ethics and Behaviour* 2: 165–184.

Barry, A., Osborne, T., and Rose, N. (1996). *Foucault and Political Reason*, London: The University of Chicago Press and UCL Press Limited.

Beauchamp, T.L. and Childress, J.F. (1989). *Principles of Biomedical Ethics*, Belmont: Wadsworth Publishing Company.

Beecher, H.-K. (1966). 'Ethics and clinical research,' *New England Journal of Medicine* 274: 1354–1360.

Bergler, J., Pennington, A., Metcalf, M., and Freis, E. (1980), 'Informed consent: how much does the patient understand?' *Clinical Pharmacology and Therapeutics* 27: 435–440.

Bogdanovic, S. and Langlands, B. (1999). *Pharmacogenomics Players*, London: Financial Times.

Boyle, J. (1992). 'A theory of information, copyright, spleens, blackmail and insider trading,' *California Law Review* 80: 1415–1540.

Burgess, M.M. (2001). 'Beyond consent: ethical and social issues in genetic testing,' *Nature Reviews: Genetics* 2: 9–14.

Cassileth, B., Zupkiss, R., Sutton-Smith, K., and March, V. (1980). 'Informed consent – why are its goals imperfectly realized?' *New England Journal of Medicine* 302: 896–900.

CERES (1999). *CERES News*. CERES (24).

Corrigan, O.P. (2000). Trial and error: a sociology of bioethics and clinical drug trials, unpublished PhD thesis, Department of Sociology, University of Essex.

Corrigan, O.P. (2002) 'A risky business: the detection of adverse drug reactions in clinical trials and post-marketing exercises', *Social Science and Medicine* 55 3: 497–507.

Corrigan, O.P. (2003). 'Empty ethics: The problem with informed consent,' *Sociology of Health and Illness* 25: 768–792.

Council for International Organizations of Medical Sciences (1993). International Ethical Guidelines for Biomedical Research Involving Human Subjects, Geneva: CIOMS.

Dresser, R. (2001). *When Science Offers Salvation: Patient Advocacy and Research Ethics*, Oxford: Oxford University Press.

Estey, A., Wilkin, G., and Dosseter, J. (1994). 'Are research subjects able to retain the information they are given during the consent process?' *Health Law Review* 3: 37–41.

Evans, W.E. and McLeod, H.L. (2003). 'Pharmacogenomics – drug disposition, drug targets, and side effects,' *New England Journal of Medicine* 348: 538–549.

Faden, R. and Beauchamp, T. (1986). *A History and Theory of Informed Consent*, New York: Oxford University Press.

Foucault, M. (1979). *The History of Sexuality*, London: Penguin.

Foucault, M. (1991). *Discipline and Punish*, London: Penguin.

Foucault, M. (1991 [1963]). *The Birth of the Clinic*, London: Routledge.

Frankenfeld, P.J. (1992). 'Technological citizenship: a normative framework for risk studies,' *Science Technology and Human Values* 17: 459.

Glass, K.C. and Lemmens, T. (1999). 'Conflicts of interest and commercialization of biomedical research', in T. Caulfield and B. Williams-Jones (eds), *The Commercialization of Genetic Research: Ethical, Legal and Policy Issues*, New York: Kluwer Academic/Plenum Press.

Gray, B.H. (1975). *Human Subjects in Medical Experimentation*, New York: John Wiley & Sons.

Hassar, M. and Weintraub, M. (1976). 'Uniformed consent and the wealthy volunteer: an analysis of patient volunteers in a clinical trial of a new anti-inflammatory drug,' *Clinical Pharmacology and Therapeutics* 29: 379–386.

Horrobin, D.F. (2000). 'Innovation in the pharmaceutical industry,' *Journal of the Royal Society of Medicine* 93: 341–345.

ICH (1998). International Conference for Harmonisation, Good Clinical Practice, Guidance for trials of medicinal products in the European Community, in C. Foster (ed.), *Manual For Research Ethics Committees*, London: Kings College, Centre of Medical Ethics and Law.

Jan, H. and DeMets, D. (1981). 'How informed is informed consent?' *Controlled Clinical Trials* 2: 287–303.

Kant, I. (1995 [1948]). *Groundwork of the Metaphysic of Morals*, London: Routledge.

Katz, J. (1992). 'The consent principle of the Nuremberg Code: its significance then and now,' in G. Annas and M. Grodin (eds), *The Nazi Doctors and the Nuremberg Code*, New York: Oxford University Press.

Kent, G. (1997). 'The views of members of Local Research Ethics Committees, researchers and members of the public towards the roles and functions of LRECs,' *Journal of Medical Ethics* 23: 186–190.

Knoppers, B.M. (1998). 'Human genetic material: commodity or gift?' in R.F. Weir (ed.), *Stored Tissue Samples: Ethical, Legal, and Public Policy Implications*, Iowa City: University of Iowa Press, pp. 226–235.

Landecker, H. (1999). 'Between beneficence and chattel: the human biological in law and science,' *Science in Context* 12: 203–225.

Marinetto, M. (2003). 'Who wants to be an active Citizen? The politics and practice of community involvement,' *Sociology* 37: 103–120.

Marks, A.D. and Steinberg, K.K. (2002). 'The ethics of access to online genetic databases: Private or public?' *American Journal of Pharmacogenomics* 2: 207–212.

Merz, J. and Sankar, P. (1998). 'DNA banking: an empirical study of a proposed consent form,' in R.F. Weir (ed.), *Stored Tissue Samples: Ethical, Legal and Public Policy Implications*, Iowa City: University of Iowa Press, pp. 198–225.

Pappworth, M. (1967). *Human Guinea Pigs*, Harmondsworth, UK: Pelican.

Rose, N. (1996). 'Governing "advanced" liberal democracies,' in A. Barry, T. Osborne and N. Rose (eds), *Foucault and Political Reason*, London: The University of Chicago Press and UCL Press Limited.

Royal College of Physicians (1990). *Research Involving Patients*, London: The Royal College of Physicians of London.

Snowdon, C., Garcia, J. and Elbourne, D. (1997). 'Making sense of randomization; responses of parents of critically ill babies to random allocation of treatment in a clinical trial,' *Social Science and Medicine* 45: 1337–1355.

Steflox, H.T., Chua, G., O'Rourke, K. *et al.* (1998). 'Conflict of interest in the debate over calcium channel antagonists,' *New England Journal of Medicine* 338: 101–106.

Taylor, K.M. (1988). 'Telling bad news: physicians and the disclosure of undesirable information,' *Sociology of Health and Illness* 10: 109–132.

Weijer, C. (2000). 'The ethical analysis of risk,' *Journal of Law, Medicine and Ethics* 28: 344–361.

World Medical Association (2000). Declaration of Helsinki: Ethical Principles for Medical Research Involving Human Subjects.

45

BIOLOGICAL WEAPONS, GENETICS AND SOCIAL ANALYSIS

Emerging responses, emerging issues – I

Brian Rappert

Source: *New Genetics and Society* 22(2) (2003): 201–213.

Abstract

Recent terrorist attacks in the USA have generated significant attention in many countries to the threats posed by biological weapons. In response to these events and the spectre of future attacks, bioscientists and professional organizations have begun or intensified asking questions about the possible malign applications of their research. As Part I of a two-part article, this paper surveys how genetics might contribute to the development of novel forms of weaponry. It is further argued that the dilemmas and difficulties facing bioscientists pose pressing and thorny questions for the hitherto agendas and orientations of those concerned with the social, ethical and political implications of genetics. Part II will examine the emerging responses initiated by biomedical organizations and spokespersons in the US and the UK. This will be done with a view to asking how scientific and medical research communities are defining and policing notions of professionalism, responsibility and accountability. On the basis of this, suggested lines for future social analysis will be offered.

Introduction

Developments in science and engineering have long provided the basis for more sophisticated weaponry. In the 20th century, that connection was perhaps most explicit in the case of weapons of mass destruction, including nuclear, chemical and biological capabilities. Attention to these topics by the public, scientists and policymakers has tended to wax and wane, depending on events and perceptions of the security situation (Balmer, 2001). While some level of concern has existed throughout the last fifty years regarding the possible use and proliferation of nuclear weapons, arguably the same could not be said of chemical and biological ones. Although certain individuals, professional bodies and policymakers have drawn attention to possible threats (e.g. Henderson, 1999; British Medical Association, 1999), generally this has not been that sustained or widespread. Whatever the complacency of the past, since 11 September 2001 and the bioterrorism that followed in the USA, substantial anxiety has been experienced across many countries about possible biological and chemical threats (e.g. Poste, 2001).

With the increasing recognition of the disturbing possibilities enabled by current scientific know-how, a wide variety of individuals and organizations within Western science and policy circles are asking what actions should be taken to prevent research from facilitating death and injury. This two-part paper surveys initial responses by biological and medical science communities to the growing societal concern with the application of their expertise. Specifically, it addresses this by considering the role of genetics research in enabling and forestalling novel forms of biological weapons targeted at humans (see Whitby, 2002; Whitby, Millet & Dando, 2002 for a discussion of animal and plant bio-weapons). The immediate policy response in many countries has been geared towards countering imminent dangers of established biological agendas by, for instance, placing further restrictions on access to dangerous pathogens (Epstein, 2001; Home Office, 2002). In contrast, this paper focuses on responses to novel prospective possibilities that may be created by genomics research. It seeks to address several questions:

- How might current genetics and genomics research contribute to the development of biological weapons?
- What is the range of responses being considered to mitigate against the threats identified?
- How are notions of professionalism, responsibility, and accountability defined within these responses?

In examining the latter two points in Part II, this paper draws on general literature within the social sciences to comment on and assess the measures being proposed. The possibilities surrounding bio-weapons have not only been of arguably marginal concern hitherto for those in the biological sciences. Despite the substantial amount of investigation into the social, legal and ethical aspects of genetics, few analysis have turned their attention to biological weapons. As argued in Parts I and II an examination of the possibilities surrounding such technologies not only poses difficult questions for the priorities and purposes of biological and medical researchers, but also for those concerned about the "social" implications of modern technoscience. This paper asks what issues biological weapons raise for the social, political and ethical studies of technology as it considers what such analyses have to say to biological weapons. It sets out a number of questions for future research, some of which will be pursued in a project by the author and Malcolm Dando (University of Bradford, UK) funded by the UK Economic and Social Research Council (see Rappert & Dando, 2002).

The threats of biological weapons

Biological warfare dates back to antiquity (Barnaby, 1997; Stockholm International Peace Research Institute (SIPRI), 2001). The fouling of water sources with animal carcasses or the catapulting of disease-infected corpses over city walls are some of the earliest recorded examples of warfare techniques designed to spread disease. During colonization, French and British forces gave smallpox infected blankets to Native Americans with the intent and effect of infecting them.

Modern biology and medicine has facilitated the production of biological weapons throughout the 20th century (Dando, 1999). On the basis of 19th century advances in bacteriology, for instance, both sides in World War I attempted to employ micro-organisms against their opponents' animals. During World War II, Japanese scientists conducted horrific experiments during their occupation of China. After the war, many industrialized countries began or furthered systematic programmes into offensive and defensive capabilities (see, e.g. Balmer, 2001). Most of these programmes were abandoned by the signing of the 1972 Biological and Toxin Weapons Convention (BTWC). The BTWC has become the cornerstone of international efforts to limit the production and proliferation of biological weapons. Article I states:

> Each State Party to this Convention undertakes never in any circumstances to develop, produce or stockpile or otherwise acquire or retain:
>
> 1. Microbial or other biological agents, or toxins whatever their origin or method of production, of types and in quantities that have no justification for prophylactic, protective or other peaceful purposes.
> 2. Weapons, equipment or means of delivery designed to use such agents or toxins for hostile purposes or in armed conflict.

While the BTWC represents one of the most significant international agreements to restrict the means of warfare, its terms have been a source of contention and negotiation. A consideration of why indicates some of the major topics of concern about biological weapons today. One major limitation of the convention is that no mechanisms exist for verifying adherence by state parties. Attempts throughout the last decade to agree such measures were halted for the time being shortly before 11 September 2001, when the US withdrew any support for inspection regimes, stating they would be both ineffective as well as reveal biodefence capabilities and commercial secrets.

Given the lack of verification measures and the widespread availability of the general materials, equipment and biological expertise necessary for weapons development, the possibility that some state parties are violating the BTWC has long been a cause of concern. The defection of high-ranking scientists from the Soviet Union post-Cold War indicated that it had an extensive offensive programme. This consisted of tens of thousands of staff and incorporated genetic engineering techniques. Likewise, Iraq had a clandestine offensive programme prior to and after the Gulf War (Wallerstein, 1998).

Outside flagrant violations, just what activities are permissible is likewise disputed. Under the BTWC, undertaking defensive measures against biological weapons is acceptable. No criteria are given though to differentiate offensive from defensive development, production or stockpiling activities. As specified by the BTWC, what counts is the intent of actions. All biological agents and toxins are banned unless otherwise justifiable. Yet, predictably, it has not been straightforward to establish intent. For instance, during the Clinton administration, the US began several projects of a questionable status. These consisted of planning to genetically engineer a potent variant of the bacterium that causes anthrax; building and testing an imitation Soviet-designed germ bomb; and assembling a mock biological weapons factory from products openly available on the commercial market (Miller, Engelberg & Broad, 2001). None was reported under the provisions of BTWC. Each was justified on the basis of needing to test current defensive capabilities against likely threats, yet each had offensive potential. Limited *offensive* developments done to test defensive measures (e.g. vaccines) have been interpreted as permissible under the BTWC (see Huxsoll, 1992). Because offensive and defensive activities are so interlinked and difficult to distinguish, there has been much disagreement regarding what counts as an infringement and what evidential basis would be needed to support particular evaluations (Wright, 1990).

Much of the most recent concern about biological weapons has centred on those outside the BTWC—specifically sub-state and terrorist groups (see, e.g. Knobler, Mahmoud & Pray, 2002). The release of the chemical nerve gas sarin during 1995 into the Tokyo subway system by the religious cult Aum Shinrikyo and its failed attempts to develop biological weapons signalled that some sub-state groups or individuals were interested in employing chemical and biological weapons. For some commentators, the ability to mask offensive programmes under civilian ones, the weight-for-weight potency of biological weapons and the relative ease of access to pathogenic agents and toxins make bio-weapons a potentially attractive option for terrorist attacks (e.g. Danzig & Berkowsky, 1997).

Whether it is feasible that individuals, sub-state groups or even many states can develop and effectively employ bio-weapons has been a topic of much debate (Broad, 2001). The ease or difficulty one believes of crafting such weapons has a direct bearing on assessments of the likelihood of attack and the necessary responsive measures. Producing a biological weapon requires obtaining a strain of an agent, culturing it in sufficient quantities, and then finding an adequate means of dispersal. Leitenberg (2000) argues there has been little or no acquisition of biological weapons by non-state actors in the 20th century because each step is highly demanding. Even though many bacteria are available from natural sources, for instance, they may not be virulent enough to act as agents in weapons.[1] Such assessments stand in sharp contrast to claims by Carl Ford, Assistant Secretary of State for Intelligence and Research in the US Department of State, who said:

> Terrorist interest in chemical and biological weapons has been growing and probably will increase in the near term. The threat is real and proven. The ease of acquisition or production of some of these weapons and the scale and terror they can cause, will likely fuel interest in using them to terrorize. The transport and dispersal techniques also are manageable and can be made effective easily . . .
>
> (Ford, 2002, cited in Leitenberg, 2002; also Pearson, 1998)

Of course some amount of disagreement may result from alternative interpretations of what counts as effective—whether that be judged by casualty figures or political, financial and psychological costs. If the latter is taken as the criterion, then heightening public attention to the dangers of biological weapons may well end up amplifying the costs of any attack by enhancing fear levels. Focusing public health attention and training into the prevention and response to bio-attacks may well end up diverting resources from the types of specific actions that are needed to address pressing health problems.[2]

Possible biological agents include viruses, bacteria, toxin, rickettsiae and fungi. Each has various strengths and weaknesses as warfare agents. *Bacillus anthracis*, the agent that causes anthrax, has long been seen as one of the principal threats because it can be found naturally, strains of it are highly potent and long lasting, an effective vaccine exists, and it is not communicable. The last two points might limit its lethality, but they also make it easier to handle and target the agent. Table 1 indicates some general characteristics of biology weapons likely to be sought for those wishing to acquire them.

Biological weapons and the new genetics

In recent years, the knowledge and techniques gained from genetics, genomics and other fields of research have grown apace. Genetic sequencing has helped to locate specific genes and to identify their associated proteins.

Table 1 Sought-after properties of biological agents

Requirements	*Desirable characteristics*
Consistently produce a given effect (death, disability or crop damage).	Possible for the using forces to protect against.
Be manufacturable on a large scale.	Difficult for a potential enemy to detect or protect against.
Be stable during production and storage, in munitions and during transportation.	A short and predictable incubation period.
Be capable of efficient dissemination.	A short and predictable persistency if the contaminated area is to be promptly occupied by friendly troops.
Be stable after dissemination.	Capable of infecting more than one kind of target (for example, man and animals) through more than one portal of entry, being disseminated by various means, producing desired psychological effects.

Source: Taken from Cohen (1997): http://www.defenselink.mil/pubs/prolif97/annex.html, which adopted the chart from US Departments of Army and Air Force (1964) *Military Biology and Biological Agents*. Departments of Army and Air Force Technical Manual 3–216/Air Force Manual 355–6, 12 March.

Combinatorial chemistry has enabled the automation of discovery process for identifying new drug candidates through the creation of large libraries of chemicals. High throughput screening provides a rapid means of finding compounds against various biological targets. Genechips generate indicators for complex multifactorial conditions. Developments in bio-informatics have supported the collection and dissemination of vast quantities of data.

As such developments become widely commercialized, the danger is that this increases the chance that new forms of bioengineered weapons can be produced to augment or replace existing capabilities. Peaceful commercial applications of genetics and genomics research can enhance the possibility that others might achieve the sought-after characteristics listed in Table 1. To use the jargon, the technology and techniques developed are of "dual-use". In 1996, the Fourth Review Conference of BTWC highlighted the danger of general advances in stating:

> The Conference, conscious of apprehensions arising from relevant scientific and technological developments, *inter alia*, in the fields of microbiology, biotechnology, molecular biology, genetic engineering, and any applications resulting from genome studies, and the possibilities of their use for purposes inconsistent with the objectives and the provisions of the Convention, reaffirms that the undertaking given by the States Parties in Article I applies to all such developments.[3]

A 1997 a US Department of Defense publication suggested a number of significant technological trends likely to influence how readily infectious agents can be used for biological warfare:

- Genetically engineered vectors in the form of modified infectious organisms will be increasingly employed as tools in medicine and the techniques will become more widely available.
- Strides will be made in the understanding of infectious disease mechanisms and in microbial genetics that are responsible for disease processes.
- An increased understanding of the human immune system function and disease mechanisms will shed light on the circumstances that cause individual susceptibility to infectious disease.
- Vaccines and antidotes will be improved over the long term, perhaps to the point where classic biological warfare agents will offer less utility as a means of causing casualties (Cohen, 1997).

Genetics and genomics figure directly or indirectly into each of these.

How might such research enable the production of new biological weapons? The possibilities are numerous (for overviews, see Drell, Sofaer & Wilson, 1999; Fraser & Dando, 2001):

- Bacterial agents such as bacillus anthracis can be made resistant to current antibiotics. Reports about the Soviet weapons' programme, for instance, indicated it was able to make the casual agent for plague resistant to numerous antibiotics (Barnaby, 1997).
- Transferring the genes responsible for pathogenic qualities into non-pathogenic organisms (such as *E. coli*) would make them so, or such genes could be transferred to already pathogenic organisms to increase their virulence (in relation to novel influenza viruses, see the *Lancet Infectious Diseases* (2002)).
- The survivability of a bacterium across a range of environmental conditions could be improved through gene splicing.
- Cells could be modified to produce greater quantities of toxins.
- Detection, mitigation and remediation from exposure to agent strains could become much more difficult if they were modified to hide telltale signatures. Russian scientists in the mid-1990s reportedly made detection tests for anthrax ineffective though genetic engineering techniques (Sunshine Project, 2002).

As elsewhere with bioengineering, there are questions about possibility and feasibility of such steps. The complexities of genetic expression and limits to understanding still hamper such efforts. The primary genes responsible for pathogenic properties might be transferred to another micro-organism but, in doing so, the virulence of the donor strain may be significantly decreased.

Two further points about the manipulations above are worth noting here. First, the know-how and tools required for these are fairly generic in genetic engineering. Attempts to develop vaccines for HIV, for instance, have inserted HIV genes into Salmonella. A major source of difference, however, is the type of materials utilized. Secondly, many of the modifications of pathogenic agents could be/are undertaken as part of efforts to develop new vaccines and diagnostic techniques or further knowledge about the basic mechanisms of virulence (see Leitenberg, 2002).

In addition to whatever possibilities are opened up for those deliberately setting out to create biological weapons, suggestions for novel approaches might arise "unintentionally". This prospect received a fairly widespread airing when Australian researchers publicized creating a particularly potent strain of mouse-pox as part of experiments into a genetically based contraceptive for mice. This was done by altering the mousepox virus through the insertion of the inter-leukin-4 gene (Jackson *et al.*, 2001; Dennis, 2001). If parallel modifications could be made to human smallpox, then it might evade existing vaccination measures.

As a further elaboration of the dilemmatic relation between "weapon" and "non-weapon" biological research, the same advances in microbial genomics used to produce bio-weapons can also set up countermeasures against them. The completion of the human genome sequence provides a new starting point for better understanding of the infectious disease process, and this has potential *against* bio-warfare. So, too, with sequence data for the casual agents of anthrax, tuberculosis and cholera (Venter, 2001). While the free availability of such data in open sources might further the proliferation of knowledge, it also enables additional prevention and treatment steps. New vaccines, compounds and diagnostics, perhaps specific to different agent strains, might be produced. Gene chips could aid in the identification of strain-specific differences in agents after an attack starts to ensure rapid and effective treatment. Each such counter-response to a potential threat in turn could bring attempts to counter it through yet a further utilization of bioengineering techniques. Such is the history of weapons development.

Novel options might be viable in the future. New bacteria or viruses might be created from component genetic parts rather than merely modifying existing agents (Fraser & Dando, 2001; Whitehouse, 2002). Viruses could be secretly introduced into a target population and then activated by a signal at some later time (Drell *et al.*, 1999). A fair amount of speculation has taken place as to whether increasing the information on the structure and function of the human genome might enable so-called "ethnic genetic weapons". In 1993, the Stockholm International Peace Research Institute warned that genetic differences between population groups might provide a basis for vectors to attack DNA sequences inside cells specific to certain sub-populations (Bartfai, Lundin & Rybeck, 1993). Whether the complexity of gene influence in disease susceptibility will allow for such targeting remains uncertain (British Medical Association, 1999).

Genetics may well offer opportunities for utilizing new agents. Consider so-called bioregulators. A US review entitled *The Biological and Chemical Warfare Threat* defined bioregulators as:

> natural substances produced in very small quantities that are essential for normal physiological functioning of the body. They control cell and body physiological functions and regulate a broad range of functions such as bronchoconstriction, vasodilation, muscle contraction, blood pressure, heart rate, temperature, and immune responses. (US, 1996; quoted from Dando, 2001)

It suggested that altering the concentration of these substances would disrupt basic body functions, with the potential for almost immediate effects. Short of inflicting outright death through, say, heart failure, bioregulators could act as incapacitants. Kagan (2001) identified several likely categories of biological warfare/terrorism bioregulators: altering levels of cytokines can cause serious illness, lethargy and fever; eicosanoids manipulations can induce asthma; and plasma proteases modifications can bring bronchoconstriction, hypertension and pain.

Neurotransmitters and hormones make up another category of bioregulators, one that has far reaching potential. Past uncertainties about the role of receptors and bioregulators in drug action have limited the possibility of devising appropriately specific drugs. As genetics and genomics research brings a greater understanding of the function of neurotransmitters in nervous system communication and the diversity of receptor types, with this comes the opportunity to affect cognition, sensory processing and mood selectively. The code for those neurotransmitters that are peptides, for instance, can be inserted into a virus in order to affect behaviour. The military applications are potentially enormous. In the 1950s and 1960s, both the UK and the US experimented with glycollates, such as the psychotropic drug BZ (3-quinuclidinyl benzilate), to see if they could be employed to cause hallucinations, disorientation or stupor. Such efforts largely failed because the biological targets were not properly characterized (see Dando, 2001). These barriers may now, or soon, be overcome.

This brief survey has indicated possible technological options as well as some of the terms for contention about biological weapons. Disagreement abounds about what threats are posed, who these come from, what should be taken as fact rather than fiction, and what steps will be required to mitigate against dangers. There are basic problems about the categorization of activities (e.g. what counts as offensive work) and the limits to empirical evidence (because of secrecy or unknowns). Those wishing to respond to the threats posed by biological weapons must operate in relation to unstable notions of anticipated future risks where the issues faced about what ought to be done are thoroughly dilemmatic. Claims about the threat of biological weapons help constitute the very issue they are commenting on. Attempts said to "talk up" or merely "highlight" potential dangers in order to unsettle complacency might well end up magnifying the social and economic damaged caused by any such attack.

The social sciences and biological weapons

What insights might be offered by past social science research to make sense of the varied and multiply conceived issues associated with biological weapons, particularly in relation to the contribution of contemporary genetics and genomics?

Most of the past interest from social scientists in biological weapons has come from those in areas aligned with international relations, such security and peace studies. The main concerns herein have been with the proliferation of production capabilities (e.g. Brauch *et al.*, 1997); the possibility of establishing effective export controls and disarmament measures (e.g. Robinson, 1998; *The CBW Conventions Bulletin*), particularly given the dual-use status of much technology (e.g. Molas-Gallart & Robinson, 1997); and the assessment of threats posed by the use of biological weapons from states, criminals or terrorist organizations (Barnaby, 1992). Practitioners in fields such as microbiology, immunology or toxicology (Hay, 1999)—many formerly part of biodefence establishments—have conducted many similar studies. Pearson (1993) has argued that any overall strategy for the reduction of the threat of biological weapons requires a "web of deterrence", consisting of such initiatives as international arms controls, export and monitoring procedures, defensive and protective measures, counter responses to the use of biological weapons, and prevention steps undertaken by the bioscience community. Outside international relations, other social analysts have examined historical transitions in major offensive biological weapons programmes (e.g. Balmer, 2001; Gould & Folb, 2002; Regis, 1999).

Despite the significant contributions made by many of the above analysts, the study of chemical and biological weapons and its links to biological and genetics research has been a rather specialized topic of study centred in certain sub-disciplines. For understandable reasons, the preoccupation of many of those concerned with this topic has been in practically examining ongoing technical and political developments rather than stepping back from the substantive area to ask what it means for wider concerns in social and political theory. In the past, no doubt due to the low public profile of bio-weapons, few outsiders to this area have asked what lessons might be learnt from its study or integrated such lessons into the mainstream of many disciplines. Even in relation to ethical issues, debates in bioethics have not been informed by the possibilities raised by biological warfare (Hooker, 1992). A sense of the marginal priority attached to this topic can be gleaned from the relatively minor attention paid to it, despite the significant sums of money being devoted to the social study of genetics. Major national their research programmes into the legal, ethical and social implications of genetics, such as ELSI, have given little attention to this topic (Greenly, 1998). This lack of attention is perhaps indicative of the general academic decline in funding priorities for defence-related studies since the end of the Cold War.

Arguably, perceptions of the importance of biological weapons have altered significantly for social scientists and funding agencies with the terrorist events of 2001. As with other areas in the past, its ascent in public discussions is likely to be followed by greater attention—as represented by this paper. In doing so, however, the problems about agendas and orientations to be faced go beyond the lack of past consideration of the topic. At stake are quite fundamental questions about what sort of response should be made. Much of past social science research into the general relation between medicine and warfare has taken a somewhat suspicious, if not down right hostile, orientation to such linkages. Sontag (1978) examined how medicine has long drawn on metaphors from warfare (e.g. "targets", "war on cancer", etc.) selectively to define proper medical responses. Others have stressed the manner in which modernity, medicine and war have been inextricably bound together in the 20th century. The mobilization of biomedicine during war has brought with it administration controls, surveillance mechanisms, regulation procedures and rationalization techniques—the overall merits of which are taken as ambivalent at best (Cooter, Harrison & Sturdy, 1998). The possible military control of genetics research has brought calls for caution from various quarters (Piller & Yamamoto, 1988; Wright, 1990; Yoxen, 1983).

Whatever the scepticism that characterized the past study of the interrelations of military, biomedicine and genetics, with the increasing regard for Western countries as targets of bioattacks and the substantial attention dedicated to biodefence measures (see the next section in Part II) the pertinent question arises of what roles those concern with the ethical, social and political implications of genetics ought to take. These might be offering critical outsider commentary that seeks to identify misrepresentations; giving practical inputs into strengthening biodefence structures; cultivating alternative responses to dominant ones; or identifying hitherto ignored security threats. Such roles are not exclusive. While questions about the way forward are hardly unique to the area of biological warfare, they are particularly acute because of the past lack of attention to this area and the speed of its rise as a matter of public concern. The points noted above about the "dual-use" of biodefence research—its potential to minimize the threat of harm from biological weapons while providing the seed of future weaponry—are destabilizing for social scientists just as they are for bioscientists. For each, it can be expected that the demarcation between offensive and defensive activities will loom large in determinations of appropriate conduct. So too will concerns about how to orientate to competing claims about what constitutes "hype" or "reality". As conducting research in this area helps constitute determination of the seriousness of the problem (albeit to varying degrees), the way forward is full of potential pitfalls.

Acknowledgements

My thanks to Brian Balmer, Malcolm Dando and Gigi Kwik for their assistance in producing this paper.

Notes

1 As an indication of this, at the time of writing, the US government considers the most likely suspect(s) for the anthrax attacks to be a biodefence insider.
2 See, e.g. <http://www.redbluffdailynews.com/display/inn_news/news65.txt>.
3 See <http://www.brad.ac.uk/acad/sbtwc/revconf/4final3.htm>.

References

Balmer, B. (2001) *Britain and Biological Warfare: Expert Advice and Science Policy 1935–65* (Basingstoke: Palgrave).

Barnaby, F. (1992) *The Role and Control of Weapons in the 1990s* (London: Routledge).

Barnaby, W. (1997) *The Plague Makers: The Secret World of Biological Warfare* (London: Vision).

Bartfai, T., Lundin, S. & Rybeck, B. (1993) Benefits and threats of developments in biotechnology and genetics engineering. In *SIPRI Yearbook 1993*, pp. 293–305 (Oxford: Oxford University Press).

British Medical Association (1999) *Biotechnology, Weapons and Humanity* (London: Harwood Academic Publishers).

Brauch, H., Graaf, H., Grin, J. & Smit, W. (1997) *Military Technology Assessment and Preventive Arms Control* (Münster-Hamburg: Lit Publishers).

Cohen, S. (1997) Technical annex. In *Proliferation: Threat and Response* (Washington, DC: US Department of Defense) <http://www.defenselink.mil/pubs/prolif97/annex.html>.

Cooter, R., Harrison, M. & Sturdy, S. (eds) (1998) *War, Medicine and Modernity* (London: Sutton Publishing Group).

Cunningham-Burely, S. & Kerr, A. (1999) Defining the "Social", *Sociology of Health & Illness*, 21, pp. 647–68.

Dando, M. (1999) The impact of the development of modern biology and medicine on the evolution of offensive biological warfare programs in the twentieth century, *Defense Analysis*, 15, pp. 43–62.

Dando, M. (2001) *The New Biological Weapons* (London: Lynee Rienner).

Danzig, R. & Berkowsky, P. (1997) Why should we be concerned about biological warfare?, *JAMA*, 278, p. 5.

Drell, S., Sofaer, A. & Wilson, G. (eds) (1999) *The New Terror: Facing the Threat of Biological and Chemical Weapons*, pp. 39–75 (Hoover Institution, Stanford).

Epstein, G. (2001) Controlling biological warfare threats, *Critical Reviews in Microbiology* 27, pp. 321–54.

Fraser, C. & Dando, M. (2001) Genomics and future biological weapons: the need for preventive action by the biomedical community, *Nature Genetics*, 22 October.

Gould, C. & Folb, P. (2002) The role of professionals in South African chemical and biological warfare programme, *Minerva*, XL (1), pp. 77–91.

Greenly, H. (1998) Legal, ethical and social issues in human genome research, *Annual Review of Anthropology*, pp. 473–502.

Hay, A. (1999) Simulants, stimulants and diseases: the evolution of the United States biological warfare programme, *Conflict, Medicine and Survival*, 15, p. 3.

Henderson, D. (1999) The looming threat of bioterrorism, *Science*, 283, 1279–82.

Home Office [United Kingdom] (2002) *Anti-terrorism, Crime and Security Act 2001 Part 6: Weapons of Mass Destruction*, HOC 16/02: see <http://www.homeoffice.gov.uk/circulars/2002/hoc16.htm>.

Hooker, M. (1992) In Zilinskas, R. (ed.) *The Microbiologist and Biological Defense Research* (New York: New York Academy of Science).

Huxsoll, D. (1992) Narrowing the zone of uncertainty between research and development. In Zilinskas, R. (ed.) *The Microbiologist and Biological Defense Research* (New York: New York Academy of Science).

Kagan, E. (2001) Bioregulators as instruments of terror, *Clinics in Laboratory Medicine* 21, pp. 607–18.

Knobler, S., Mahmoud, A. & Pray, L. (eds) (2002) *Biological Threats and Terrorism* (Washington, DC: National Academy Press).

Lancet Infectious Diseases (2001) Too much learning is a dangerous thing, *Lancet Infectious Diseases*, 1, p. 287.

Leitenberg, M. (2000) An assessment of the threat of the use of biological weapons or biological agents. In Martellini, M. (ed.) *Biosecurity and Bioterrorism* (Washington, DC: Landau Network Centro Volta).

Leitenberg, M. (2002) *Biological Weapons in the Twentieth Century: A Review and Analysis*: see <http://www.fas.org/bwc/papers/21 centurybw.htm>.

Molas-Gallart, J. & Robinson, J.P. (1997) *Assessment of Dual-use Technologies in the Context of European Security and Defence*, p. 58 (Brussels: European Parliament Scientific and Technological Options Assessment).

Pearson, G. (1993) Prospects for chemical and biological arm control, *The Washington Quarterly*, Spring, pp. 145–62.

Pearson, G. (1998) *The Threat of Deliberate Disease in the 21st Century* <www.bradford.ac.uk/acad/ sbtwc/other/disease.htm>.

Piller, C. & Yamamoto, K. (1988) *Gene Wars* (New York: Beech Tree Books).

Poste, G. (2001) A rude awakening to the forces of good and evil, *Financial Times*, 27 November, IV.

Rappert, B. & Dando, M. (2002) *Accountability and the Governance of Expertise: Anticipating Genetic Bioweapons* <http://sbs-xnet.sbs.ox.ac.uk/scisoc/html/projects/project_detail_08.asp>.

Regis, E. (1999) *The Biology of Doom* (New York: Holt).

Robinson, J.P. (1998) The chemical weapons convention verification regime: implications for biotechnology. In *A Strengthened Biological and Toxin Weapons Convention: Potential implications for Biotechnology*, pp. 15–28 (Vienna: Institute of Applied Microbiology).

Sontag, S. (1978) *Illness as Metaphor* (New York: Farrar, Straus & Giroux).

Stockholm International Peace Research Institute (2001) *Biotechnology and the Future of the Biological and Toxin Weapons Convention* (Stockholm: SIPRI).

Sunshine Project. (2002) *An Introduction to Biological Weapons, Their Prohibition, and the Relationship to Biosafety*, Backgrounder Series No. 10 (Austin, TX: Sunshine Project).

United States. (1996) *The Biological Chemical Warfare Threat: A Tutorial* (Washington, DC: US Government) In Dando, M. Genomics, bioregulators, cell receptors and potential biological weapons, *Defense Analysis*, 17, pp. 239–58.

Venter, C. (2001) Understanding the genome for pathogen detection, identification and treatment presentation at american association for the advancement of science. Paper presented at the 2001 Annual Meeting Bio-technology and Bio-Weapons, 17 February.

Wallerstein, M. (1998) Responding to proliferation threats, *Strategic Forum*, 13.

Whitby, S. (2002) *Biological Warfare Against Crops* (Basingstoke: Palgrave)

Whitby, S., Millett, P. & Dando, M. (2002) The potential for abuse of genetics in military significant biological weapons, *Medicine, Conflict and Survival*, 18, pp. 138–56.

Whitehouse, D. (2002) First synthetic virus created, *BBC News*, 11 July <http://news.bbc. co.uk/1/hi/sci/tech/2122619.stm>.

Wright, S. (ed.) (1990) *Preventing A Biological Arms Race* (Cambridge, MA: MIT Press).

Yoxen, E. (1983) *The Gene Business* (London: Pan).

46

DNA IDENTIFICATION AND SURVEILLANCE CREEP

Dorothy Nelkin and Lori Andrews

Source: P. Conrad and J. Gabe (eds), *Sociological Perspectives on the New Genetics*, Oxford: Blackwell, 1999, pp. 191–208.

Introduction

A wad of spit, a spot of blood, a semen stain, or a single hair is all that is necessary to create a DNA 'fingerprint'. DNA profiles can be extracted not only from blood or sperm at a crime scene, but also from objects touched by a person's hands, and from saliva used to lick stamps. From a tiny sample of body tissue, a forensic laboratory can create an image on an autoradiogram; a cluster of horizontal bands form a pattern, resembling a bar code. Despite its imperfect accuracy rate, law enforcement officials argue that DNA fingerprinting can be a unique way of identifying people and thus should be used as the gold standard.

DNA analysis had been developed in a medical context as a technique to identify the markers that indicate familial disorders. But in 1983, a British geneticist used the technique to identify a rapist. Subsequently, DNA testing spread out of the medical sphere into the sphere of public surveillance. As a non-intrusive and easy procedure, DNA fingerprinting has been used in many nonmedical contexts. It appeals to military, law enforcement and other governmental authorities: those seeking evidence to establish the identity of a dead body, a missing person, a biological relative, or the perpetrator of a crime. In 1990, the US Congress authorised and funded a military programme mandating the collection of blood and tissue for DNA testing of all military personnel. The FBI and the law enforcement agencies in every state require convicted felons to have a sample of their body tissue banked and tested for purposes of future identification. In some states, non-violent offenders and misdemeanants are included. Some countries require immigrants to provide a DNA sample as a condition of entry. DNA samples are collected and stored for the identification of missing children, or elderly

Alzheimer's victims, or babies switched at birth, or genetic fathers in child support and paternity disputes. Men have brought their children to genetics clinics and had them secretly tested to determine if they were the 'real' fathers or if their wives had had an affair. The collection of body fluids for DNA identification is an expanding enterprise. At least 50 commercial DNA testing centres have been established in response to identity quests. A mail order company is selling a 'Defender DNA' device. If a perpetrator comes close, the potential victim jabs a pin into the attacker, activating an alarm and gathering an identifying DNA sample.

There are, of course, many reasonable purposes for DNA identification. Why not facilitate crime control by having records of recidivists? Why not develop accurate means of identifying missing persons or the remains of soldiers killed in war? To the military and law enforcement officials who collect body fluids for DNA identification, body material is an efficient means to implement legitimate policy goals. But the expanding use of DNA identification also reinforces a pervasive trend towards increased surveillance.

The early development of computer data systems attracted sociological attention to the possibilities and problems of growing surveillance. James Rule (1974) warned of institutional trends in which powerful bureaucracies would collect and store information on private persons to control their behaviour. He predicted a future of increasingly efficient mass surveillance and control. Michel Foucault conceptualised surveillance as a means of 'normalization' and focused attention on the production of knowledge in the service of power. Tests, he wrote, have become a means to 'compare, differentiate, hierarchize, homogenize, exclude' (1979: 183–4). Responding to the expansion of national information systems in the 1980s, Kenneth Laudon (1986) warned about the implications of a 'dossier society' for personal privacy and civil rights. Gary Marx focused on the use of these powerful data collection and data sharing tools in undercover police work, warning of 'almost imperceptible surveillance creep' marked by subtle, invisible, involuntary forms of social control (1988: 2). Criminologist David Garland (1995) takes a different view, arguing that surveillance technologies are essential and inevitable in complex societies that require formalised systems of data gathering. In his view, they are not in themselves the problem; rather, the issue is how they can be regulated to avoid abuse.

However, the development of DNA tests and banking systems has intensified concerns about surveillance technologies. For DNA samples, more than just a source of identification, can reveal information about a person's health and predispositions. How will organisations and political systems use DNA identification? Who will have access to the data? How can those in control of DNA data appropriately balance the identification benefits of this technology while protecting the social values of individual freedom and privacy? DNA databanking may be a useful tool for fighting crime and meeting military exigencies, but it is also subject to abuse for political or economic ends.

This chapter describes the gradual expansion of mandatory genetic testing for DNA identification. We focus on the disputes that develop when those required to provide DNA samples raise concerns about loss of benefits, psychological harm, and discrimination based on the information revealed by their DNA. And we use these disputes to analyse the problems of 'surveillance creep' as growing numbers of people have their DNA on file (McEwen and Reilly 1994).

The military DNA collection programme

In January 1995, Corporal Joseph Vlacovsky and Lance Corporal John Mayfield III, two Marines stationed at the Kaneohe base near Honolulu, were ordered to provide blood and cheek epithelial cell samples as part of the military's mandatory genetic testing programme. Their DNA samples would be stored at the Department of Defense DNA Repository in order to facilitate the efficient identification of the remains of soldiers killed in battle. They refused to comply and were court-martialled for violating an order from their superior officer (Mayfield v. Dalton 1995). They became 'the first DNA conscientious objectors'.

The Department of Defense (DOD) began collecting blood and tissue from every person in the military services in 1992. Included in this programme are all active duty and reserve personnel as well as civilian employees and contractors, even though these persons are much less likely to become unknown soldiers. The tissue is collected, analysed, and then stored in the DOD's DNA Repository in Gaithersburg, Maryland. Each individual is provided with two sealed plastic cards that have a fingerprint, signature, blood stain and oral swab, and a bar code; the cost of this kit, pencil included, is $3.00 (not including the cost of analysing the DNA). DNA samples are vacuum sealed and frozen to ensure their survival for 40 years. When needed, bits of bodies can be identified by matching their DNA to the samples kept on file. By 2001, over 4 million tissue samples from military personnel will be in the Maryland repository. It is the largest DNA bank in the world.

Military authorities regard the identification of remains as a compelling interest for soldiers and their families. In every war, the military has established departments to identify the dead, and have developed techniques to assist in identification. Lt. Col. Victor Weedn, the initial programme manager of the Armed Forces DNA Identification Laboratory, is a forensic pathologist—a lawyer, MD, and expert on DNA analysis. He had used DNA techniques to identify the skeletal remains of Czar Nicholas II, the victims recovered from TWA Flight 800, and the members of the Branch Davidian cult killed in the fire in Waco, Texas.

Weedn explains the importance of a mandated military testing programme: 'It's an issue to the soldiers, sailors and airmen. They want to know that—if they pay the ultimate sacrifice—they will be remembered' (Weedn 1996). He talks about the effectiveness of DNA identification and the time it would save in the slow and painful task of identifying human remains. He hopes to make the Tomb of the Unknown Soldier a thing of the past.

The problem of identifying remains was a major issue during World War I, with its devastating death toll and enormous number of bodies that had been dismembered beyond identification (Bourke 1996). The dead body of the 'unknown soldier' became a potent symbol of the horror of war. But Vlacovsky and Mayfield cared less about the identification of their dead bodies than their ability to control the integrity of their living bodies. Explaining his willingness to risk a court-martial, Vlacovsky said: 'This won't destroy the rest of my life. When this is over, I will still have control over my DNA'.

The Marines argued in court that the taking of their tissue violated their Fourth Amendment right to privacy. They regarded the requirement as unreasonable search and seizure: 'I expected to give up some privacy when I joined the military', said Mayfield. 'But,' he added, 'It doesn't say we hereby waive our constitutional rights'. 'It is our God-given right to maintain possession of our genes' (Mayfield 1995). Having little trust in military authorities, they also suspected that the tissue samples would be misused or used without their knowledge or consent for purposes other than the identification of their remains. Samples could be used to assess soldiers' predisposition to homosexuality (Hamer and Copeland 1994), or to a genetic disease and then used against their personal interests. Or their DNA could be used for purposes they oppose, such as the development of biological weapons (Cole 1996). And the information collected from their DNA samples could be available to law enforcement authorities in criminal investigations.

The Council for Responsible Genetics and a clinician, Paul Billings, submitted affidavits supporting the objecting Marines. They documented cases where access to genetic information revealing predisposition to genetic disease had resulted in discrimination. 'Thousands of tests could be done on these samples', Billings said: 'The military may have kept the door open as a way to counteract rising benefits costs by excluding coverage for those with pre-existing conditions that can be discovered in DNA samples' (Billings 1995).

Weedn dismissed such concerns: 'When you've licked a stamp on your tax return you've sent the government a DNA sample' (indeed, FBI had tested the saliva on postage stamps to link a suspect to the World Trade Center bombing and to identify the Unabomber). Weedn insisted that 'each specimen is treated as a medical specimen with confidentiality and respect'. Questioned about the initial DOD policy of keeping specimens after service members were discharged from the military—a policy later changed—he replied that it would be 'extremely costly and time consuming to return or destroy the specimens' (Weedn 1996). Instead, Weedn called for greater trust; he said he would not abuse the information in the DNA files.

Weedn's plea for trust did not convince the Marines. 'With nuclear testing, they just handed people some dark glasses and said "here, watch the bomb go off" . . . In the 1950s they gave LSD to army troops . . . In the 1970s they sprayed Agent Orange over their troops' (Chadwin 1996). Trust could hardly have been strengthened when Assistant US Attorney Theodore Meeker dismissed the importance of informed consent: 'If the military's use of unproved drugs on its personnel does not require informed consent, collection and storage of blood samples and oral swabs for possible use in identifying human remains does not require consent' (Meeker 1995).

Though both Vlacovsky and Mayfield had exemplary service records, they were threatened with court-martial proceedings, incarceration, fines, and dishonourable discharge for refusing to obey orders., However, a military judge dropped the charges in light of the fact that there were no existing regulations dealing with the consequences of failing to comply with the programme. Thus, both Marines received honourable discharges with veterans benefits—and they kept their DNA.

By the time of the decision, however, they had become interested in the general issues raised by mandatory testing. So Vlacovsky and Mayfield filed a civil suit in the District Court in Hawaii against the DOD on behalf of all service personnel claiming that the military programme ignored existing protocols for confidentiality and consent. The court dismissed their suit, contrasting the 'hypothetical' arguments of the Marines to the 'compelling interest' of the military to account for its troops. The court held that the taking of a blood sample was not unreasonable seizure and thus did not violate the Fourth Amendment. Taking blood was, after all, legally considered a minimal intrusion. And, because there was no immediate plan to use the specimens for research, the Nuremberg Code's requirements for informed consent were not relevant.

The DOD did amend the rules for its original banking programme. They reiterated that the samples would only be used for the identification of human remains unless they were subpoenaed for a criminal investigation or 'other uses compelled by other applicable law'. The programme was accelerated—testing of all current military personnel, including civilian employees, was to be completed by 1999.

Other cases quickly followed. In April 1996, Sgt. Warren Sinclair, age 33, a 14-year Air Force veteran and medical equipment repairman, refused to submit blood samples for genetic testing. Vlacovsky and Mayfield generally mistrusted military motives, but Sinclair, an African American, had specific political concerns about the use of his body tissue. He was convinced that DNA samples would be used to support racist claims. 'Would we ask Jews to give their genes to Germans? No. . . . Until the issue of racism is resolved, Afro-Americans should maintain possession of their genetic material' (Sinclair 1996). Sinclair recalled the use of genetic testing in the 1970s when blacks in the Navy were tested for sickle cell carrier status. Though no scientific evidence suggests this would affect a person's health (only reproductive decisions), those found to be carriers were disqualified from certain jobs. Black servicemen interpreted the exclusion as one more way to restrict their opportunities.

The Air Force court ruled against Sinclair, arguing that the interest of the government in assuring the identification of remains outweighed the intrusiveness of taking blood. Sinclair was convicted by court-martial on 10 May 1996 and sentenced to 14 days of hard labour and a two-grade reduction in rank (US v. Sinclair 1996).

In April 1996, Donald P. Power, a 1st Class Petty Officer and Navy nuclear technician, refused to give a DNA specimen because it violated his religious principles as a member of a Native American Lodge. Power said: 'My body is a sacred recipe to me, and I didn't think I should share it . . . They were not holding a part of me on a shelf. . . . You find personal power in knowing who you are' (Hinde 1997). For his refusal, Power lost a stripe, his security classification, and 40 per cent of his income. He applied for a waiver on grounds of religious freedom and it was accepted 18 months later. But few members of the armed services will be able to make use of the narrow religious exception, and moral objections are not enough to avoid military rules.

Those who refused to comply with mandatory genetic testing were challenging longstanding assumptions about the authority of the military over the bodies of its men. The military, after all, sends bodies into battle and soldiers cannot refuse assignments that threaten their bodily integrity. Refusal was in effect a declaration of rights based on a view that DNA holds special meaning for the individual; it was beyond the usual domain of military intrusion. As one of the Marines put it: 'It's your genetic blueprint, how you were created. . . . Your body is one of the few things that you have control over'. Moreover, the Marines mistrusted the promises of confidentiality; they believed that their samples would become a useful and efficient resource not just for identification purposes, but for decisions about promotion, health insurance, and law enforcement.

DNA dragnets

Since World War II, law enforcement agencies have been expanding their information systems. But until the 1960s, many record-keeping systems functioned on the basis of local tradition, and the management of criminal justice information was uneven. By the late 1960s, the increasing sophistication of computers converged with growing fear of crime to encourage experiments with identification and surveillance systems (Marx 1988). In 1968, President

Johnson's Commission on Law Enforcement and the Administration of Justice declared that information and systems technology was the most important tool for controlling crime. The Commission also proposed the creation of a national computerised criminal history repository.

But the efforts to implement a system faced public opposition. Attitude surveys suggested that Americans had little confidence in institutional leaders, mistrusted centralised information systems, and feared the implications for liberty and privacy (McClosky and Brill 1986). The idea of a national identity centre raised fears that growing government surveillance would extend well beyond the criminal justice system. Critics emerged to warn of the potential for abuse, the unwarranted tracking of 'suspicious' persons, the selective surveillance of particular groups, the harassment of political activists, and the leakage of information to private organisations seeking information for employment or credit ratings.

Although proposals to collect and bank human tissue for DNA identification raise similar issues, there has been remarkably little protest against them. Indeed, they have been welcomed as an effective means to lower the cost of criminal investigations. People may worry that computer banks store information on their economic status or credit rating, but few believe that the collection of DNA samples will affect their personal interests. And the aura of science underlying DNA technologies contributes to the legitimacy of testing and overrides privacy concerns.

However, those who are directly affected by mandatory genetic testing have responded by using the courts to challenge the programmes. In 1991, six inmates from Virginia's Tazewell Correctional Unit Number 31 challenged the state's mandatory DNA testing programme (Jones v. Murray). Virginia, in 1989, had been the first state to require the collection of blood samples from convicted sex offenders and felons for use in a state DNA database. Law enforcement officials attempted to justify the programme by citing a study indicating high recidivism rates.

The inmates claimed that the Virginia programme was unconstitutional; that in the absence of individualised suspicion, mandatory extraction of DNA samples violated their Fourth Amendment right against search and seizure. Also, they argued, imposing blood test requirements as a condition of release would impose additional punishment for their crimes and interfere with the right to due process by putting extra conditions on possibilities of parole.

Like soldiers, prisoners relinquish certain rights. But the prisoners defined their right to bodily integrity in a distinctive category; their body fluids, their genetic blueprint, should not be violated even in the context of the prison system. The Virginia Court disagreed. It conceded that the state could not meet the standards of probable cause or individual suspicion. But the court balanced the government interests in deterring and detecting crime against the privacy interests of inmates, and found the law to be reasonable.

Convicted felons, said the court, already lose the right to privacy from routine searches of the cavities of their bodies and their prison cells. Most searches, however, are conducted to determine whether the inmates present a current danger—by, for example, concealing weapons. In contrast, the collection of DNA is to protect against a remote future risk.

Dissenting, Circuit Judge Murnaghan concurred with the decision as applied to violent felons, but lambasted the idea of collecting blood from prisoners who had been convicted of nonviolent crimes and who thus were unlikely to commit violent crimes in the future. He questioned whether the lack of reasonable expectation of privacy in a prison cell should extend to permit searches of the body fluids of every felon, violent and nonviolent alike. Perhaps most important, Judge Murnaghan suggested that 'The only state interest offered by the Commonwealth for including non-violent felons is administrative ease'. Reviving the 1980s concerns about the social implications of creating computerised criminal history systems, Murnaghan expressed 'a deep, disturbing and overriding concern that . . . the Commonwealth may be successful in taking significant strides towards the establishment of a future police state, in which broad and vague concerns for administrative efficiency will serve to support substantial intrusions into the privacy of citizens' (Murnaghan 1992).

Other states also began to develop DNA identification programmes, and in 1993, the FBI implemented CODIS, a national programme to assist federal, state, and local law enforcement agencies, support development of a population statistical data base, improve DNA forensic analysis methods, and serve humanitarian purposes such as the identification of missing persons or the human remains from mass disasters. The FBI promoted CODIS on the grounds of 'productivity and efficiency'. Former Director of the FBI crime laboratory, John Hicks, had described the computer databank as 'nothing more than an information management and screening tool' (Hoeffel 1990). He expected that 'It will save time and effort, and courts will have fewer cases to process because investigations can be better focused and coordinated' (FBI 1991). CODIS links the DNA profiles of convicts gathered by scattered law enforcement DNA labs, encourages uniform standards, and pools DNA data to facilitate identification of criminals across borders. The 1994 Crime Control Act reinforced these efforts through a provision for coordinating nationwide DNA databank systems. A report, commissioned by the Justice Department to implement the Act, announced an award of $8.75 million in grants to states and city crime agencies to improve their DNA testing capabilities (Butterfield 1996). As a result of that incentive, all 50 states have adopted laws requiring specified offenders to provide blood samples for forensic DNA testing.

State statutes vary. Most statutes initially required that saliva and blood samples be obtained from sex offenders on their release from prison, or as a condition of probation or parole. Statistically, sex offenders do have a high rate of recidivism. Strategically, selecting a group with such a negative public image for mandatory DNA testing was unlikely to provoke objections. Once in place, the DNA programmes expanded to cover a range of both violent and nonviolent crimes. In New York, blood samples are taken from defendants convicted of felony, sex offence, felony assaults, incest, or prison escape. In Virginia, those convicted of mail fraud must also store samples, and Mississippi law includes persons convicted of elder abuse. In Wisconsin, if a judge determines that a crime 'might have escalated to a sex crime' he can require registration. At least seven states test misdemeanants.

Provisions for access to criminal data banks also vary; the most restrictive statutes allow access only for law enforcement purposes. Maintaining confidentiality is also problematic: there are over 19,000 law enforcement agencies in the US and over 51,000 additional criminal justice agencies worldwide, which means over 600,000 employees have direct access to the National Crime Information Center maintained by the FBI.

Law enforcement agencies also enjoy access to many other DNA sources. The military had admitted it is willing to release its data to law enforcement officials (Gill 1997: 185). A proposed federal law, the Human Genome Privacy Act, would allow police officers to have access to hospital diagnostic DNA collections without patient authorization (Pear 1996). It is easier for police investigators to gain access to medical records than to bank records, e-mail information, or video-rental receipts—all of which are protected by federal privacy statutes.

Yet law enforcement purposes can be very broadly defined. Indeed, what is a law enforcement purpose? Would it include the identification of a man who failed to provide child support? Ohio explicitly allows its databank to be used pursuant to a court order for proceedings establishing paternity or maternity. Laws in New Jersey and Maryland allow their DNA banks to be used to find genetic parents where the party seeking to search the databank has obtained a court order.

DNA data could also be used to explore whether an individual has a genetic profile that might predispose him to aggressive acts (Andrews 1998). Both political parties have made crime a priority issue and state legislatures are granting money to build more and higher security prisons in the hopes that this will reduce crime. There is increased discussion in the popular and policy media about predicting and preventing crime by identifying those people thought to have 'criminal genes' (Nelkin and Lindee 1995). Information that purports to identify those 'predisposed' to violent behaviour holds considerable policy appeal. Yet many innocent people could be snared in the law enforcement net if their genes suggested they were potentially violent.

The military and prisons are self contained, or what sociologist Erving Goffman called 'total institutions' that operate under special rules with respect to social control and the right to intrude on the privacy of individuals (Goffman 1961). But the attraction of efficiency has also encouraged a wider use of DNA fingerprinting as DNA dragnets are used to search for suspects in serious crimes. These may involve many innocent people.

In 1990, the San Diego police department collected blood samples from 800 men during a search for a serial killer. They selected men who matched the description of a 'dark-skinned male'. In 1993, the two-year-old daughter of an American Army Sergeant stationed in Germany was kidnapped, raped, and murdered. The murderer was identified after an eight-month dragnet investigation that included DNA screening of 1900 men who had been near the military housing complex (Atkinson 1995). In 1995, police in Prince George County, Florida, searching for a serial rapist, collected saliva samples from more than 2300 men, stopping them at random on a road near the scene of the crime. In 1998, 50 hospital employees were asked to provide saliva samples in a dragnet search for the strangler of a popular nursing administrator. The police chief, John S. Farrell explained his use of the DNA dragnet: 'It is a way to focus the investigation efficiently . . . in a more business like fashion. . . . It would save time and money [each DNA test cost only $30] . . . It is an extremely cost-effective tool' (Pan 1998). But the African Americans on the hospital staff felt they were targets of discriminatory suspicion.

In his dissent in Jones v. Murray (1992), Judge Murnaghan contended that efficiency is not a legitimate interest and could be used to justify the testing of any citizen on grounds that it might reduce administrative work-loads. He worried that arguments for administrative efficiency could justify DNA testing of all citizens at birth simple because of the likelihood of some future manifestation of violence; even when there is no specific evidence this will occur. Murnaghan also pointed out that, if the state was willing to allow the collection of blood from nonviolent offenders, the same logic would allow 'the testing of other discrete populations, e.g. racial minorities or residents of underprivileged areas'. Under the majority's logic, the state could go into the inner city and demand blood samples.

A similar argument was effective in a Massachusetts case in which a court halted collection of DNA from inmates, probationers, and parolees on grounds that it violated Fourth Amendment Rights. Benjamin Keehn, a Boston Public Defender, had pointed to the slippery slope: 'Why not round up poor people? Poor people are more likely to have their DNA on file. Of course there are benefits every time you get a cold hit. There are going to be dramatic success stories. But where does it stop? Why not take DNA samples at birth?' (Keehn 1998).

Refusal to comply with requests to submit a blood sample in a DNA dragnet is bound to imply guilt. Submission to testing is not necessarily voluntary. In addition, the collection of body tissue for DNA testing presents a distinctive set of problems, for unlike fingerprints, tissue samples expose individuals to the risk that the cells will be used for purposes other than identification. They can reveal information about predisposition to disease or physical traits, a fact that becomes increasingly problematic as public authorities responsible for social control in an expanding range of situations—such as immigration—are attracted to DNA testing as a means to extend surveillance and facilitate investigations.

Surveillance creep

In 1989, the Thatcher government in Great Britain instituted a policy allowing officials to use DNA fingerprint tests on immigrant applicants seeking to prove they have relatives in Britain. Over the subsequent few years, 18,000 tests were carried out on immigrants. Most testing is done in UK consulates in the country of origin. DNA testing is considered a cheap and more effective alternative to hours of questioning. But the British testing programme was criticised as racially discriminatory, and as creating 'a bureaucratic barrier and financial barrier' to immigration. Officially, applicants have a right to refuse to be tested, but rarely do so. Many do not understand why British officials want their blood, and, as in DNA dragnets, the implications of refusal lead most people to comply (Evans 1995).

The practice of testing immigrants spread to Canada in 1991. The purpose of the Canadian policy was to help immigrants with inadequate documents reunite with their families. But the cost, borne by the families sponsoring immigrants, is high: the government charged $975 for the applicant and $325 for each relative sponsored. A national committee on the status of women called the immigrant testing programme a way to discourage Third World immigration. High priced tests resulted in differential application of immigration policies, thereby 'adding fuel to a growing tide of racism and anti-immigrant sentiment . . . It would be more fair for Canada to come out publicly and say we don't want family sponsorship any more rather than put up all kinds of ridiculous obstacles which cost so much money' (Rinehart 1995).

The United States has never required DNA testing for immigrants, but in the anti-immigration fervour of the 1990s, pilot projects were set in motion to develop worker ID cards that would be linked to an electronic data base. They would include fingerprints, voice prints, and DNA sequences—to assure that only citizens and legal aliens hold jobs (Davis 1995). Under the US Immigration and nationality Act of 1952, aliens can be excluded from immigration for mental or physical defects, diseases, or disabilities. Information about diseases, disabilities, and predispositions could be gleaned from the applicant's DNA and therefore used as a basis to deny the application. Should genetic testing, then, be required of all immigration applicants?

In an odd extension of genetic testing, an Israeli researcher applied DNA techniques to the identification of the true Cohanim—the Orthodox Jews who trace their lineage 3300 years back to the first high priest. He found a genetic pattern on the Y chromosome that is shared by the descendants of the Cohanim (Grady 1997). These descendants are accorded higher status and are the only rabbis to perform certain religious duties. A DNA test could be used to validate a presumption of priesthood.

As DNA replaces other technologies such as HLA blood typing or fingerprints, the collection and banking of body tissue for DNA analysis is becoming mandatory in more and more contexts. It is the preferred technology for identifying recidivists and remains, but its use is expanding.

Increasingly, it is not doctors or public health officials who collect tissue samples for identification, but government, law enforcement agencies, the military, and immigration authorities. Private firms are increasingly involved as collecting tissue becomes a growing business. One company, Identigene, advertises in taxis, subway cars, and bill boards (Call 1–800 DNA TYPE). It collects tissue for DNA identification (at $600) that can establish paternity in child support disputes or family relationships for immigration purposes. A phone call to the company suggests they have not considered questions about the privacy of stored samples or who should have access to them.

Greater efficiency and reduced costs are also encouraging the 'surveillance creep' that Marx (1988) had predicted. In 1998, a senior member of the British police force called for a national DNA testing data base of the entire population, arguing that it would cut the time and cost of investigating crimes (Gammon 1998). Techniques of DNA are improving and costs are declining. The Department of Justice expects that the average cost of a DNA test can be reduced to less than $10. Technological developments have also increased the feasibility of manufacturing DNA testing dogtags. A one-centimetre-sized chip can contain all the genetic information needed to identify 10s of thousands of genes at a time and to store them on a DNA card.

Why worry?

At first glance, DNA identification programmes seem an efficient way of solving crimes, preventing immigration fraud, and identifying soldiers who die serving their country. The increased collection of DNA data and the expansion of centralised DNA banks have evoked little public response. It may seem comforting to law-abiding citizens who view the technology as resolving social problems with little relevance to themselves. But the cases described above suggest reasons for concern. Just as the targets of DNA fingerprinting have expanded, so too have the uses of DNA information. Yet, there are possibilities for error in the collecting and banking of DNA samples and many potentials for abuse. Innocent people can be snared in the DNA identification net.

Possibilities of error

DNA fingerprinting has been called the 'gold standard' of identification. The technology is premised on the assumption that DNA fingerprints are unique for each individual. Indeed, a printout of a person's entire genome of more than three billion base pairs would be a unique identifier, except for identical twins. However, forensic tests look at only a small subset of a person's genome, and segments may be shared by other individuals especially within an ethnic group. Statistically, the chances that DNA from a crime scene actually comes from a particular suspect depends on how many other individuals could share that DNA pattern. Laboratories use a reference population of samples to assess the probability of a random match. However, selecting reference populations that are appropriately representative is often difficult (Lander 1994).

In addition, there is potential for laboratory error. Proficiency tests, used to check laboratory capabilities, have revealed many problems. In one proficiency test, a Cellmark analyst, instead of loading the two samples in different lanes, loaded the same sample in both lanes—and unwittingly declared a match where there was none. In a 1993 study, 45 laboratories were asked whether particular DNA samples matched. The labs were undoubtedly using their best techniques since they knew they were being tested. Yet, in the 223 tests performed, the study identified matches in 18 cases where matches did not exist (Koehler et al. 1995). If these had been real trials, innocent people would have been convicted.

In an actual case, the bloodstain found on the watch of a defendant seemed to match the blood from the victim, both showing three DNA bands. But the DNA from the watch showed two extra bands, not mentioned in the laboratory report. Rather than concluding that the accused individual might be innocent, the company decided that the sample probably included non-human contaminants. But, in an unusual move, genetic experts for both the prosecution and defence analysed the forensic DNA work and issued a joint statement that it was unreliable: 'If these data were submitted to a peer-reviewed journal in support of a conclusion, they would not be accepted', said the experts (Lander 1994: 197).

The interpretation of DNA fingerprints is also prone to bias. In a sociological study of forensic scientists, William Thompson documented ways in which the institutional context in which these scientists work influences 'the development and validation of new testing procedures, the interpretation of results, and their presentation in court' (Thompson 1997, 1117). Forensic scientists have professional incentives to adopt the goals of their clients and this may compromise scientific detachment. Keen on justifying the value of their services, they may be reluctant to question the reliability of tests. And in ambiguous situations, they may make interpretive errors (Thompson 1997).

In a study of responses to DNA evidence, University of Texas researchers set up a team of mock jurors (actually university students) and found they did not appreciate the importance of laboratory error rates. Jurors are often faced with 'misleading assurances from forensic science experts that laboratory errors are impossible or nearly impossible'. The study indicated to the researchers that DNA evidence 'could lead to conviction where *acquittals* might otherwise result' (Koehler et al. 1995).

Errors can also be a problem in the military context. To justify the use of DNA identification in the military, Lt. Col. Weedn argued that 15 to 30 per cent of previously collected fingerprint cards could not be used for identification because the cards were smudged. But DNA samples are no less prone to error and sloppy handling.

Changing methods of analysis open new possibilities of error, but neither the laws regulating forensic programmes nor those allowing court admission of DNA evidence provide adequate protection. 'Providing blanket admissibility for any type of DNA analysis whatsoever (as these loosely drawn statutes might appear to do) is an invitation for mischief', says geneticist Eric Lander (Lander 1994).

Potentials for abuse

In the cases we have described, plaintiffs feared their samples would be used for other than identification purposes, that military or forensic investigators might also look at health status and other information that could lead to genetic discrimination. To date, forensic DNA bankers have been able to claim they are testing only for identification purposes and that their samples could not be used for revealing health problems. But there are possibilities for abuse in the future: certain markers that were not thought to indicate medical risks may turn out to reveal health information. Some state forensic departments are buying used gene sequencers from companies, putting within their control the ability to sequence genes including those that indicate health risks.

Abuses are likely to increase as interest turns to behavioural genetics, for predictive information about behaviour may be useful in both military and criminal contexts. The uses of such data are likely to reflect existing stereotypes. The scenarios Judge Murnaghan presented are realistic. If genes associated with aggressive or criminal behaviour were identified, it would be easy to envisage selective testing of black men in light of current stereotypes associating crime with race. Numerous incidents of selective investigation suggest the way stereotypes might influence data banking practices. In Pennsylvania, state police instructed bank employees to photograph suspicious looking blacks, in effect, creating a criminal profile applying to specific race groups. Airport security profiles have also used race as an indicator of potential terrorism or drug smuggling crimes.

If genetic predispositions were identified for antisocial acts, social interests could encourage measures to prevent crime by circumscribing the rights of people thought to have criminal genes. This might include identifying those with antisocial genes, keeping them under surveillance, or preventively detaining them. Their profiles might be kept on file to be consulted when a crime is committed. In the military context, identification of genetic predisposition to homosexuality could—as the Marines suspected—have devastating consequences. Though the military may not directly ask, the genes may tell. Soldiers with that genetic makeup could simply be discharged even when there was no evidence they engaged in homosexual behaviour.

Concerns about such misuse of surveillance technologies are not without basis. In the 1960s and 1970s the FBI and local law enforcement officials kept tabs on thousands of citizens who were active in the civil rights and anti-war movements, and in some cases harassed innocent people. Today, the tools of surveillance are improving with the growing capacity of central data banks that include DNA. The Fourth Amendment might protect against secondary use of samples that were collected for purposes of identification, but this is a matter of dispute. Indeed, the concerns about privacy that had been raised by critics in the early days of centralised computer data banks are increasingly urgent when the data are DNA.

Violations of privacy

To the marines, prisoners, and immigrants who challenged mandatory testing, body tissue holds religious, social and political meanings, and privacy concerns were critical. For Native American Donald Power, the military's taking of DNA violated his religious beliefs. Mayfield and Vlacovsky defined their DNA in terms of personal identity. And even prisoners, whose privacy rights are compromised, defined the taking of DNA as different from the searching of prison cells or body cavities.

While convicted felons would have lesser rights than other individuals, the potential uses of forensic DNA banks affect more than just criminals. Those tested in a DNA dragnet because they happen to be in an area will then have their DNA samples on file. Victims also have their DNA tested at forensic labs and their samples may be banked. Family members related to the offenders are also affected because health information about the offender (say, a genetic predisposition to cancer) indicates genetic risks to relatives as well.

Collecting tissue samples from an individual who has not been charged or convicted of a crime—as in a DNA dragnet—could violate the person's Fourth Amendment right to be free from unreasonable searches and seizures. However, persuaded by the 'scientific' nature of 'profiling', courts have allowed the random stopping of individuals thought to fit criminal profiles of hijackers or drug smugglers. One judge, referring to a hijacker profiles as 'elegant and objective' was convinced that hijackers had characteristics 'markedly distinguishing them from the general traveling public' (US v. Lopez 1971).

As a controlling tool for modern institutions and complex organisations, the use of DNA is appealing. But in light of problems of privacy and potentials for abuse, testing should be closely regulated. The Fourth Amendment could be used to prohibit non-consensual tissue collection so as to limit the taking of DNA samples (Krent 1992). Informed consent doctrines could be more widely enforced to limit possibilities of abuse. Or the property interests in the body could be recognised, giving people greater control over what is done with their tissue (Andrews 1986). There could be greater limits on access to DNA information: Vermont legislation, for example, includes language in a databanking bill to prohibit the use of genetic information as a basis for employment and insurance decisions.

Though the practice of testing and banking DNA is extending to a widening range of people—from soldiers who go to battle to chaplain's assistants, from violent to nonviolent felons, from immigrant families to foreign adoptees—there has been little public concern about the practice. The possibilities of error are deflected by faith in science and, especially, the promise of genetics. Potential abuses of DNA data are deflected by perceptions that surveillance pertains to 'others'—the soldier, the criminal or the illegal immigrant—and a belief that DNA identification is an efficient means to maintain social order.

Moreover, Americans these days have few expectations of privacy, accepting surveillance in many spheres. Shoppers accept television surveillance in department stores, strollers accept camera surveillance in public parks (Nelkin 1995). The dossier society that Laudon and Rule predicted years ago has crept up on us; facilitated by the ability to gather, store, and access information—not just about finances, credit rating, or consumer preferences, but about the body, identity, and health.

In 1972, a legal scholar wrote that the social security numbers assigned to us at birth have become a 'leash around our necks, subjecting us to constant monitoring and making credible the fear of the fabled womb-to-tomb dossier' (Miller 1972). Could DNA identifiers eventually replace social security numbers, requiring every person to have DNA on file? Today, according to Janet Hoeffel (1990), 'It is not merely paranoia to imagine the incremental steps the current government would take that would lead from a data bank with DNA profiles on criminals to a data bank with profiles on each of us'. Indeed, molecular biologist Leroy Hood has predicted that within 20 years all Americans will carry a credit card type plastic stripe that contains computer readouts of their personal genomes: 'Your entire genome and medical history will be on a credit card' (Hood 1996).

Acknowledgements

We wish to acknowledge the support of the National Science Foundation EVS Program grant #SBR-9710345, the assistance of Michelle Hibbert, and the comments of David Garland and David H. Kaye.

References

Andrews, L. (1986) My body, my property, *Hastings Center Report*, October, 28–38.

Andrews, L. (1998) Predicting and punishing anti-social acts: how the criminal justice system might use behavioral genetics. In Rothstein, M. and Carson, R. (eds) *Behavioral Genetics and Society: the Clash of Culture and Biology*. Baltimore, MD: John Hopkins Press.

Atkinson, R. (1995) DNA samples catch American killer of toddler in Germany, *The Washington Post*, 1 January, A27.

Billings, P. (1995) Amicus in Mayfield v. Dalton 95–00344.

Bourke, J. (1996) *Dismembering the Male: Men's Bodies, Britain and the Great War*. Chicago: University of Chicago Press.

Butterfield, F. (1996) US has plans to broaden availability of DNA testing, *New York Times*, 14 July.

Chadwin, D. (1996) The DNA war, *The Village Voice*, May, 14.

Cole, L. (1996) The specter of biological weapons, *Scientific American*, December, 62.

Davis, A. (1995) Digital IDs for workers in the cards, *National Law Journal*, 10 April, 17.

Evans, K. (1995) Targets of the genetic inquisition, *The Guardian*, 29 March, 12.

Federal Bureau of Investigation (1991) Legislative guidelines for DNA databases, US Department of Justice, November.

Foucault, M. (1979) *Discipline and Punish*. New York: Vintage Books.

Gammon, P. (1998) cited in UK police chief calls for national DNA data base, *Nature*, 393, 14 May, 106.

Garland, D. (1995) Surveillance and society, *CJM*, 20, Summer, 3–4.

Gill, S. (1997) The military's DNA registry: an analysis of current law and a proposal for safeguards, *Naval Law Review*, 44, 175–222.

Goffman, E. (1961) *Asylums*. New York: Anchor Books.

Grady, D. (1997) Who is Aaron's heir? *New York Times*, 19 January, 4.

Hamer, D. and Copeland, P. (1994) *The Science of Desire*. New York: Simon and Schuster.

Hinde, J. (1997) Their hands on your genes, *Times Higher*, 7 March, 19.

Hoeffel, J. (1990) The dark side of DNA profiling, *Stanford Law Review*, 42 January, 465–538.

Hood L. (1996) quoted in Garrett, Laurie The dots are almost connected, *Los Angeles Times Magazine*, 3 March, 49.

Keehn, B. (1998) quoted in Goldberg, Carey DNA databanks giving police a powerful weapon, and critics, *New York Times*, 19 February, A1, A12.

Koehler, J., Chia, A. and Lindsey, S. (1995) The random match probability in DNA evidence: irrelevant and prejudicial? *Jurimetrics*, 35, 201–9.

Krent, H.E. (1992) Of diaries and DNA banks: use regulations under the Fourth Amendment, *Texas Law Review*, 74, 49–100.

Lander, E. (1994) DNA fingerprinting, science law and the ultimate identifier. In Kevles, D. and Hood, L. (eds) *The Code of Codes*. Cambridge: Harvard University Press.

Laudon, K. (1986) *The Dossier Society*. New York: Columbia University Press.

Marx, G.T. (1988) *Undercover: Police Surveillance in America*. Berkeley: University of California Press.

Mayfield, J. (1995) quoted in Essoyan, S. Two marines challenge Pentagon order, *Los Angeles Times*, 27 December, A1.

McCloskey, H. and Brill, A. (1986) *Dimensions of Tolerance*. New York: Russell Sage Foundation.

McKewen, J.E. and Reilly, P. (1994) A review of state legislation on DNA forensic data banking, *American Journal of Human Genetics*, 54, 941–58.

Meeker, T. (1995) quoted in Essoyan, S. Two marines challenge Pentagon order, *Los Angeles Times*, 27 December, 5.

Muller, A. (1972) *The Assault on Privacy*. New York: New American Library.

Murnaghan, J. (1992). In *Jones v. Murray*, 962 F.2d. 302, at 313 (dissenting).

Nelkin, D. (1995) Forums of intrusion: comparing resistance to information technology and biotechnology. In Bauer, M. (ed) *Resistance to New Technology*. Cambridge: Cambridge University Press.

Nelkin, D. and Lindee, M.S. (1995) *The DNA Mystique: the Gene as a Cultural Icon*. New York: W.H. Freemen.

Pan, P. (1998) Prince George's chief has used serial testing before, *Washington Post*, 31 January, B01.

Pear, R. (1996) Clinton would broaden access of police to medical records, *New York Times*, 10 December.

Rinehart, D. (1995) DNA tests help not hinder relatives. *Montreal Gazette*, 10 May, A16.

Rule, J. (1974) *Private Lives and Police Surveillance*. New York: Schocken Books.

Sinclair, W. (1996) Memorandum to convening authorities, 27 June.

Thompson, W.C. (1997) A sociological perspective on the science of forensic DNA testing, *University of California Davis Law Review*, 30, 4, 1113–36.

Weedn, V. (1996) Stored biological specimens for military identification, Oberman Seminars, 14 June.

Cases

Jones v. Murray (1992) 962 F 2d 402 (4th Cir.), cert. denied. 506 US 977.

Mayfield v. Dalton, 901 Fu. Supp. 300 (d. Hawaii 1991), vacated and remanded, 109 F. 3d 1423 (9th Cir 1997).

US v Lopez (1971) 328 F Supp. 1077, 1081.

US v. Sinclair (1996) (Central Judicial Circuit, USAF Trial Judiciary, 10 May).

Part 3

PATENTING AND OWNERSHIP

47

GIFT RELATIONSHIPS IN GENETICS RESEARCH

Richard Tutton

Source: *Science as Culture* 11(4) (2002): 523–542.

Over the last few years, regulators and commentators in Britain have become increasingly interested in the question of how the voluntary donation of human biological samples to genetics research should be conceptualized. This can be attributed to growing scientific demand for the creation of tissue collections and databases of personal genetic information in support of biomedical research (Coghlan, 2000; Fears and Poste, 1999).

In the autumn of 1999, the Medical Research Council (MRC) and the Wellcome Trust announced their joint proposal to set up UK Biobank (which at the time was called the UK Population Biomedical Collection) (McKie, 2000). This will be a repository for biological samples, National Health Service medical records, and lifestyle data collected from a national cohort of 500,000 volunteers that will represent the ethnic make-up of British society (Meade, 2000). According to the Wellcome Trust, it will help to elucidate 'the role genetic variation plays in disease, health and illness', and produce 'deeper understandings of the complex interplay between genes, environment and lifestyle' (Wellcome, 1999).

One of the more notable aspects of the increased interest in samples emerged from the ethical guidelines published by the MRC (2001) on the collection, use and storage of human tissue. The authors recommend that blood or saliva donated by members of the public should be treated as gifts. They evoke Richard M. Titmuss' (1997) work on blood transfusion when they suggest that this is 'preferable from a moral and ethical point of view, as it promotes the "gift relationship" between participants and researchers, and underlines the altruistic motivation for participation in research' (MRC, 2001, p. 8). The appearance of the 'gift relationship' in the context of this discourse of professional ethics invited a number of questions: can a conceptualization of public participation in the British blood donor system of the 1960s provide an insight into the debates around contemporary genetics? Can it be relevant for an initiative such as UK Biobank? After all, its organizers acknowledge that it will depend on commercial involvement. And why should a group of geneticists, ethicists and medical legal scholars, drawing up a set of guidelines for the MRC, evoke Titmuss' celebrated and criticized concept at this particular time?

It is not my aim in this paper to examine in any detail these nonetheless pertinent questions. I want to consider instead how the public might conceive of the biological samples that they donate to genetics research, and whether the concepts of 'gift' and 'gift relationship' help to illuminate this question. In the first part of this paper, I outline the work of Richard M. Titmuss in his classic study of blood donor systems, *The Gift Relationship*, and begin to address its relevance to the contemporary social and scientific moment. In the second part, I turn to my own research and to evidence from my fieldwork in Orkney, the site of a genetic study conducted under the rubric of the 'European Human Genome Diversity Project'. While it represents an example of research on human genetic variation, which is at once distinctive from the biomedical concerns of the organizers of the MRC/Wellcome initiative, it will, nonetheless, begin to provide an insight into the complexities of the issues surrounding tissue donation for the purpose of genetics research. Moreover, my discussion of the study will indicate that the real value of the blood given by donors resides not in the physical substance itself, but in the genetic information derived from it.

The 'blood gift'

Richard Titmuss was professor of Social Administration at the London School of Economics for over two decades, and one of the most influential figures in that emerging discipline in the post World War II period (Marshall, 1973). In *The Gift Relationship*, Titmuss makes two significant contributions to public policy on blood transfusion. The first of these is related to the conclusions that he draws from his comparative analysis of the commercial system operated inthe United States and its voluntary counterpart in Britain. As Oakley and Ashton (1997) observe, in the 1960s some right wing thinkers associated with the Institute for Economic Affairs in Britain, had advocated the merits of a market-based donor system. However, Titmuss demonstrated that the US commercial system was more administratively inefficient, involved higher operating costs, and was more likely to supply contaminated blood than the British system. Oakley and Ashton credit these conclusions with having been an important influence on the US Government in its policy to adopt a voluntary system in the 1970s (Oakley and Ashton, 1997). Although more recent events, such as HIV/AIDS, have undermined some of the claims that Titmuss made about the benefits of the voluntary system in terms of its safety, his overall argument continues to inform the direction of public policy on blood transfusion in the twenty-first century (Berridge, 1997). The second of Titmuss' contributions arose from his use of the French anthropologist Marcel Mauss' (1997) work on gift exchange to inform his account of the British blood donor system.

In what is widely regarded as the classic exposition of gift theory, *The Gift: The Form and Reason for Exchange in Archaic Societies*, Mauss argues that gift was an important force for solidarity. He observed that in the indigenous cultures of Polynesia, Melanesia, and the American Northwest 'what they exchange is not solely property and wealth, movable and immovable goods, and things economically useful. In particular such exchanges are acts of politeness: banquets, rituals, military services, women, children, dances, festivals, and fairs, in which economic transaction is only one element' (Mauss, 1997, p. 5). Mauss notes that, although these exchanges appeared to be voluntary, in reality, they were founded on a set of interlocking obligations. He argues that 'to refuse to give, to fail to invite, just as to refuse to accept, is tantamount to declaring war; it is to reject the bonds of alliance and commonality' (p. 13). This has been supported by other writers, such as the anthropologist Mary Douglas who concludes: a 'gift that does not enhance solidarity is a contradiction' (Douglas, 1997, p. vii), while Lewis Hyde suggests that: 'any exchange, be it of ideas or of goats, will tend toward gift, if it is intended to recognize, establish and maintain community' (Hyde, 1983, p. 78).

A 'free human gift'

It is precisely in this understanding of the sociality of gift that Titmuss finds an attractive analogue to his own proposition that the voluntary donor system in Britain could also be conceptualized as a kind of gift exchange. Titmuss may not have been the first to see the potential relevance of Mauss' account for thinking about the donation of human tissue, but his use of it is notable for the way he departed from Mauss on certain key points.[1] Most crucially, he suggests that the British donor system was not defined by a set of obligations to give or reciprocate the 'blood gift' (see Figure 1). This related to his conception of the system as one organized around the principle of voluntary donation. People who gave blood did so from a sense of altruism with no expectation of a return. In support of this, he constructs an ideal type, the 'voluntary community donor', which he claimed constitutes the majority of those who gave blood in the British system. He observes:

> This type is the closet approximation in social reality to the abstract concept of a 'free human gift'. The primary characteristics of such donations are: the absence of tangible immediate rewards in monetary or non-monetary forms; the absence of penalties, financial or otherwise; and the knowledge among donors that their donations are

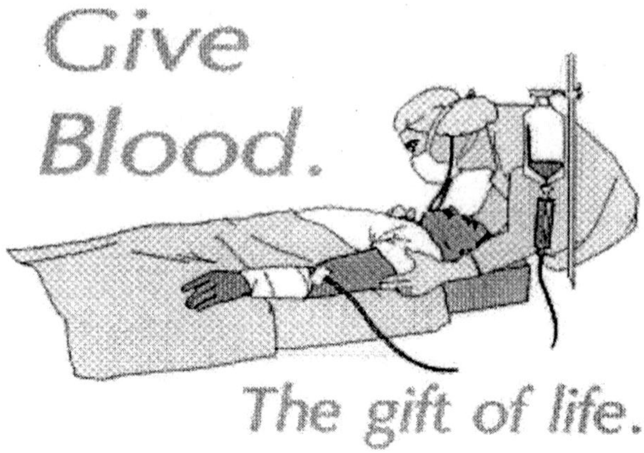

Figure 1 The blood gift.

Source: PPR Garden's blood donation website, www.geocities.com/pprgarden/blood.html.

> for unnamed strangers without distinction of age, sex, medical condition, income, class, religion or ethnic group.
>
> (Titmuss, 1997, p. 140)

The characteristics of this ideal type reflect many of the themes with which Titmuss was concerned throughout his research. As Hilary Rose notes in her essay 'Re-reading Titmuss', he worked within a framework that was defined by the 'conception of the good society as community' (Rose, 1981, p. 481) and by the understanding that 'altruism describes the character of the social relations of community' (p. 481). Moreover, as Ann Oakley and John Ashton note, Titmuss 'replaced the traditional conceptual framework for policy analysis focussing on administrative aspects with one that instead seeks to understand the underlying objectives of different social policies' (Oakley and Ashton, 1997, p. 8). In other words, he argues that it should be the role of social institutions, such as the National Health Service of which the blood donor system was a part, to encourage the expression of altruism in order to foster the relations of community in British society (Lafitte, 1972).

Reflecting on the previous three decades since the establishment of the welfare state, Titmuss concludes that:

> The most unsordid act of British social policy in the twentieth century has allowed and encouraged the sentiments of altruism, reciprocity and social duty to express themselves; to be made explicit and identifiable in measurable patterns of behaviour by all social groups and classes. In part this is attributable to the fact that, structurally and

> functionally, the Heath Service is not socially divisive; its universal and free access basis has contributed much [. . .] to the social liberties of the subject in allowing people the choice to give, or not to give, blood for unseen strangers.
>
> (Titmuss, 1997, p. 292)

The conceptualization of the blood donor system in the context of the modern welfare state through gift theory is not unproblematic. One reviewer of the *Gift Relationship* suggests that Titmuss' 'use of gift theory from anthropology is more token than actual' (Leach, 1971, p. 114), a point that relates particularly to the absence of obligation in the blood donor system, which was a key dimension to gift theory. Others since then have been stronger in their criticism. The literary theorist John Frow argues that the 'the state is not a "gift" domain because its forms of sociability do not involve the magical and dangerous ties of personal obligation. At this level, obligation is an abstract matter [. . .] in any strict sense, the concept of the gift is irrelevant to the structural understanding of modern societies, with the exception of the micro-level of everyday life' (Frow, 1996, p. 108).

While it may not be applicable in any strict sense, imaginatively applied, I suggest that Titmuss' notion of the gift has achieved a certain metaphorical resonance as part of a broader political discourse on the values of social equality, altruism and community. These were the values he saw embodied in the liberal welfare state. It is against this background that the 'gift relationship' should be read as essentially a political conceptualization of the British donor system, defending the welfare state, and attacking the commercialism represented by the US system of market-based healthcare. It helped frame the voluntary and commercial donor systems within a gift/commodity opposition that Frow recognizes to be 'one of the most powerful and illuminating ways of thinking about the pattern of relations between persons' (Frow, 1997, p. 102). It also enabled Titmuss to argue that the British system was characterized by a fundamental social equality in which donors gave blood to help anonymous others irrespective of social, cultural or ethnic differences.

Investigating the 'gift relationship'

Given my argument that *The Gift Relationship* should be read as primarily a political account, the empirical grounding of Titmuss' research is of particular note. What evidence was there, from what donors themselves reported, that they gave blood from this sense of altruism? And, in doing so, shared the values of community and equality that Titmuss saw upheld by the system? To conclude *The Gift Relationship*, he included empirical data from a survey of the socioeconomic status, demographic composition, and attitudes and motives of 3,818 donors in England.[2] One of the questions that he asked the respondents in this survey was why they first became donors. Titmuss classifies the wide range of different reasons that were reported into 14 categories. The most popularly reported of these categories was altruism, with Titmuss grouping 26% of the responses under this heading. He describes them as expressions of the general desire to help, with some respondents identifying the National Health Service or specific groups in society as the recipients of their donation. The next most popular category was general appeal, which referred to those who had been influenced by an appeal for donors in the media (18.0%); followed by personal appeal, which referred to those who had been encouraged by friends or family to donate (13.2%); and reciprocity, which relates to those who had either personally received, or knew of family or friends who had received transfusions (9.8%) (see Titmuss, 1997, pp. 293–302).

Titmuss comments that the 'vividness, individuality and diversity of these responses add life and community to the statistical generalities' (Titmuss, 1997, p. 292). At the same time, he sounds a note of caution on whether these responses could be read as revealing individuals' *motives* for giving blood. He thought that analysing the responses in these terms was fraught with conceptual and analytical problems.[3] Instead, he suggests that they be taken at 'face value'. Perhaps this accounts for why Titmuss failed to subject this data to any detailed analysis, although he was certain to point out how, at 'face value', what these donors reported appeared to support his conceptualization of the voluntary donation of blood as a 'gift relationship'. This is despite the fact that none of the respondents actually described the blood which they gave as being a 'gift'. I argue that this data suggests a more complex understanding of how different members of the public might conceive of blood donation than is readily apparent in the concept of the 'blood gift', although it is not my intention to revisit the analysis of this data here.

In the next part of this paper, it is this question of public conceptions that I want to address with reference to my own work on public participation in research on human genetic variation. I consider the relevance of the 'gift relationship' as a theoretical framework for the conceptualization of the voluntary donation of biological samples. I also address the important differences between blood donation and genetics research in terms of the uses to which the blood is put.

Gift relationships in genetics research

In my research, I have focused on a study organized by the geneticist Walter Bodmer and his research team—his wife Julia and daughter Helen—from the Cancer and Immunogenetics laboratory at the Imperial Cancer Research Fund (ICRF). This study was conducted under the rubric of the 'European Human Genome Diversity Project'. It took place in Orkney, an archipelago of some 70 islands off the north coast of Scotland (see Figure 2). Bodmer was involved in the production of a BBC science documentary *Sir Walter's*

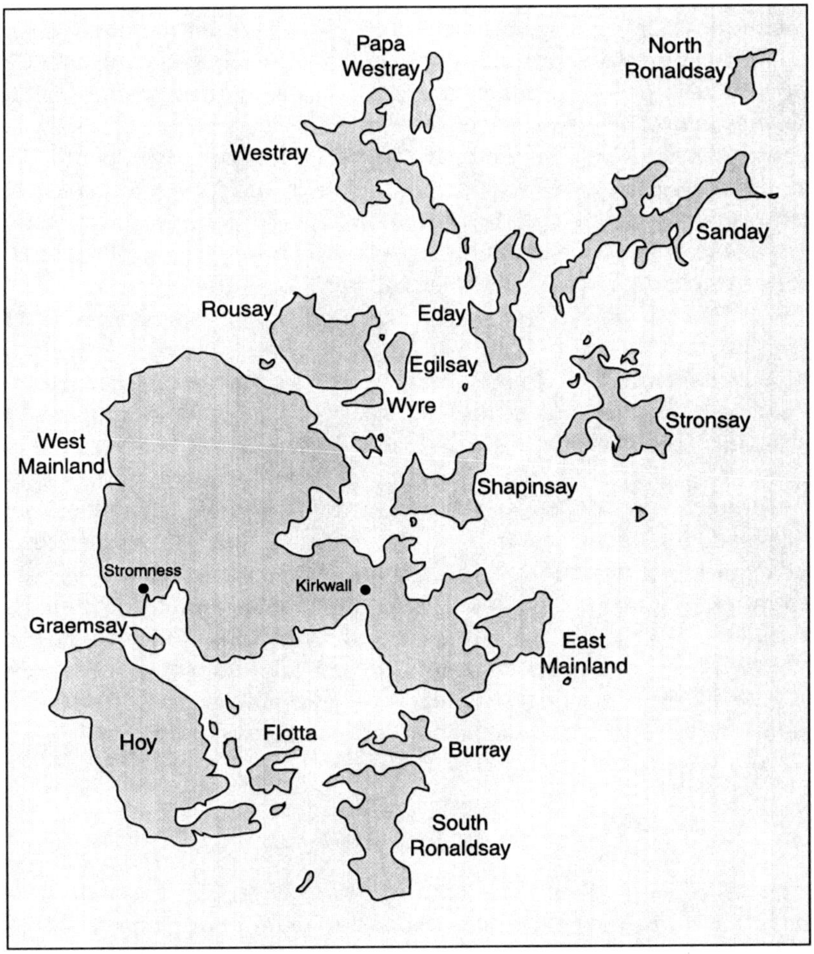

Figure 2 A map of Orkney.

Source: Mardi Jenkins, 1998, from www.orkney.org.

Journey (Horizon, 1994), and the geneticists were filmed collecting blood samples from 157 volunteers recruited from the Orkney public.[4]

The aim of this research was to delineate the genetic traces of the population who inhabited this part of the British Isles before the arrival of the Neolithic agriculturists 6,000 years ago. It would also make a contribution to the understanding of the population history in Europe. Orkney was a Norse colony until the fifteenth century. Therefore, the research team had first to determine whether the genetic impact of the Norse was as significant as their cultural influence, for which there is ample evidence in the form of archaeological remains.[5] The preliminary results of the collection were presented in the documentary: Walter Bodmer reported that the Orcadian samples, when compared with other populations in Scandinavia and the 'Celtic Fringe', indicated that the Viking heritage is likely to be more 'cultural' than 'genetic'. There were also some unique markers suggestive of an older, more isolated indigenous population in the islands (Horizon, 1994, p. 15). These results were based on an analysis of the Human Leucocyte Antigen (HLA) system and other unspecified DNA markers. In addition to making a contribution to the genetic understanding of the population history of Europe, the samples were also relevant for the Bodmers' biomedical research funded by the ICRF on the distribution of HLA polymorphisms across different populations.[6]

In the summer of 1999, I visited Orkney to research public participation in this study. I conducted unstructured interviews with 15 Orcadians, eight women and seven men, a 10% sample of the original cohort. It is very unlikely that this self-selecting sample is representative of that cohort, dependent as I was on people responding to my appeals in the local media for assistance with my research. In any event, I was less concerned with achieving a representative sample than with exploring something of what it meant to participate in this collection for these 15 individuals. I should also note that my analysis of their responses is conducted with the knowledge that these interviews were conducted five years after the collection was completed, and with the understanding that they may be indicative of how respondents chose to represent themselves to me, a Southern English, white, male social researcher.[7]

Giving blood in Orkney

Titmuss asked respondents in his survey to report on why they first became blood donors. Likewise, the first question I asked my respondents was what first interested them in participating in this collection. One notable aspect of my interview data was that responses could not easily be categorized in the way that Titmuss had done with the data from his survey. I found that many of the respondents reported more than one interest for their participation in the collection, and this indicated the complexity that is to be encountered when attempting to investigate issues of donation. However, certain respondents did emphasize particular reasons. For example, some reported that they participated in this collection from a desire to help either the researcher or the community of Orkney.

The first of these responses comes from Jane, who reports the following when asked why she was interested in participating in the collection:

> I think I took part because I thought, well, I'm a true Orcadian and thought, you know, I ought to do something being a true Orcadian

> [. . .] I mean I just did it because, I thought, it was helping, doing something for Orkney by going to do this, and doing my little bit.

Jane perceives the study as an opportunity to make a contribution to the community of Orkney, although she does not specify the nature of this contribution, or even how it would serve to 'help' Orkney. This lack of specificity is reflected in the fact that she has no expectations regarding the outcomes of the study, in either general or personal terms. This resonates with the second response from Thomas, who reports that his participation would help the researcher:

> Well I am also interested in Orkney and . . . if somebody is to offer genuine research on the subject, well if he can't get people to help them, it's no use, so it's reasonable to help, if someone's doing genuine research.

Thomas' involvement seemed to have arisen from the sense that his contribution would help to ensure that Bodmer's research, which he obviously considers a 'genuine' attempt to illuminate some aspect of Orkney, is successful in doing so. Like Jane, he has no expectations for himself from the study; the contribution he perceives he made to 'genuine' research on Orkney is probably an outcome in itself.

In contrast to these two respondents, Rachel, when reflecting on different reasons for her participation, was rather more specific about what she could have helped through her participation:

> I think I did it because my mother was a [ ] which is a very old Orkney name. In fact it goes right back to St Magnus [ ], and we thought that we might maybe find out more. We have a family tree that goes back probably to the 15-somethings I think, but there's a gap then, and then we come to the St Magnus period and we can fill that in but there's a gap between the two. But um . . . and I think maybe, not only that, I had in the back of my mind that I might help towards research as well, because my mother was an MS sufferer, because there's an awful lot of that in Orkney.

Rachel expresses a very specific concern about the high incidence of Multiple Sclerosis (MS) in Orkney of which she has had some personal experience. Her hope is that she could have contributed towards research to benefit those with MS, particularly in Orkney, which raises the interesting question of what she understood were the aims of the study. This may illustrate how some participants have perceptions of research as very open-ended, in which the samples are used for different purposes despite what they are informed of at the time of donation. This is not necessarily an issue of informed consent, but one of the public understandings of the research process. None of these respondents reported any expectation of a personal return from their participation in the research. Therefore, in terms of Titmuss' conceptualization of the 'gift relationship', these responses could be read as expressions of altruism. However, it is interesting to note the different recipients or beneficiaries they had in mind from their participation in the collection. Rachel considers that she was helping 'research' with a particular relevance for Orkney, Thomas identifies the researcher, although again he related this to his own interest in Orkney, while Jane explicitly states that the wider community of Orkney would be the beneficiary of her participation.

Through the different recipients indicated by each of the different respondents, this data also reveals an important distinction between the contexts of blood transfusion and genetics research. Although personal contacts can be an important element in blood transfusion, overall it is characterized by an institutional anonymity. Indeed, Titmuss argued that an essential feature of the 'gift relationship' was the fact that the donor and recipient of the 'blood gift' not be made known to each other. If they were, Titmuss warns that either of them could refuse to give or accept this gift on ethnic, religious or other grounds. In this collection in Orkney, the dynamics are somewhat different. The geneticists had appeared in the local media to appeal for participants. While this does not mean that there is a personal relationship between the participants and researchers, it does mean that some people may have identified with the researchers and with what they were trying to achieve through the collection.

There is also the social specificity in which this research took place. As a society, Orkney is arguably constituted by different social relations to the anonymous society of 'unconnected strangers' within which Titmuss conceptualized the 'gift relationship'. This is not to say that it is characterized by face-to-face relations like those societies about which Mauss wrote (except perhaps for some of the smaller islands), but there are distinct interconnecting networks of extended families, and a strong sense of belonging, which are important social dimensions to this study.

Jane's response indicates that her sense of desire to help the community of Orkney by participating in the collection appears to be related to her identification as a 'true Orcadian'. This particular kind of research on human genetic variation, with its aim of elucidating questions of origins and descent, particularly highlights the issues of social and cultural identities. This is not to say that the donation of blood for transfusion or for biomedical research does not also involve these issues. After all, Titmuss' comment that if donors and recipients were made to each other, then they could refuse to participate in the donor system on the basis of ethnic, social or cultural differences, suggests that a consideration of these issues in relation to blood transfusion would be an interesting line of inquiry.

Donation and identity

One particular way in which to think about identity that is relevant for my research is suggested by Stuart Hall (1990). He argues that 'identities are the names we give to the different ways we are positioned by, and position ourselves within, the narratives of the past' (Hall, 1990, p. 225). The research that Walter Bodmer conducted offered a particular narrative of Orkney's past to its participants, one based on the analysis of the distribution of HLA polymorphisms in the local population. However, some of the respondents in my research already position themselves within other narratives of the past, derived from genealogies and family histories. They expect to learn more about their own ancestry from the results of the research.

In part, this can be attributed to the offer made by the geneticists at the time of the collection to provide participants with information specific to their individual samples. Walter Bodmer remarks that: 'we did actually say to people if they wanted their own types we could let them know what they were, and a few people did. Generally, people are not concerned at that level. They want the general information, they're not so much concerned . . . well, I think they're interested in the general information, of course some people might want to know well are they really Vikings or Celts or whatever' (Bodmer, 1999). In other words, the geneticists offered to tell them about their own HLA types in terms of what these indicated about their own ancestry. Given this, it is evident that some respondents conceive of their participation in the collection as a reciprocal relationship based around the exchange of blood samples for information about their family genealogies.

Prentice, a retired health worker, reports:

> I was interested in my own, because they used to say that there was quite a bit of Spanish blood in me [. . .] Well, it's just hearsay, they used to say that my father was a pure Spaniard, you see, he had the aquiline nose but I don't know whether that . . . if there's really Spanish blood in him or no. It's just . . . just a hearsay [. . .] Whether that was true or not I can't tell you, but I thought it might have come out in this test.

Michelle expresses a similar expectation:

> I know some . . . there was a person used to tease me husband about us being related to the American Indians. But I don't really see it . . . I've never tried to trace that, but there is a lot of folk who went to Hudson Bay, working Hudson Bay, and I know some of me forebears was out there, but I've never traced through the records yet. I just thought they might get more individual, you know, that they might come back and say no, it canna be that, you know, they might

> come and say, oh yeah, you've this sort of blood, or you're prone to that, or you could be, you know, ancestors of, you know, but it's more sort of just population movements in Europe and that.

These family genealogies were inextricably linked with local history and folklore. Prentice referred to stories that circulate in Orkney of how, after being routed in the English Channel, surviving ships of the Spanish Armada were forced to return to Spain around the north coast of the British Isles to escape destruction. According to these stories, some ships foundered on the coast of Orkney and the survivors, apparently welcomed by the indigenous population, settled amongst them. In support of his possible Spanish descent, Prentice notes how others had perceived some of his father's physical characteristics as typically 'Spanish'. Likewise, Michelle's interest in discovering her possible 'American Indian' ancestry was also prompted by remarks made about her father's physical appearance, as well as her knowledge that a relative had been in the employment of the Hudson Bay Company in Canada.[8] Both Prentice and Michelle anticipate that the information from their individual sample would provide some evidence of these possible lines of descent. It is also of interest to note that, in light of the academic nature of the collection, Michelle suggests that she expected to receive information with medical relevance in terms of what she might be 'prone to'. Arguably, they both conceive of their sample as part of a reciprocal relationship between researcher and volunteer. It is their expectation that the sample they donated to the collection would illuminate something about their family, and by extension, the local history of Orkney. These responses indicate an added complexity to the use of the 'gift relationship' as a conceptual model for the voluntary donation of samples to genetics research.

From blood to genetic information

In this section I want to develop the idea that the respondents who participated in this study from an interest to discover more about their family genealogy, conceived of a reciprocal relationship with the researcher. John Frow (1997) remarks that: 'there is nothing inherent in objects that designates them as gifts; objects can almost always follow varying trajectories. Gifts are precisely not *objects* at all, but transactions and social relations' (original emphasis) (Frow, 1997, p. 124). However, I think that the use of human tissue in genetics research calls for some reflection on the nature of the gift itself. In Titmuss' account of the 'gift relationship', the gift resides in the physical substance of the blood donated for transfusion. The biological material that participants in the Orkney study donated was also blood. Therefore, the potential contrast lies in the difference between blood donated for therapeutic purposes of transfusion, and the donation of blood for genetics research.

Despite the common substance, the nature of the gift is rather different in the two cases. In the Orkney study, it can be argued that the gift resides not in the physical samples of blood donated by participants, but in the genetic information that can be derived from them. This highlights an important difference between the 'corporeal' economy of blood transfusion, in which the blood itself is of immense value and is used to assist someone in emergency medical need, with the informational economy of research, in which blood is merely an easy way of getting to the DNA. Clearly, more complications arise from the use of blood in this way because the outcomes are less certain and probably less transparent to the participant. This raises questions around the attitudes people may hold towards scientific research, how it is conducted and by whom. However, these questions are for another paper.

The evidence from the responses I have discussed suggests that the study in Orkney could be conceptualized as a reciprocal exchange of information: participants provided samples from which the geneticists extracted valuable genetic information, and, in return, gave those participants information about their own genetic ancestry. Yet is this reflected in how participants conceive of the blood they have donated to the study? It is clear that some of them understand that they could receive information, which would come from some kind of 'test' on their sample, but it is not at all evident that they appreciate the idea that genetic information is derived from the physical sample of blood. Only a few of my respondents even mentioned the terms genes or genetics in the course of the interviews, preferring instead to speak about what was 'in the blood' as that which connected them with other people, either in Orkney or more widely. Also, if we consider the experiential dimension of participation in this collection, then the process by which the blood was collected would not have been too dissimilar to that experienced by donors giving blood for transfusion. Therefore, the actual scientific process to which the donated blood is subject for the purposes of obtaining genetic information is not necessarily apparent from the donor's perspective.[9]

Conclusion

In this paper I have been concerned with the relevance of the 'gift relationship' for the conceptualization of tissue donation to contemporary genetics research. In current social, legal and ethical debate, the 'gift' has been articulated in the context of the MRC/Wellcome Trust proposal to create UK Biobank that will support biomedical research. The Orkney study, on which I have focused, represents a number of important differences from both this initiative and Titmuss' work on blood donation.

To begin with, the primary purpose of this study was to elucidate population history rather than to undertake specific biomedical research, and it took place within a particular social context. Arguably, the background for Titmuss' 'gift relationship' was that of a rather abstract and generalized national population, which was in principle characterized by class, gender, religious and ethnic differences, although these were ultimately elided by Titmuss in his desire to champion the fundamental social equality he claimed was embodied in the donor system. In contrast, I have sought to highlight the social specificity within which the Orkney study was conducted. Located on the geopolitical periphery of Britain, Orkney represents a notable difference from the generalized population of the *Gift Relationship*. It is a community structured by interconnecting extended families, a distinct identity, and a strong sense of belonging. The questions of population history investigated by Bodmer were relevant to many in Orkney because family genealogies here (as to some extent elsewhere) are important 'narratives of the past' in which people position themselves.

However, this is not to say that my findings do not have wider implications. I have discussed the issue of how blood donated for the purposes of genetics research has informational rather than corporeal value. As my fieldwork indicates, this may shape people's perceptions of the blood they are giving to researchers. The interview data also suggests that, as a conceptual model, the 'gift relationship' is only likely to provide a partial account of why people participate in contemporary genetics research. The responses of those interviewed indicate the complex and multilayered reasons people are likely to have for participation in different forms of genetics research, either for population history or biomedical purposes. It is my contention that the issues of identity, on which I have touched here, will surface as a significant feature of what other researchers may uncover when investigating public participation in other genetic studies and tissue collections in Britain and beyond.

Acknowledgements

The research reported on in this article was undertaken as part of my doctoral research funded by an ESRC studentship. I would like to acknowledge the 'gift relationship' that characterized my own research in Orkney, by thanking those Orcadians who shared their words, hospitality, and time with me; I would like also to thank Walter Bodmer, Anne-Marie Fortier, Sarah Franklin, John Frow, Georgina Haarhoff, Klaus Lindgaard Hoeyer, Maureen McNeil, and Jim Wilson for their comments on earlier versions of this paper.

Notes

1 In the preface to the first edition of *The Courage to Fail: A Social View of Organ Transplants and Dialysis*, Fox and Swazey (1978) recorded that the term 'gift' had gained a certain currency in popular and medical discourses around organ transplantation during the 1960s: 'it cannot be said that we discovered the gift dimension in transplantation and dialysis [. . .] However, the notion of the gift had not been conceptualized in a way that yielded insights into its broader medical and social significance' (Fox and Swazey, 1978, p. xii).

2 Titmuss conducted this survey in 1967 with the assistance of the Ministry of Health and the National Blood Transfusion Service. A total of 3,818 donors completed the questionnaires: 'of this total, 65 per cent were to be drawn from the general public, 32 per cent from institutions and 3 per cent from the Defence Services to accord broadly with national proportions. On grounds of cost, organization and administrative convenience, three hospital regions were selected: Manchester, Birmingham and South East Metropolitan' (Titmuss, 1997, p. 179).

3 From his extensive review of research on blood donor motivation, Oswalt (1977) found that expressions of altruism and replacement (for blood taken) are among the reasons for giving blood most often reported to researchers. However, he had some reservations about the validity of these responses. He referred to one psychology researcher whose motivational analysis 'indicates that the true motivation tends to be more selfish. He indicates that giving blood makes a person feel superior, more virile, and it makes him proud of himself, thus heightening his own self-esteem [sic]' (Oswalt, 1977, p. 125). Given the gendered nature of this observation, with its references to 'virility', it is unclear whether this researcher also considered women who donate blood to react in a similar fashion.

4 See Cross (2001) for an analysis of this fascinating documentary, which remains one of the only popular representations of the 'European Human Genome Diversity Project' in a specifically British context.

5 In the ninth century Orkney was colonized by the Vikings, and was ruled by Norway and Denmark until 1472 when, with Shetland, it was transferred to Scotland as security against an unpaid dowry of Princess Margaret of Denmark to King James III of Scotland. The interested reader can consult Linklater (1984) and Berry and Firth (1986) for absorbing accounts of the history of Orkney.

6 From research in the late 1950s, scientists realized that leucocytes (white blood cells), essential to the immune system, had antigenic specificities in a similar fashion to red blood cells. Due to the enormous amount of individual variation, with billions of combinations of the 200 HLA types so far identified, the HLA system is the most polymorphic marker in population genetics. See Bodmer and McKie (1994, pp. 164–176) and Lewontin (1995, p. 36) for further explanation. Visit the ICRF website for details of the research on HLA conducted by Bodmer's laboratory at <www.icnet.uk/research/prospectus/bodmer.html>.

7 I could have considered other aspects of my identity with respect to my fieldwork in Orkney, and the interviews that I conducted. However, my regional accent, my ethnicity, my skin colour, along with my gender were the most obvious markers apparent to respondents. There was also the added factor that Orkney has seen much immigration from England in the last 20 years, which has been the source of some social conflict (see Forsythe, 1982). Therefore, I felt a certain conspicuousness about my Englishness during the time of my fieldwork in Orkney. In terms of my skin colour, it is also of interest to note that in the 1991 census, the local population was reported to be 99.7% white.

8 The potential 'American Indian' connexion is supported by historical accounts of how employees of the Hudson Bay Company married Inuit and Cree-Indian women. It is reported that some of the children of these marriages were taken to Orkney for their education (cf. Miller, 1986).

9 Richards (1996), in his research on public understandings of inheritance, noted that there is an uncertain relationship between blood and genetic substance in the context of kinship. Interestingly, he documented how his research suggested that 'a minority of the population believes that genetic diseases can be transmitted via blood transfusions. This conclusion may be encouraged by the use of genetic tests involving blood samples' (Richards, 1996, p. 221).

References

Berridge, V. (1997) 'AIDS and the gift relationship in the UK', in A. Oakley and J. Ashton (Eds), *R.M. Titmuss, The Gift Relationship, From Human Blood to Social Policy*, pp. 15–40. London: London School of Economics and Political Science.

Berry, R.J. and Firth, H.N. (Eds) (1986) *The People of Orkney*. Kirkwall: The Orkney Press.

Bodmer, W. (1999) Interview, Hertford College, Oxford, 1 June.

Bodmer, W. and McKie, R. (1994) *The Book of Man: The Quest to Discover our Genetic Heritage*. London: Abacus Books.

Coghlan, A. (2000) 'Land of opportunity', *New Scientist*, 2263(4 November): 31–33.

Cross, K. (2001) 'Framing whiteness', *Science as Culture*, 10(3): 411–438.

Douglas, M. (1997) 'No free gifts', foreword to *The Gift, The Form and Reason for Exchange in Archaic Societies*, translated by W.D. Halls, pp. vii–xviii. London: Routledge.

Fears, R. and Poste, G. (1999) 'Building population genetics using the UK NHS', *Science*, 284: 267–268.

Forsythe, D. (1982) *Urban–Rural Migration, Change in an Orkney Island Community*. London: Social Research Council.

Fox, R.C. and Swazey, J. (1978) *The Courage to Fail: A Social View of Organ Transplants and Dialysis*. London: University of Chicago Press.

Frow, J. (1996) 'Information as gift and commodity', *New Left Review*, 219: 89–108.

Frow, J. (1997) *Time and Commodity Culture, Essays in Cultural Theory and Postmodernity*. Oxford: Clarendon Press.

Hall, S. (1990) 'Cultural identity and Diaspora', in J. Rutherford (Ed.), *Identity, Community, Culture, Difference*, pp. 222–237. London: Lawrence & Wishert.

Horizon (1994) *Sir Walter's Journey*, produced by Tim Haines, edited by Jana Bennett, broadcast BBC TV 28 March 1994.

Hyde, L. (1983) *The Gift, Imagination and the Erotic Life of Property*. New York: Vintage Books.

Lafitte, F. (1972) 'Review of Richard M. Titmuss *The Gift Relationship*', *Journal of Social Policy*, 1: 81–84.

Leach, E. (1971) 'The heart of the matter', *New Society*, 17(434): 114–115.

Lewontin, R. (1995) *Human Diversity*. New York: Scientific American Library.

Linklater, E. (1984 [1965]) *Orkney and Shetland, An Historical, Geographical, Social and Scenic Survey*, revised by James R. Nicolson. London: Robert Hale.

Marshall, R.M. (1973) 'An appreciation—Richard M. Titmuss', *British Journal of Sociology*, 24: 137–139.

Mauss, M. (1997) *The Gift, The Form and Reason for Exchange in Archaic Societies*, translated by W.D. Halls. London: Routledge.

McKie, R. (2000) 'The gene map of Britain, and how it could save your life', *The Observer*, 13 February, pp. 16–17.

Meade, T. (2000) 'After the Human Genome Project: the UK Population Biomedical Collection', Millennium Lecture Series, St Catherine's College, Oxford, 18 May.

Medical Research Council (MRC) (2001) *Human Tissue and Biological Samples for Use in Research*, April. London: MRC.

Miller, R. (1986) 'Who are the Orcadians?', in R. J. Berry and H. N. Firth (Eds), *The People of Orkney*. Kirkwall: The Orkney Press.

Oakley, A. and Ashton, J. (Eds) (1997) 'Introduction to the new edition', *The Gift Relationship, From Human Blood to Social Policy*, pp. 3–14. London: London School of Economics and Political Science.

Oswalt, R.M. (1977) 'A review of blood donor motivation and recruitment', *Transfusion*, 17(2): 123–135.

Richards, M. (1996) 'Lay and professional knowledge of genetics and inheritance', *Public Understanding of Science*, 5: 217–230.

Rose, H. (1981) 'Re-reading Titmuss: the sexual division of welfare', *Journal of Social Policy*, 10(4): 477–502.

Titmuss, R.M. (1997) *The Gift Relationship, From Human Blood to Social Policy.*

Wellcome Trust (1999) *Biological Sample Collections for Medical Research (Including Genetic Research): Research Specification for an Initial Public Consultation.* London: Wellcome Trust.

48

LEGAL, ETHICAL AND SOCIAL ISSUES IN HUMAN GENOME RESEARCH

Henry T. Greely

Source: *Annual Review of Anthropology*, 27 (1998): 470–502.

Abstract

In the past several decades, biological sciences have been revolutionized by their increased understanding of how life works at the molecular level. In what ways, and to what extent, will this scientific revolution affect the human societies within which the science is situated? The legal, ethical, and social implications of research in human genetics have been discussed in depth, particularly in the context of the Human Genome Project and, to a lesser extent, the proposed Human Genome Diversity Project. Both projects could have significant effects on society, the former largely at the level of individuals or families and the latter primarily at the level of ethnic groups or nations. These effects can be grouped in six broad categories: identity, prediction, history, manipulation, ownership and control, and destiny.

Introduction

> It has not escaped our notice that the specific pairing we have postulated immediately suggests a possible copying mechanism for the genetic material.
>
> Watson & Crick (1953)

Watson & Crick did not announce the discovery of DNA in their famous, understated paper. Nor did they invent molecular biology or human genetics. But the now-familiar double helix they discovered, with its embedded sequence of nucleotides coding for proteins, has combined with the so-called

"Central Dogma"—that genetic information passes from DNA to RNA to protein—to transform biology (Judson 1979). The result has been a vision of DNA molecules, and the genomes they constitute, as the "blueprints" of life with vast consequences for science.

Nor did it escape the notice of others that the science of DNA, when applied to humans rather than bacteria or fruit flies, immediately suggested many and varied applications to human affairs. As the tools for analyzing and understanding the human genome improved, the possible value of those tools for understanding human lives captured the attention of many publics, academic, political, and popular. Forty-five years later, the implications of human genome research for human societies, though still largely in the future, are becoming clearer.

These implications have also become the subject of what is nearly a discipline: ELSI—ethical, legal, and social implications. The social effects of increased knowledge of human genetics probably have been speculated about—and studied—more than those of any ongoing social change in history. Some of this attention derives from the fascination, sometimes tinged with horror, humans have with any technology that offers—fairly or not—to teach us about our pasts, our futures, and our essences. Although substantial scholarly attention was paid to these issues before 1990 (Andrews 1987; Fletcher 1988; Milunsky & Annas 1975, 1980, 1985; President's Comm. Study Ethical Probl. Med. Biomed. Res. 1982, Reilly 1977), some of the attention stems from the decision, apparently made on the spur of the moment by James Watson as director of the United States Human Genome Project, to commit at least 3% of the project's budget to ELSI (Cook-Deegan 1994). The results include myriad conferences, symposia, books, and articles, as well as grants for anthropologists, lawyers, philosophers, physicians, psychologists, sociologists, students of religion, and others (Annas 1992, Cranor 1994, Frankel & Teich 1993, Kevles & Hood 1992, Kitcher 1996, Murray & Lappé 1994, Murray et al 1996, Pollock 1994, Weir & Lawrence 1994, Wertz & Fletcher 1989).

This chapter does not seek to summarize or synthesize all that work. Instead, it seeks to point out the more interesting legal and social issues raised by human genome research and to group them into some analytically useful categories. The Human Genome Project (Natl. Res. Counc., Comm. Mapp. Seq. Hum. Genome 1988; US Congr., Off. Technol. Assess. 1988a; Cook-Deegan 1994) and the proposed Human Genome Diversity Project (Weiss 1998; Natl. Res. Counc., Comm. Hum. Genome Divers. 1997) are especially useful for this effort because their effects fall largely at different levels of social organization. The Human Genome Project (HGP), with its effort to find and understand the full set of human genes, speaks primarily to the implications of genetic knowledge for individuals or families. The more controversial proposed Human Genome Diversity Project (HGDP) (Greely 1997b), with its effort to collect and assess samples of human genomes from throughout the world, speaks primarily to the implications of genetic knowledge for larger human groups (and, of course, for the people within them).

This chapter first examines specific concerns arising from human genetic research, which are grouped into six categories:

1. Human genomes and identity
2. Human genetics and predicting the future
3. Human genetics and revealing the past
4. Manipulating human genomes
5. Ownership and control of human genes and genetic information
6. Genomes, souls, and destiny

The chapter then points out some common strands in many of these issues. Some of those strands point to ways in which human genetic research has social implications similar to those of other kinds of science or technology. Other strands identify ways in which those implications are unique. Both sets are crucial to viewing the possible consequences of human genetic research in perspective.

And perspective is vital. Discussions of the ethical, legal, and social implications of human genetic research are usually alarming. The foreseen social implications are usually threatening to some or all people, and the likelihood that they will occur is unknown. The foreseen benefits, although also largely speculative, fall mainly in the medical and scientific realms. Both the uncertainties and the differences in the nature of these kinds of implications make it difficult to weigh the long-term human benefits against the long-term human costs. Most discussions of the social consequences of human genetic research do not even make the attempt. This chapter is no exception, so, like similar discussions, it could be read as a litany of plausible horrors. That is not my goal. I want to alert the reader to the consequences I think are most plausible and most significant.

Ethical, legal, and social issues: specific concerns

The six categories discussed below do not begin to exhaust the implications of human genome research. The possible direct consequences of genome research for medicine are ignored, yet genomics-based interventions that either increase the average life span or decrease morbidity during life could have enormous social implications. Issues arising within the research community itself—corporate, university, and public—are diverse and fascinating (Rabinow 1996, Greely 1995) but are ignored here. Instead, I want to look at the possible implications of these technologies for ordinary people in nonmedical situations.

The following discussion attempts both to expose the reader to many of those implications and to classify them into a few analytically useful categories. The first three categories focus on the effects of information available as a result of genetic research on individuals, groups, and societies. The fourth category deals with the intentional manipulation of human genomes. The fifth deals with issues relating to the acquisition, control, and use of genetic information. The final category deals with the possibly profound implications of human genetic research for our perceptions of ourselves.

Human genomes and identity

The use of genetic information for identification gives rise to concerns in at least three areas: individual identification, cloning, and group membership.

Individual identity

The use of DNA tests for individual identification has been most discussed in the context of criminal law. DNA testing has been used since the mid-1980s to try to match suspects to DNA-containing residues left at the scene of a crime. The forensic uses of DNA technology for identification have prompted concerns about competence, overimpressiveness, ethnic discrimination, privacy, and implicit coercion (Lander 1989; US Congr., Off. Technol. Assess. 1990a; Billings 1992; Natl. Res. Counc. 1992, 1996).

The first concern is a straightforward fear that the technology may not be effective or that it may not be done well by average practitioners. This concern is similar to that faced with any new forensic procedure, but it is a continuing worry, as new and more sensitive types of DNA analysis are regularly introduced as the technology improves. The second concern is a fear that jurors would be overly swayed by the apparent scientific proof implied by DNA testing.

A third concern has been with ethnic discrimination. The markers used to analyze the DNA may not be distributed randomly across the population. Certainly, when a suspect has an identical twin, the odds of someone matching are good. Similarly, the odds that family members will match are higher than the odds of a random match. And it is possible that the variations that were examined in a suspect's DNA are found more commonly in the suspect's ethnic group than among the rest of the population. Failure to take that into account might produce overly low estimates of the possibility of a random match. This issue was debated extensively in the late 1980s and early 1990s; the controversy appears to have been resolved with the techniques now being used by forensic labs to calculate the odds of a match (Natl. Res. Council 1996).

Another concern revolves around privacy (McEwen 1997; Am. Soc. Hum. Genet., Ad Hoc Comm. DNA Technol. 1988). The possible value of DNA for forensic purposes has led to the creation of databases that could invade privacy. Some states have created databases of DNA "fingerprints" for those imprisoned or those convicted of certain crimes. These databases are used to investigate other crimes. The US military is creating a database of DNA samples for all uniformed personnel for use in identifying remains if necessary. These kinds of databases raise concerns about who may get access to the DNA or its analysis, the uses to which the samples might be put, and the coercive nature of such laws. (Of course, if the database contains only the analysis of DNA samples, checked against a panel of common markers, the possible abuses of the database are much more limited because those markers usually do not have health consequences.)

Finally, coercion may be present apart from mandatory participation in databases. The first use of DNA testing in a criminal context occurred in a rural area of England after a rape and murder. Young men in the nearby villages were asked to volunteer to provide DNA samples to be matched against DNA left on the victim. One man paid an acquaintance to provide a sample in his name; the acquaintance told the police, leading to the suspect's arrest and conviction (Cook-Deegan 1994). It may be worth worrying about how voluntary participation in such a screening program would truly be, when failure to participate could reasonably be expected to draw the attention of the police. Without sympathizing with the guilty, one may be concerned for all those not guilty who are effectively forced to provide a DNA sample.

Convicting the guilty does not normally raise ethical concerns, but the issues raised above can affect the rights and interests of both the guilty and innocent. As the technologies improve and their implementation becomes more routine, some of the concerns about false matches will fade, but that increased accuracy may only increase the pressure for invasions of individual privacy. It is, however, worth noting that DNA testing rules out suspects about as often as it incriminates them—it is much easier to say that there is no match between two samples than to compute the odds that a false match has occurred by chance (US Dep. Justice 1996).

Also, DNA identification has been used for other purposes, such as paternity testing, reuniting separated families, and identifying human remains. The latter two functions have been used in recent years in situations of political interest. DNA analysis has brought together the parents and children of people "disappeared" by the military during the junta's rule in Argentina, and it has been used in Central America to identify the remains of victims of government massacres.

Cloning

The birth of Dolly has made it seem possible that new human babies might be created with genes taken solely from one adult parent. Rather than resulting from the random mixing of genes from mother and father, the child would have the entire nuclear genome of one parent. It remains unclear whether humans can be cloned from adult cells—thus far, no one has even cloned a second sheep from adult sheep cells. But the possibility of human cloning has raised concerns about the physical, and psychological, identity of the resulting child (Silver 1997, Kolata 1998, Pence 1998).

Some of these concerns may have stemmed from a naive, Hollywood understanding of clones as photocopies—one 30-year-old person suddenly becoming five identical 30-year-olds. Clones would begin life as babies. The clone and the "donor" would be no more identical than, at most, identical twins. If the donor nucleus came from an adult cell, the clone would necessarily grow up in an environment different from that of his or her "older twin." Perhaps more important, the clone would develop before birth in a different environment, either an entirely different womb or the same womb but of a much older mother. These are reasons to suspect that clones would be substantially less similar than identical twins, but at this point that remains speculative. That the confusion of identical genomes with identity (Brock 1992)—making a copy of one's self—is a major thread in the negative reaction to cloning seems clear.

Group membership

Population genetics and the HGDP raise another possible concern about identity. It is possible that membership in various tribal or ethnic groups might be defined on the basis of an individual's genetic variations (Cavalli-Sforza et al 1994). Thus, for example, the Basques or Euskadi, who constitute the population with the highest known frequency of the Rh-negative blood type, could conceivably require Rh-negative blood as a condition for recognition as a community member.

Such a method for determining group membership would be extremely arbitrary. Groups generally are defined culturally, frequently with provisions for "adoption" into a group. Even for group members who share recent common ancestry, genetic variations will not provide a sensible method for defining a group. It is highly unlikely that any genetic variation is found, in the entire human species, in only one particular population. It is even less likely that it is found in all members of that population. Population genetics deals, as its name indicates, with populations, not with individuals. Although Basques are the people most likely to be Rh negative, most Rh-negative people are not Basque and most Basques are not Rh negative. The group frequencies, unless they are absolutely 100% or 0%, cannot be used logically to determine whether an individual is a group member.

That a membership criterion is arbitrary does not, of course, mean that it might not be used. The ethical implications of the use of such an arbitrary criterion are likely to hinge on the circumstances—including whether the criterion is chosen by the group or imposed externally—regardless of whether the criterion is genetic. Use of a genetically based group membership criterion could, however, have the negative effect of reinforcing false ideas that ethnic groups or races can be scientifically defined through genetics.

Human genetics and predicting the future

Probably the most important immediate issues concerning these new technologies revolve around their ability to predict future expressed characteristics. The strength of the link between the human gene sequence, or genotype, and the functioning person, or phenotype, is often exaggerated, but in some cases genotype does determine phenotype and in other cases strongly influences it. Genetic tests look for genotypes that help predict phenotypes. To the extent that such tests allow medical interventions that benefit the tested person, they are rarely controversial. But the predictions from genetic tests can be problematic in at least four respects: (*a*) through parental decisions about childbearing, (*b*) through individual and familial reactions to prediction, (*c*) through private third-party discrimination based on the predictive information, and (*d*) through government action. Many of these effects could take place as a result of research on either individual or population genetics.

Childbearing

Genetic counseling for any patient raises complex issues (Capron & Lappé 1979; President's Comm. Study Ethical Probl. Med. Biomed. Res. 1983; Rapp 1988; Holtzman 1989; Bosk 1992; Caplan et al 1993; Inst. Med., Natl. Acad. Sci. 1993; Nelkin & Tancredi 1994; Wertz et al 1994), perhaps especially in childbearing. Genetic tests are already in use to predict a couple's chances of having fetuses with particular characteristics or the actual characteristics of an already created embryo or fetus. Pre-conception analysis of the possible parents can lead to decisions about marriage or bearing children; analysis of embryos or fetuses can be used to determine which embryos or fetuses will be allowed to survive to birth.

Prospective parents can be tested to determine if the couple is at risk for bearing a child with some genetic diseases. Such pre-conception testing for the risk of Tay-Sachs disease is common in the Ashkenazi Jewish community. This kind of testing can raise conflicts between an individual's possible duty to disclose genetic risks to a potential partner and his or her own privacy and protection from discrimination.

Prenatal fetal diagnosis evokes a spectrum of responses. At one end, some question whether parents should allow a genetically diseased infant to be born; at the other, some attack the morality of elective abortion. An intermediate position, allowing or encouraging abortions for some kinds of genetic diseases but not others, raises difficult questions of drawing lines. Abortion for Tay-Sachs disease, in which children inevitably die after short, unpleasant lives, might raise fewer concerns than would abortion for late-onset Alzheimer's disease, with which people lead normal lives until old age, or for BRCA1 mutations, which confer a 50–80% chance of a breast cancer diagnosis sometime in life. Although under current US constitutional law the issue would not be whether abortion for a given disease would be allowed, one might regulate whether certain genetic tests could legally or ethically be used for prenatal testing (Capron 1990; Robertson 1990, 1994; Rothenberg & Thomson 1994; Rothman 1986).

One twist on this issue could come from parents with unusual preferences for their children, preferences for what many would view as a genetic disease. Thus, two people with achondroplasia, one of the common forms of very short stature, might want to use prenatal diagnosis to select a fetus with achondroplasia. If society allows prenatal testing for genetic diseases, who defines "disease?"

These latter issues may pose even greater concerns if and when genetic tests can predict phenotypic characteristics that are not generally viewed as diseases. Thus far, only one such characteristic is readily tested—sex. The ethics of parental sex selection by abortion, and the ethics of telling parents fetal sex when sex-selective abortion may be a possibility, are controversial. Some other nondisease characteristics of interest to some parents, such as skin, hair, and eye color and possibly potential height, seem likely to be heavily influenced by genetics and hence eventually testable.

It remains unclear whether other, more complex—and more interesting—traits, such as various aspects of intelligence or personality, will be amenable to genetic testing. If so, they could raise not only issues of the propriety of parents using such traits to abort fetuses (or to decide not to implant embryos), but also broader social issues of "parental eugenics." The ability to select children for particular characteristics may confer on those children cultural advantages. If such selection is allowed but restricted by financial considerations or religious beliefs, some parents—but not others—would be able to give their children a prenatal, genetic advantage or even just a valuable *perception* of genetic advantage. Such a result may have effects significantly greater than the effects of different environments available to children based on their family's wealth, class, or beliefs, but it would be an extension of those kinds of advantages, with possible implications for class structure and distributive justice (Duster 1989).

Individual and familial reactions

The personal, psychological, and familial consequences of predictive genetic information are often overlooked, but they may well raise ethical or legal concerns. Apart from the actual health risks revealed by a genetic test, new knowledge of a risk or a perceived flaw can affect a person's self-perception and happiness (King et al 1993, Natl. Soc. Genet. Couns. 1997). Similarly, genetic tests can have significant effects within a family by affecting relationships with spouses, parents, siblings, and children. Also, when one family member tests positive for a disease-related gene, that person's parents, siblings, and children all have a 50% chance of carrying that version of the gene. Whether a physician or genetic counselor has legal or moral obligations to inform other family members of their risks remains uncertain (Andrews 1997).

Private third-party discrimination

Concerns about discrimination in health insurance and employment are often reported in the United States as major deterrents to genetic testing. These issues have also been raised for disability insurance, life insurance, and adoption (as to genetic tests both of the child and of the potential adoptive parents). Only a few anecdotal cases of employment and insurance discrimination have been reported (Billings et al 1992; US Congr., Off. Technol. Assess. 1988b, 1990b), but concern is nonetheless high.

These situations share the characteristic that private actors seek to use genetic information about a person to make decisions concerning that person. The information may be used in a rational manner, where the genetic information really does provide data significant for the decision. In other situations, the information could be used inaccurately and irrationally, on the basis of misunderstandings about the nature of the genetic condition.

The legal questions surrounding employment and health insurance discrimination are murky (Gostin 1991; Greely 1992; Alper & Natowicz 1993; Rothstein 1993; Task Force Genet. Inf. Ins., Natl. Inst. Health/Dep. Energy Work. Group Ethical, Legal, Soc. Implic. Hum. Genome Res. 1994). First, it is not clear whether the federal Americans with Disabilities Act prohibits employment discrimination on the basis of genetic characteristics. The issue would turn largely on whether a genetic characteristic actually is a disability, defined by the statute as a mental or physical impairment that significantly affects a major life activity, or is merely regarded as a disability by the employer. No judicial decisions have interpreted that language with regard to genetic characteristics; the Equal Employment Opportunity Commission, which has the power to enforce the employment discrimination provisions of the act, has stated that genetic characteristics may be disabilities, but that conclusion does not have the force of law (Equal Employ. Oppor. Comm. 1995). Many states have statutes similar to the Americans with Disabilities Act; the application of these statutes to genetic characteristics generally also remains unclear.

The situation surrounding health insurance is even more confusing. The issue revolves around medical underwriting and preexisting condition limitations. Insurers and health maintenance organizations medically underwrite consumers when they decide whether to issue coverage based in part on information about consumers' predicted future health (and hence future covered health costs). Most Americans with health coverage are not subject to medical underwriting, as they receive coverage from large employers, who rarely use medical underwriting, or from Medicare and Medicaid, public programs that do not medically underwrite. People purchasing their own health coverage or receiving it from a small employer, however, may be subject to medical underwriting with respect to their past health and known health risks. Similarly, individual coverage, small-employer coverage, and some large-employer coverage can include preexisting condition limitations that exclude, for a fixed period of time or forever, payments for conditions that the insured had at the time the coverage began.

In insurance markets where they are used, medical underwriting and preexisting condition limitations can be used to reject for insurance people with a record of many conditions, such as cancer, hypertension, diabetes, obesity, or smoking. Genetic testing would add another class of factors (and another group of people) to those affected by these practices: genetic factors that predict future illness without causing substantial present symptoms.

Federal law does not generally prohibit either medical underwriting or preexisting condition limitations (Greely 1992, Hudson et al 1995). The Americans with Disabilities Act largely exempts health insurance from its coverage, even when the insurance is related to employment. The recently passed Kassebaum-Kennedy bill, however, prohibits most employers from providing health coverage for some employees but not others based on medical conditions. It does not, however, regulate what prices can be charged and what conditions can be covered. The act also puts a complicated time limit on the length of preexisting condition limitations imposed by most employers and restricts the ability of covered employers to classify genetic characteristics as preexisting conditions.

About 20 states have now prohibited insurers from using genetic information in medical underwriting, although those statutes generally do not deal with genetic characteristics as preexisting conditions. The meaning of "genetic information" in these statutes is variable and somewhat uncertain. The scope of these statutes is limited because state jurisdiction over employer-provided health coverage is, to a large extent, prohibited by a federal statute called the Employee Retirement and Income Security Act (ERISA) (Rothenberg 1995, Greely 1992).

Beyond the surprisingly unclear questions of the current legal status of employment and health insurance discrimination based on genetic information lies the normative question of whether such discrimination should be legal. Employers, insurers, and others discriminate among people on many grounds; are genetic grounds inherently inappropriate? Some argue that they are, analogizing them to other characteristics over which individuals have no control, such as their race, sex, or age. Others argue that the propriety of genetic discrimination depends on the importance of the good that could be denied. Under that kind of analysis, discrimination might be banned in health insurance but allowed in life insurance, which could be seen as less important. One could also point out that singling out genetic characteristics for protection tends to emphasize a view of genetic characteristics as inherently immutable, which is often not true. Thus, a person with a genetic predisposition to a given condition might be able either to avoid the condition altogether or to mitigate its effects, through, for example, using eyeglasses to counteract myopia.

All of these concerns can be the result of either individual or population genetic research. Although individuals carry alleles, some kinds of genetic conditions, and hence some varieties of alleles, are more common in some groups than others. Sickle-cell anemia is found at unusually high levels among people from parts of Africa and the Mediterranean; Tay-Sachs disease is disproportionately common among Jews and French Canadians; cystic fibrosis is unusually frequent in people from northern Europe. Research that revealed a high rate of a particular genetic characteristic in certain populations could be used to stigmatize the entire population, leading insurers to reject all its members without examining the genetic makeup of individuals in that group. Of course, to the extent that this kind of stigmatization of an entire group exists already, based on epidemiological evidence, the development of individual genetic tests could be used to protect individuals from the discriminatory consequences. Even if some individual members of the group escape discrimination, however, the stigma may still have negative consequences for the group as a whole (Duster 1989).

Finally, one must note that these issues will present themselves to different people and in different cultural settings. Americans over the age of 65 almost always can receive health coverage through Medicare, with no medical underwriting. Health insurance discrimination should be of little concern to them, as, if they are retired, should employment discrimination. In the United Kingdom, health coverage is universal through the National Health Service, but medically underwritten credit life insurance is important for those seeking home or automobile loans. The use of genetic tests in health insurance is not a major issue there; the use of genetic tests in life insurance is.

Government action

Governments could also use genetic predictions to make distinctions among individuals or groups. The history of the eugenics movement looms over this issue. This movement viewed with alarm what it considered the increase in "genetically defective" people caused by what it perceived as a greater rate of reproduction among the "genetically inferior" (Kevles 1986, Müller-Hill 1988, Proctor 1988, Paul 1992). Many countries, including the United States, Germany, Sweden, and the United Kingdom forcibly sterilized people on grounds of genetic predispositions to such traits as criminality and feeble-mindedness. The US Supreme Court, in an opinion by Justice Holmes, upheld one such statute (Buck v. Bell 1927) with the famous words "three generations of imbeciles are enough," although it later applied constitutional restrictions to a similar statute (Skinner v. Oklahoma ex rel. Williamson 1942). Arguments from the eugenics movement were also used to support immigration legislation that greatly favored northern Europeans.

State-sponsored eugenics is not simply a matter of history. The People's Republic of China recently adopted legislation that requires all couples to receive a medical certificate before being allowed to marry; the legislation also requires prenatal testing for pregnancies involving those with "serious" genetic diseases (Gewirtz 1994). The statute at the very least encourages long-term contraception, sterilization, and abortion for those affected. In other countries, the concern that the genetically less favored are "over-reproducing" continues to resonate.

The original eugenics movement collapsed with the revelations of the lengths to which Germany took these arguments during the Third Reich, but its demise was aided by an increased scientific understanding of the complex genetic and environmental roots of the conditions at which eugenicists took aim. As knowledge of genetic influences on those kinds of complex conditions increases, some form of more scientifically based, and hence more limited, eugenics can take shape. Required abortion or sterilization is one extreme, but incentives short of requiring abortion might be imposed, such as a refusal to provide health coverage for infants born after such a prenatal diagnosis. Such efforts would raise constitutional questions in the United States and ethical and moral questions everywhere.

Government action based on the predictive power of genetics would not necessarily be limited to eugenics, however. The press occasionally reports the discovery of a "violence gene." A government might decide to take preventive criminal action against someone carrying that genotype. Similarly, a government might use genetic evidence of low intelligence or learning disabilities to limit the education and life opportunities of certain children. On the other hand, such information might be useful in identifying children who would benefit from special attention.

Finally, although the discussion above has been about government action against individuals, data from population genetics could be used to cast an entire ethnic group as genetically inferior. There is no evidence from existing genetic studies that such an outcome is scientifically plausible. All human populations seem to contain nearly all the same genetic variants, albeit at somewhat different frequencies. Levels of particular genetic disorders higher than the overall species average can be found in almost any group. Nonetheless, the absence of good arguments to be drawn from population genetic data does not mean that bad but politically attractive arguments will not be made. Misused data might be cited to argue that governments should take action against "genetically inferior" peoples, action up to and including genocide.

Human genetics and revealing the past

The "fortune-telling" aspects of human genetics have received the most attention, but in some circumstances that information can look backward as well, to provide evidence about the past. These uses of genetic data can raise less obvious social concerns, with respect to both individuals and groups.

Individual ancestry

One of the most straightforward uses of human genetic information is to trace family relationships. Parenthood can usually be established readily, as can more distant relationships. Although such analyses have long been done with blood types (based on proteins that are themselves genetically determined), additional genetic data can make the conclusions even more certain.

Whether establishing "true" genetic family ties is a good or bad thing may depend on the circumstances. When used to reunite the children of victims of the Argentinean junta with their grandparents, its advantages seem clear. But if it is used without the consent of those involved, it could have serious implications for their privacy, their social relationships, and even their lives. Genetic counselors can encounter this problem when testing for a familial disorder. They have often discovered that the putative father was not, in fact, the genetic father. In that situation, they had to decide whom, if anyone, to tell about the true parentage. Similarly, use of genetic tests to establish paternity for child-support purposes, though justifiable, would violate the privacy of both a non-consenting father and a nonconsenting mother, with possibly serious effects on their lives.

Historical figures

Writers have long speculated on the medical and psychological ailments and characteristics of famous people. DNA testing has provided another method to learn more about individuals who have died. Genetic tests on remains can sometimes be used to determine whose remains they really are, as with the Romanovs and Jesse James. Some have urged that remains from Abraham Lincoln be analyzed to determine whether he had Marfan's syndrome, as has often been speculated. Genetic tests, in such circumstances, provide just one more way of analyzing the past, but one that may garner greater attention because of the high status of genetic information. The spread of such testing has implications for the privacy of the dead historical figures, with surviving family members probably experiencing the strongest effects. It also could have broader social implications, if, for example, the identification and reburial of the Romanov remains were used as weapons in Russian politics.

The history of peoples—and people

The HGDP raises another set of issues about the past, centering on particular populations. With enough data, population geneticists can estimate how closely related different populations of humans are to each other. This information could then be used, with linguistic, anthropological, archeological, and historical analysis, as one line of evidence to help trace the history of human migrations across the globe (Cavalli-Sforza et al 1991, Cavalli-Sforza et al 1994, Weiss 1998).

For some cultures, that evidence might be unwelcome. Many societies have their own creation stories, one of which is the Garden of Eden. Although belief in the literal truth of the Bible's creation story remains strong in the United States in spite of much scientific evidence to the contrary, for some peoples additional inconsistent scientific evidence might disrupt their own historical beliefs or self-image. In some situations, such information could disrupt an entire culture.

Genetically based historical information could have modern political implications. In the American Southwest, for example, the Hopi and Navajo peoples have long disputed ownership of certain land. Genetic evidence as to whether the Navajo are relatively recent immigrants to the region, as the Hopi maintain, could have political consequences for that land dispute. Of course, the genetic evidence would be only one additional strand of evidence, but it could be an important strand.

More broadly, evidence from population genetics could help confirm, or rebut, understandings of overall human evolution. The "out of Africa" and multiregional hypotheses certainly could be affected by more genetic data, which in turn could change scientific and popular views about human creation or evolution.

Even more broadly, one of the most striking findings from research into genomes is how much genetic material is shared by very different kinds of life. Genes found in humans can almost always be found in substantially similar form in mice; they can often be found in fruit flies, nematodes, and even yeast. "Human" genes are often just slight variants on "mammalian" genes or "vertebrate" genes or genes that are universal to all life on earth. A broad public appreciation of this reality might affect our understanding of nature and humanity's place in it. Not only would all humans be cousins, but all life would be demonstrably related.

Manipulating human genomes

Thus far, the implications discussed have arisen from information derived from existing (and inferred past and future) human genomes. The manipulation of human genomes may also raise ethical issues.

Moving DNA bearing genes from one species into another species in order to create chimeras—organisms that, genetically, are part one species and part another—is a mainstay of research in molecular biology and is also used extensively in biotechnology. Moving human DNA into non-human animals strikes some as a profanation of humanity; mixing DNA from non-human species can be perceived as a violation of a perceived natural order. The creation of mixes of humans and other animals that are apparent at a non-molecular level—chimpanzees with near-human brains, humans with feline fur—remains in the realm of science fiction, but if accomplished would raise serious questions about the meaning of "human" for various laws and cultural norms. Finally, supporters of animal rights would also raise concerns about the purposeful use of members of other species for human ends, especially when the human gene caused harm or discomfort to the animal or when the animal ultimately was killed for human use, as would be the case with organ transplants.

Human gene therapy

Moving genes from one human to another for therapeutic purposes is less controversial. The paradigm of gene therapy, as originally conceived, was to insert a "proper" gene into the relevant cells of a person with a genetic disease so that the gene would produce a protein necessary for health. Thus, gene therapy for cystic fibrosis would provide lung cells with genes making functional copies of the CTFR1 gene (Lyon & Gorner 1996).

Because of its direct connection to human health, gene therapy aimed at inserting functioning genes into the relevant cells of the body has caused little ethical controversy. This approach to treatment, however, has produced little success in spite of major funding; the appropriate level of support has thus become controversial. Some have also questioned the ethics of human gene therapy research in offering highly experimental treatments to often desperate patients.

Human germ line gene manipulation

The somatic cell gene therapy discussed above would affect only the patient who received it; it would not normally have any effects on the genes of the patients' descendants. Particular variants of genes could also be inserted into eggs and sperm, thus making intentional changes in the genomes of the patients' descendants indefinitely. This could be done for the purpose of avoiding a genetic disease or, possibly, for "enhancing" the characteristics of the child. The alleles that are added might come from the parents or from other humans. In theory, at least, they could also come from other species or come, newly made, from laboratories. It is not clear how realistic these scenarios are. All these technologies would be technically difficult and of uncertain safety for the fetus; the desire for alleles found in the parents or in other humans could also be met by preimplantation diagnosis and sperm or egg donation.

Germ line genetic manipulation raises all the issues of prenatal genetic selection, through preimplantation diagnosis or otherwise. It is not clear whether it raises others. Some might argue that the affirmative act of inserting alleles not otherwise present is different from the sorting through of genetic combinations produced by the merger of eggs and sperm, perhaps paralleling the action/inaction distinction. The social implications of the two technologies, whether voluntarily chosen by parents with the opportunity (and the money) to make the choice or imposed by the state, seem largely identical.

Ownership and control of human genes and genetic information

The next set of issues moves from how human genetic material and information could be used to who should control those materials and their uses. Control might be exercised in several ways: through property rules, through privacy protections, through informed consent, or through direct government regulation.

Ownership of human biological materials and human genetic information

Human genetic research raises issues as to appropriate legal and ethical ownership of both human biological materials and the information they contain. Human bodies and their parts have long been subject to special treatment under the law of property. Anglo-American law, at least after the abolition of slavery, does not recognize property interests in human bodies, living or dead. The next of kin of a decedent have a long-recognized quasi-property right over the corpse, but only for the purpose of arranging for its proper disposition. While living, people are viewed as having control over their bodies and bodily integrity, but not as a result of the laws of property. Thus, in the United States, competent adults are not legally able to exercise one of the prime characteristics of a property right with respect to their organs: They are not allowed to sell them.

On the other hand, some kinds of body parts can be sold: human hair, blood and blood products (in some states), and eggs and sperm. Other organs may be donated and, where they come from a living donor, may be designated for donation to a specific person. The donors are entitled to get their expenses but no price for the organs themselves.

Biological materials sought primarily for their genetic information do not fit easily into this framework. Who owns—or should own—chromosomes or genes, or, indeed, cells contained within human tissues, once they have been separated from the person in whom they grew? (US Congr., Off. Technol. Assess. 1987; Knoppers 1997; Knoppers et al 1996a; Gold 1996). Only one court decision has addressed this issue—*Moore v. Regents of the University of California*—and *Moore*'s importance is unclear (Moore v. Regents of the Univ. Calif. 1991). The California Supreme Court in *Moore* rejected a claim that a patient owned a cell line derived from his surgically removed tissue, although it allowed Moore to try to prove at trial a claim that he had not given true informed consent to the procedure. Because of its facts, *Moore* may not apply to anything other than cell lines or other substantially modified human biological materials; in any event, at this point it has not been adopted—or rejected—by any other court. It remains the express law in California only, leaving the property status of human biological materials in other contexts legally uncertain.

The ethical discussion of ownership of human biological materials has two major aspects. The most prominent argument revolves around the effects such ownership would have on our views of humans, on whether humans would become commodities. This view has been expressed forcefully with respect to surrogate motherhood and the selling of organs for transplantation purposes; the connection between individual cells, molecules, or parts of molecules from human bodies to humans seems more tenuous but can still be asserted.

The second argument, raised clearly by John Moore's suit, is the fairness of the distribution of benefits from human tissue: If any profit is made as a result of research with human biological materials, how should it be shared between the source of the materials and the researchers? Individual donors of materials rarely will be able to show that their contributions played a crucial role, as most human genetic research depends on contributions from many individuals. The argument, however, becomes stronger when the entire group that has contributed to the research—extended families, people who share a genetic disease, or communities—asserts such a claim (Greely 1997a; North Am. Reg. Comm., Hum. Genome Divers. Proj. 1997).

The legal status of intellectual property in the *information* in human genetic material seems clear, except at the margins. The ethical status of that kind of property is much more controversial (Eisenberg 1990).

The United States, Europe, and Japan have all issued patents for "inventions" that include human genetic information. These patents follow from the long-standing patent treatment of complex organic chemicals. Even though the chemical might be found in nature, a method for producing a useful compound at greater purity or higher effectiveness than found in nature can be patented. For the patent offices, human DNA has largely been just another complex organic chemical.

Patents issued on human DNA confer on the holder a monopoly on the right to use that invention for a limited amount of time, now usually 20 years from the date of the patent application. The patent holder cannot claim control over naturally occurring DNA with that sequence—it could not demand royalties from all humans for their normal use of "its" DNA—but it can control commercial uses of that DNA sequence during the patent's term. (There is a statutory exception for some research uses in the patent statutes of many countries; the United States has no statutory research exception but there is some unclear support for one in court opinions.)

Some "inventions" involving human DNA, such as expressed sequence tags, may not qualify for patent protection because they lack some of the essential attributes of patentability, notably utility, enablement, and novelty. And in many countries, but not the United States, the patent statutes contain an exception for inventions that are "against public morality;" this might be used against some patents on biotechnology. The core concept that useful stretches of human DNA sequence can be patented is not, however, in substantial legal doubt in the United States, Europe, or Japan.

Whether such DNA patents are ethical is, however, controversial. Patents on human DNA have attracted opposition from many directions. Some opponents object, on religious grounds, to any claim of human ownership to divinely created creatures or their DNA. Others view the human genome, in particular, as a common heritage of mankind that should belong to all humanity. This position, adopted by the UNESCO International Bioethics Committee in its Universal Declaration of Rights in the Human Genome, rejects the idea that some should profit from the human genome. Others object to the possibility of genetic engineering generally and view the elimination of DNA patents as a useful instrument for slowing or stopping the entire technology. Still others, equating parts of the human genome with humans themselves, see gene patents as a commodification of humanity. Finally, even some biotechnology companies and academic researchers object to certain patents or the scope of some patents, such as patents on expressed sequence tags or on single nucleotide polymorphisms, as likely to interfere unduly with research.

Thus far, the movement against gene patents has led to some relatively minor modifications in a European patent directive; otherwise it seems to have had little influence on the constantly increasing number of patents issued on DNA sequences.

Privacy

The concept of genetic privacy has great appeal (Andrews & Jaeger 1991; Annas et al 1995a,b; Rothstein 1997; Gostin 1993a). It is the subject of much proposed and enacted legislation throughout the United States, in part as a means of avoiding the kinds of employment and insurance discrimination discussed above and in part as an end in itself. When examined closely, however, genetic privacy becomes complex and difficult (Greely 1998b).

First, privacy itself has many meanings. Privacy can refer to the ability to make particular decisions without governmental intrusion, as in abortion. It can mean a claimed right to choose with whom to associate, as in marriage or certain residential situations. It can refer to a right to avoid unwanted publicity, even if accurate. It can denote a right against unwanted physical intrusion into one's body or one's home, as in the Fourth Amendment's protection from unreasonable governmental search and seizure. Finally, it can mean confidentiality—the right to insist that information, conveyed to one party for a particular purpose, not be retransmitted to another. Most of these forms of privacy are subject to some degree of constitutional or statutory protection in the United States. Discussion of genetic privacy has focused on intrusion and confidentiality, with some discussion of a decisional privacy right in making decisions about childbearing (Robertson 1990, 1994).

Protection against intrusional privacy would involve prohibitions on the collection of genetic information on people without their consent. At one level, this is not difficult. One cannot normally force someone to provide a blood sample, cheek swab, or other source of genetic material. On the other hand, people constantly shed biological materials that contain cells from which DNA can be recovered, in sloughed skin, fallen hairs, saliva, urine, and feces. There seems to be little legal protection for such, often unintentionally, discarded materials; nor does there seem any practical way to enforce a prohibition on the collection of such materials.

The intrusion, however, could be viewed not as the collection of biological materials but their analysis for genetic information. In that case, a ban on analysis for genetic information of materials collected without consent might be plausible. Such a rule would, however, raise problems both in the definition of genetic information and in the situation of partial consent: If a person authorizes the collection and analysis of a blood sample for medical purposes, would all genetic information have to be the subject of specific consent? At least one US court has ruled that testing voluntarily provided blood samples for "intimate" medical information—information on pregnancy, syphilis infection, and carrying an allele for sickle-cell anemia—without specific consent may violate the law (*Norman-Bloodsaw v. Lawrence Berkeley Lab. 1998*).

That kind of limitation on the unconsented genetic analysis shares a second problem with attempted protection of the confidentiality of genetic information. "Genetic information" may be a meaningless term because a great deal of information, medical and otherwise, provides evidence, conclusive or weak, about genes. For example, a person's ABO blood group is determined by a test of certain molecules found on the surface of red blood cells. Those molecules are determined by the variations in specific genes; knowing that a person is blood type O gives an observer specific information about the person's genetic sequence. Knowing that a couple had a child with sickle-cell anemia provides the observer with definite knowledge of the sequence of a hemoglobin gene found in each parent. Even knowing that people are male or female provides information about their genomes. If genetic information is defined narrowly, as the result of analyses of a gene's sequence, it misses family history or protein tests that would provide strong evidence about the person's genome. If genetic information is defined broadly, it encompasses nearly all medical information.

Medical information is currently protected by various legal doctrines, but that protection is generally considered woefully inadequate (US Congr., Off. Technol. Assess. 1993; Gostin 1993b, 1995). The difficulty of defining "genetic information" could argue for more effective protection of all medical information. Unfortunately, trends in both medicine and the financial management of health care in the United States are pushing for broader collection and easier dissemination of medical information. The practical scope for legislation improving the privacy of medical information is unclear (Holtzman 1995, Reilly 1995, Greely 1998b).

Genetic privacy also has a corollary in population genetics. Communities studied by population genetic researchers may not want to be identified for fear of various negative consequences, such as discrimination or weakening a political argument. It may be possible to blur the identity of such groups by referring to them only as members of a broader group of people. Thus, instead of identifying research as being done in a particular village of a particular Apache nation, published accounts might speak of research in an Apache village or, more broadly, a village of speakers of a language in the Na Dene language family. This technique, if chosen by the group, is feasible to some extent, but the greater the imprecision in the description of the group, the less useful the published data would be for fine-scale research.

Informed consent

In the past 40 years, the requirement of informed consent of a patient to medical intervention has become an important part of US law governing the doctor-patient relationship. During the same period, the principle of requiring informed consent of human participants in research has become enshrined in international and US law. Both kinds of informed consent have important implications in human genetics.

In the clinical setting, the law of informed consent may well be insufficient to protect patients adequately with respect to genetic testing. States have taken two approaches to determining when informed consent is necessary. In some states, informed consent is required only when the standards of professional practice require it—that is, doctors have to get their patients' informed consent for a particular intervention only if doctors usually get informed consent in that situation. Other states have adopted a "reasonable patient" standard: Doctors have to get their patients' informed consent when the intervention involves risks that a reasonable patient would want to know about.

Under either standard, it is not clear that informed consent would be required for genetic tests. There is no established standard for most genetic tests, and most kinds of clinical tests are not currently the subject of informed consent. More fundamentally, the risks to a patient from genetic tests are not the kinds of direct medical risks, such as death or paralysis, that informed consent usually covers. Indeed, the risks of the "procedure" could be said to be limited to the trivial risks of drawing a blood sample.

If informed consent were required, what kind of information should be given? The main risks of genetic tests are usually the risks of psychological consequences, familial disturbances, and employment and insurance discrimination that test results might bring. These are not the kinds of risks that physicians normally view as the consequences of medical interventions, but they could well be much more important to patients than a minute risk of a negative medical effect of, for example, a spinal tap.

Beyond the legal and ethical questions about the existence and scope of any informed consent obligation in the clinical use of genetic tests lies the question of *whose* informed consent. Should informed consent be required from the patient or the entire family that might be affected by the information? Knowledge of the genetic status of one relative provides powerful, sometimes conclusive, information about the genetic status of parents, siblings, and children. It could be argued that the obligation of clinical informed consent should be expanded to encompass all those strongly affected by the resulting information. Such an expansion might give other family members a veto over the patient's decision or merely give them information about the test and an opportunity to discuss the decision with the patient.

For research, US law generally requires informed consent from human subjects participating in research at institutions receiving federal funding or in research that will be used in a submission to the Food and Drug Administration (FDA). These rules, and similar rules in other countries, raise at least four important legal and ethical issues.

One currently heated battle involves the use of previously collected human biological materials for genetic research. Hospitals, universities, blood banks, states, and others have vast collections of human tissue samples, some with associated clinical information. These kinds of samples are potentially invaluable resources, particularly when tied to clinical data. On the other hand, the patients involved did not give informed consent for this kind of research use, and where their identities could be linked to the sample, the analysis could harm them—for example, by identifying them as at high risk for a disease (Clayton et al 1995). The US National Bioethics Advisory Commission is currently studying this issue; it is expected to report on it during 1998.

A second concern deals with the scope of uses that may permissibly be requested through informed consent. Some argue that research subjects can give consent only to research on specific and narrowly defined topics. Subjects could not be fully informed of the risks and benefits of research with their samples unless the research plan has been formulated. This position would arguably make it impossible for subjects to agree to allow their samples to be used as general resources for genetic studies, as the HGDP contemplates, or even, perhaps, to be part of the sequencing of the entire genome planned by the HGP. The research that might be conducted with that information, and the possible consequences for the subject, cannot be specified in advance.

A third issue focuses on the use of informed consent to restrict, in a binding way, the uses made of a research subject's samples. Research subjects, as part of informed consent, might be able to authorize use of their samples for research into diabetes but forbid their use for research into alcoholism. They might require the destruction or return of their samples within a specified period or might put conditions, including financial conditions, on any subsequent use of their samples for commercial purposes.

It is not clear under existing law whether the informed consent process, as currently constituted, leads to a binding contractual agreement between the researcher and the subject. There seems no reason, however, to think that binding contracts could not be achieved in connection with informed consent. Such contracts could provide research subjects with potentially valuable ways to control the use of their samples, as long as the subjects had the ability, as well as the legal right, to enforce such contracts (Greely 1997a,b; North Am. Reg. Comm., Hum. Genome Divers. Proj. 1997).

Finally, for some kinds of research, as for some kinds of clinical genetic tests, one might ask whose informed consent should be required. This is particularly true for population genetics, where the object of the study is primarily the population as a whole and not the individual participants, but it could also apply to research with individuals or families affected with genetic diseases. When a group is the actual subject of the research, it might be appropriate to require the informed consent of the entire group, acting through whatever authorities it recognizes. A study of the genetic constitution of Icelanders, for example, affects all people from Iceland, whether they are among the individual research subjects or not.

If there are entities that the group recognizes as having authority over them as a group, such as a national or tribal government or an important religious or cultural body, such group consent might be demanded. This consent would have the beneficial effect of allowing the group, as a whole, to determine whether to run the risks of genetic research to their history, political standing, insurability, and so on. It would also, of course, be difficult to implement: What are the boundaries of a group, and what are the culturally appropriate authorities that speak for them? It could also be viewed as infringing on individual liberty by preventing willing group members from participating in research that they found worthwhile (North Am. Reg. Comm., Hum. Genome Divers. Proj. 1997).

Social regulation

Apart from rules of property, privacy, and informed consent, other forms of regulation could be imposed on genetic technologies through, among other possibilities, professional organizations, governments, or international organizations. These regulations could cover many kinds of human genetic research or its applications. For example, many of those groups have already attempted to ban human cloning. Perhaps the most interesting case, at least in the United States, is the commercial availability of genetic testing.

In the United States, neither drugs nor biologicals can legally be sold or used without FDA permission, which is based on strong proof of their safety and efficacy as well as close monitoring of the conditions under which they are produced. The FDA also limits the advertising, marketing, and labeling of drugs. Medical devices are subject to similar, though somewhat less stringent, regulation. The practice of medicine, however, is subject to no such limitations. While a new drug, vaccine, or device can be used commercially only after government permission, a physician can use a new medical procedure at any time, with fear largely of only possible malpractice litigation.

Depending on how they are offered, genetic tests may be treated like medical procedures and not like drugs or medical devices. If the test is being offered as a service, by a clinical laboratory, the FDA has not asserted jurisdiction over it as a medical device. If, however, the test is packaged for resale to clinical laboratories, physicians, or individuals, it is a medical device and must be shown to be safe and effective before it can be sold. Not surprisingly, firms marketing genetic tests have chosen to market them as services and not as test kits, thus avoiding the stringent FDA approval process or real review of the value of test. The Clinical Laboratory Improvement Act Amendments of 1988 regulate the credentials and working conditions of the technicians who perform clinical laboratory analyses and provides for regular tests of laboratory competence, but it does not limit what tests can be performed. Some states have imposed more stringent limitations on clinical laboratories, but they have not required rigorous proof of the safety and efficacy of genetic tests. The extent to which the state or federal governments should, or will, impose such regulation remains in doubt.

Genes, souls, and destiny

The most important social implications of human genetic research may lie in its possible deep cultural effects. Two consequences seem particularly plausible: the "sacralization" of human DNA and an increasing belief in determinism. Both derive from attaching a seemingly exaggerated importance to a set of very long molecules.

It has been argued that, for some people, a human's genome has taken on attributes of the traditional Christian soul. It is uniquely individual (except in identical twins), it is a person's essence, and it is, at least in the form of information, potentially immortal. Nelkin & Lindée even note an effort to sell bits of celebrity DNA, which, they point out, parallels medieval practices with Christian relics (Nelkin & Lindée 1995).

In some cultures, the sacred nature of human biological materials already imposes some constraints on genetic research; for example, when DNA samples may be taken from cheek swabs or other tissues but not from blood. A broadening sense that human DNA is somehow sacred has implications for research, for genetic manipulation, and for privacy. At the same time, the demonstrably close connections between human genes and non-human genes could diffuse some of this sense of sacredness to other life.

The second possible consequence is a greater belief that individuals' abilities, and fates, are determined by their DNA. Reports of "gay" genes or "risk-seeking" genes, like phrenology, astrology, and other variants of fortune-telling, attempt both to predict and to provide explanations for human behavior and human outcomes (Hamer & Copeland 1998). The war between free will and determinism has been played out in different contexts during different eras in Western culture (and probably many other cultures as well), but the tension remains (Degler 1991). Increasing associations of genetic variations with particular traits may well increase the appeal—and the perceived "scientific truth"—of determinism.

In fact, much human genetic research does not support such a deterministic view (Lewontin et al 1984, Lewontin 1992, Hubbard & Wald 1993). Only a few, usually rare genetic conditions are solely and completely determined by a known gene or genes. Variation in the severity of the phenotype exists for many major genetic disorders, such as cystic fibrosis and sickle-cell anemia. Variation in the penetrance of other disease-related genotypes, the percentage of people with the genotype who will get the disease phenotype, is common, with examples from Alzheimer's disease to breast cancer to colon cancer. Many conditions with strong genetic components, such as phenylketonuria, and weaker ones, such as myopia, can be treated successfully. And most human traits, disease-related or otherwise, seem to be a result of a complex interaction of many genes and many environments, including perhaps prenatal environment. For some unfortunate humans, such as children born with Tay-Sachs disease, genes are destiny. For most of us, they are only one more influence in the contingent histories of our lives.

I believe that the paragraph above is correct. I also want to live in a society that believes it is correct. The possible serious undercutting of such a belief by research in human genetics would be, to me, its most negative consequence.

Ethical, legal, and social issues: common strands

Three common strands run through most of the issues discussed above. First, much human genetic research simultaneously raises many quite different issues. Second, the social consequences of human genetic research are similar to those of many other kinds of research. Third, the *perception* that genetic research is different is powerful. These latter two themes are important in themselves and combine to create a policy dilemma.

This chapter has carefully assigned different concerns arising from human genetics research to one of six categories, yet any given research project is likely to raise many issues from different categories. For example, research on genetic links to some forms of intelligence would spark concerns about, among other things, discrimination, prenatal testing and abortion, eugenics, germ line genetic manipulation, genetics, property, and determinism.

The proposed HGDP, whose North American ethics subcommittee I chair, provides another example. This proposal for collection, storage, and analysis of DNA samples from a broad spectrum of human populations has been attacked as "biopiracy," as reinforcing racism, as violating informed consent principles, as breaking down culturally important myths, and as leading to discrimination against ethnic groups (if not biological warfare), among other things (Mead 1996; Rural Advancement Found. Intl. 1993, 1995, 1997; Cultural Survival Q. 1996). The breadth of the issues involved both confuses the analysis and guarantees that there will be something for almost anyone to worry about. The resulting debates have led to some working out of ethical principles (Human Genome Diversity Project 1994, UNESCO Intl. Bioethics Comm., Subcomm. on Bioethics and Population Genetics 1995, Human Genome Organisation 1996, Knoppers et al 1996b, North Am. Reg. Comm., Hum. Genome Divers. Proj. 1997, Natl. Res. Council, Comm. Hum. Genome Divers. 1997) and, perhaps, to some improved understanding between, and among, the researchers, activists, and populations involved. But the very breadth of the issues involved has made those understandings more difficult to reach, as well as underlining the necessity of some continuing ethical oversight of this and other such endeavors. The issues are too complex, the individual contexts are too important, and the participants are too human to create "the" solution to the issues this kind of effort raises; only a continuing and vigilant process can minimize negative consequences from such research.

The second common strand is that the issues discussed above do not arise uniquely from genetic research. They are consequences of information about people or peoples; in some respects, human genetic research just makes possible another kind of information.

Thus, the forensic use of DNA raises issues largely identical to the earlier introduction of fingerprints or blood group tests. Genetic tests merely extend the reach of discrimination in employment and individually underwritten health insurance from, for example, people with a history of breast cancer to people with a genetically high risk of getting breast cancer in the future. Any advantages wealthy parents could confer on their children from genetic selection are probably dwarfed by the educational and cultural advantages money buys during their lives. The historical use of population genetics just adds one more line of evidence to findings that may undermine common beliefs about ethnic, national, or human origins. Privacy is threatened whenever information people care about has value to others, whether it is genetic information, medical information, or credit information. Many clinical tests and not just genetic tests should probably have expanded informed consent, as a "mere" test is not necessarily an entirely benign procedure.

The consistent thread is a new sort of information of value. Even some of the wilder issues, such as part-human chimeras, have parallels outside genetics, as in speculation about the legal and moral status of truly intelligent computers.

The third strand is that genetic information is often different from other kinds of information in its scientific origins and is always different in its perceived "essential" nature. Genetic information comes from examining invisible molecules. It is the result of the work of high-status scientists, laboring in clean laboratories in major universities. It is threatening (Rifkin & Howard 1981, Rifkin 1998), and it comes from one's "blueprint," the "Code of Codes," the stuff that "makes us what we are." And, perhaps as a result of those facts, research in human genetics, whether speculative or confirmed, is often on the front page of newspapers and the cover of news magazines (Nelkin & Lindée 1995).

This elevated status for the human genome is not only deeply inappropriate but also dangerous. At the same time, as a social phenomenon, it is very real. This sets up a serious dilemma that runs through many policy issues in genetics. Greater public knowledge of, interest in, and concern about genetic research gives this kind of research unusual power. Whether genetic information gives insurers anything more than medical information may be unclear, but if insurers believe it is powerful, it will have a greater effect on people's lives. And if people believe that, their anxiety concerning the uses of genetics will rise.

Thus, the mere perception of a peculiar power in human genetics may cause heightened risks that could justify special intervention for genetic information. Or, alternatively, the heightened concern might provide enough political support to ban, for example, medical underwriting for genetic susceptibilities even though sufficient support does not exist to ban medical underwriting for current or past health conditions.

But, on the other hand, providing special legislation or regulation may just feed the cultural belief that genetics truly *is* special. If that is the case, one might win a small tactical victory against the misuse of genetics in ways that harm people while reinforcing dangerous misperceptions of the power of genetics (Wolf 1995). This deep dilemma is perhaps the greatest ethical, legal, and social challenge posed by human genetic research.

Conclusion

This article is an attempt to forecast the major ethical, legal, and social implications of ongoing research into human genetics. It is, necessarily, a somewhat idiosyncratic vision of those implications. Only two conclusions seem certain. The first is that the implications will be important, leading to changes in day-to-day matters like medicine and in concepts of our world and our humanity. This revolutionary expansion in our knowledge of the molecular biology of life, especially human life, will have major effects on all cultures that partake of it. The second conclusion is that this article's discussion will prove to be, in some important parts, wrong. Issues will not play out as expected, unforeseen problems will arise, and time and chance will have their effects. Most of the effects of DNA on human society will prove to be no more predetermined than most of its effects on individual humans.

References

Alper JS, Natowicz MR. 1993. Genetic discrimination and the public entities and public accommodations titles of the Americans with Disabilities Act. *Am. J. Hum. Genet.* 53:26–32

Am. Soc. Hum. Genet., Ad Hoc Comm. DNA Technol. 1988. DNA banking and DNA analysis: points to consider. *Am. J. Hum. Genet.* 42:781–83

Andrews LB. 1987. *Medical Genetics: A Legal Frontier*. Chicago: Am. Bar Found.

Andrews LB. 1997. Gen-etiquette: genetic information, family relationships, and adoption. In *Genetic Secrets: Protecting Privacy and Confidentiality in the Genetic Era*, ed. MA Rothstein, pp. 255–80. New Haven: Yale Univ. Press

Andrews LB, Jaeger AS. 1991. Confidentiality of genetic information in the workplace. *Am. J. Law Med.* 17:75–108

Annas GJ, ed. 1992. *Gene Mapping: Using Law and Ethics as Guides*. New York: Oxford Univ. Press

Annas GJ, Glantz LH, Roche PA. 1995a. Drafting the genetic privacy act: science, policy and practical considerations. *J. Law Med. Ethics* 23:360–66

Annas GJ, Glantz LH, Roche PA. 1995b. *The Genetic Privacy Act and Commentary*. Boston: Boston Univ. Sch. Health

Billings PR, ed. 1992. *DNA on Trial: Genetic Identification and Criminal Justice*. Plainview, NY: Cold Spring Harbor Lab.

Billings PR, Kohn MA, Decuevas M, Beck-with J, Alper JS, Natowicz MR. 1992. Discrimination as a consequence of genetic testing. *Am. J. Hum. Genet.* 50:476–82

Bosk C. 1992. *All God's Mistakes: Genetic Counseling in a Pediatric Hospital.* Chicago: Chicago Univ. Press

Brock DW. 1992. The human genome project and human identity. *Houston Law Rev.* 29: 7–22

Buck v. Bell. 1927. 274 U.S. 200, 47 S.Ct. 584, 71 L.Ed. 1000

Caplan AL, Bartels DM, LeRoy BS, eds. 1993. *Prescribing Our Futures: Ethical Challenges in Genetic Counseling.* New York: Aldine de Gruyter

Capron AM. 1990. Which ills to bear? Re-evaluating the "threat" of modern genetics. *Emory Law J.* 39:665–96

Capron AM, Lappé M, eds. 1979. *Genetic Counseling: Facts, Values and Norms.* New York: Liss

Cavalli-Sforza LL, Menozzi P, Piazza A. 1994. *The History and Geography of Human Genes.* Princeton: Princeton Univ. Press

Cavalli-Sforza LL, Wilson AC, Cantor CR, Cook-Deegan RM, King MC. 1991. Call for a worldwide survey of human genetic diversity: A vanishing opportunity for the human genome project. *Genomics* 11: 490–91

Clayton EW, Steinberg KK, Khoury MJ, Thomson E, Andrews L, et al. 1995. Informed consent for research on stored tissue samples. *JAMA* 274:1786–92

Cook-Deegan R. 1994. *The Gene Wars: Science, Politics, and the Human Genome.* New York: Norton

Cranor CF, ed. 1994. *Are Genes Us? The Social Consequences of the New Genetics.* New Brunswick, NJ: Rutgers Univ. Press

Degler CN. 1991. *In Search of Human Nature: The Decline and Revival of Darwinism in American Social Thought.* New York: Oxford Univ. Press

Duster T. 1989. *Backdoor to Eugenics.* New York: Routledge

Eisenberg RS. 1990. Patenting the human genome. *Emory Law J.* 39:721–45

Equal Employ. Oppor. Comm. 1995. *Order 915.002. Definition of the Term "Disability."* Reprinted in *Daily Labor Report.*, March 16, 1995, pp. E1, E23

Fletcher JC. 1988. *The Ethics of Genetic Control: Ending Reproductive Roulette.* Buffalo, NY: Prometheus

Frankel MS, Teich AH. 1993. *The Genetic Frontier: Ethics, Law, and Policy.* Washington, DC: Am. Assoc. Adv. Sci.

Gewirtz DS. 1994. Toward a quality population: China's eugenic sterilization of the mentally retarded. *NY Law Sch. J. Int. Comp. Law* 15:147–62

Gold ER. 1996. *Body Parts: Property Rights and the Ownership of Human Biological Materials.* Washington, DC: Georgetown Univ. Press

Gostin LO. 1991. Genetic discrimination: the use of genetically based diagnostic and prognostic tests by employers and insurers. *Am. J. Law Med.* 17:109–44

Gostin LO. 1993a. Genetic privacy. *J. Law Med. Ethics* 23:320–30

Gostin LO. 1993b. Privacy and security of personal information in a new health care system. *JAMA* 270:2487–93

Gostin LO. 1995. Health information privacy. *Cornell Law Rev.* 80:451–527

Greely HT. 1992. Health insurance, employment discrimination, and the genetics revolution. In *The Code of Codes: Scientific and Social Issues in the Human Genome Project*, ed. DJ Kevles, L Hood, pp. 264–80. Cambridge, MA: Harvard Univ. Press

Greely HT. 1995. Conflicts in the biotechnology industry. *J. Law Med. Ethics* 23: 354–59

Greely HT. 1997a. The control of genetic research: involving "the groups between." *Houston Law Rev.* 33:1397–430

Greely HT. 1997b. The ethics of the Human Genome Diversity Project: the North American Regional Committee's proposed model ethical protocol. See Knoppers 1997, pp. 239–56

Greely HT. 1998a. Informed consent, stored tissue samples, and the Human Genome Diversity Project: protecting the rights of research participants. In *Ethical Issues in Stored Tissue Samples*, ed. RF Weir. Iowa City: Univ. Iowa Press. In press

Greely HT. 1998b. Problems in protecting "genetic privacy." *Chicago-Kent Law Rev.* In press

Hamer D, Copeland P. 1998. *Living with Our Genes: Why They Matter More Than You Think*. New York: Doubleday

Holtzman NA. 1989. *Proceed with Caution: Predicting Genetic Risks in the Recombinant DNA Era.* Baltimore, MD: Johns Hopkins Univ. Press

Holtzman NA. 1995. The attempt to pass the Genetic Privacy Act in Maryland. *J. Law Med. Ethics* 23:367–70

Hubbard R, Wald E. 1993. *Exploding the Gene Myth.* Boston: Beacon

Hudson KL, Rothenberg KH, Andrews LB, Kahn MJE, Collins FS. 1995. Genetic discrimination and health insurance: an urgent need for reform. *Science* 278:391–93

Human Genome Diversity Project. 1994. *Summary Document.* Available at http://www.stanford.edu/group/morrinst/HGDP. html.

Human Genome Organisation. 1996. HUGO statement on the principled conduct of genetics research. *Genome Digest* 2 (May 1996). Also available at http://hugo.gdb.org:90/conduct.htm.

Inst. Med., Natl. Acad. Sci. 1993. *Assessing Genetic Risks: Implications for Health and Social Policy*. Washington, DC: Natl. Acad. Press

Judson HF. 1979. *The Eighth Day of Creation: The Makers of the Revolution in Biology*. New York: Simon & Schuster

Kevles DJ. 1986. *In the Name of Eugenics: Genetics and the Uses of Human Heredity.* Berkeley: Univ. Calif. Press

Kevles DJ, Hood L, eds. 1992. *The Code of Codes: Scientific and Social Issues in the Human Genome Project*. Cambridge, MA: Harvard Univ. Press

King M-C, Rowell S, Love SM. 1993. Inherited breast and ovarian cancer: what are the risks? What are the choices? *JAMA* 269:1975–80

Kitcher P. 1996. *Lives to Come: The Genetic Revolution and Human Possibilities*. New York: Simon & Schuster

Knoppers BM, ed. 1997. *Human DNA Sampling: Law and Policy—International and Comparative Perspectives*. The Hague: Kluwer Law Int.

Knoppers BM, Caulfield T, Kinsell TD, eds. 1996a. *Legal Rights and Human Genetic Material.* Toronto: Montgomery Publ.

Knoppers BM, Hirtle M, Lormeau S. 1996b. Ethical issues in international collaborative research on the human genome: the HGP and the HGDP. *Genomics* 34: 272–93

Kolata G. 1998. *Clone: The Road to Dolly and the Path Ahead.* New York: Morrow

Lander ES. 1989. DNA fingerprinting on trial. *Nature* 339:501–5

Lewontin RC. 1992. *Biology and Ideology: The Doctrine of DNA*. New York: Harper Collins

Lewontin RC, Rose S, Kamin LJ. 1984. *Not in Our Genes*. New York: Pantheon Books

Lyon J, Gorner P. 1996. *Altered Fates: Gene Therapy and the Retooling of Human Life*. New York: Norton

McEwen JE. 1997. DNA data banks. See Rothstein 1997, pp. 321–51

Mead A. 1996. Genealogy, sacredness, and the commodities market. *Cult. Surviv. Q.* 20(Summer):46–52

Milunsky A, Annas GJ, eds. 1975. *Genetics and the Law*. New York: Plenum

Milunsky A, Annas GJ, eds. 1980. *Genetics and the Law II*. New York: Plenum

Milunsky A, Annas GJ, eds. 1985. *Genetics and the Law III*. New York: Plenum

Moore v. Regents of the Univ. Calif. 1990. S. Cal. 3d, 120, 793, p. 2d 479. 271 Cal. Rptr. 146

Müller-Hill B. 1988. *Murderous Science: Elimination by Scientific Selection of Jews, Gypsies, and Others 1933–1945*. Oxford: Oxford Univ. Press

Murray TH, Lappé M, eds. 1994. *Justice and the Human Genome Project*. Berkeley: Univ. Calif. Press

Murray TH, Rothstein MA, Murray RF, eds. 1996. *The Human Genome Project and the Future of Health Care*. Bloomington: Ind. Univ. Press

Natl. Res. Counc. 1992. *DNA Technology in Forensic Science*. Washington, DC: Natl. Acad. Press

Natl. Res. Counc. 1996. *DNA Technology in Forensic Science: An Update*. Washington, DC: Natl. Acad. Press

Natl. Res. Counc., Comm. Hum. Genome Divers. 1997. *Evaluating Human Genetic Diversity*. Washington, DC: Natl. Acad. Press

Natl. Res. Counc., Comm. Mapp. Seq. Hum. Genome. 1988. *Mapping and Sequencing the Human Genome*. Washington, DC: Natl. Acad. Press

Natl. Soc. Genet. Couns. 1997. Predisposition genetic testing for late-onset disorders in adults. *JAMA* 278:1217–20

Nelkin D, Lindée MS. 1995. *The DNA Mystique: The Gene as a Cultural Icon*. New York: Freeman

Nelkin D, Tancredi L. 1994. *Dangerous Diagnostics: The Social Power of Biological Information*. Chicago: Univ. Chicago Press. 2nd ed.

Norman-Bloodsaw v. Lawrence Berkeley Lab. 1998. 135 F.3d 1260 (9th Circuit)

North Am. Reg. Comm., Hum. Genome Divers. Proj. 1997. Proposed model ethical protocol for collecting DNA samples. *Houston Law Rev.* 33:1431–73

Paul D. 1992. Eugenic anxieties, social realities, and political choices. *Soc. Res.* 59: 663–83

Pence GE. 1998. *Who's Afraid of Human Cloning?* Lanham, MD: Rowman & Littlefield

Pollock R. 1994. *Signs of Life: The Language and Meanings of DNA*. Boston: Houghton Mifflin

President's Comm. Study Ethical Probl. Med. Biomed. Res. 1982. *Splicing Life: A Report on the Social and Ethical Issues of Genetic Engineering with Human Beings*. Washington, DC: The Commission

President's Comm. Study Ethical Probl. Med. Biomed. Res. 1983. *Screening and Counseling for Genetic Conditions: A Report on the Ethical, Social, and Legal Implications of Genetic Screening, Counseling and Education Programs*. Washington, DC: The Commission

Proctor R. 1988. *Racial Hygiene: Medicine Under the Nazis*. Cambridge, MA: Harvard Univ. Press

Rabinow P. 1996. *Making PCR: A Story of Biotechnology*. Chicago: Univ. Chicago Press

Rapp R. 1988. Chromosomes and communication: the discourse of genetic counseling. *Med. Anthropol. Q.* 2:143–57

Reilly PR. 1995. The impact of the genetic privacy act on medicine. *J. Law Med. Ethics* 23:378–81

Reilly PR. 1977. *Genetics, Law, and Social Policy*. Cambridge, MA: Harvard Univ. Press

Rifkin J. 1998. *The Biotech Century: Harnessing the Gene and Remaking the World*. New York: Tarcher/Putnam

Rifkin J, Howard T. 1981. *Who Should Play God? The Artificial Creation of Life and What It Means for the Human Race*. New York: Dell

Robertson JA. 1990. Procreative liberty and human genetics. *Emory Law J.* 39: 697–719

Robertson JA. 1994. *Children of Choice: Freedom and the New Reproductive Technologies*. Princeton, NJ: Princeton Univ. Press

Rothenberg KH. 1995. Genetic information and health insurance: state legislative approaches. *J. Law Med. Ethics* 23:312–19

Rothenberg KH, Thomson EJ, eds. 1994. *Women and Prenatal Testing: Facing the Challenges of Genetic Technology*. Columbus: Ohio State Univ. Press

Rothman BK. 1986. *The Tentative Pregnancy: Prenatal Diagnosis and the Future of Motherhood*. New York: Viking

Rothstein MA. 1993. Genetic discrimination in employment and the Americans with Disabilities Act. *Houston Law Rev.* 29: 23–84

Rothstein MA, ed. 1997. *Genetic Secrets: Protecting Privacy and Confidentiality in the Genetic Era*. New Haven, CT: Yale Univ. Press

Rural Advancement Found. Intnatl. 1993. *Communiqué: Patents, Indigenous People, and Human Genetic Diversity* (May 1993). Also available at http://www.rafi.ca

Rural Advancement Found. Intnatl. 1997. *Communiqué: The Human Tissue Trade* (Jan../Feb 1997). Also available at http://www.rafi.ca

Rural Advancement Found. Int. 1995. *Indigenous Person from Papua New Guinea Claimed in US Government Patent* (Oct. 4, 1995). Available at http://bioc09.uthscsa.edu/natnet/archive/nl/hgdp.html.

Silver LM. 1997. *Remaking Eden: Cloning and Beyond in a Brave New World*. New York: Avon Books

Skinner v. Oklahoma ex rel. Williamson. 1942. 316 US 535, 62 S. Circuit. 1110, 86 L. Ed. 1655

Task Force Genet. Inf. Ins., Natl. Inst. Health/Dep. Energy Work. Group Ethical, Legal, Soc. Implic. Hum. Genome Res. 1994. *Genetic Information and Health Insurance*. Bethesda, MD: Natl. Inst. Health

UNESCO Intnatl. Bioethics Comm., Sub-comm. on Bioethics and Population Genetics. 1995. *Bioethics and Human Population Genetics Research*. Available at http://www.biol.tsukuba.ac.jp/~macer/PG. html

US Congr., Off. Technol. Assess. 1987. *New Developments in Biotechnology: Ownership of Human Tissues and Cells*. Washington, DC: US Gov. Print. Off.

US Congr., Off. Technol. Assess. 1988a. *Mapping Our Genes—The Human Genome Project: How Big, How Fast?* Washington, DC: US Gov. Print. Off.

US Congr., Off. Technol. Assess. 1988b. *Medical Testing in the Workplace*. Washington, DC: US Gov. Print. Off.

US Congr., Off. Technol. Assess. 1990a. *Genetic Witness: Forensic Uses of DNA Tests*. Washington, DC: US Gov. Print. Off.

US Congr., Off. Technol. Assess. 1990b. *Genetic Screening in the Workplace*. Washington, DC: US Gov. Print. Off.

US Congr., Off. Technol. Assess. 1992. *Cystic Fibrosis and DNA Tests: Implications of Carrier Screening*. Washington, DC: US Gov. Print. Off.

US Congr., Off. Technol. Assess. 1993. *Protecting Privacy in Computerized Medical Information*. Washington, DC: US Gov. Print. Off.

US Dep. Justice. 1996. *Convicted by Juries: Exonerated by Science: Case Studies in the Use of DNA Evidence to Establish Innocence After Trial*. Washington, DC: Natl. Inst. Justice

Watson JD, Crick FHC. 1953. A structure of deoxyribose nucleic acid. *Nature* 171: 737–38

Weir RF, Lawrence SC, eds. 1994. *Genes and Human Self-Knowledge*. Iowa City: Univ. Iowa Press

Weiss KM. 1998. Coming to terms with human variation. *Annu. Rev. Anthropol.* 27: 273–300

Wertz DC, Fanos JH, Reilly PR. 1994. Genetic testing for children and adolescents: who decides? *JAMA* 272:875–81

Wertz DC, Fletcher JC, eds. 1989. *Ethics and Human Genetics: A Cross-Cultural Perspective*. New York: Springer-Verlag

Wolf SM. 1995. Beyond "genetic discrimination": toward a broader harm of geneticism. *J. Law Med. Ethics* 23:345–53

49

ENCLOSING THE GLOBAL PLANT GENETIC COMMONS[1]

Robert W. Herdt[2]

Source: Paper prepared for delivery at the China Centre for Economic Research, 24 May 1999, pp. 1–19.

Abstract

Property rights are assured by the collective standing behind one's claim to the benefits stream generated by property. Changing technology and institutions have interacted throughout history to create property rights from what had previously been public goods. The discovery of knowledge about DNA and the useful products that can be created through applications of that knowledge has generated conditions that have led to intellectual private property claims on many new processes and products. Large multinational life sciences companies seeking to capitalize on these developments have purchased many heretofore independent seed companies, leading to a high level of concentration in the seed industry. These issues are treated differently is various countries and many developing countries have limited capacity to deal with them. Policies to address the challenges created by this set of events are outlined.

Though all the fruits [the earth] naturally produces, and beasts it feeds, belong to mankind in common, as they are produced by the spontaneous hand of Nature, and nobody has originally a private dominion exclusive of the rest of mankind in any of them, as they are thus in their natural state, yet being given for the use of men, there must of necessity be a means to appropriate them some way or other before they can be of any use, or at all beneficial, to any particular men.

John Locke, Concerning the True, Original, Extent, and End of Civil Government.

Property can be defined as assets that generate a benefit stream. Property rights in turn, can be defined as "the capacity to call upon the collective to stand behind one's claim to a benefit stream."[3] We generally think of property as land, buildings and financial instruments, but since the eighteenth century, technology has probably created property of far greater value than all existing real estate. Compare, for example, a technology-based corporation like Microsoft with a physical asset-based one like General Motors. In late December, 1998, Microsoft corporation was valued by the stock market at $353 bill while General Motors was valued at $48 bill. Microsoft had annual sales of $16 bill while GM had annual sales of $150 bill.[4] The huge asset represented by Microsoft was generated simply through technology as embodied in software (and market power), while land, building and machinery are an insignificant part of the total.

A second example provides further illustration. Prior to the development of the telegraph in 1837, the electromagnetic spectrum existed but had no economic value. Visible light from the sun was the primary member of the electromagnetic spectrum that provided services recognized by humanity. Sunlight stands as one of the few remaining examples of a public goods – its use by one person does not compete with or rival its use by another, and no person can exclude other persons from its use (although blockage of natural light in cities is changing this, and physical constraints like imprisonment can deny individuals access to sunlight). In economics jargon, sunlight is a pure public good because its consumption is non-rival and non-excludable.

Technological developments have transformed other members of the electromagnetic spectrum from unrecognized public goods to private assets of immense value. Radio waves were unknown in the eighteenth century. They certainly were not private property. Since then the world has seen their conversion from unknown phenomena, into public goods and then into private goods through the application of science to create technologies and then the creation of laws and other social institutions by which to appropriate that property and allocate the income streams generated therefrom. Further, the nature of electromagnetic waves – their ability to travel infinite distances through space, and the applications invented for their use – long distance communications – means that to be effective, the laws or agreements about their use must cover large distances. Electromagnetic waves quickly became the subject of international agreements.

As the telegraph spread, it became apparent that without some standards and regulations its use would be limited. In 1865 the International Telegraph Union was created with the adoption of the first Convention by twenty countries.[5] The invention of the telephone in 1876 was followed by agreements on its use in 1885; the first voice radio transmission in 1902 was followed in 1906 by the Berlin International Radiotelegraph Conference; the first broadcast radio in 1920 by international agreements on its use in 1927; and so forth through television, the launching of Sputnik, communications satellites, cable TV, and cellular telephone.[6] The Iridium system, which claims the ability to provide instant voice and data communications from any spot on the globe however remote, to any other, is the latest step in this series of technological developments. With each innovation, regulations allocated the relevant electromagnetic waves. In the beginning, allocations had little value, but as the value of communications for advertising became apparent, broadcast licenses became important assets. The transformation of the ether from a valueless, ephemeral, universal presence into private property, through interrelated innovations in technology and institutions, is complete.

As might be expected, the creation of this wealth has generated conflict, with different individuals and groups each seeking to maximize their own benefits. The nature of the conflict is not always obvious. In 1963, the Hat Creek radio telescope of the University of California detected unexpectedly strong transmissions from deep space that were discovered to be the "first known natural maser, an intense blast of laserlike, organized radio waves unleashed by molecules excited by cosmic radiation."[7] This unusually strong signal, localized at 1612 megahertz (MHz), has led to an important scientific tool that enables astronomers to estimate the diameter of stars and their distance from earth. But while this particular signal is quite distinct and localized, other signals close to 1612 MHz can interfere with the scientific applications. For the past forty years astronomers have sought to protect the area around 1612 MHz from radio signals. In 1992, the International Telecommunications Union awarded radio astronomers primary rights to the spectrum from 1610 to 1613.5 MHz to protect them from harmful interference. However, the Iridium system has rights to an adjacent band, from 1616 to 1626.5, and stray radiation from Iridium is claimed to degrade the quality of the astronomers' signals.[8] Technological and regulatory solutions to the conflict are being explored.

Similarly to the electromagnetic spectrum, which has existed from the beginning of time, DNA has existed from the beginning of life as we know it, providing the essential service of transmitting all biological traits from one generation to the next. It was generating great value, but was not an asset. It was like sunlight, a part of nature whose use was non-rival and non-excludable. The discovery of DNA and the invention of ways to visualize it, characterize it, map it and move it from one organism to another has generated a new class of asset whose ownership is now being contested by multi-billion dollar companies. What was a global genetic commons is being enclosed. The pace is rapid, and at the end of the Twentieth Century intellectual property rights over plant genomic information is being concentrated in the hands of a few highly capitalized companies. "Monsanto along with a handful of other major corporations now holds most of the primary biotechnology-related assets needed for doing business in the agricultural sector."[9] The potential implication of these developments for the public interest is the focus of this paper.

Property rights in crop plants

Varieties of vegetatively propagated plants have been patentable in the United States since 1930, and sexually propagated plants (i.e. those reproduced through ordinary seeds), have been eligible for intellectual property protection under the Plant Variety Protection Act since 1970. But there was little public attention to the issue of intellectual property in plants. The seed industry consisted of hundreds of small companies producing seeds for relatively local conditions. Plant breeders in universities and government agencies continued to be the main creators of new varieties, which were provided to seed companies at little or no cost. The technical nature of seed was unchanged, despite the legal ability to claim patent protection.

Open-pollinated seeds (i.e., seeds other than hybrids), reproduce the traits of their parents with high fidelity, so once a farmer has even a small quantity of seed of a particular variety, she can obtain additional seed simply by growing the crop and harvesting the grain, which is seed for the next planting. The seeds of desirable crop varieties have historically spread quickly from farmer to farmer. In the low-income countries of Asia this is how the "Green Revolution," spread, and not withstanding many reports to the contrary, spread quickly and widely to farmers of all sizes and characteristics.[10] Not only do seeds self-reproduce, but it is difficult or impossible to distinguish varieties by looking at their seeds (sometimes even by looking at the growing plant because appearance is inevitably affected by growing conditions). Thus, farmers could easily reproduce seeds of new varieties and companies would have no way of proving the seed was their variety. The incentive for private companies to invest in developing new plant varieties was limited because open-pollinated seeds are like a public good – once they exist it is difficult to prevent any farmer from using them (non-excludable), and because they self-reproduce, their use by one farmer does not compete with their use by another (non-rival).

Hybrids provide the contrasting example and were the primary driver of private investment in the seed business before the 1990s. Hybrids are produced by crossing two distinctly different parental lines. Planting the seed harvested from a hybrid crop typically gives a range of different-looking, different-quality, different-maturity offspring. Farmers who want to grow the original hybrid have to obtain new seed for each planting. Companies that were able to keep the identity of the parental lines secret were able to prevent others from producing their hybrids, thereby generating continuing sales potential. By creating hybrids with improved production traits that were attractive to farmers, companies built businesses based on hybrid seeds beginning in the 1940s. Hybrid seeds are not public goods because it is possible for companies to exclude farmers from producing their seed. But of all the major crops, only corn was well-suited to hybrid seed production.

The development of DNA-based techniques, generally known as biotechnology, and their application to plants opened new possibilities for seed businesses. These techniques permit one to routinely distinguish any individual living organism from any other, regardless of how similar they may look. This makes it possible to identify the ancestors of any plant variety and hence the developers of a variety with near certainty, just as in the animal world a sample of DNA can be matched to any individual with near certainty. That certainty provides a way to exercise property rights over seed varieties because now one can prove who created the original variety. DNA techniques also promise to make possible the development of a wide range of new plant traits that will increase the economic value of plants.

On the institutional side, the 1980 *Diamond vs. Chakrabarty* Supreme Court decision permitted patenting of biological organisms, traits and genes set off an explosion of gene-related investment, with a significant amount focused on plant genes.[11] Still, as recently as 1989, it was unclear "whether widespread patenting of agriculturally important genes is a likely prospect," partly because broader property claims based on a variety or trait rather than genes were also possible.[12]

Most new non-hybrid varieties were produced by plant breeders working for land grant universities in the United States or public institutions in other countries. But that is changing. It is estimated that in 1994, private companies accounted for 66% of the approximately 2241 scientist years devoted to plant breeding in the United States, up from 63% in 1990.[13] Carl Pray reports that now in the United States "more food and agricultural research is conducted in the private sector than in the public sector . . . public sector research is expenditure is increasing more rapidly than public expenditure . . . and "private research is particularly strong in some of the fastest growing and most important areas of applied agricultural research such as biotechnology, plant and animal breeding, precision agriculture, and animal health."[14]

The full consequences of the DNA revolution are still to be played out, but a number of trends of interest have emerged:

1. Intellectual property rights are being rapidly extended to plants and their smallest genomic components in the industrialized countries.
2. Intellectual property rights in plants are being pressed on developing countries through international agreements.
3. A few companies have acquired most of the world's seed production capacity.
4. Agribusiness, pharmaceutical and chemicals companies are merging to create a new, highly concentrated, life sciences industry.
5. These new companies are providing unprecedented levels of funding to public research universities.

Extension of intellectual property rights

Until recently, intellectual property rights (IPRs) were of limited relevance to plant breeding. In both industrialized and developing countries public-sector plant breeding programs were the chief producers of improved crop varieties which were created by freely sharing genetic resources and research findings across regions and countries. Public sector plant breeders took discoveries from the academic world, combine them with local knowledge and genetic resources and distribute the resulting varieties through local seed production systems. The success of such collaborative research depends on a willingness to share discoveries and materials for the common good. In the genetic improvement of plants, each enhancement (whether through breeding or genetic engineering) is built on preceding generations so the process of adding value requires physical use of the earlier plant material itself.

Taking advantage of the technical capability made possible by DNA, companies have applied for and been granted patent rights to plants and animals as well as biotechnology research methods and cloned genes. Intellectual property rights extend not only to plant traits and genes but also to the smallest fragments of genetic material. In the United States DNA constructs that meet "all the conditions for patentability (such as statutory subject matter utility, enablement, written description, novelty, and non-obviousness)" can be patented, although "in order for DNA sequences to be distinguished from their naturally occurring counterparts, which cannot be patented, the patent application must state that the invention has been purified or isolated or is part of a recombinant molecule or is now part of a vector."[15] In particular, short fragments of DNA used as genome markers like expressed sequence tags (ESTs) and single nucleotide polymorphisms (SNPs), although they may not directly identify genes, "may still be extremely useful and thus satisfy the utility requirement," and hence be patentable.

The extension of patents to such short sequences of DNA raises some scientific and intellectual issues which are illustrated by Gelbart's analogy between an instruction book and the genome. He points out that if every set of information needed to deploy a given protein or polypeptide in the correct set of cells at the proper developmental times and in the requisite quantities were "analogous to one sentence in the instruction manual that we call the genome, a reasonable current assessment is that we have partial but still quite incomplete knowledge of how to identify and read certain nouns (the structures of the nascent polypeptides and protein-coding exons of mRNAs). Our ability to identify the verbs and adjectives and other components of these genomic sentences is vanishing low. Further, we do not understand the grammar at all – how to read a sentence, how to weave the different sentences together to form sensible paragraphs describing how to build multi-component proteins and other complexes."[16]

In such an instruction book, is it appropriate that all elements – nouns, verbs, adjectives, grammar, phrases, sentences, paragraphs, sections, and the whole – be subjected to intellectual property rights? Is there not a distinction between the basic elements of DNA and their combination something like the distinction between ideas and the expression of ideas in copyright of conventional books, or like the distinction between words and particular sequences of words?

Might it be preferable for much DNA information to be placed in the public domain by its inventor and remain there? It is certainly possible, for if a sequence or "invention has been described in a patent or printed publication anywhere in the world, or if it has been in public use or on sale in the United States for more than 1 year before the date on which an application for patent is filed in the United States, a patent cannot be obtained. Thus any SNPs or ESTs that have been available in a public database for more than one year prior to the filing date of the application cannot be patented."[17] Placing basic DNA information and materials in the public domain would allow property rights enforcement to focus on traits of utility to farmers, which is after all where value is generated in agriculture.

Extending intellectual property rights to developing countries

The principles of sharing germplasm and research results among participants were at the center of the plant breeding work of the centers supported by the CGIAR and were enshrined in an International Undertaking of the Food and Agriculture Organization of the United Nations.[18] But this tradition of sharing conflicts with several other international agreements. The Convention on Biological Diversity (CBD) which emerged from the 1992 UN Conference on Environment and Development established national sovereign rights over indigenous germplasm. While it was motivated by a desire to protect germplasm that might have pharmaceutical value, it applies to crop germplasm as well.[19] A third international agreement, the Trade-Related Aspects of Intellectual Property Rights, which forms part of the World Trade Organization apparatus, is requiring many developing countries to establish intellectual property rights for plants if they wish to participate on an equal basis in international trade in any products.

Pros and Cons of IPR in developing countries

Pros

- Provides incentives to create knowledge
- Encourages greater research and development within the country
- Discloses information other researchers can use to make further advances

- Encourages greater research and development investments in industrialized countries on issues of concern in developing countries
- Encourages transfer of knowledge to developing countries
- Creates a market for knowledge by providing a legal basis for technology sales and licensing

Cons

- Increases costs of protected products and may price them beyond the reach of the poor
- Shifts bargaining power toward producers of knowledge rather than users
- Broad patents may discourage follow-on inventions and slow the overall pace of innovation
- The knowledge gap between industrial and developing countries may increase
- Industrial-country firms may take advantage of indigenous knowledge and natural products without compensation to local communities

Although some believe that patents and other IPR may raise costs or discourage innovations in the developing world, or shift power unfairly to industrialized country firms away from developing country organizations, it appears that the momentum behind the international agreements will make the introduction of intellectual property rights in developing countries inevitable. These initiatives, driven by industrial intellectual property concerns have engulfed agricultural interests with a speed and intensity that has caught many in the developing world unaware and unable to deal with intellectual property rights.

Consolidation of the seed industry

The consolidation of seed producers into a few mega-seed companies has proceeded at an astounding rate over the past 15 years. In 1982 Monsanto, now the second largest seed seller in the world, was a pure chemical company until about 1982 when it entered the seed business by acquiring DeKalb's wheat research program. In the '90s they increased their seed business by obtaining AgriCetus, Britain's Plant Breeding Institute, Cargill's seed operations in Africa, Asia, Central and Latin America and Europe, and others. They purchased Calgene Inc, a leader in plant biotechnology in 1995 and Asgrow Agronomics, a leader in soybean and corn seeds, in 1996. In May, 1998 Monsanto announced that it had reached agreements to acquire two seed companies – DeKalb Genetics Corporation and Delta & Pine Land Company,[20] and in mid-year announced a merger with American Home Products. The merger, however, was abandoned, and in November Monsanto announced plans to cut its workforce by 2500, possibly to help cover the $4.2 bill cost of DeKalb and Delta Pine.[21]

DuPont, another traditional chemical company, in 1997 established a $400 million joint venture with Pioneer Hi-Bred, the world's largest seed company. The venture aimed at producing feedgrains designed to optimize weight gains in livestock, and other products tailored to meet specific markets. Subsequently, DuPont acquired controlling ownership of Pioneer. Novartis, the third largest seed company, was created through a combination of Ciba-Geigy and Sandoz. It owns Northrup King, Funk Seeds, Rogers Seed Company, and six other seed companies and has sales of almost $1 bill annually.

Thus, Monsanto, along with a handful of other giant corporations – Pioneer/DuPont, Novartis, AgrEvo, Zeneca, and Dow – are coming to dominate the world seed industry. The ten largest firms have 30% of global seed sales and have attracted the attention and opposition of groups that claim to represent farmers.[22] The business plans of the megaseed companies seem straightforward: control everything from genetic engineering of seeds to the selling of seeds to farmers, to marketing plant-grown drugs, modified foods, and industrial products. They aggressively employ patents to claim intellectual property and defend those claims equally aggressively, as illustrated by the case among Monsanto, Novartis and Mycogen over rights to corn with "built-in insect resistance.[23]" They are developing new technologies to protect intellectual property technically, such as a gene that makes the seed produced by their varieties sterile. Within the next few years many, if not most, of the potentially valuable genes from major crops are likely to be the intellectual property of these corporations. They can be expected to seek profits from new high-quality seeds for farmers who are capable of paying a premium price. But they have little incentive for producing crops that are important to the poor and disadvantaged farmers who primarily save their own seeds.

The life sciences industry

DNA contains all the information that distinguishes a living organism and many believe it holds the key to creating all kinds of biologically active materials that can cure or prevent diseases in humans, animals and plants. This makes DNA technology of great interest to pharmaceutical companies which have reached out to acquire biotechnology capacity even as they were acquiring seed companies. At the same time, because many chemicals are used to affect biological processes such as pest control in agriculture, chemical companies are interested in them. The possibility of modifying the biochemical properties of grains and oilseeds used as inputs in processed foods has attracted the attention of food companies, who envision the possibility of lowering input costs or creating foods with properties highly desired by consumers. The demonstrated possibility of producing pharmaceuticals through genetically engineered plants and animals reinforces the interests of drug companies in agricultural biotechnology.

A new life sciences industry is emerging from the combinations of chemical, agricultural and pharmaceutical companies. Dealing with such masses of data and making sense of them requires big computers and outstanding scientists. It is estimated that the maize genome contains about 30,000 genes and about 3 billion base pairs, "which makes it comparable in size to the human genome."[24] Companies are becoming ever bigger to secure as much of the talent and information as possible and are aggressively patenting processes, traits, and sequences. The US Patent and Trademark Office received 4000 patent requests for nucleic acid sequences in 1991 and 500,000 in 1996.[25]

Mergers and acquisitions across pharmaceuticals, chemicals, and agribusiness has created a few mega-companies like Novartis, valued at over $100 billion.[26] Monsanto began its biotechnology work in 1981 before it bought its first seed company. Its biggest agricultural venture has been the creation of crops engineered to be resistant to its Roundup herbicide. Selling seeds of such crops through its subsidiary companies or licensing the Roundup Ready gene has been the primary reason why herbicide resistant crops are planted on nearly 20 million hectares of land in 1998.[27] In 1997 Monsanto took a major step in transforming itself into a DNA-based company by splitting into biotech and chemical divisions and then selling off its chemical division.[28] DuPont, another traditional chemical company, in early 1998 split its life sciences work into one of three business units and shortly thereafter announced plans to sell off its energy company in order to invest the proceeds in life sciences.[29] Hoechst sold its basic chemicals divisions and invested in AgrEvo, an agribusiness biotechnology venture with Schering, bought Plant Genetic Systems for $600 million, acquired Cargill's North American seed business for $650 million and recently announced a merger of their agricultural, pharmaceuticals and veterinary medicine activities with Rhone-Poulenc.[30]

Dow was a chemical company until 1989 when it absorbed Eli Lilly as a wholly-owned subsidiary and changed its name to DowElanco. In December, 1996 the company expanded its access to biotechnology when it became a majority shareholder in Mycogen, a diversified biotechnology company with technology for insect-resistant crops. In April, 1998 the company acquired a cotton seed breeding program located in the Chaco province of Argentina and a Brazilian hybrid maize seed company. In July, 1998 Mycogen and Rhône-Poulenc Agro, the crop protection subsidiary of Rhône-Poulenc SA, signed a letter of intent to pool plant biotechnology assets to develop and market genetically modified plants and seed products containing multiple traits, initially focusing on modifying cotton and sugarcane with Mycogen's insect resistance gene derived from the natural insecticide Bacillus thuringiensis (Bt) and Rhône-Poulenc Agro's gene sequences which provide tolerance to various herbicides.[31]

Research investments in public sector organizations

At the same time as the life sciences industry has been consolidating small seed companies into mega-seed companies, they have been investing huge sums in research, not only in their own laboratories but in new ventures with public sector organizations. In July of 1998, a new private-public consortium announced the creation of a \$145 million basic plant research center in St Louis. With a staff of 80 and an annual budget of \$15 million, the new center will be run as a not-for-profit organization. The sponsors include the Danforth Foundation, Washington University, the Missouri Botanical Garden, the University of Missouri, the Monsanto Fund and Monsanto itself. The center will receive its own patents and any income from licensing. Monsanto will benefit through close association with the center, but the center is intended to operate independently from its backers.[32] In July of 1998, Novartis announced the planned investment of \$600 million to fund initiatives in plant biology. The first step was the creation of the Novartis Agricultural Discovery Institute (NADI), to be located in San Diego, California with about 180 researchers.[33] In November 1998 the University of California at Berkeley announced a \$25 million deal with NADI for the support of basic research in agricultural genetics in return for first rights to negotiate for the findings. The basics of the deal provide \$5 million a year, two-thirds for direct support of projects selected on a competitive basis and one-third for overhead and infrastructure costs to the University. NADI may also "supply proprietary tools, technology or tangible research materials under appropriate confidentiality and/or material transfer agreements. As it becomes developed, a database of NADI genomics bioinformation may also be made available to program participants who choose to access a NADI facility (under a separate "access" agreement) where a suite of workstations will be available." A six member Advisory Committee is responsible for the relationship between the University and Novartis while a five member Research Committee selects and oversees projects.[34]

There is little doubt that more deals like this between major public sector institutions and the big private companies will be struck in the future. Other initiatives will also be generated, in part as reactions to such deals. Recently, the Director of the Cornell University Agricultural Experiment Station promulgated a "Statement of Intent to Form the Consortium for Genomics Research in the Public Sector." The national public sector plant research organizations of Brazil, China, India, and the United States, as well as five of the CGIAR research centers (IRRI, ICARDA, CIMMYT, ICRISAT, CIAT) have indicated their willingness to subscribe to the consortium,[35] which seeks to serve the world community through the development, application and dissemination of genomics and computer technology relevant to world germplasm resources in plants. Cornell University has established an Institute for Genomic Diversity devoted to developing and applying genomic technologies and computational tools to the conservation, evaluation and utilization of plant genomic resources worldwide and intends to support the consortium through information resources, knowledge, and training.

What are the challenges?

The speed and economic muscle behind recent developments in private company plant genetics are viewed with pride and hope by some and with apprehension by others. The mega-seed companies' views are reflected in their aggressive corporate advertising campaigns and high-tech home pages on the World Wide Web.[36] Financial market forces have generated immense investment and shareholder value. But the trends also raise a number of concerns of public interest. With such power directed at applications, what will happen to basic science and who will teach the next generation of plant biologists? With market forces directing priorities, how will the needs of those without market power be met? How will developing countries adjust to the rapidly changing plant germplasm realities? With such a great degree of market concentration, how will abuses of power be avoided?

Basic research and education

Basic research is conducted with little or no thought of a product other than increase of knowledge; it encourage researchers to simply learn as much as they can about whatever they care to investigate. Scientists employed by private companies are sharply focused on products. The more money companies pour into research, the greater the proportion of work directed toward technology development and the less toward basic research for knowledge building. "But technological progress cannot continue without the input of basic research and the conceptual breakthroughs it makes possible . . . Apart from a handful of exceptions, the new technological companies in the Silicon Valley mold do not invest in long-term research."[37] Basic research has no role in corporate research or in educational programs driven by sharply defined end-products.

Considerably more knowledge is needed to achieve the promises of biotechnology. We still know very little about how to make use of DNA sequence information. The task of matching data with function or with means of affecting the function is huge. Even with the full DNA sequence of an organism, "with regard to understanding how to make sense of the A's, T's, G's, and C's of genomic sequence, by and large we are functional illiterates . . . For now the genomic sequence of an organism is written in a language we barely comprehend. However, through the work of the scientific community, we can attach biological meaning to limited regions of the sequence."[38] Scientists working toward highly applied goals will not generate this kind of more basic knowledge. It is short sighted to believe that simply because the life sciences industry is making large investments in plant biotechnology that the knowledge base for the next fundamental breakthrough beyond DNA will be built.

A related problem will occur with educating the next generation of plant biologists. If the most talented scientists are employed in companies and not in universities, the quality of education will suffer, graduating scientists will know more technology specifics but be less able to build new knowledge. Likewise, because training in biology so much a hands-on matter, unless students can use the latest processes, tools and materials, which may be restricted if produced in corporate labs, knowledge will be restrained.

In most cases owners of intellectual property permit its use for teaching and research, but simply obtaining appropriate permissions takes time and resources, thereby slowing the process. Recognition of these concerns may be one of the factors motivating the funding of university biology departments by Novartis and Monsanto. The unrestricted nature of the research conducted under Monsanto's funding to the Danforth lab and of Novartis' funding to Berkeley suggests this may be the case. Of course, Novartis has members on both the Advisory Committee and the Research Committee that are responsible for the Berkeley grant, so skeptics may claim that enables sufficient direction by the company. Still, the funding does provide some recognition of the value of undirected research.

Needs poorly expressed in the market

According to Evenson, "Clearly there is an economic role for the public sector even when the private sector is actively inventing new usable technology. The standard public good argument applies to plant germplasm."[39] While this is possibly an overstatement, as plant germplasm may not be a pure public good, he identifies three elements in the case for public sector investment in crop germplasm variety development:

1. Clear identification of the problems and limitations of existing technology in order that germplasm can be tailored to overcome these problems may not be done completely by market-driven mechanisms,
2. The need to provide technology for markets too small to interest the existing firms,
3. When public sector investment is needed to ensure technological competition.

Small markets exist for many minor crops in the United States where the opportunity for profit is limited. This may include crops that (1) are cultivated on a limited acreage, (2) provide relatively limited gross revenue, (3) for which there is no existing plant breeding base. "A National Plan for Promoting

Breeding Programs for Minor Crops in the U.S. has been developed by a coalition of private and public plant breeders (the "Buckwheat Coalition").[40]

Two mitigating factors suggest some of these concerns may be overdrawn for industrialized countries. As the plant breeding activities of the big companies expand to fully meet the needs for the major crops there seems no obvious reasons why they would not turn to the minor crops. Second, according to the theory of "synteny" there is great similarity among genes and where they lie in relation to one another along the chromosomes, at least within families of plants. To the extent that this holds, research on one plant will have great value in addressing the challenges in other plants.[41] These mitigating factors do not, however, offset major concerns of developing countries.

Developing countries concerns

Crop productivity is determined by genetics and environment and by their interactions, so plant biotechnology addressing plant production challenges of the industrialized North have limited relevance to the tropics. If the genes that control a plant's reactions to elements of tropical production environments are not identified, the gains from synteny will be limited. Further, because over half the investment in plant biotechnology is directed toward herbicide resistance or product quality,[42] which may not be the highest priority for developing countries, the potential for spillover is limited.

Developing country crop variety development systems are poorly equipped to deal with the rapid changes that are occurring because they have depended so extensively on international free exchange of germplasm. As plant research in the industrialized world has come to be dominated by private companies this process has slowed. But public-sector plant breeders don't know how to respond, and when they try are handicapped by the huge disparity in resources and negotiating power between themselves and the companies.

Among the most important assets accumulated by previous public investments are the plant germplasm collections of developing countries and the CGIAR. These logically should form the basis for variety development efforts for those markets. It is now possible for a company to take an existing variety, insert a few genes, take out a patent, and try to sell the seeds wherever they are adapted. Of course, the value of any variety is determined by the combinations of genes that determine height, leaf density, drought tolerance, disease resistance, grain type, maturation period, and so forth. The addition of one or two traits adds value, but clearly there was value in the preexisting line or variety. Without IPR on that preexisting line any compensation to plant breeders or institutions who developed the previous variety is unlikely. All varieties are built on preexisting varieties, and as recently as 50 to 100 years ago, all seeds were selections by farmers of the most attractive plants in their fields. These so-called landraces underlie all plant breeding.

The argument is made that the public interest demands some compensation by private companies who add a few genes to existing varieties. One possibility is market segmentation to allow the poorest developing-country farmers to obtain improved varieties at minimum cost while imposing higher charges to others. To ensure this is at least a possibility, public-sector plant breeding organizations must gain the capability to deal with IPRs and negotiate with private companies to preserve the public interest in assets accumulated with public funding.

Company power and freedom to operate

The ownership of a large fraction of seed production capacity by a few firms is beginning to attract public attention. One only need recall how domination of the American automobile industry by the Big Three allowed them to set prices and determine product characteristics. Only when foreign firms introduced new ideas was price and quality competition regained. The current court case against Microsoft is providing many claims about the anticompetitive nature of a highly concentrated market.

Given the existing intellectual property rights on the tools, processes, materials, and cloned genes that are inherent components of plant biotechnology, a company has rights to products made using those components. However, there are disputes about how far rights should "reach through" to subsequent discoveries. For example, Cre-loxP is a tool used to create research animals to study the function of certain genes. A gene has been identified that appears to play a role in a particular disease. Further research has identified a molecule that serves as a drug target. Drug development and screening is pursued without use of the Cre-loxP. Can the drug developer enter into licenses for the drug's production and sale without involving the company that owns the IPR on Cre-loxP, the original tool?

Some commentators claim the company is leveraging "its proprietary position in upstream research tools into a broad veto right over downstream research and product development," while representatives of the company that owns the tool claim they "reserve neither right in our license agreements with academic and other not-for-profit institutions."[43] Subsequently representatives of another academic institution report having approached the holders of IPR on proprietary research tools such as promoters and transformation systems for discussions on commercializing applications of the technology in crop plants. They report with "surprise and dismay" that "the holders of these proprietary "upstream" technologies "have effective veto power over whom universities can and cannot approach with their technologies for commercial development."[44]

Examples of power plays by companies are making the rounds of the academic plant biology community, and they extend to the developing world. A developing country research institution working with a CGIAR research center using public support seeks agreement to use a proprietary gene in the development of technology to address a major problem confronting smallholder farmers in Africa. The company holding the IPR agrees to license the gene only if it and it alone can determine how commercialization will proceed in the event the research proves successful.[45] Many researchers are discovering that it is easy to get a material transfer agreement or other license to use proprietary technology for research purposes, but when the research turns up a potentially useful technology, some owners of IPR don't hesitate to use their power to dictate how and even whether commercialization can proceed. The freedom to conduct research is different from the right to negotiate with the owner of IPR when it comes time to apply research results in farmers fields.

Where does the public interest lie?

In a time of rapid change, it is not easy to see where the public interest lies. Most might agree it is in the public interest to promote those social desirable objectives that may be neglected without deliberate action and to discourage or prohibit socially undesirable results that might occur if deliberate public action is not taken, or that intellectual property "rights should be shared so that the maximum benefit eventually accrues to the maximum number of people."[46] But these idealistic concepts provide little concrete policy guidance.

The discovery of DNA and the invention of techniques using DNA have made the previously invisible global genetic commons visible. The interaction of technological change and intellectual property rights is allowing the transformation of DNA from a public good whose consumption was non-rival and non-excludable (and perhaps not even recognized) into private property. If one believes that such claims on the global genetic commons are necessary in order to make the "fruits of nature" usable, one may view these developments positively. On the other hand, the privatization of the genetic commons might simply redound to the benefit of some particular persons at the expense of the larger society.

Despite the great wealth creating power of market forces, it is clear they do not resolve all issues in a socially desirable way. Collective social action is needed to set the institutional framework within which market forces operate, and the institutional solutions may differ for industrialized and developing countries. Increasingly, as cross-national flows of capital, information, goods, and people have reaches massive proportions, national institutional frameworks are becoming inadequate. Multinational companies can shift operations and funds across national boundaries with ease and impunity, and as a consequence, there is a need to examine plant biotechnology issues not only from national viewpoints, but also from an international perspective.

Market power

Markets in which there is a single seller (monopoly) or several sellers (oligopoly) tend to operate in ways that follow from their structure. In order to maximize their profits, sellers in such markets have the incentive to increase prices and reduce quantities relative to prices and quantities that would occur in a competitive market. The concentration of seed production in the hands of a few companies will inevitably lead in those directions. The classical cases of "natural monopoly" that have existed in earlier eras, for example, with public utilities led some countries to establish nationalized monopolies or alternatively to establish regulatory bodies that set prices charged to consumers.[47] Alternative public responses have been dissolution of monopolies. Consolidation of seed producers has reached such an extent that representatives of farmers in the United States are becoming alarmed and are beginning to call for some such remedies.

Many developing countries are at a significant disadvantage in controlling monopolies, having relatively few persons with the knowledge required to regulate monopolies, especially in a relatively esoteric field like plant biotechnology. Many countries have historically provided for a controlled monopoly or a government monopoly in crop research and seed production. The experience has been mixed, however, and in recent years the World Bank has urged borrowing countries to divest themselves of operating companies like those that produce and sell seeds. Typically, such companies are purchased by one of the big multinationals, but the monopoly position is preserved. A more preferable policy would be to allow local production or importation and sales of seeds by any company, subject only to quality standards. In addition, the large multinational seed companies are actively negotiating with developing countries for rights to import and sell their products under special arrangements, which amount to government-blessed monopolies. All monopoly and oligopoly situations should be resisted because competitive markets in seed development, production and sales will generate the greatest public benefit.

Patents and plant variety protection

Patents are a special case of monopoly, one endorsed by society for a limited time period and requiring disclosure of the invention. Many argue that patents function to stimulate economic activity by granting exclusive rights to the inventor to economically exploit the invention. However, others argue that for small developing countries, patents raise the costs of new goods and restrict the ability of local, low-capital innovators to improve on existing technology, especially in the plant area.

A key question with implications for building knowledge in plant biotechnology is what shall be patented. The United States Patent Office allows the patenting of SNPs – single nucleotide polymorphisms or single base pair polymorphisms in the DNA of organisms. Companies specializing in the identification of SNPs and other molecular markers anticipate selling that information to others who would commercialize the application of the knowledge. SNPs are, however, essentially existing constructs of nature, discovered but not invented, and it is reasonable to ask where the line should be drawn between public and private ownership of the genome. This is an issue requiring further public debate and, when resolved, perhaps further legislation.

The United States is the only major country to allow the use of patents to protect crop varieties, and U.S.-based companies have pressured the United States government to urge, in international negotiations, that patents be the system of choice for all countries. Ironically, the patenting of varieties is currently being challenged in a case before the United States District Court.[48] There is a chance that in the U.S. patenting of varieties will be disallowed because there exists legislation providing an alternative system of protecting the intellectual property embodied in plant varieties. If the practice of patenting varieties is overturned, intellectual property in crop varieties in the United States would be governed by the plant variety protection (PVP) system, which is similar to protection provided in many other countries and is internationally recognized under the UPOV treaty.

The plant variety protection system allows farmers to save seed for their own reuse and plant breeders to use them in research designed to produce further varietal improvement. Both these provisions are prohibited by patents but seen to have significant public welfare advantages, certainly for developing country agriculture. Regardless of what happens in the United States, the use of the PVP system, which prevails in most of Europe, is preferable for developing countries. With little competitive loss, the big seed companies could agree to use PVP in developing countries in cooperation with public breeding agencies.

Biotechnology and IPR capacity in developing countries

The capacity of governments or the science sectors of many developing countries to understand, deploy and negotiate regarding biotechnology is quite limited. They simply have very few people with the education and experience required to understand these matters. Training to build national capacity in the science and management of biotechnology, intellectual property rights, biosafety, and international negotiations is an overwhelming need. This will require considerable investment by individuals, countries and donors. Life science companies have contributed to building the needed capacity through projects that link specific technologies with training of scientists, such as with the sweet potato in East Africa or potato in Mexico. But these arrangements have involved fewer than 10 scientists. What is needed is a massive effort to create a cadre of national scientists so countries themselves can make determinations about what standards are appropriate and what technology to use.

Companies could establish a fellowship program for training developing country scientists in crop biotechnology, intellectual property and international negotiations. Participating companies could contribute to a fellowship fund administered by a neutral fellowship agency, something like the Fullbright program. Fellows would be chosen on a competitive basis according to criteria that might include national agricultural needs and academic excellence. Training 500 individuals to the PhD level would require about $75 million. If the fund and the award process is insulated from the companies, fellows would be seen as national resources and would not be obligated to the commercial interests.

Biotechnology tools and techniques

To speed the development of biotechnology capacity in developing countries, companies that have intellectual property rights over certain techniques or materials might agree to license these for use in developing countries at no cost. The agrobactrium transformation system is one possible candidate for such treatment. Another candidate might be all the privately held genomic data about a key crop species in which much genomic data is already in the public domain, say rice. Making these key tools available, not just for research, but for commercial applications in developing countries might give a boost to developing country scientists.

Another set of actions would relieve apprehension in the developing world over what they perceive as unreasonable intellectual property rights claims on crop varieties or crop traits of distinct national origin such as South Asian basmati rice or Thailand's Jasmine rice. The U.S. patent on "Basmati rice lines and grains" specifically claims the invention of "novel rice lines with plants that are semi-dwarf in stature, substantially photoperiod insensitive and high yielding, and that produce rice grains having characteristics similar or superior to those of good quality basmati rice grains produced in India and Pakistan."[49] An agreement to share financial rewards from such inventions or grant free licenses to use such lines in breeding programs in the country of origin of the said trait might gain the appreciation of developing country researchers. One might also foresee the possibility of legal actions in the United States and Europe against companies that claim IPR on such traditional crop traits.

Gene protection systems

Efforts to induce farmers to purchase all the seeds they need are probably not socially optimal in developing countries because seed production systems are not adequate to supplying all the needed seeds and farmers have limited cash liquidity for such purchases. This can be seen from the experience with hybrids. They provide a certain natural inducement to purchase new seeds each year but in many developing countries farmers choose not to do so but instead to accept the lower yield associated with reusing the grain from hybrid maize as seed. The idea of using in developing countries technologies designed to prevent germination of grain as seed, thereby blocking farmers from saving their own seed (called "terminator" by some), has engendered strong opposition in most quarters. The possible consequences if farmers who are unaware of the characteristics of such seed purchase it and attempt to reuse it are certainly negative and may outweigh any social benefits of the seeds thereby protected. In any case, there is widespread negative public reaction to encouraging poor farmers to use such technology.

An alternative approach to protecting genetic technology might be to develop a "lock and key" gene control system based on the idea that genes would be freely available for incorporation by any breeder, but would be silent until they are turned on by farmers through a deliberate process, say the application of an environmentally friendly, bio-degradable "key" compound. All the "base" traits in such seeds would be reproduced when it was saved and replanted unlike with the terminator gene, but only when the key compound is applied would the additional, special traits be expressed.

Food and environmental safety

Over the past six months an intense controversy has erupted in Europe, especially in England, about genetically modified organisms (GMOs) in foods that threatens the entire agricultural biotechnology process. One focus of challenge is the safety of GMOs as food. A second challenge is the possible negative environmental effects of the introduction of GMOs into the crop production environment. These issues have been the subject of much scrutiny during the development of genetically engineered crops and are a number of procedures have been put in place by national authorities to ensure safety. Another objection is the moral, ethical issue of meddling with nature or "playing God." These issues are being contested, but many believe they could be resolved if genetically engineered foods were labeled.

If instead, policies are implemented in Europe that impose such strict constraints on GMOs as to stop or significantly slow the deployment of GMOs in developing countries, either through regulations or because of spillover of opinion, it could have significant negative consequences on using biotechnology in developing countries and hence on future food supply. Already many countries lag behind with implementing procedures under which such GMOs could be tested and approved for production. Meanwhile, in 1998, over 60 million acres of GMO corn and soybeans were grown, mainly in the United States and Canada.

Need for a public dialogue

It is clear that the wide range of suggestions made here will not simply be accepted by companies or governments. In many cases there is no clearly preferable position. In some cases agreement by a single company would disadvantage that company. What is needed is a wide-ranging, broadly inclusive dialogue among all concerned parties — civil society, governments, companies, scientists, ethicists, and intergovernmental agencies to consider these issues and agree on a framework for their resolution in the future. It is certain they will continue to become ever more complex.

Notes

1 Prepared for delivery at the China Center for Economic Research, May 24, 1999, based on a paper delivered at the Institute for International Studies, Stanford University, January 14, 1999.
2 Director for Agricultural Sciences, The Rockefeller Foundation, 420 Fifth Ave, New York, NY, 10018. The views presented here are those of the author and not necessarily those of the Rockefeller Foundation.
3 Bromley, D.W. Environment and Economy: Property Rights and Public Policy, Cambridge, Mass., Basil Blackwell, Inc. 1991.
4 http://www.marketguide.com/mig/
5 http://www.itu.int/itudoc/about/itu/history/landmrks_e_5517.txt.
6 *Ibid.*
7 David Malakoff, "Iridium Accelerates Squeeze on the Spectrum," Science 282, p. 34–5, October 2, 1998.
8 *Ibid.*
9 Jeffrey L. Fox, "Monsanto unaffected by merger halt? Nature Biotechnology 16 December 1998 p. 1307.
10 See for example, "Who Benefits from the New technology? Ch 10, The Rice Economy of Asia, Randolph Barker, Robert W. Herdt with Beth Rose, Resources for the Future, Washington D.C. 1985
11 (C.O. Qualset, E.C.A. Runge, J.J. Mortvedt, Foreward, Intellectual Property Rights Associates with Plants, ASA Special Publication Number 52, Crop Science Society of America, Inc., American Society of Agronomy, Inc., Soil Science Society of America, Inc., Madison Wisconsin, USA, 1989)
12 Stephen A. Bent "Patenting Genes that Encode Agriculturally Important Traits," *bid*, p. 118
13 Frey, K.J., National Plant Breeding Study-I: Human and Financial resources devoted to Plant Breeding and Development in the United States in 1994. Iowa Agricultural and Home Economics Experiment Station, Iowa State University, Ames, Iowa.
14 Carl E. Pray, Role of the Private Sector in Linking the U.S. Agricultural Scientific and Technological Community with the Global Scientific and Technological Community, unpublished paper, March 30, 1999.
15 John J. Doll, The Patenting of DNA, Science, v.280, 1 may 1998, p. 689–90.
16 William M. Gelbart, Databases in Genomic Research, Science, v. 282, 23 October, 1998.
17 *Ibid.*

18 See the documents reproduced in Genetic Resources, Biotechnology and Intellectual Property Rights, International Potato Center, Lima, Peru, June 1998.
19 Frisvold, George B. The Convention on Biological Diversity and Agriculture: Implications and Unresolved Debates. World Development V. 26 (4), p. 551–70, 1998.
20 http://www.monsanto.com/ag/articles/98–05–11acquireDeltaPine.htm
21 Ed Schafer, Associated Press, November 12, 1998
22 Data assembled by Rural Advancement Foundation International, http://www.rafi.ca
23 http://www.monsanto.com/monsanto/media/current/980630-btpatent.html
24 Jon Cohen, Please Pass the Data, Science V. 276, 27 June, 1997, p.1961.
25 US Patent and Trademark Office, Hearing and Request for Comments on Issues Relating to Patent Protection for Nucleic Acid Sequences," University of California, San Diego, April 16, 1996.
26 Enriquez, Juan, Genomics and the World Economy, Science, V. 128, 14 August, 1998, p. 925.
27 James, Clive, Global review of Commercialized Transgenic Crops: 1998. ISAAA Briefs No. 8–1998, Ithaca, NY.
28 http://www.monsanto.com/Monsanto/About/Past/default.htm
29 Enriquez, *op. cite.*
30 John Hodgson, Nature Biotechnology v. 17 January 1999, p. 15.
31 http://www.seedquest.com/News/Companies/Mycogen.htm
32 Kaiser, Jocelyn, Plant Biologists Score Two New Major Facilities, Science, V. 281, 17 July 1998, p. 317.
33 http://www.prenewswire.com/cgi-bin
34 http://www.urel.berkeley.edu/urel_1/CampusNews/PressReleases/releases/Agreement.html
35 W. Ronnie Coffman, "Future of Plant Breeding in Public Institutions," a paper presented to the American Seed Trade Association, December, 1998.
36 http://www.monsanto.com/monsanto/index4.html
37 Nathan Myhrvold, Supporting Science, Essays on Science and Society, Science, v. 282, 23 October, 1998.
38 William M. Gelbart, Databases in Genomic Research, Science v. 282, 23 October, 1998, p. 659–61.
39 Evenson, R.E. Intellectual property Rights and Agribusiness Research and development: Implications for the Public Agricultural Research System. American Journal of Agricultural Economics, 65 (5) December 1983.
40 W.R. Coffman, *op. cite.*
41 Gale, M.D. and K.M. Devos, Plant comparative genetics after 10 years, Science v. 282 23 October, 1997, p. 656–8.
42 As reflected in the proportion of field tests of genetically engineered crops in the United States. See Clive James, Global Status of Transgenic Crops in 1997, ISAAA Brief No. 5–1997, Ithaca, NY.
43 See the exchange of letters between David S. Block and Daniel J. Curran of DuPont Merck and Rebecca S. Eisenberg and Michael Heller of the University of Michigan Law School, Science v. 282, 20 November 1998.
44 Colm Lawler and Fred Erbisch, From Mice to Maize, Science v.283, 1 January, 1999.
45 Joe deVries, personal communication.
46 See the editorial in New Scientist, 17 April 1999 on information and property rights.
47 In recent years technological developments have undercut the natural monopoly of the telephone company, first with alternative long-distance carriers made

possible through cheap fiber optic or high capacity satellite-based service, and more recently through local cellular telephone service.

48 Kilman, Scott, Biotech Industry Shivers at Threat to Seed Patents, Wall Street Journal, March 3, 1999.

49 Sarreal et. al. United States Patent 5,663,484, September 2, 1997.

50

THE INSCRIPTION OF LIFE IN LAW

Genes, patents and bio-politics

Alain Pottage

Source: R. Brownsword, W. R. Cornish and M. Llewelyn (eds), *Law and Human Genetics: Regulating a Revolution*, Oxford: Hart, 1998, pp. 148–173.

'Bio-colonialism' in New Guinea

In 1989, Carol Jenkins, an American medical anthropologist attached to the Institute of Medical Research in Papua New Guinea, extracted a set of blood samples from some members of the Hagahai, a small group of hunter-gatherers living in an inaccessible mountain range: 'I told them we wanted to see a "binitang" — an insect — in their blood'.[1] The suspicion that this peculiar 'insect' existed had arisen from research carried out by scientists from the US National Institutes of Health (NIH), who were conducting a survey of the distribution of variants of the HTLV-I virus in Melanesia.[2] In her own (retrospective) accounts of the transaction that was taking shape, Jenkins identified herself as a disinterested intermediary, whose primary concern was to ensure that the Hagahai received the proper commercial or therapeutic return for their participation. Her original research interests had little to do with the genetic properties of their blood. In a paper describing the methods and objectives of her medical-anthropological survey, written before these particular blood samples had been extracted, Jenkins heralded 'a rare opportunity to document the effects of contact on a previously isolated population in Papua New Guinea'.[3] Apparently, the Hagahai had avoided direct contact with 'globalised' society until 1983, when the effects of various illnesses became so severe that they came to a missionary outpost in search of assistance. The missionaries had their own ambitions for the Hagahai, which were not necessarily in the best interests of the health of the group,[4] and the first effective measures of medical assistance arrived only when Jenkins and her colleagues began their study. Indeed, Jenkins' published account states that 'the ethical obligation to intervene with medical aid was immediately obvious to the researchers'.[5] However, although the protocols of research and publication did not exclude an active concern for the health of the Hagahai,[6] it remains the case that the governing objective of Jenkins' programme was 'to investigate the causes of their decline and to monitor the consequences of increased contact to their biological and cultural survival'.[7] The extensive array of tests carried out on the members of the group facilitated various therapeutic objectives, but they also provided a criterion against which the course of their decline could be plotted. The prognosis was far from encouraging. Jenkins' research paper states, somewhat tersely, that 'the net population increase is negative and the Hagahai are dying out'.[8]

Jenkins' medical-anthropological publications are written in the language of traditional ethnography; they betray none of modern anthropology's ethical or epistemological anxieties about the status of its objects. Medical anthropology deals with subject profiles fashioned from parasitological surveys, mortality rates, comparative assessments of growth rates, cranial measurements, and so on. These techniques were entirely appropriate to the research programme, which identified the Hagahai population as a site of epidemiological negotiation, where cultureborne viruses and antibodies were exchanged or cancelled. The question of how culture was reproduced, or of how 'biology' was bound to 'culture' in that process of reproduction, was no part of this programme, whether one takes the 'culture' in question to be that of the Hagahai or that of the researchers. Similarly, for the NIH scientists the Hagahai were interesting only as the human carriers of a rare viral strain and its correlative antigens. Both perspectives condensed the social or cultural agency of the group into a set of biological ciphers. Given that these reductive accounts supply the basic fabric of even the most 'critical' interventions, it is not surprising that most versions of the story adopt the notion of genes as 'patrimony', a notion that obscures the most vital issue in the patenting of 'life', namely, the question of how societies (re-)invent the distinction between the social and the biological. In this context, it is particularly important that these processes of (re-)invention should be made explicit, rather than being allowed to condense into inarticulate ciphers, because the most vital aspect of biotechnology patents is their role in forcing the re-negotiation of traditional elaborations of the distinction between nature and culture.

The precise nature of the extracted 'insect' was identified in the laboratories of the NIH, where analysis of one particular culture, taken from a young man, revealed the existence of antibodies to a variant of the HTLV-I leukaemia virus.[9] This culture was used to develop an immortal T-cell line, 'persistently infected' with the HTLV-I strain, which was the basis for a patent application relating to the cell line, the infecting virus, and a set of ancillary diagnostic test kits.[10] Carol Jenkins was identified as one of the five inventors of the PNG-1 cell line, a status which seems to have been attributed almost entirely in recognition of her role as the agent of 'her' population. According to Jenkins' account, the terms of the agreement which she made with the Hagahai were considerably more sophisticated than the euphemistic reference to 'insects' might suggest. Not only did the Hagahai 'leaders' have 'a clear understanding of the concept of ownership',[11] but they were sophisticated enough to appreciate the advantages of patenting the cell line.[12] The patent was supposed to secure an appropriate return for their participation: as Jenkins put it, 'despite what exploitation may take place around the world, in this case the rights of the Hagahai people have been specifically safeguarded'.[13] Apparently, the terms of the 'agreement' were that the Hagahai should receive some (unspecified) share of the commercial profits of the invention. However, in late 1996, the patent was officially disclaimed by the NIH, after a heated international controversy during which Jenkins was apparently threatened with deportation from Papua New Guinea.[14] The NIH Office of Technology Transfer explained its abandonment of the patent in general terms; the Institutes would retain patents only in relation to those inventions with significant commercial potential, and they would no longer file patent applications on 'primary research tools'.[15] Although Jenkins and her advisors had considered assigning the unwanted patent to the Hagahai themselves, the administrative costs would have been entirely disproportionate, given that the invention had almost no prospects of commercial exploitation.

In these 'public' accounts of the PNG-I patent, the Hagahai themselves are absent other than as the passive objects or mute clients of their observers and advocates. Of the latter, the most significant was a Canadian organisation called the Rural Advancement Foundation International (RAFI), which dramatised their 'exploitation' as a story of 'bio-colonialism':[16]

> Pieces of indigenous and remote rural peoples' very bodies are now, without any doubt, the 'intellectual property' of the corporations and governments researching them in the North. The Papua New Guinea patent dramatically underscores why groups like the Human Genome Diversity Project can no longer hide behind claims to be collecting only for historical, cultural, or medical research. It is now clear that they are also potentially shuttling the cells, DNA, and other biological materials of people into the intellectual property portfolios and cashboxes of the life industries.[17]

Coloured with this polemical theme, the story of the PNG-1 patent becomes an exemplary account of the methods which globalised industry has used to appropriate the genetic wealth of indigenous peoples. In some cases their bodily tissues are themselves the raw materials for some commercial process, but in other cases the exploitative relation is more oblique. For instance, traditional endogamous family structures make it considerably easier to identify and isolate the genetic polymorphisms which are thought to be active in producing certain illnesses.[18] 'Bio-colonialism' encompasses processes of commodification which appropriate not only the 'bodily' genetic resources of indigenous peoples, but also their traditional cultural artefacts. Multinational pharmaceuticals companies have identified the 'biotechnological' bases of those therapeutic or agricultural benefits which indigenous traditions have attributed to plants such as the turmeric root, the neem tree, or basmati rice.[19] The basic mode is to (re-)attach traditional applications to an 'invented' molecular structure; or, in other words, to stretch 'the veneer of [technological] modernism over traditional prior art'.[20]

'Bio-colonialism' discloses an asymmetrical process of exchange in which the wealthy but genetically impoverished nations of the North exploit the genetic patrimony of the vulnerable South. One might ask whether this version of events is as 'true' to the agency and identity of indigenous peoples as it claims to be. For example, it is sometimes argued that (even where they are 'successful') claims to cultural property reconstruct traditional cultures in the 'inauthentic' language of property and patrimony.[21] However, given that traditional cultures are always already included or implicated in globalised society, it is not clear who would be placed to adjudicate on authenticity.[22] For that reason, it may be more productive to ask how accounts are settled within globalised society; either quite literally, by considering how the commercial profits of bio-colonialism should be (re-)distributed, or, more abstractly, by considering the ways in which these stories of exploitation express global society's own anxieties about the 'commodification of life'. This article begins with the second, more abstract, question. Therefore it has little to say about the dynamics of particular local disputes, or about the broader, technical, question of how international regulatory frameworks structure economic asymmetries. Rather, it explores the ways in which themes such as 'biocolonialism' articulate a set of productive misunderstandings about the workings of the patents process and of property processes in general. The element of misunderstanding arises because arguments about the 'commodification of life' reckon with concepts that afford only a limited understanding of how property norms bind events, but these conceptualisations are nevertheless productive because they facilitate the very processes that they misrecognise. It is instructive to consider how critical themes such as 'biocolonialism' are metabolised by the very technological and economic processes which they claim to hold at some sort of critical distance.

Patents and bio-politics

Despite its unpromisingly technical and bureaucratic tone, patent law is politically and socially significant because it constitutes a nexus in which some deep-running anxieties about the relation between 'nature' and 'culture' are negotiated. To the extent that one might treat generalised ideas about life, body, or even property, as the components of an 'indigenous (Euro-American) social theory', which facilitates social transactions by articulating a traditional, local, version of the distinction between nature and culture,[23] the dramatisation of bio-technologies as a peculiarly novel and transgressive form of science poses the sharpest and most persistent challenge to this 'indigenous' theory. And, because biotechnology patents are so often understood as proprietary claims to 'life itself' – claims that breach the boundaries set by 'indigenous' social-theoretical distinctions between nature and culture – the legal and bureaucratic administration of biotechnology patents has become the most intensive site of this confrontation. The threat to generalised conceptions of 'life' is that they might be forced into a mode of reflexivity to which they are quite unaccustomed; that is, they are increasingly being forced to address the question of how or why they draw the distinction between nature and culture in this way rather than another. Moreover, these generalised, indigenous, interpretations of the distinction between nature and culture are complicit with legal doctrine, in the sense that while indigenous simplifications are readily absorbed into legal doctrine as resources for thematising the attribution of disparate rights and duties, legal doctrine repays the favour by lending its institutional and ideological *imprimatur* to some of these cultural simplifications. So although many social anxieties about biotechnology are cast as complaints against law, the complicity between doctrine and indigenous theory is such that these complaints betray a confusion within law itself. Legal doctrine is faced with the problem of how to re-invent the motifs that have traditionally explained or justified the attribution of rights. The ethical or ecological arguments which give voice to social anxieties are apt to seem extravagant. For example, an application to patent the expressed nucleotide sequence for the protein relaxin, which is produced during pregnancy, was opposed by the Green Party on the grounds that it enslaved the women from whom the original samples had been extracted; it involved 'the dismembering of women and their piecemeal sale to commercial enterprises throughout the world'. More abstractly, the application was 'immoral' within the terms of the European Patent Convention because 'the patenting of human genes means that human life is being patented'.[24] Yet, despite their eccentricity,[25] and although they have made no obvious or immediate impression on legal doctrine, these objections have identified a structural failing in patents doctrine. The familiar conception of property, whether it is reconstructed in formal legal doctrine or elaborated in the looser languages of 'indigenous' theory, is unable to account for biotechnological manipulations of 'life'.

The 'bio-colonial' dimensions of patents such as the PNG-1 cell line offer the most immediate and compelling dramatisation of the problems that biotechnology poses for doctrinal and indigenous understandings of property. The ownership of 'genes' is represented as a form of appropriation that is even more threatening than other contemporary forms of bodily 'enslavement'.

Whereas slavery perverts property by treating a person as a thing, biotechnology undoes the very distinction that makes property, and hence slavery, possible. The basis of most critiques of 'bio-colonial' commodification is the familiar distinction between those entities which lose nothing essential in the process of legal and economic exchange, and those which are somehow 'degraded' by the medium of money. This distinction may yield little more than the familiar idea of gene ownership as a form of slavery. However, even this banal interpretation is significant because it shows how both legal doctrine and its critics have responded to the challenges of biotechnology by attempting to re-institute the old idea of property. The proprietary distinction between persons and things serves as a basic criterion for the interpretation and evaluation of biotechnological manipulations. And, precisely because they begin with this criterion, ostensibly critical themes such as 'bio-colonialism' or 'commodification' bind patent law and its opponents to the very distinctions between nature and culture, or discovery and invention, that biotechnology has rendered transparent or implausible.

Most critiques of biotechnology patents seek to restore the proper bounds of property; the common basic practical response consists in the 'juridification' of human life, granting it a form of legal immunity from commodification. In political argument, this juridical model surfaces only in metaphorical form. For example, the quite common notion of genetic or ecological 'patrimony' implicitly draws on the technical connotations of the concept of patrimony in civilian jurisdictions, where it is conceived as an abstract, indivisible, and inalienable fund, possessed of a juridical personality that secures it against appropriation by any particular individual. Hence the objections made by Greenpeace to genetically-modified plants on the ground that 'plant genetic resources are the heritage of mankind, and should remain available to all without restriction and be preserved intact for future generations'.[26] When the concept of patrimony is re-appropriated within legal theory, these political metaphors do seem rather simplistic. Instead, the virtues of 'patrimony' are attributed to its status as a 'hybridised' legal form,[27] that allows for a 'dialectical' oscillation between local and global, present and future expectations, or property and personality, and which therefore allows nature to be treated as a commodity for some purposes, a non-commodifiable fund for other purposes, and the substance of human personality for yet others. This is a more complicated adaptation of the juridical relation between persons and things, but the distinction is nevertheless preserved. For some tastes, patrimony is too feeble and equivocal a barrier to the commodification of life; in effect, it makes nature a potential commodity which economic forces would soon colonise. 'Dignity' may seem to be a less fallible criterion. So, for example, Bernard Edelman defends French legal principle against the 'cynical pragmatism' of Anglo-American law[28] by arguing that French legal tradition, which instituted the body as the inalienable foundation of 'legal individualism', offered a surer protection for personal dignity than a regulatory framework premised on the 'liberal' principles of informed consent:

> The body is the means of existence of the human person, and as such it cannot be sold, leased, or given away without immediately reviving slavery. In other words, a contract relating to the human body would be inconceivable in law, because it would offend against human dignity: in that respect, the inalienability of the body is of the same order as the inalienability of freedom [or] citizenship . . . In short, we can enjoy our freedom, but we cannot dispose of it.[29]

This is the most uncompromising argument for 'juridified' life. The body as the substrate of individuality is a legal artefact; or, more precisely, the dimension occupied by or incarnated in the 'body' is a medium of subjective existence that is instituted and normalised by law, even before the law begins to attribute rights and liabilities.

These conceptions of juridified life reinstate the boundaries of property, the better to preserve human life from commercialised science. They do so by reaffirming the distinction between persons and things (even if that distinction is complicated by the quasi-Hegelian notion of a domain of things that is so thoroughly infused with 'personality' as to be unavailable to economic transactions). This is a somewhat conservative manoeuvre. Doubly so, since the objective is not only to re-institute property, but also to re-institute law itself. The dimension of 'life' or 'nature' that critiques of commodification seek to protect is bound to a particular definition of legal power, so that the project of rescuing nature is implicitly associated with the project of (re-)instituting a form of legal 'sovereignty'. If, in terms of the familiar theoretical distinction between 'sovereignty' and 'bio-power', the patents process is almost the paradigmatic example of a bio-political programme, the model of 'juridified' life is obviously on the side of sovereignty. Although 'sovereignty' so readily connotes repression, nevertheless it implicitly guarantees the imperilled distinction between persons and things, life and law, nature and society, or invention and discovery.[30] Only within the theme of sovereignty is it possible to sustain the idea of a pre-given natural substrate from which society or personality can be distinguished.

Sovereign power, as the precursor of bio-power, based itself upon a fundamental difference between law and nature, or between norms and the 'life' to which they refer. The association between sovereignty and repression arises from sovereignty's representation of itself as a normative power ranged against 'life', but this very asymmetry also secures the status of life or nature as a domain that is radically distinct from law or society. Sovereignty was not what it took itself to be. Although it wholeheartedly identified itself with one term of the distinction between 'law' and 'life', this distinction was sovereignty's own creation. 'Normalisation' — interpreted both etymologically and conceptually — consists less in a process of intervention in nature, than in the process of establishing the very perspectival horizon within which such interventions are conceived.[31] In effect, there was no such thing as nature until it was instituted as such by the exercise of sovereign power. For that reason, each exercise of sovereign power effectively intervened in its own representations. This form of self-referentiality was plausible and sustainable only for so long as forms of institutional aesthetics allowed it to be reconstructed as a brazen tautology, dignifying the simple say-so of the sovereign as an unquestionable source or cause of order. However, this tautological mode of self-reference unravelled as the institutions of power began to address their own conditions of existence, prompting a transformation in the way in which power was rationalised; more specifically, a shift from the blind investment in excess to a bio-political economics of income and expenditure.[32] 'Sovereignty', and the distinction between life and law that it sustained, was therefore dissolved by a set of bio-political or governmental techniques which undramatically addressed the problem of operational self-reference. As suggested further on, bio-political programmes do not identify themselves with one side of the distinction between norm and nature; rather, they treat the distinction itself as a provisional programme which serves to govern their own operations. In these terms, the anxieties surrounding biotechnology patents stem from the fact that implosion of sovereignty — and its correlative distinction between law and nature — has only belatedly become transparent. The insistence on a proprietary distinction between persons and things as a defence against commodification might be seen as an attempt to re-institute law as sovereignty, and, by definition, nature as the correlate of sovereign legal power.

Critiques of commodification assume that the language of property norms is descriptive of its own actualisation. In other words, the old legal-philosophical account of how 'things' are subjected to legal norms is mistaken for a quasi-sociological account of the performativity of norms. Of course, themes such as 'sanctity' or 'dignity' are routinely deployed in legal argument, but their functional value consists in processes of expression or actualisation that escape juridical thematisations of power. The ontological assumptions of traditional property semantics, assumptions about scarcity, exclusivity, or distinctions between form and matter, have been overtaken by social evolution.[33] The language of property norms does not command or describe the processes by which those norms are actualised in the world. So, whereas critiques of commodification represent biotechnological sciences as a threat to the integrity of property, or as a peculiarly delicate problem of proprietary regulation, property has already lost the integrity which it is supposed to have. In fact, biotechnological models of gene expression offer an instructive analogy for the processes or strategies through which property norms are socially (or sociologically) actualised in modern society.

Biotechnological practices resemble proprietary practices in that they slice across the distinctions between form and matter, or difference and identity, that are essential to the lawyer's conception of property. It is true that molecular biology has reliably identified the molecular structure of many expressible DNA sequences; precisely those structural sequences that are the basis of many biotechnology patents. For that reason one might be tempted to identify 'genes' with these determinate molecular structures, in which case the basic operations discovered or invented by biotechnology would be referable to precisely the sorts of things — or 'thing-analogues' — that are presupposed by traditional theories of property. However, the identity or quiddity of 'genes' becomes increasingly elusive as the specification of molecular structures improves. The difficulty is that a gene, however it is defined, consists not in a structure but in the process that 'expresses' or actualises a given molecular strand. The specification of a particular gene's molecular composition, or the question of what counts as an integral part of that composition, can be addressed only by referring to the process of gene expression. If attention is shifted from structures to the processes that animate them, the question 'what is a gene?' becomes somewhat more problematic. For example, because the function of a particular transcription unit is determined by regulatory sequences that may be quite remote from the transcription unit itself, it is inadequate simply to identify the molecular structure of genes with transcription sequences. More interestingly, the same DNA sequence can be cut and spliced in different ways to produce functional RNA. For example, the structural composition of an expressible nucleotide sequence may be the result of a selection made by the process of protein expression, so that the same element of a 'raw' DNA sequence can function either as an intron or as an exon depending on the way the sequence is spliced in the process of gene expression. Similarly, the process of protein manufacture itself is often regulated by a cybernetic feedback loop, which retroactively adjusts the 'input' codes of gene expression to their actual 'output'.[34] The molecular structures that are commonly identified as 'genes' are a highly selective interruption of these processes. And, although lawyers or scientists might for that reason prefer to deal with an expressed sequence — that is to say a functional nucleotide sequence that reliably yields a given protein — rather than a 'gene', it is not possible entirely to sever these smaller and more determinate mechanisms from the broader set of environmental or ecological conditions in which they are embedded.

For an enterprise such as population genetics, the tension between local structures or mechanisms and their general conditions of existence surfaces as the problem of how to bind genetic types or variations to their human carriers (see below). The conditions of existence of genetic traits are understood as natural-historical processes, and the work of binding genes to carriers is pursued through techniques such as the statistical sampling of selected populations. For commercial biotechnology, the question of how to bind molecular structures to social actors presents itself in a different form. The conditions of existence of biotechnological artefacts are not 'natural-historical'; rather, they consist in those economic, legal and political processes that shape the environment of biotechnological inventions. And, although molecular biology can specify structures or determinate mechanisms of expression, these elements are charged with the value or 'potential' that they are attributed by the reproduction of these processes. The story of the application made by the NIH for a patent relating to a number of cDNA sequences nicely illustrates, from the perspective of the patents process, this continuous re-invention or re-valuation of molecular structures. The NIH application was rejected on the basis that it failed to satisfy the utility requirements of US patent law. And although the NIH in America and the Medical Research Council in Britain decided not to pursue their applications, the decision by the US Patents and Trademarks Office generated some uncertainty as to the patentability of gene sequences.[35] Many of these uncertainties had to do with the identification of 'artefactual' gene sequences. Did a patent relate to the molecular structure of the gene sequence or to some functional characterisation of that structure? Or rather, given that structure is never sufficient 'in itself' to warrant a patent, how was one to distinguish structure and function, or how was one to decide what sort of functional supplement would make a gene sequence patentable?[36] These questions suggest how the quiddity of a gene — or even a particular nucleotide sequence — is not reducible to a determinate mechanism or structure, but that these structures are variables that acquire different values according to the expectations that are invested in them. The role of the patents process is to bind these expectations in such a way as to facilitate the reproduction of the 'patents vector'.[37]

There are two ways of relating property to biotechnology. First, there is the sense in which biotechnology serve as an analogy for property practices. 'Hybridity' as the *modus operandi* and *opus operatum* of recombinant DNA technologies — which splice 'homologous' DNA sequences into 'heterologous' expression vectors — is a sort of working model for sociological concepts of 'hybridity', which describe how social artefacts are characterised by the processes which machine them rather than by any prior 'essence'. Or, in simpler and more abstract terms, biotechnological artefacts are 'hybrid' objects; 'objects whose essence is not prior to their existence'.[38] From that perspective one might say that the semantic elements of property norms are articulated by operations that are just as liquid and productive as the practices of biotechnology. Secondly, however, there is a different sort of relation, in which property practices are not merely analogical to biotechnological processes, but are operationally bound to biotechnology. If one provisionally takes the side of property, this relation might be described as a parasitic or symbiotic bond through which proprietary strategies exploit the indeterminacy of biotechnological practices. For example, the most basic manoeuvre of litigation strategies is to dissolve scientific artefacts (back) into their constituting processes, which are in turn filtered through doctrinal themes such as novelty, obviousness, or utility. In this way, legal concepts are strung together by a process of argumentation which 'parasitically' abstracts the energy or complexity of biotechnological operations by using the latter as a dynamic, re-vivifying, expression vector for legal language. In semantic terms, these proprietary strategies may not be especially 'novel', given that they deploy the familiar legal conception of ownership as priority or exclusivity. The novelty or complexity lies in the process by which lawyers adapt legal semantics to the dynamic relationships between political, scientific, and economic expectations. These semantic elements make up the basic repertoire of patent law but, again, their functional value is given only by the processes which adapt them to economic and legal expectations. Although the particular example of DNA sequences may no longer be commercially significant, given that gene sequences with no assigned function are now rarely seen as economic commodities, it does suggest how the indeterminacy of 'genes' — here expressed as a tension between structure and function — allows patent law to fold itself into the complexity of biotechnological programmes. Indeed, it is also the way in which the semantic energy of biotechnological models is abstracted by political debates. The ethical or ecological debate surrounding the European Union's 'Biotech Directives' exploited the indeterminacy of genes in a similar manner, emphasising either structure or function depending on the objectives of the argument.

An invented virus?

Patent law is bound to its own version of the traditional, 'proprietary' distinction between law and nature. Indeed, the distinction itself is considered sufficient to disqualify any ethical or ecological objection to the 'commodification of life'. According to Henry Greely, a member of the Ethics Committee of the Human Genome Project, and legal advisor to Carol Jenkins in the latter phase of the controversy over the PNG-1 patent, ecological activists had failed to grasp the simple distinction between a cell line and the materials from which it was derived:

> [T]he patent doesn't patent a person. It doesn't even patent human genetic material. It's the cell line, a viral preparation derived from the cell line, and three different bioassays to see whether people are infected by this virus. And the idea that the US government owns this person or his genetic material is absolute rubbish. [T]he donors involved can continue, obviously, to use their own DNA to run their bodies. They could also, if they chose, patent anything they wanted to patent that was an 'invention' from their DNA.[39]

The legal claims of the Hagahai were on no different or better footing than those of John Moore, who unsuccessfully claimed a share of the profits of a cell line abstracted from his diseased spleen. Moore failed because the medical and commercial value of the Mo-1 cell line was entirely attributed to the inventive art of the scientists who had identified and isolated his 'golden cells'.[40] This use of a banal doctrinal distinction to disqualify political or ethical objections is quite routine in patent law. Perhaps because the constitution of the European Patent Office has exposed it to politically-motivated interventions, the decisions of the EPO frequently resort to this manoeuvre. For example, faced with the Green Party's claim that the Relaxin application did violence to 'life' in general, and more specifically to the donors from whom the source tissues had been taken, the Opposition Division of the EPO responded in terms that amounted to little more than the proposition that an invented artefact is not the same thing as a natural substance.[41] The same strategy informed the Commission of the European Union's attempts to resolve the uncertainties created by the initial rejection of its long-contested Biotech Directive.[42] A personal, and somewhat more polemical, account of this bureaucratic strategy suggests that many objections to biotechnology patents are the result of 'technical misunderstandings which arise from a wilful refusal to understand the difference between discovery and invention'.[43] The first chapter of the Commission's revised, and successful, proposal for a Biotech Directive corrected these misunderstandings.[44] Taking the view that political oppositions had arisen only because the original formula — according to which 'the human body as such or parts of the human body as such' was not to be patentable — was ambiguous, the revised formula specifies that 'the human body and its elements in their natural state shall not be considered patentable inventions'.[45] Supposedly, this formulation improved on the first because it clarified the point that, as knowledge of the human body in its natural state could only qualify as 'discovery' rather than 'invention', natural processes are automatically secured against commodification.[46]

The story of the PNG-1 cell line makes these doctrinal 'clarifications' seem quite implausible. The text of the PNG-1 patent application implicitly concedes the point that, despite their technical sophistication, recombinant DNA and hybridoma technologies remain dependent on the extraction and exchange of 'raw' human materials.[47] A good part of the application is dedicated to the task of differentiating the PNG-1 virus from neighbouring or potentially antecedent strains of HTLV-I. The object is to establish the exclusivity of the PNG-1 viral strain and the peculiarity of the population(s) which might benefit from the various bio-assays in which the invention was embodied:

> Unlike strains of HTLV-I from Japan, the West Indies, the Americas, and Africa, which share a 97% sequence homology, the PNG-1 variant is only about 92% identical to a Japanese prototype HTLV-I.

> [The] PNG-1 variant, in turn, differs by approximately 4% from that of the variants from Melanesian Solomon Islanders, indicating the existence of a new HTLV-I quasispecies.[48]
>
> Due to the genetic variability between HTLV-1 isolates from Melanesia and other geographical locales, the widespread screening for infection in human populations in Melanesia can be best served by using a virus strain which is indigenous to that area.[49]

First, the designation of Carol Jenkins as one of the five 'inventors' of the cell line indicates that the cell line was not conceived as an entirely 'scientific' artefact; it could hardly have been argued that Jenkins had made a direct contribution to the formulation of a conception of the molecular structure.[50] Her formal status highlights the function of the PNG-1 patent — and perhaps patents in general — in concretising a fluid and open-ended process of co-operation between various social actors into a singular scientific 'fact'.[51] Secondly, given that the innovative content of this 'invention' seemed to owe more to the natural components of the extracted materials than to any technological process of extraction and purification,[52] the idea that the production of the cell line involved a radically inventive process seems especially implausible.[53] Unlike the more complex and precarious achievements of recombinant DNA technology, the production of an immortal cell line demands little more of the 'inventor' than the mastery of a routine scientific technique.[54] The 'inventive' process seems merely to transcribe a natural code into a new medium. In these cases, the dogmatic distinction between discovery and invention gives way to a more flexible and more plausible distinction between form and substance, which characterises biotechnological programmes as either novel or natural by referring to the use which they make of the body. Hence the argument that biotechnological inventions should not be patentable if they treat the human body 'as a structured entirety of cells, tissue and organs', and, more specifically, that cell lines and hybridomas should be patentable only if 'both the individual and the point in time of the extraction of the cells were selected by chance, ie not for the mere or predominant purpose of obtaining valuable substances'.[55]

There is something quite unsatisfactory about these endless permutations of 'nature' and 'artefact'. Each perspective exposes its opponent's particular 'fetish'[56] by trading on a distinction between two types of causal genealogy — natural mechanism and human intervention. This same distinction allows political arguments to undo patent law's fetishistic attachment to scientific idols, while legal doctrine repays the compliment by disclosing the artefactual sinews of ecological 'nature'. This process of mutual misunderstanding is interesting and instructive only because the exchange exposes a common 'blind spot'. On the one hand, in attempting to press a political or ethical distinction between 'commodifiable things' and 'non-commodifiable things' into a narrow doctrinal distinction between discoveries and inventions, the strategy behind the second European 'Biotech Directive' misrepresented the nature of the problem. Political oppositions are not a function of doctrinal confusion. Rather, although their force is blunted by legal proceduralisation, political arguments show how legal doctrine mistakes its own interested definition of 'nature' for nature 'as such'. From this perspective the doctrinal distinction between inventions and discoveries is itself a form of violence against nature.[57] On the other hand, the technical definitions of patent law reveal the unacknowledged complexity of ecological 'nature'. The technology involved in the production of cell lines seems unremarkable only because science is all too successful in reproducing itself as 'nature'. The apprehension of 'natural facts' is now mediated by multiple layers of social representation, and the more deeply embedded these layers, the more readily they congeal into a substance that is misrepresented as 'nature'. The common failing of both perspectives is their addictive attachment to the idea of nature 'as such', a pre-given substrate that guarantees the existence of artefacts, and which, depending on one's point of view, defines the proper sphere of human invention. This attachment obscures the fact that there are no such things as inherently 'natural' things, no more than there are entirely 'invented' things. Interestingly, that lesson might have been learned from the Hagahai themselves, whose hybridised social relations disclose the contingency of any distinction between nature and culture.[58]

The distinction between the 'natural' and the 'social' is itself a social artefact, continuously re-invented by discursive programmes that emerge from a point 'between' the terms that they distinguish. Recombinant DNA technologies are 'transgressive' not because they intervene in natural processes more violently or more effectively than other forms of technology, but because they begin 'before' any distinction between nature and technology. In that respect biotechnologies are again representative of discursive processes generally. The incorporation of the 'natural' into the 'social' is not the result of a process of normative authorship, in which 'social' forces are imprinted upon a 'natural' substance. Bio-political rationalities employ the distinction between the natural and social to regulate themselves, but they cannot be identified with either term. Bio-political programmes are sustained by a reflexive process of intervention in their own distinctions, which serve as criteria by which to steer operations and by which to observe the effects of those operations. They are neither norm nor nature, occupying instead the indeterminate point 'between' those terms, and consisting in nothing more than this process of recursive self-observation.[59] This proposition does not depend on the claim that there is no such thing as material reality.[60] Rather, it confines itself to a description of how events are thematised within discursive programmes. 'Natural' events, just like their 'social' counterparts, are ascriptions which these programmes use as prompts or occasions for their reproduction.[61] Nor does the proposition imply that the distinction betwen nature and artefact, or the social and the biological, is entirely illusory.

Scientists, lawyers, sociologists, and politicians routinely assign events to one or the other of its terms. The world is semantically 'materialised' by the reproduction of social programmes which may deploy some version of the distinction between nature and artefact, or things and persons, in their interpretation of observed events. The more interesting question concerns the way in which these distinctions work within the reproduction of bio-political norms.

The theme(s) of authorship or invention might be seen as the components of a legal technique of 'purification'.[62] In this sense, 'purification' describes the way in which discursive processes manage social contingency. It is, so to speak, a way of 'interrupting' or 'patterning' the flow of events. Themes such as 'discovery and invention' turn processual events into localisable facts by fastening them to a conceptual structure which grants them a determinate identity and a stable pattern of causal co-ordinates. The purification of events therefore transcribes contingency into a stable, manageable, topography. However, because there is a radical distinction between these 'purifying' themes and the operations in which they are expressed, so that a theme cannot account for its own expression, 'purification' merely reproduces the processes which it seeks to interrupt. It is therefore a particular technique of hybridisation.[63] For example, the patents process purifies scientific and economic events by characterising them in terms of concepts such as obviousness, sufficiency of disclosure, or industrial application, which are in turn re-characterised by economic and scientific strategies. This version of 'purification' emphasises the thoroughly productive capacity of proprietary themes. It also suggests why biotechological artefacts are so troubling from the perspective of traditional property law. The problem is not that property is losing its mastery of things, quite simply because it has never enjoyed the mastery imagined by legal philosophy. Rather, it is becoming more and more obvious that the old purifying themes have outlived their usefulness, so that the relationship between themes and operations has to be re-negotiated. For example, there is a basic incompatibility between ethical and political arguments and the scientific-rational character of the patents system, which can accommodate 'ethical' themes only if they are translated, or betrayed, by some form of risk calculus.[64] The questions asked of property law increasingly require it to adjudicate on this sort of incompatibility. That may be why both law and science have converged on themes such as informed consent which seem to have the ability to bind scientific or economic characterisations of persons and events to the legal conception of the autonomous individual. For example, returning to the tension between ethical themes and impersonal calculations of risk, questions of risk can be treated as criteria by which to assess the adequacy of disclosure (or the extent to which a particular act of consent is properly 'informed'), thereby preserving the legal fiction of the consenting subject from the more complex representations of agency that are implicit in scientific or economic assessments of risk.

Commodification as co-variation

If one focuses on processes rather than their purification, what kind of objects or artefacts do biotechnological inventions become? If appropriation is seen as a form of discursive inscription rather than a mastery of 'things themselves', how does the patents process function in the reproduction of these artefacts? One answer is suggested by the concept of patents as 'vectors':

> [A]s patent law functions to draw boundaries around information, it allows research to be treated as a form of property or as an asset. In so doing, it enables research to become part of the commercial currency; for it to be mortgaged, insured, included on the balance sheet, and used as a basis upon which decisions about investment can be made. In this sense patents act as vectors which enable operational links to be forged between higher education institutes and industry, and between public and private spheres of research.[65]

There are several versions of the argument that patents regulate the flow of scientific information. One might, for example, ask to what extent the norms of patent law work as the functional equivalents of Merton's norms of scientific culture.[66] However, this question still represents the agency of property norms in conventional legal or regulatory terms; the relation between patent laws and scientific processes is understood as a relation between a normative template and the world to which it is applied. And, although scientific practices might be allowed a capacity to resist, or to evolve forms of 'creative compliance', these processes of negotiation are understood in terms of the prescriptive content of norms rather than the discursive operations that articulate prescriptive themes. In other words, the process of regulatory negotiation is deciphered by measuring evasive activity against the propositional content of regulatory norms. From the start, the field of regulatory events or activities is 'framed' or delimited by these propositional contents rather than the more complex and heterogeneous tissue of discursive operations. The model of intellectual property rights as 'vectors' is distinctive because it represents the nexus between law, economy, and science as an emergent bond, in which no participant enjoys a privileged perspective from which to observe and intervene in the others. Hierarchical distinctions such as the distinction between norms and facts are embedded in the discursive programme which employs them. In other words, the social plane is folded into a set of local hierarchies, none of which, not least that of the sociologist or social theorist, can claim any sort of epistemological or practical priority. So, although science has always been dependent upon particular institutional or economic configurations, and upon the ability of scientists to develop and exploit these networks,[67] the idea that patent law has been reduced to an 'expression vector' for commodified science is entirely misplaced. The nexus between law, economy and science is emergent in the sense that each of these practices exists only in and through its relation to its partners, so that money, norms, and scientific facts are already bound into relations of mutual representation and mutual convertibility. This more fluid theoretical model is not too far removed from the perspective of the participants in the patents process. For example, the research director of SmithKline Beecham, defending the company's use of patents to structure access to its projected map of the human genome, observed that: 'there's absolutely no reason why our genes can't be used as "currency" in negotiating creative alliance frameworks with various partners'.[68]

The concept of an emergent process implies that (patented) artefacts are produced by the ongoing articulation of discursive processes. Some sense of how this might work is suggested by Marie-Angèle Hermitte's interpretation of the requirement that a plant variety right — a *droit d'obtention végétal* — should be granted only in relation to those plant varieties that are distinguishable from all previously known varieties by at least one significant characteristic.[69] There is no closed categorisation of distinguishing characteristics, which allows the process of identification to be responsive to economic and scientific imperatives:

> If varieties are classified according to a limited range of characteristics it is relatively difficult to create a new variety but each existing variety enjoys a broad and stable market position. To extend the list of classificatory characteristics, something which the examiners consider they are entitled to do even in the course of an examination, makes it much easier to create new varieties, even if these enjoy a less stable and much narrower market position.[70]

Hermitte's argument is that economic interests infiltrate the taxonomic categories of intellectual property, causing them to become more or less rigid in response to market strategies. If that were so, the role of patent law and the law relating to plant variety rights would indeed be to serve as nothing more than a receptive medium for the active conversation between economic and scientific criteria. The role of law would simply be to facilitate this exchange rather than to impose its own definition of what should qualify as a sufficient distinguishing characteristic. According to Hermitte, intellectual property laws therefore become a form of 'soft' law, which retains the external institutional attributes of a system of legal norms, but whose internal mechanisms have been thoroughly colonised by an economic rationality. There are various problems with this analysis, which have to do with the priority accorded to economic programmes,[71] but the basic formulation points beyond itself, towards a model of how patentable inventions are constructed by the 'co-variation' of legal, scientific, and economic programmes. A patent is a legal artefact, referable in the first instance to law's particular perspective on this process of co-variation. However, even if one privileges law to this limited extent, the essential point is that a patented invention is produced by the legal characterisation of a fluid process, in which law is itself implicated, rather than the legal description of an identifiable fact or mechanism. The general category of the invention, which emerges from the articulation of various criteria of inventive priority, is a sort of virtual entity, sculpted by the legal characterisation of economic and scientific expectations. As this representation of the nexus of expectations is progressively woven into the language of inventive step, obviousness, industrial application, and so on, it gradually comes to be expressed as an intangible 'thing'. However, the materiality of this intangible artefact consists in nothing more substantial than the legal apprehension of a fluid nexus of expectations. More precisely, the intangible *res* is the referent of propositions or prescriptions that consist in the co-variation of legal discourse and its companion discourses. In that sense, the patents process 'commodifies' not by binding pre-constituted things or mechanisms, but by representing its discursive binding of expectations as an intangible 'thing'. In these terms, a model that Hermitte uses to argue for the concept of intellectual property law as a form of 'soft law' instead becomes the model of a relation in which each participant patterns the nexus of expectations in accordance with its own purifying themes; a proprietary 'vector' is reproduced as a dynamic process in which scientific programmes accommodate themselves to an economic accommodation of legal criteria, and so on, in a circular movement of overlapping observation.

The general category of a patentable invention — or the notion of 'the addition of a new idea to the existing stock of knowledge'[72] — therefore precipitates from the various themes which the law uses to decipher scientific processes, and more especially to observe the complex relations between scientific and economic processes. This process of legal invention involves a relentlessly corrosive mode of purification, in which scientific commodities are decomposed into competing accounts of their emergence.[73] If science turns artefacts into facts, patent law turns scientific facts back into artefacts. Litigation strategies deploy the familiar proprietary themes of identity and continuity[74] — which are here diffracted into various sub-themes such as (non-)obviousness, (in)sufficiency of disclosure, or, in the context of biotechnological inventions, the distinction between biological and micro-biological processes[75] — so as to unravel the patents vector, reconstructing between scientific and economic parts of the bundle in legal terms. So, for example, the distinction between 'obvious' and 'obvious to try' allows litigators to shift between a genealogy based on a 'purely' scientific representation of the process and a representation of the scientific process as it is mediated by economic or corporate expectations. More important, these legal themes have to take account of law's immediate implication in the patents 'vector'. For instance, concepts such as 'sufficiency of disclosure' allow lawyers and judges to advertise or accommodate economic concerns about the 'breadth' of a proposed patent, and thereby to bind scientific and economic expectations in a way that anticipates the longer-term effects of patent law upon the systems that it binds. 'Commodification' consists in this process of unravelling and recomposing the fabric of expectations.

The uses of 'bio-colonialism'

In functional terms, the patents 'vector' is a technique that binds events through processes of bio-political normativity rather than 'juridical' power. This theoretical distinction is hardly novel, but the example of biotechnology patents offers a clearer specification of the concepts of 'bio-power' and 'governmentality'. The concept of 'purification' is an essential part of this specification. For example, an important implication of the preceding arguments is that concepts of 'biocolonialism' or 'commodification' constitute a form of political or legal purification; they press complex processes of co-variation into simplifying templates fashioned from traditional conceptions of property and political power. The interesting observation about these templates is not that they are selective or out-dated, but rather that they serve to reproduce the very processes that they criticise. Thus, critical representations of bio-colonialism have been incorporated into the commercial strategies of multinational pharmaceuticals corporations, not just because they facilitate the process of exploitation, but more importantly because they ease the articulation of commercial vectors.

Complaints about commodification are often met by the argument that biotechnology patents do not place human individuality at stake: 'the effect of a patent on human genes is restricted to the exclusion of others from using the genes in an industrial process and, thus, does not affect at all the personality of human beings'.[76] Superficially, this is quite correct. However, examples such as that of the PNG-1 cell line patent reveal the sort of difficult questions that biotechnology patents ask of modern conceptions of human individuality. Even if the basic notion of embodied individuality remains unquestionable,[77] the reproduction of this conception of individuality has been complicated by the manner in which biotechnology undoes the distinction between nature and culture, (re-)opening the 'socio-biological' question of how to attach human individuals to their genetic components.

Scientific studies of genes are not directly interested in individual human beings; instead they deal with a very different kind of individual. Genetic types or traits have biographies that unfold through particular human lives, and those biographies are shaped by processes of fertility and mortality which exceed the temporal or geographical limits of embodied human existence. 'Genes' exist primarily as laboratory artefacts or statistical figures, and as such they occupy a dimension that has only an oblique and problematic relation to the dimension of social individuality. Indeed, recombinant DNA technologies have so thoroughly appropriated the molecular matter of genes that the notion of an 'original' human source is of little or no relevance. DNA fragments have acquired a malleable individuality of their own, and one might well ask whether, for example, it is appropriate to 'identify a gene which expresses in the human liver with the label human'.[78] In this respect genetic technologies have amplified a problem that had already become apparent in relation to the simpler question of the legal status of human tissues. As medical-technological resources, human tissues such as blood, gametes, organs, or hospital waste have been assigned to different medical sectors and accorded different regulatory statuses.[79] Each of these statuses implies a particular representation of the body, and a particular technique for (re-)attaching tissues to bodies. The bond between each kind of tissue and its body of origin is reconstructed as a specific compound of social or technological expectations, which determine who is authorised to give consent, what makes consent actual or effective, whether or not some form of payment is appropriate, what is in the best interests of research, and so on. The difficulty then is to find some way of stitching these different regulatory statuses (back) together, as a coherent representation of individuality. 'Informed consent' is a particularly successful answer to this difficulty because it works with a model of individuality that is so vacuous and pliable as to be almost infinitely adaptable. In themselves, the questions of what counts as consent, whether 'actual' or 'counterfactual', or of what counts as adequate disclosure, allow for the construction of quite different categories of consenting subject.

In one sense, the tension between social individuality and genetic materiality is simply the modern form of a very old question: how does one reconcile physical materiality, bodily integrity and social individuality?; or, more abstractly, how can embodied individuality be defended against 'process'?[80] However, the difficulty of the question is acutely accentuated by the fact that life and death have become bio-political concepts, so that the issue is no longer that of how to attach 'social' individuality to 'biological' processes, but rather how to construct a model of individuality that takes account of the radical contingency of the bond between the social and the biological.

Legal themes, and specifically the notion of informed consent, are quite successful in addressing these difficulties. In fact, they intervene even before one reaches formal legal procedures such as patent applications, in such a way as to reproduce the basic scientific material of 'genetic patrimony', not least viruses such as the PNG-1 strain. These more discreet functions are evident in the way that population genetics — exemplified in the work of the Human Genome Diversity Project — has come to style itself as a form of 'sincere' science, conscious of the broader ethical implications of its activities. The sincerity of population genetics is advertised by its basic normative programme, which is to establish a historical geography of the human family that begins with the egalitarian and apolitical substance of genes rather than a divisive and asymmetrical category such as 'race'. In practical terms, this involves the construction of a genetic 'family tree' that shows how humanity became differentiated into its existing (genetic) populations, or of how 'genes, peoples and languages have diverged . . . through a series of migrations that apparently began in Africa and spread through Asia to Europe, the New World and the Pacific'.[81] In order to retrace the evolution of genetic types, population geneticists begin with existing populations, and calculate their proximity to one another — or the 'genetic distance' between them[82] — by calibrating the variations in the molecular sequences of certain genes. In other words, affinities inferred from variations in genetic sequences are traced to a point of division or divergence: '[w]hen other matters are equal, genetic distance increases simply and regularly over time. The longer two populations are separated, the greater their genetic distance should be'. It should therefore be possible to identify the temporal and geographical point at which two existing populations diverged, and thence to reconstruct the entire genealogy of the human family. However, these affinities are difficult to infer where one is dealing with heterogeneous populations in which the pure lines of evolutionary drift have been complicated or contaminated by intensive cultural exchange.[83] For that reason, population geneticists prefer to begin with populations that have retained a 'pristine genetic constitution'.[84] If the great virtue of 'pristine' populations is that they constitute the only reliable co-ordinates for the evolution of genetic types Papua New Guinea, for the population geneticist as much as the doctoral student in anthropology, is the very ideal of purity and complexity.[85] Indeed, the investigative programme which resulted in the 'invention' of the PNG-1 virus and its correlative antibodies implemented the basic schema of population genetics; so much so that the virus itself was identified by its own genealogical 'dendrogram', which discreetly constituted the 'genes' that formed the 'genetic patrimony' of the Hagahai, and an essential part of their identity.[86]

The Human Genome Diversity Project is faced with the problem of how to (re-) incorporate the molecular substance of genetic types into human society. There are two aspects to this problem. First, how does one persuade the members of pristine populations that their participation in the project would be of benefit to themselves and to the 'human family' in general? Secondly, how can the abstract elements of genetic evolution — statistical populations and molecular sequences — be bound to social personality in such a way as to produce a plausible story of 'human' evolution? In other words, if there is indeed some 'evolutionary' correlation between genes and languages as the representatives of, respectively, biology and society, how are people to recognise and orient themselves by reference to this abstract distinction? These difficulties are resolved by the Human Genome Diversity Project's ethical protocols, which also take full advantage of the indeterminacy of the concept of informed consent:

> It is important to identify the most appropriate persons with whom to communicate, the persons from whom clearance should be obtained, and the appropriate content and media of communication. Research will need to take account of the group's social organisation, goals and aspirations, cultural values and mores, and laws (both statutory and customary).[87]

These protocols allow indigenous peoples to be represented 'as partners in the work rather than merely subjects of it'.[88] This is, of course, a particularly inequitable kind of 'partnership',[89] which has evolved more as a way of ensuring the survival of the project than as a way of recognising indigenous peoples.[90] Moreover, in representing the participation of individuals in these terms, ethical protocols yield an unexpected scientific or cognitive benefit; the contractual bond of informed consent is essential to construction of a 'scientific' relation between evolutionary genetics and individuals.

In the population geneticist's story of genes, peoples, and languages, the reference to 'peoples' is somewhat opportunistic. It represents 'people' as the vital medium of a relation between genes and languages. First, the hypothesis of a causal connection between the transmission of genes and the transmission of languages supposes that people are the common 'carriers' of both codes. Secondly, 'people' themselves are supposed to derive their cultural identity from this compound of genes and languages; indeed, in one sense they simply are this compound of genes and languages (which is why the story of how genes are reproduced can be represented as a story of how humanity itself is reproduced and differentiated). The trouble is that because population genetics reckons only with molecular structures, linguistic classifications, and statistical populations, each of which is a theoretical abstraction with no self-evident relation to 'real' people, it is unable to explain just how genes characterise 'people'. Population genetics takes on the difficult task of determining how genes and culture are bound together by some process of co-evolution or co-variation; how, that is, genes affect social individuality, or of how personality serves as a medium through which 'culture' is re-absorbed into 'genes', particularly in a culture that has come to thematise itself in terms of its genetic components. Yet it does no more than retrace two similarly-shaped developmental 'curves', without explaining the mechanisms by which each affects or responds to the other. Obviously, 'people' are supposed to constitute the medium or mechanism of this exchange, but the nature of this role is inferred rather than explained. And, although this inference gains explanatory consistency by being fashioned into a scientifically-constructed genealogy of human populations — and, by implication, people — one immediately encounters the problem that, like all family trees, the population geneticist's 'dendrogram' is a highly selective model. For example, the populations from which the genetic history of humanity is inferred are highly selective groupings, referable not only to the geneticist's preference for 'pristine' populations, but also to the statistical sampling techniques which allow these populations to serve as co-ordinates for the evolution of humanity as a whole. In other words, the genetic past which is supposed to characterise modern humanity as a whole is just the retrospective elaboration of an arbitrary representation of contemporary humanity.

Population genetics creates a role for 'people' that it cannot explain within its own frame of reference, and then compounds the problem by turning a weak hypothesis into a grand genealogy. The ethical practice of requiring informed consent functions in such a way as to reinforce this fragile genealogical characterisation of human beings. Because consenting participants are already identified in terms of the genetic population to which they belong or which they represent (simply by virtue of the fact they are recognised as the appropriate individuals or representatives to give consent to a particular project) their subscription to the project already makes of them what the project would like them to become, namely, the 'people' who ensure the articulation of genes and languages, or the 'social and the biological'. Thus, a legal, contractual, relation binds participants into a vital scientific or theoretical role that is presupposed by scientific explanations but which cannot be accounted for by the science of population genetics itself. This is only one of the ways in which 'ethical', legal, protocols are constructively re-appropriated by the processes they are supposed to regulate, thereby reconstructing the individuals or populations who are ostensibly accorded recognition or protection. The theme of 'bio-colonialism' has also suffered this fate, and informed consent is again the principal vehicle of reappropriation.

There is of course a somewhat crude sense in which ethical or ecological objections to biotechnology patents are appropriated within the patents process itself. The existence of ethical opposition to a patent is a factor that can be mobilised in legal argument. Plainly, the options are somewhat less interesting for the patentee, who can only attempt to minimise the long-term damage caused by adverse publicity,[91] than they are for the infringer, who can turn publicity to strategic advantage.[92] However, informed consent facilitates a rather more subtle and interesting technique for the incorporation of ethical themes in the patents process. For example, the case of the PNG-1 patent has been superseded by commercial practices such as those adopted by Merck or Hoffmann LaRoche, in which industrial corporations have negotiated the assignment of some of the anticipated profits of pharmaceutical inventions to indigenous groups in return for their approval or endorsement.[93] Even from the perspective of those organisations who have done most to expose 'bio-colonialism', informed consent is considered to be an effective remedy to the violence of bio-colonial commodification. For instance, the Rural Advancement Foundation International readily accepts informed consent as an approved procedure for commercial science; researchers should disclose 'any intention to seek patents and the prospects of commercial gain', and 'prior informed consent is required from both the subjects of medical research and their communities or governments before materials or information arising from the research can be commercialised in any form'.[94]

Informed consent, and the contractual procedures that it has inspired, constitute a particularly effective way of integrating the theme of 'bio-colonialism' within bio-colonial transactions. And again, as with the Human Genome Diversity Project's ethical protocols, this process of integration involves more than the simple conscription of indigenous participants. As the extended debate over the European Biotech Directive suggests, commercial enterprises cannot entirely insulate themselves from political opposition; this has less to do with the force of political arguments than with the fact that biotechnologies have corroded the different purifiying themes which facilitate the reproduction of proprietary 'vectors'. Ethical or ecological arguments against 'bio-colonialism' betray tensions or incompatibilities existing within the compound structure of 'bio-colonial' transactions. Here again, the virtue of 'informed consent' is that it is a highly 'deformable' theme, which readily negotiates the complexity produced by the increasingly transparent differentiation of concepts of individuality and body. Because it consists in an open-ended relation between 'subjects' and 'information', each of which is already a complex, indeterminate, category, the procedure of informed consent is sufficiently malleable to reconcile the quite different conceptions of agency that are found in law, science, politics, and economics. And, because the category of 'information' is based upon the translation of complex economic or scientific assessments of 'cost' and 'benefit' into a set of criteria that would make sense from the perspective of the fictitious consenting subject, it performs a dual mediating function. First, it makes abstract conceptions of agency or individuality commensurable with the familiar juridical notion of subjectivity; secondly, and perhaps more importantly, it makes these abstractions commensurable with each other, relegating the juridical notion of subjectivity to the role of a simple common denominator. And, as with the role played by informed consent in constituting an ostensibly scientific account of the relation between genes, peoples, and languages, the act of consenting characterises the subscriber in terms of a prefabricated role. If one asks who, for example, is qualified to give consent to a 'bio-colonial' transaction, the protocols adopted by the HGDP and advocated by RAFI suggest that consent should be given by the population as a whole, or the government, or the relevant individuals, or by all of these 'actors'. The indeterminacy of these specifications means that from the start they are apt to incorporate and reproduce precisely those scientific and economic processes which informed consent is supposed to regulate. Indeed, definitions of populations and individuals are so thoroughly infused with economic and scientific expectations that the options available to the consenting actor have already, even before consent is given or withheld, been configured by commercialised science.

Again, the act of consent yields some unexpected benefits. For example, the judicial arguments in the case of *Moore* v *Regents of UCLA* reveal how procedures of informed consent (which were ostensibly developed to secure the interests of patients) ensure the reconciliation of economic, scientific, and therapeutic rationalisations of medical practice. The rise of commercialised biotechnologies has accentuated the differentiation of medicine, producing some sharp divisions between 'Hippocratic' therapy, laboratory research, and commercial profit. Increasingly, the function of legal regulation and adjudication is to mediate between these overlapping representations of the patient. By drawing on the almost inexhaustibly adaptable resource of legal subjectivity, the procedure of informed consent manages to bind these representations to a common denominator. The patient's consent is the common term of each separate transaction; the fiction of the informed patient translates competing representations into a basic calculus of costs and benefits, thereby unifying or absorbing the tensions between rationalities. Of course, informed consent is not a property law concept, and these observations might be taken as evidence for the proposition that human dignity and autonomy should be granted a more fundamental form of protection. Whereas informed consent does no more than impose conditions on the commercialisation of human beings, the more radical argument for dignity would quite simply prohibit commercialisation. The most obvious practical means of granting protection would be to create an exceptional category akin to the Roman law concept of a *res extra commercium*. However, to create or defend an exception is to concede the claims of the rule. In other words, the technique of exemption is plausible only if one accepts the traditional account of how property norms bind things. As soon as concepts such as dignity or autonomy are treated as purifying themes rather than effective descriptions, the question is not whether dignity should be enforced, but how the theme of 'dignity' would be absorbed into commercial vectors. Here, somewhat crudely, it might be observed that the claims of 'dignity' are somewhat less plausible than those of 'informed consent'. The 'cynical pragmatism' of Anglo-American law is especially troubling to Bernard Edelman because French law has come to adopt informed consent as the means of proceduralising its conceptions of dignity. However, the undoing of dignity is not entirely the fault of informed consent. Conceptions of dignity presuppose some mode of proceduralisation, if only to determine what counts as 'pure' research, how conflicts between research and therapy should be addressed, under what conditions it would be appropriate for patients to take part in randomised drugs trials, how the donation of tissues should be regulated, and so on. These unavoidable practical questions require a mode of proceduralisation which shares the flexible mediating skills of the concept of informed consent. Again, the simple point is that legalistic representations of genetic technologies have to take account of law's immediate implication in biotechnological processes.

Acknowledgements

I am grateful to the participants in a research seminar at Griffith University for their comments on a version of this article.

Notes

1 Carol Jenkins, quoted in *The Detroit News* 20 April 1996.
2 Jenkins and two other members of the Papua New Guinea Institute of Medical Research were co-signatories, together with researchers from the National Institute of Neurological Disorders and Stroke (a branch of the US National Institutes of Health) of a letter to a medical journal describing the isolation of the strain of the virus (Richard Yanagihara *et al*, 'Isolation of HTLV-I from Members of a Remote Tribe in New Guinea' (1990) 323 *New England Journal of Medicine* 993).
3 Carol L. Jenkins, 'Health in the Early Contact Period: A Contemporary Example from Papua New Guinea' (1988) 26:10 *Social Science and Medicine* 997, 997.
4 The missionaries' first response was to send a helicopter bearing sweets and toys. More important, in attempting to convert the Hagahai to a more settled, sedentary, life style, the missionaries rendered them more vulnerable to infection by unfamiliar viruses and bacteria.
5 Jenkins, n 3 above, 1004.
6 In fact, through her publications Jenkins actively raised some of the funds needed to buy essential vaccines and medical supplies for the Hagahai.
7 Carol L. Jenkins, 'Medical Anthropology in the Western Schrader Range, Papua New Guinea' (1987) 3:4 *National Geographic Research* 412, 413.
8 *ibid* 412.
9 Yanagihara *et al*, n 2 above.
10 'The present invention relates to a human T-cell line (PNG-1) persistently infected with a Papua New Guinea (PNG) HTLV-I variant and to the infecting virus (PNG-1 variant). Cells of the present invention express viral antigens, type C particles and have a low level of reverse transcriptase activity. The establishment of this cell line, the first of its kind from an individual in Papua New Guinea, makes possible the screening of Melanesian population using a local virus strain. The present invention also relates to vaccines for use in humans against infection with and diseases caused by HTLV-I and related viruses. The invention further relates to a variety of bioassays and kits for the detection and diagnosis of infection with and diseases caused by HLTV-I and related viruses' (Abstract, US Patent Application No 05397696, assigned to the National Institutes of Health, granted 14 March 1995. The full text is available at <http://www.rafi.ca/pp/patent.txt>).
11 'Scientists Attacked for "Patenting" Pacific Tribe' (1995) 270 *Science* 1112.
12 'Anthropologist Cleared in Patent Dispute' (1996) 380 *Nature*. 374.
13 Carol Jenkins, quoted in 'Bio-pirates Face Challenge in Pacific Blood Feud' Gemini News Service Report, <http://www.oneworld.org/gemini/wk42-gemini-bio.html>.
14 *The Independent* 29 March 1996; 'Anthropologist Cleared in Patent Dispute' (1996) 380 *Nature* 374.
15 'US Drops Patent Claim to Hagahai Cell Line' (1996) 384 *Nature* 500.
16 '[A]ccusations came from a small Canadian-based group known as the Rural Advancement Foundation International, which says it is dedicated to the socially-responsible development of "technologies useful to rural societies". By distributing the release via the Internet – a medium prized by scientists for its ability to

disseminate information, but one proving equally adept at spreading misinformation – RAFI ensured a wide and rapid airing' ('Scientists Attacked for "Patenting" Pacific Tribe' (1995) 270 *Science* 1112).

17 RAFI Communique, Mar/Apr 1996, 'New Questions About Management and Exchange of Human Tissues at NIH' (text available at <http://www.cptech.org/ip/rafi.html>). See also Aroha Te Pareake Mead, 'Resisting the Gene Raiders' (1997) 293 *New Internationalist* 26.

18 'Tribal Groups Attack Ethics of Genome Diversity Project' (1996) 383 *Nature* 208; 'Gene Hunters Home in on India' (1996) 381 *Nature* 13.

19 'GATT adds Fuel to Patents Controversy' (1993) 366 *Nature* 625; 'US Patent Office Withdraws Patent on Indian Herb' (1997) 389 *Nature* 6; 'India to Challenge Basmati Rice Invention' (1998) 391 *Nature* 728. Of these, the last example is by far the least plausible.

20 Correspondence, (1995) 378 *Nature* 532.

21 Rosemary Coombe, 'The Properties of Culture and the Politics of Possessing Identity: Native Claims in the Cultural Appropriation Controversy, (1993) 6(2) *Canadian Journal of Law and Jurisprudence* 249.

22 On these issues, see Lisa Wilder, 'Local Futures. From Denunciation to Revalorisation of the Native Other' in Gunther Teubner (ed), *Global Law Without a State* (Aldershot: Dartmouth, 1997).

23 See Marilyn Strathern, 'New Knowledge for Old? Reflections following Fox's Reproduction and Succession' (1994) 2:3 *Social Anthropology* 263; esp at 270–276.

24 See the decision of the EPO Opposition Division in *HOWARD FLOREY/Relaxin* [1995] EPOR 541, point 6.1. of the reasons.

25 As in, for example, the curious proposition that to permit the patenting of gene sequences would be to 'privatise humanity in the European Union' (representative of the German Green Party of the European Parliament, comments reported in (1994) 372 *Nature*, at 310). It may be that this sort of argument will not survive the 'modernisation' of the Green Party's perspective on biotechnology (see for example (1998) 392 *Nature*, at 213).

26 *PLANT GENETIC SYSTEMS/Glutamine synthetase inhibitors* [1995] EPOR 357, 363.

27 Francois Ost, 'Le patrimoine, un concept dialectique adapté à la complexité du milieu' (forthcoming).

28 Bernard Edelman, 'Génétique et liberté' (1991) 13 *Droits* 31; *idem*, 'Expérimentation sur l'homme: une loi sacrificielle' (1991) 235 *La Recherche* 1056.

29 Edelman, 'Expérimentation sur l'homme: une loi sacrificielle' *ibid* 1061.

30 Appropriately, the common form of these distinctions is most convincingly demonstrated by Hegel, for whom the acquisition of property and accession to language are quintessentially juridical processes.

31 'Law is normative in character; it functions as a norm (in the etymological sense of a set square) not because it commands or prescribes but because in the first instance it establishes the framework of its own reference to real life — it normalises life' (Giorgio Agamben, *Homo Sacer. I: Le pouvoir souverain et la vie nue* [French translation of *Homo Sacer. I: Il potere sovrano e la nuda vita*] (Paris: Seuil, 1997) 34).

32 This transformation is especially well described by Alain Guery in 'Le roi dépensier. Le don, la contrainte, et l'origine du système financier de la monarchie française d'Ancien Régime' (1984) 39 *Annales ESC* 1241.

33 See my (somewhat compressed) account of this shift in 'Instituting Property' (1998) 18 OJLS 331.

34 See François Jacob, *La logique du vivant. Une histoire de l'hérédité* (Paris: Gallimard, 1970) esp at 299–306.

35 'US and British Researchers Agree not to Seek Gene Fragment Patents' (1994) 367 *Nature* 583.

36 If anything, these uncertainties were only amplified by the US Patent Office's subsequent determination that expressed sequence tags (short nucleotide sequences which are used to identify transcription units that have been expressed as functional RNA) were patentable.

37 See below for an extended discussion of this latter concept.

38 Bruno Latour, *Nous n'avons jamais été modernes* (Paris: La Decouverte, 1996) 118.

39 'Scientists Attacked for "Patenting" Pacific Tribe' (1995) 270 *Science* 1112.

40 *Moore* v *Regents of University of California* 111 S Ct 1383 (1991); Bernard Edelman, 'L'homme aux cellules d'or' (1989) 34 *Receuil Dalloz Sirey* 225. A simpler example of the distinction between invention and discovery is found in *Diamond* v *Chakrabarty*, where it was held that an engineered bacterium 'plainly qualified as patentable subject matter' because the patentee's claim was 'not to a hitherto unknown natural phenomenon, but to a non-naturally occurring manufacture or composition of matter — a product of human ingenuity' (*US Supreme Court Reports* 65 L Ed 2d (1980) 144, 150). In *Chakrabarty* the distinction between nature and artifice is unhelpfully stark, not least because the majority in the Supreme Court chose to reduce the case to a formalistic exercise in textual interpretation.

41 First, the application related to a DNA sequence which was structurally different from the sequence as it was found 'in nature', because a 'natural' sequence will express a given protein only if functionally redundant introns have been spliced out. Secondly, even if the application had claimed a molecular structure that was identical to the nucleotide structure in its 'natural' form, the technological interventions which allowed the sequence to be identified and isolated were sufficient to turn a discovery into an invention. More fundamentally, having already characterised the nucleotide sequence in question as a 'DNA fragment' rather than a 'gene', the Opposition Division asserted that '. . . DNA is not "life", but a chemical substance which carries genetic information and can be used as an intermediate in the production of proteins which may be medically useful' (*HOWARD FLOREY/Relaxin* [1995] EPOR 541, point 6.3.4 of the reasons).

42 'Proposal for a Council Directive on the Legal Protection of Biotechnological Inventions', COM (88) 496 final, OJ 1989 C 10/3. Another, less explicit strategic reason is the desire to maintain the fiction that the implementation of the directive would not alter established patent laws or practices, but that it would merely harmonise existing practices.

43 Dominique Vandergheynst (Commission of the European Union, DG XV), with reference to the argument that gene sequences or derivatives should be patentable only if they were essentially different from sequences or elements as they were naturally found in the body ('Brevet sur le vivant: une nouvelle invention juridique?' in Libois and Strowel (eds), *Profil de la création* (Brussels: FUSL, 1997) 65, 88).

44 '[I]n particular, to determine the difference between inventions and discoveries with regard to the patentability of certain elements of human origin' ('Proposal for a European Parliament and Council Directive on the Legal Protection of Biotechnological Inventions' COM (95) 661 final, OJ 1996 C 296/4, 5). At the time of going to press, the text of the Directive as approved by the European Parliament had not appeared in the *Official Journal*. However, this formula reappears in recital 14 of the approved text.

45 ibid, article 3(1). The text of the Directive makes a sharper distinction between 'discovery' and 'invention'. See article 5:

> 1. The human body, at the various stages of its formation and development, and the simple discovery of one of its elements, including the sequence or partial sequence of a gene, cannot constitute patentable inventions.
> 2. An element isolated from the human body or otherwise produced by means of a technical process, including the squence or partial sequence of a gene, may constitute a patentable invention, even if the structure of that element is identical to that of a natural element.
> 3. The industrial application of a sequenced or a partial sequence of a gene must be disclosed in the patent application.

See also Alain Gallochat, 'Directive on Legal Protection of Biotechnological Inventions: Another (or Last) Chance for Europe?' (1996) 82 *Patent World* 13, esp at 15: 'This improvement should be wholeheartedly approved. It relates back to a fundamental principle in the patents field, a principle that has been somewhat forgotten, namely, that a discovery cannot be patented while an invention can . . .' (The author is counsel for the Pasteur Institute, one of the more aggressive European public sector holders of biotechnology patents).

46 See the account of the directive's ambitions in Vandergheynst, n 43 above, 80–82.

47 See generally RAFI Communique, Jan/Feb 1997, 'The Human Tissue Trade' <http://www.rafi.ca/communique/fltxt/19971.html>.

48 Abstract, PNG-1 Patent Application, n 10 above.

49 *ibid.* Some sense of the significance of these figures can be gathered by referring to the contest between Luc Montagnier of the Institut Pasteur and Robert Gallo of the National Cancer Institute as to who enjoyed priority — both 'intellectual' and 'proprietary' — in the discovery of the AIDS virus. One of the evidential facts in this contest was the small degree of sequence variation between the two discovered viruses. The nucleotide sequences of LAV (the Pasteur isolate, which had been forwarded to the NCI) and HTLV-III (the strain isolated by Gallo) varied by only 1.5 per cent, a factor of difference which was so minimal as to suggest that the NCI strain was identical to the Pasteur strain, any divergence being attributable to spontaneous mutations in culture (see Rebecca S. Eisenberg, 'Proprietary Rights and the Norms of Science in Biotechnology Research' (1987) 97:2 *Yale Law Journal* 177, 226–229, and esp n 260).

50 In US law, an 'invention' is predicated upon the conception of a molecular structure: 'conception of a DNA sequence, like conception of any chemical substance, requires definition of that substance other than by its functional utility' (*Fiers* v *Sugamo* (Fed Cir) (1993) 25 USPQ 2d 1601). For an illustration of how this notion of invention can become an issue where the commercial stakes are significant, see the dispute over the BRCA1 patent and specifically the question whether NIH scientists who had participated in the inventive programme should be recognised as inventors (reported in (1995) 373 *Nature* 649).

51 The nature and the limits of Jenkins' role are implicitly disclosed in the assertion made by NIH scientists, that 'at no time did NIH investigators have direct interactions with the Hagahai' (response to a Freedom of Information Act request, made by RAFI, cited in RAFI Communique Mar/Apr 1996, n 17 above).

52 In some cases there may be a technical or doctrinal answer, which is that the process itself is the invention. One example is the story of the 'polypeptide expression' application. One of the reasons given by the EPO's Examining Division for rejecting the patent was that this particular recombinant DNA process 'could not be identically repeated since the source of DNA in humans varied with the individual'. In response, the Technical Board of Appeal emphasised that the patent

was a process, which could be used to transform various materials — a 'general methodology which is fully applicable with any starting material' — and that this process was sufficiently described, and repeatable, so long as the methodology itself functioned reliably: 'generally applicable biological processes are not insufficiently described for the sole reason that some starting materials or genetic precursors therefor, eg. a particular DNA or plasmid, are not readily available to obtain each and every variant of the expected result of the invention provided the process as such is reproducible' (*GENENTECH I/Polypeptide expression* [1989] EPOR 1, point 3.3.3 of the reasons).

53 '[T]he element isolated is the result of the technical processes used to identify, purify and classify it and to reproduce it outside the human body, techniques which human beings alone are capable of putting into practice and which Nature is incapable of accomplishing by itself' (COM (95) 661 final, n 44 above, 5).

54 Of course, some elements of the 'routine' might consist in a form of 'tacit' knowledge, which is tied to the circulation of laboratory technicians and post-doctoral researchers. For example, an intuitive 'feel' for the development of a cell culture, or a sense of just when and how to shake a flask or test tube, often plays an important part in even the most sophisticated procedures (see, for example, A. Cambrosio and P. Keating, 'Going Monoclonal: Art, Science, and Magic in the Day-to-Day Use of Hybridoma Technology' (1988) 35 *Social Problems* 244, and also the role of laboratory procedures in the 'discovery' of PCR (Paul Rabinow, *Making PCR. A Story of Biotechnology* (Chicago: University of Chicago Press, 1996).

55 Observations reported in Rainer Moufang, 'Patenting of Human Genes, Cells and Parts of the Body? — The Ethical Dimensions of Patent Law' (1994) 25:4 *IIC* 487, 510–511. The more precise formulation consists in a distinction between 'inventions' relating to lymphocytes extracted from 'sensitised test persons', which would not be patentable, and those which arise from a 'one-time and incidental extraction' of myeloma cells, which would be patentable.

56 See generally Bruno Latour, *Petite réflexion sur le culte moderne des dieux faitiches* (Paris: Les empêcheurs de penser en rond, 1996).

57 See, for example, the argument of a 'bioethicist' quoted in a RAFI text: 'The product of nature doctrine has been rendered vacuous by allowing that the isolation, purification, or alteration of an entity or substance from its natural state turns it into something "not found in nature" ' (see RAFI Communique, Jan/Feb 1994, 'The Patenting of Human Genetic Material' (text available at<http://www.rafi.ca.communique/fltxt/1994 1.html>).

58 Jenkins' account of the Hagahai contains the observation that the women of the group would breast feed both their children and their piglets, the piglets being the more privileged of the two classes of dependant. For an account of how these 'transgenic' social practices 'hybridise' the distinction between nature and culture, see generally Marilyn Strathern, *The Gender of the Gift* (Berkeley: University of California Press, 1988).

59 For this shift from tautology to recursivity, see my argument in 'Power as an Art of Contingency: Luhmann, Deleuze, Foucault' (1998) 27:1 *Economy and Society* 1.

60 '[M]eaning could not outlast the destruction of life or of its chemical and physical basis [but] this dependency is not an operative premise of meaning events themselves' (Niklas Luhmann, *Social Systems* (Stanford: Stanford University Press, 1955) 66).

61 See 'Power as an Art of Contingency', n 59 above.

62 This concept is borrowed from Bruno Latour, but it should be noted that my appropriation diverges significantly from Latour's use, in that the model

of differentiation presupposed by my argument does not allow the various 'purifications' to be referred to a single, common, hybrid object (cf Latour, n 38 above, 197).

63 *ibid*, 107.

64 The question of risk is excluded just as readily as it is included, by the argument that 'potential risks in relation to the exploitation of a given invention for which a patent has been granted cannot be anticipated merely on the basis of the disclosure of the invention in the patent specification' (decision of the EPO Technical Board of Appeal in *PLANT GENETIC SYSTEMS/Glutamine synthetase inhibitors* [1995] EPOR 357, point 18.4 of the reasons). In other words, the risk is accepted as a legitimate criterion in principle, but one which is inapplicable in practice because the risk falls to be assessed during the lifetime of the patent rather than at the point at which the application is made. Consequently, any form of risk assessment would be the proper concern of regulatory bodies other than the EPO. See also *HARVARD/Onco-mouse* [1991] EPOR 525, point 4 of the decision, and the Commission's new Biotech Directive, n 44 above, para 23, at 6. Certainly, the proposition that 'ethical considerations permeate the entire normative structure of the patent system' seems somewhat naive (Moufang, n 55 above, 514).

65 Brad Sherman, 'Governing Science: Patents and Public Sector Research' (1994) 7(3) *Science in Context* 515, 525.

66 Rebecca S. Eisenberg, n 49 above.

67 Or, borrowing again from Bruno Latour, one might say that a scientific concept is 'a strategic entity' that allows one to 'interest' and hold on to one's allies (*Le métier de chercheur – regard d'un anthropologue* (Paris: INRA, 1994) 28).

68 Quoted in (1994) 371 *Nature* 365.

69 For example, the length of the stem, the number of seeds, the time of fruiting, and so on.

70 Hermitte, 'Le rôle des concepts mous dans les techniques de déjuridicisation. L'exemple des droits intellectuels' (1985) *Archives de Philosophie du Droit* 334, 344.

71 cf Alain Pottage and Brad Sherman, 'Création et appropriation' in Libois and Strowel (eds) *Profils de la création* (Brussels: FUSL, 1997).

72 *Biogen Inc v Medeva* [1997] RPC 1.

73 '[I]t is less the facts of science that are at issue in a patent trial [sic] than the scientific process itself' (Alberto Cambrosio *et al*, 'Scientific Practice in The Courtroom: The Construction of Sociotechnical Identities in a Biotechnological Patent Dispute' (1990) 37:3 *Social Problems* 275, 275).

74 For an excellent analysis, see Cambrosio *et al, ibid*.

75 This distinction was essential to the HARVARD/Onco-mouse claim; the engineered mouse was not the product of a biological process because the decisive interventions or manipulations happened at a 'micro' level: 'the oncogene is inserted by technical means into a vector (eg a plasmid), which is then microinjected at an early embryo stage' ([1990] EPOR 510, point 4.9 of the reasons). Of course, this simpler distinction became more complicated when these microbiological interventions were supplemented by biological processes of reproduction (see *PLANT GENETIC SYSTEMS* [1995] EPOR 357).

76 Joseph Straus, 'Patenting Human Genes in Europe — Past Developments and Prospects for the Future' (1995) 26:6 IIC 920, 927.

77 'It goes without saying that one must assume that individuality in the sense of a self-propelling, psycho-physical unity, and above all in terms of each person's individual death, is something accepted by all societies' (Niklas Luhmann; *Love as Passion. The Codification of Intimacy* (Cambridge: Polity Press, 1986) 14).

78 See for example the observations of one German scientist, quoted in Straus, n 76 above, 927: 'the human genome has a label which I think is appropriate because man is defined in a certain way, at least with respect to its macroscopic structure, by the total structure of a genome. But we are talking here and in all other cases about little bits of DNA and if the source comes from plants, or does it come from bacteria or from humans, then we give an etiquette which is not appropriate. . . . I think when we talk in terms of biology, then, of course, its only the genome which defines a certain species'.

79 See Marie-Angèle Hermitte, 'Le corps hors du commerce, hors du marché' (1988) *Archives de Philosophie du Droit* 323, esp at 329–341.

80 See Caroline Walker Bynum, *The Resurrection of the Body in Western Christianity, 200–1336* (New York: Columbia University Press, 1995).

81 Luigi Luca Cavalli-Sforza, 'Genes, Peoples and Languages' (1991) 265 *Scientific American* 72, 72.

82 'If we know that there exist different genetic types of a specific protein or other strictly inherited character, we can count individuals carrying one type or the other and establish the proportions of that type in the population being examined. These proportions vary from one population to another because they change over time in each population in a relatively unpredictable manner. The change in proportions of these types over time is the *evolutionary process* itself (L.L. Cavalli-Sforza *et al, The History and Geography of Human Genes* (Princeton: Princeton University Press, 1994) 4).

83 Therefore, the larger part of the theoretical resources of population genetics is devoted to the work of isolating the pure diachronic line of 'evolution itself' from a skein of social relationships in which the diachronic is enmeshed in the synchronic. The great advantage of genes is that they promise a pure measure of human history: 'Genes, always transmitted from parents to children, describe a vertical path through the generations. Culture can also pass vertically from generation to generation, but unlike genes, it can also be transmitted horizontally, between unrelated individuals' (Cavalli-Sforza, 'Genes, Peoples and Languages' n 81 above, 78). The trouble with this approach is that it overlooks the fact that any diachronic purification is the result of understandings and selections that are embedded in society as it is constituted here and now.

84 Luigi Luca Cavalli-Sforza, *ibid*, 76.

85 'In no other place can one find almost intact ways of life that disappeared from Europe and most of Asia thousands of years ago, and from Africa centuries ago . . . From the point of view of population structure, information on tribal customs, demographic properties, and especially migratory changes can therefore be extremely interesting. The basic problem, what fraction of genetic variation is due to differential selection or random genetic drift, requires knowledge of these factors' (Cavalli-Sforza *et al*, n 82 above, 351–352). See also the Report of the Sub-committee on Bioethics and Population Genetics, UNESCO International Bioethics Committee, section 1.2.3: 'Samples could be, and frequently are, analysed from all over the world. However, for genetic studies the more isolated homogeneous human groups are thought to be the most informative'.

86 The uniqueness of the PNG-1 cell line was established by recourse to many of the techniques used in population genetics. For example, the claim that 'alignment and comparison of the nucleotide sequence of each provirus with the published sequence of a prototype [HTLV-I virus] revealed the existence not only of highly divergent variants of HTLV-I in Melanesia but of new quasi-species within this HTLV-I variant' (see the text of the patent application, n 10 above) was supported by a viral genealogy, a sort of family tree; or, in the more specialised language of

the patent application, a 'dendrogram' allowing one 'to deduce the evolutionary relationships among members of the HTLV-I family'. For an explanation of 'phylogenetic tree analysis' see Cavalli-Sforza *et al*, n 82 above, 30–39).

87 Report of the UNESCO Sub-committee on Bioethics and Population Genetics, n 85 above, section 2.2.1. (See also the comment of one of the members of the UNESCO Sub-committee: 'The latest measures produced by the HGDP planners contain some of the most ethically sophisticated and detailed procedures for obtaining informed consent from individuals and groups in population genetics research' (1996) 379 *Nature* 11).

88 Cavalli-Sforza, cited in *ibid* section 3.1.4.

89 The violence of this is plainly expressed in a photograph of Professor Cavalli-Sforza, which, according to the caption, shows the author taking 'a genetic sample' from 'a member of the Aka tribe of African pygmies' ('Genes, Peoples and Languages', n 81 above, 78). The population geneticist occupies the focal spot of the photograph; he smiles directly at the camera, while the woman subject from whom he is about to take a sample sits with her arm limply extended over a table, bound by tourniquet, her facial expression signalling only utter, abject, resignation.

90 See Donna J. Haraway, *Modest_Witness @ Second_Millenium. FemaleMan ©_Meets_Oncomouse™* (New York: Routledge, 1997) 249: 'the people to be sampled might give or withhold permission, to be more or less carefully sought and thoroughly explained, but they were not regarded as partners in knowledge production who might have ends and means of their own in such an undertaking'.

91 'Opposing the involvement of a visible and respected public interest group can frequently generate adverse reactions in the market' (William S. Feiler, 'Devising and Implementing an Effective Biotechnology Litigation Strategy' (1995) 72 *Patent World* 34, 35).

92 'When the client is a defendant in a biotech suit, enlisting the participation of high profile public interest groups can be a potent stratagem' (*ibid*).

93 See generally (1998) 392 *Nature*, at 525 and 535–540. Although it concerns a group or nation that is less obviously 'indigenous', the partnership between Hoffmann LaRoche and the Icelandic genomics company deCode Genetics, under the terms of which 'the Icelandic nation' is to receive free of charge any pharmaceutical products of this corporate collaboration, offers a practical example of how 'pristine populations' might be compensated or enriched by these new forms of 'bio-colonial' transaction. See the account given by deCode at its website (http://www.decode.is/news.html).

94 RAFI Communique Mar/Apr 1996, n 17 above.

51

PATENTING CULTURE IN SCIENCE

Reinventing the scientific wheel of credibility

Kathryn C. Packer and Andrew Webster

Source: *Science, Technology and Human Values* 21(4) (1996): 427–453.

Abstract

This article discusses the emergence of a patenting culture in university science. Patenting culture is examined empirically in the context of the increasing commercialization of science, and theoretically within debates over scientific "credibility." The article explores the translation of academic credit into patents, and vice versa, and argues that this process raises new questions for our understanding of scientific recognition and of scientists networks. In particular the analysis suggests that scientists must move between two distinct social worlds to manage the rewards that academic and patent cultures carry.

Introduction

Typically, research on the locus of scientific practice has concentrated on the academic research community. The industrial context within which much of scientific research and development (R&D) is located and the growing interface between corporate and public sector research have received relatively little sociological attention. It is only recently that a number of contributions (Webster 1994; Etzkowitz 1993) have sought to explore what Cambrosio (1994) has called "inter-laboratory life," or the connections between the private and public contexts of the production and exchange of scientific knowledge. This article contributes to this new work by focusing on the emergent patenting culture in public sector science that reflects the general trend toward the increasing commercialization of university and government laboratory research (Mackenzie, Keating, and Cambrosio 1990; Bozeman and Crow 1991). The growing commercialization of science is said to pose a threat to the long-established conventions and norms of public sector, especially university, research (Seashore Louis, Anderson, and Rosenberg forthcoming). A commonly held view is that commercialization may compromise scientific impartiality by introducing the profit motive into the direction and evaluation of scientific knowledge claims (Brooks 1993; Krimsky 1991). Were this to happen, the normal processes of scientific reward would be undermined since claims would be less likely to be subject to the disinterested process of objective peer review.

The article examines how scientists determine what is patentable and how this differs from their traditional conceptions of novelty in science. It considers the implications of the patenting culture for the way scientific work is recognized and rewarded—the "cycle of credibility" through which scientists' reputation and position are secured—and suggests a new way of conceiving of the boundaries and networks between "public" and "private science" (see Rappert 1995). We thus paint a picture of the social world(s) of scientists whose credibility within academic and patenting networks is secured in different ways. Our analysis raises a number of criticisms of the approach most closely associated with network analysis in STS, the actor network theory (ANT).

Scientific rewards clearly depend on the dissemination and evaluation of scientific knowledge claims inscribed in scientific papers, conference presentations, grant proposals, and end-of-grant reports to funding agencies. Patents might also be seen as part of the dissemination process, and indeed, Narin (1994) argues that for bibliometric purposes, patents and papers can be treated in much the same way. A patent can thus be regarded as merely another form of scientific claim that generates scientific capital and, like a scientific article, brings its own rewards.

Earlier work on the forms and patterns of rewards in science has included Hagstrom's (1966) functionalist analysis of "the scientific community" engaging in "gift-exchanges," Bourdieu's (1974) image of a more competitive world of science characterized by competitive investments in "symbolic capital," and, as an extension of this, the concept of the "credit cycle" developed by Latour and Woolgar (1986). Latour and Woolgar argue that credit is more than simply recognition or reward for the work that scientists do; instead, "scientists' behavior is remarkably similar to that of an investor of capital" (p. 197). Scientists build their stock of scientific capital through making investments that will secure their *credibility* as scientists: "the receipt of reward is just one small part of a larger cycle of credibility investment" (p. 197). The credibility of scientists is measured in terms of the trust and reliability *others* can invest in them: the more this is secured, the quicker the cycle of credibility turns and the stronger the stock of capital held by the scientist.

How would patents as a particular form of scientific investment fit into such cycles? Do they contribute significantly to scientists' credibility? Do scientists judge them to be successful or not to the degree that they can be converted into other forms of scientific capital in the credit cycle? Does failure of a patent claim based on a particular piece of scientific research necessarily weaken the credibility of that research, and vice versa? What counts as a *trust*worthy patent claim, and whose trust is to be secured? If it is found that there are differences in the way in which scientific investment in patenting relates to credit and credibility, can we simply incorporate patents into the reward system? If we cannot do so easily, might this suggest that there are different social worlds scientists occupy as they move between patenting and scientific research? Like Hagstrom (1966), Latour and Woolgar's (1986) account of credibility assumes a *singular* world for scientists insofar as "reward and credibility originate essentially from peers' comments" on their and other scientists' work (p. 198). As this article suggests, this network of peers forms only one social world, one that is much less significant as a source of credibility when it comes to investing in patent claims. The scientific capital or currency that patents generate may not be easily exchanged for other forms of capital that scientists seek to hold. Patenting requires engaging in a different form of investment with distinct criteria, and the sorts of skills and credibility needed for securing reward are different from those needed in academic research.

Patents, competence, and novelty

The research reported here is part of a project that examines the culture of patenting in the organizational contexts within which scientists work. It focuses specifically on how patenting works (or fails) in the context of public sector research. Our project began (in 1993) with the first national survey of all British universities' technology transfer offices. Fifty-seven institutions that replied to the survey (44 percent response rate) showed a wide range of patenting across universities. For example, whereas 572 patents were held in total, some universities had none, and one held as many as 60; 30 percent of the universities in the survey had no formal policy on intellectual property, and 60 percent of respondents said that patenting activity was *not* self-financing.

From these returns, we then selected ten universities that had varying levels of patenting activity in terms of filing and grants secured and that also had strong pharmacology and biotechnology research teams. These two fields were chosen because they were seen to represent, on the one hand (in pharmacology), a well-established research tradition in a mature field within which judgments about novelty and scope were guided by long-established conventions and, on the other hand, a newer field (biotechnology) in which novelty and scope of claims had yet to stabilize. We also checked whether these groups had links with industry. The overall intention was to secure respondents who were more likely to have engaged with the patent system and who might, thereby, have research experiences more akin to industrial scientists. We also visited all the research laboratories of one of the U.K. research councils to determine whether the universities faced similar problems as other public sector research institutions. In addition, we visited a smaller number (five) of corporations in the pharmaceutical and biotechnology sectors. Overall, we conducted in-depth interviews with fifty respondents: the majority of these (thirty-five) were from academic laboratories, but we also interviewed industrial liaison officers, patent agents, and examiners at the U.K. Patent Office. The results reported here relate specifically to the sample of university scientists.

Central themes of the research are the use and construction of concepts such as "novelty" and "scope," which are both defined as technical, legal terms but may also be understood in different ways by academic scientists who are patenting research work. We are interested in how this interaction between legal and scientific definitions of novelty—*and the distinct sense of "credibility" they carry*—affects the practice of scientific research and the increasingly important drive to push technology out of the laboratory and into companies. This article examines two aspects of the patenting process. We ask, first, how scientists acquire patents, by learning and applying what we call specific "sociotechnical competencies," and second, what scientists do with patents once they have been granted.

Exploiting research to produce patentable results

Translation of scientific findings into patents has to be understood as a *process* that involves a number of different social actors: scientists, local university-based technology transfer officers, patent agents, patent examiners, and, typically, companies that may be funding the research upon which the patent claim is based. Each of these actors judges the merits of the claim according to different local criteria: the scientific worth of the claim, its relationship to other existing claims, who holds existing patents in relation to the claim (e.g., other companies), the potential revenue it might bring, and so on. This indicates the range of different constituencies that need to be addressed in pursuing a claim. The list is not exhaustive. A wider constituency of international patent offices, international regulatory agencies and their policies (such as those associated with the European Commission's Biotechnology Directive), and groups opposed to patenting may all have a role to play in determining the ultimate career of the claim. Thus the identification and consolidation of claims as patentable is *never* a singular event—despite the "eventful" image of "discovery"—but rather an iterative *process*. Not surprisingly, it can take some considerable time to persuade others of one's claims (Myers 1995), and patenting can be exceedingly costly (see Figure 1). These financial costs of patenting are unlikely to be borne by individual scientists or even by their departments; it is much more likely that they would be met by a firm that had received licensing rights from a

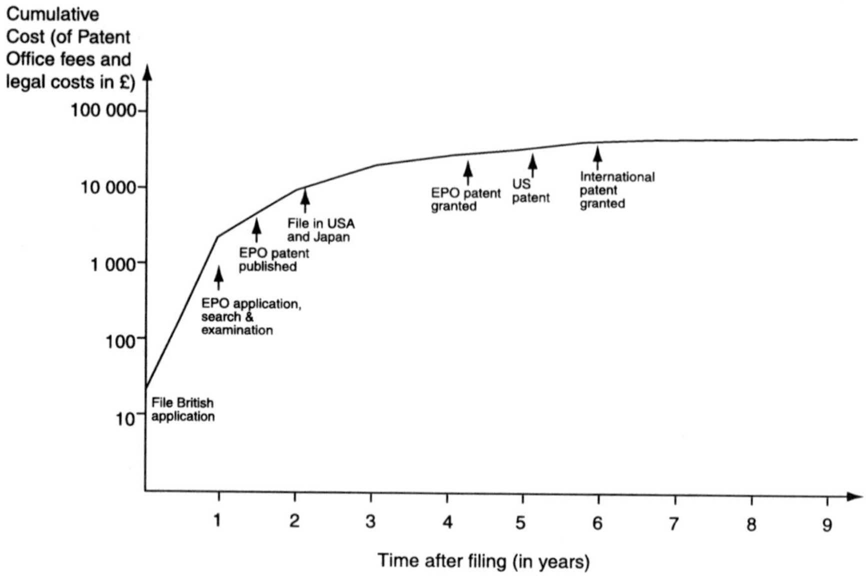

Figure 1 Costs of international patent filing.

Note: EPO = European Patent Office.

university. Nevertheless, to secure the needed financial support, scientists must present themselves as credible patentees.

If they are to have a chance of securing a claim to knowledge as a patentable claim, scientists must successfully complete a range of social and technical tasks. Failure to complete any one of these tasks could ultimately mean the collapse of the claim. From the perspective of Latour and Woolgar's (1986) credibility cycle, these social and technical tasks of interpretation are crucial to being a "competent" ' or credible patent filer.

Sociotechnical competencies

To secure patentable ideas and then the patents filed on them, research scientists must have access to a sophisticated and both materially and socially well-resourced *cultural infrastructure*. This infrastructure consists of the range of different interpretive skills, networks, information sources, and professional services that a scientist or group of scientists rely on. It will enable them to develop or draw on the sort of technical and social competencies (see Figure 2) they will need to negotiate the hurdles of patenting. We can distinguish two kinds of such competencies, although we are not suggesting that this categorization reflects hard and fast differences in the activities and skills involved.

Sociotechnical Competencies

1. Production work
   - Using legal terminology and making judgments in light of it
   - Searching patent literature
   - Writing patent examples and helping formulate claims
2. Articulation work
   - Performing new experiments to embody a patent claim
   - Planning research around existing patents
   - Altering or delaying publication to avoid damaging patent applications

Figure 2 Competencies required for successful patenting.

First, scientists who are experienced academic researchers but unfamiliar with the patent system (see Myers 1995) need to acquire (through their own hard work or through purchasing someone else's services) competence in intellectual property issues. They must develop an understanding of legal terminology and practices and understand how to manipulate them. They must be able to distinguish between scientific and legal "novelty" and have the capacity to search patent literature, to formulate patent claims and relate them to a portfolio of sister claims, or to make interpretive judgments about wider conventions of being "skilled in the art." Such competencies also include the skills needed to produce a certain type of science-based text that can be credibly presented to the patent system. Such texts are similar in form but very different in content from articles submitted to journals. In Fujimura's (1987) terms, all these skills might be described as "production" work.

The other type of sociotechnical competence can be likened to Fujimura's (1987) notion of "articulation" work. Meeting the requirements of the patenting process requires changes in the experimental and social work of the laboratory. Scientists seeking patents must be prepared and able to provide an "embodiment" of the invention, work around existing patents, perform necessary research, and be willing and able to delay publication (in light of other restrictions on finishing a doctorate or producing a certain number of publications).

Fujimura (1987) shows that articulation and production tasks must be successfully completed *at different levels*. She argues that scientists will only succeed when they align the different demands made of them by their own immediate experimental practice, the wider laboratory, and the much wider "social world" (R&D managers, sponsors, and so on) that their work addresses. She does not, of course, relate this process to the social world of patenting. In patenting, we find not only that scientists must engage with different social worlds but also that the "alignment" process is rather different. Fujimura's model assumes (in a way that is similar to ANT) that scientists can enroll the support of those from other social worlds through translating their claims to scientific novelty into scientific facts at each "level" so as to align successfully the distinct "levels." But when scientists in our study translated scientific claims into patentable claims, they were *not* thereby aligning the worlds of scientific experiment and patenting. The determination of novelty and credibility in the first was not easily translated or exchanged for novelty and credibility in the second. Instead, scientists moved between distinct social worlds.

Determining novelty in science

The most important technical competence of research scientists involved in patenting is the ability to recognize and distinguish between scientific and legal versions of the novelty of their work. Even the most well-developed and funded industrial liaison office is unlikely to be able to monitor scientists' work for patentable inventions, certainly not to the extent that company R&D managers do. This limitation stems partly from the lack of specialist technical and market knowledge across the wide range of research areas that would be necessary to monitor university research and partly from the traditional distrust of central university administration. Typically, therefore, scientists carry the primary responsibility for recognizing and developing a patentable invention.

Although Eisenberg (1987) discusses disclosure and enablement in some depth, she pays scant attention to novelty and nonobviousness after initially recognizing their importance. On the surface, the black letter law of patenting and the rhetoric of scientific norms appear to treat novelty in similar terms. However, judgments of the novelty and nonobviousness of biotechnology patents have been crucial in causing and deciding disputes. The comments of the scientists we interviewed also suggest that scientists perceive the legal concepts of novelty and nonobviousness very differently from the way they judge the originality of academic work.

The legal concept of novelty in patenting is quite a simple one. For an invention to be novel, it must not have been publicly disclosed by the applicant or anyone else before the patent application has been filed (or one year prior to the filing in the United States, where a system different to that operated under the EPO exists; the rest of this discussion relates to the first-to-file system since it is under its auspices that all of the scientists that we spoke to would file their initial claims). Disclosure includes any sort of publication (an article in an academic journal, a patent, a newspaper article, or book), a conference presentation, a commercial use of the invention, and any form of oral disclosure not covered by a confidentiality agreement (for example, see warnings in *Bio/technology* about the dangers of disclosure while chatting on a coach on the way to a scientific meeting; Williams 1994). A patent application is "tested" for novelty by searching (usually using keywords) through electronic databases of academic publications and patents (published and unpublished). When scientists in U.K. universities first started to patent their research, their unfamiliarity with the system meant that a number of applications were turned down because they had published their findings before filing a patent application. However, it is now rare for such an application in the biological sciences to be disallowed because of failure to meet the novelty requirement.

This does not, however, mean that the issue of originality is unimportant in the examination of patent applications. Nearly all patents are initially referred back to inventors, and many of the patent examiners' queries ask whether the invention disclosed in the application meets the requirement of being nonobvious, a concept closely related to originality. In judging whether the patent makes an inventive step and thereby fulfils this criterion, the patent examiner assesses the claims not only in relation to the letter of the law and to prior court rulings but also, and more specifically, in relation to an imaginary character, the "person skilled in the art." To be inventive, a patent application must demonstrate how its contents enable others to repeat a procedure or make an object that would not have been an *obvious* extension of the existing knowledge or practice (prior art) available to someone with a reasonable knowledge of the field at the time the application was filed. The person skilled in the art is someone conversant with all the relevant publications and experimental procedures of a specialism but without any capacity to engage in original thought (that is, not someone most scientists would be keen to employ or take on as a student).

Making judgments about the mental capacity of someone skilled in the art is a difficult task, even for experienced examiners, since the conception of the person skilled in the art not only must change to accommodate different technologies relating to different patent applications but also must be firmly grounded in the time of the patent application. Owing to delays in the processing—especially of biotechnology applications—patent examinations or legal disputes may take place months or even years after the initial application has been filed. Research will have moved on considerably since that time, but the examiners must avoid using hindsight. They must base their judgments on what prior art was available at the time of filing (patents and scientific publications are cited by examiners in their search reports that accompany published patents and correspondence with applicants) and avoid being inventive themselves in its interpretation. As one examiner commented, "You have to avoid being an inventive examiner. You have not to look at things with hindsight; you have to look at the situation there and then."

Although the legal arena of patenting is more bureaucratic than the invisible college of the scientific community, *interpretation* of the tacit "rules" is just as important in this arena as the laws themselves. In talking to scientists about their experience of patenting, we wanted to find out how they dealt with this mixture of definitive law and flexible interpretation and how they integrated the requirements of patenting with the requirements of academic institutions and communities concerning the production and publication of original or "novel" research.

The scientists we spoke to generally believed in gradual advancement of knowledge in small steps and did not evaluate research programs simply in terms of the march of intellectual history. New genes were sequenced, novel strains of bacteria were isolated from newly acquired samples, improved and ingenious methodologies were developed, and reviews of previous work led to more perceptive theoretical insights. Although some advances were seen as more significant than others (and therefore easier to get published), they were all considered novel and contributing to progress in the field. This complicated notion of progress in science makes it very difficult to generalize about the matter in which these differing types and conceptions of scientific novelty are relevant to patenting. However, most of the scientists we spoke to recognized that there was a difference between their conceptions of what novelty was in terms of academic science and what it meant for the legal requirements of patentability.

The criteria of novelty and nonobviousness in the patent system were generally seen as *less* stringent than the standards by which originality is judged in academic science. One scientist felt some embarrassment at being a named inventor on a patent that she regarded as obvious:

> I am in bit of an awkward position because I had to be an inventor on this because I suggested they do it, but it is so obvious from the literature and it is so derivative that I am absolutely surprised it issued.

All the scientists recognized that the concept of the "inventive step" was difficult to locate within their social world. Two reasons were cited. First, "discoveries" are legally nonpatentable. So although coming across a new, naturally occurring species, disease, or chemical entity might be considered a significant advance in science, in patenting terms it would carry no reward. It would not count as intellectual property to which one could secure legal rights (unless one could show how the effort of purification or preparation of this discovery made it into a product of human labor). Second, patenting judgments are based on a *universalistic* notion of a virtual scientific community sharing a common stock of knowledge, whereas scientists' own judgments are based on a sense of *specific* communities, which, although geographically dispersed, are local in the sense of being based on subject specialties and interests. As Star (1986) wrote, "Each actor, site or node of a scientific community has a viewpoint, a partial truth consisting of local beliefs, local practices, local constants, and resources, none of which are fully verifiable across all sites" (p. 46).

Unlike the judgment of novelty by the "partial truth" of a specific scientific community, novelty in patenting is judged against all possible forms of publication: as members of specific communities, scientists would probably only expect to read what they are sent by colleagues or find in their department libraries. Similarly, patenting judgments about nonobviousness are made against the baseline of someone skilled in the art, who can draw on all the relevant literature. This combination means that the decision characterized as universal actually applies only to someone with peculiarly good access to publications and active involvement in whatever research project one cares to name. In the social world of science, as one of our respondents noted, it would be ridiculous to consider an invention as nonobvious if only a handful of people had ever had the inclination or opportunity to provide a solution to the question at hand: "[Something is only really inventive if] you ask ten people and only one person comes up with the solution. They get round it by essentially only asking one person."

How scientists relate to the patent system depends on the sort of judgments they make about the virtual community of science, represented by the invisible skilled-in-the-art person, and whether they perceive intellectual property rights as something awarded only to major advances or to seemingly petty advances. Scientists who had the least involvement with patenting thought that only major advances can be patented:

> Basically the question is that there has never been anything that I'm really convinced that is absolutely wonderful. . . . There are a lot of people working in the field so you read their papers, and they read yours, and you make a little step forward on the basis of that. So there's nothing like the invention of PCR, that's such a fantastic idea which you say I must patent it. If something like that happened, and you say this is going to revolutionize molecular biology, then you obviously would patent it.

This view of patenting is perhaps not surprising given the tendency in the science press to characterize scientific breakthroughs as revolutionary or "eureka" events. Such events play a central part in the rhetoric surrounding milestones in research. For example, this is how Kary Mullis described his own work on Polymerase Chain Reaction, a widely used molecular biology technology:

> a revelation came to me one Friday night in April 1983, as I gripped the steering wheel of my car and snaked along a moonlit mountain road into northern California's redwood country . . . that was how I stumbled across a process.
>
> (Mullis 1990, 36)

The patent system generally deals with incremental changes. Yet the impression often given in the reporting of patenting in journals such as *Nature*, *Science*, and *New Scientist* mirrors the "breakthrough" story of

Mullis's (1990) account. Such reporting has focused on major advances and disputed patents in the area of biotechnology. In *Nature*, five biotechnology patents accounted for over half of the news items dealing with individual patents (as opposed to changes in patent law, etc.) between 1974 and 1992. Although pharmaceutical patents are not so widely publicized, the rhetoric surrounding the need for strong patent protection of new chemical entities reinforces the association between patenting and single, distinct, immensely important inventions.

Scientists with experience of patenting recalled that initially they had also thought patents were associated with major advances in science, though they now believed this to be a naive view. Remembering his early experience, one scientist claimed,

> [the main reason for not proceeding with the patent application was that] we couldn't convince ourselves that what we'd report would be regarded as new in scientific terms.

It might seem surprising that those who thought that patenting requires less originality from the scientist also found it more difficult to present their work as inventive in patenting terms than to write scientific articles. This admission, however, merely confirms that when scientists make patent claims, they must move between distinct social worlds.

The experienced difficulty is not just the shock of the new: even those who are fully conversant with the patent system still perceive a definite boundary between patenting and the rest of their work as academic scientists. They have not ended up integrating the distinct audiences of the "invisible college" of science and of the virtual universe of patentable knowledge *that has no "membership" status even when one participates in it*. There is no one unified process of evaluating and presenting their work in these two arenas. The differences are expressed not only in the scientists' initial feelings of unease at venturing into new territory but also in the mental, literary, and discursive actions that scientists undertake to facilitate their movement. Unlike in the alignment process described by Fujimura (1987), translation here works across two very different social worlds that are relatively independent of each other. This helps explain why in some specific areas, a translation or exchange of "credit" between the two is closed off.

Some scientists coped with these two worlds and managed their contrasting conceptions of novelty and audience by treating the experience (and perhaps the rest of their scientific and administrative work) as a game. Different rules apply to patenting than to academic publication, but one is not transgressing any sort of ethical code by conforming to these rules. Moving between the different games, one is merely doing what is necessary to remain a player. Others were less cynical but still maintain a sense of two distinct cultures:

> When I go into a legal way of thinking then I use novelty (2) instead of novelty (1). I think people can cope with that. It's just a question as to whether *lawyers* will agree amongst themselves as to what constitutes novelty. (emphasis added)

For this scientist, the credibility of claims is *not* judged by the constituency of scientific peers but by "lawyers" represented as having their own local debate about what, according to their criteria, can be credibly considered novel. This particular scientist also said that he had come to be seen by others in the laboratory as someone to whom they could turn for advice on legal issues such as patenting. His status as an advisor was not due to any prior laboratory experience in dealing with intellectual property rights but to the fact that he had worked for his *church* in helping lawyers draw up an agreement with another church about shared use of a building. In this case, patenting was firmly perceived as part of the legal rather than the scientific arena, and *any* experience of the law was regarded as helpful to the uninitiated in the world of the laboratory.

Scientists involved in patenting not only have to think differently about novelty but also have to change some of their activities. Clearly, they have to write up their research for a new type of audience, using legal terminology with which they may be unfamiliar. They have to position their work in relation to that of their colleagues and peers differently from the way they do so in a scientific article. One of the respondents explained:

> Papers are written in a more scientific way—in that your references are made to papers and you have to explain other people's work and discuss it with relation to your work, whereas a patent is just a write up of *your* work.

In scientific articles, citations to earlier work can be used to lend support to the direction of thinking and experiment reported in the article. They can position the new research as the next logical step for all concerned. In a patent, on the contrary, such contextualization of claims might lead to the work being seen as obvious and not patentable. One scientist explained why it had taken a long time to decide whether his invention was patentable. His "invention," as he discovered, was a melange of discrete findings:

> In the BTG[1] patent the novelty was taking three completely different, separate, possibly published situations and linking them together to create a new system. It wasn't the type of novelty of sitting down with a piece of paper and saying, here is a type of chemistry that nobody has ever thought about before.

The ideal way to use citations in patents is to recognize work that the applicants know the examiners will be likely to find and that can lend support to certain assumptions or techniques. These references, however, cannot be too closely related to the disclosed invention because they should not jeopardize the patent claim. A "credible" claim in this context is built by constructing nonobviousness.

Changes in the scientists' practice are not limited to their patent writing activities. Literature searches may need to be improved and new experiments performed to support the patent claims or dispute those of others. Such activities may be needed to support a current patent application or to improve the possibility of obtaining patents in the future. One laboratory made relatively large investments in staff time and equipment to perform work that would not have been necessary if patenting were not a goal:

> It has been one of the stumbling blocks of the [new technology transfer activities] whether or not it should be associated with a reduction to practice laboratory and I think we decided that we didn't know basically whether that was a good thing to do, [but] it is a very expensive thing to do.

In general, although both for patenting and for academic research scientists develop what is initially *articulation* work into *production* work, there is still a divide between the concepts, actions, and constituencies—the social worlds—associated with patent production work and those associated with academic production work. Scientists translate the novelty of academic production into patent production by trying to map scientific novelty claims onto the patent system: the experiences this gives them—especially with regard to the amount of articulation work they have to perform—will determine the process of translation. In the biosciences, for example, academics have traditionally had little experience with commercializing their research; now they are increasingly given the opportunity, or are expected, to take part in technology transfer. As they become more familiar with the practice and rules of the patent system, they adjust the claims of novelty that they make. Or, as one bioscientist put it, "now I keep it to the clean and simples."

When scientists have established themselves as credible patentees—whether on the basis of "the clean and simples" or by patents of more ambitious scope—to what extent can they use the credibility established through a patent in the world of academic research? If so much translation was required to enter the distinct social world of patents in the first place, is it possible then to use the currency and credibility of the patent in the very different context of academia? Scientists do use patents within their research environment—the world of the laboratory, their peers, their universities—but in a way that has much less to do with enrolling others and more to do with *controlling* others. Holding patents is not so much a means of enhancing the credibility of scientists in their research world but a means of defending prior investment in the area.

Exploiting patents in academia: controlling scientific research

To illustrate these processes in more detail, we can return to Fujimura's (1987) account of the social world of science and examine the relations between Fujimura's three levels: the laboratory, the institution, and the wider corporate world.

Within the laboratory

As sources of scientific and technical information, one would expect patents to serve in the laboratories as resources for experimental and theoretical work. However, very few of the scientists we spoke to used patents as a source of information for their research work. Even scientists who had patents taken out in their names had never considered using patents in this manner. In its attempts to educate researchers about the patent system, the U.K. Patent Office's promotional material often stresses that much of the information that appears in patents is never published elsewhere and could, therefore, serve as a potentially important source of information. For a number of reasons, our interviewees were skeptical about these claims.

Patents are much more difficult and more expensive to get hold of than articles, particularly papers sent by colleagues or given out at conferences. Offprints, preprints, or references sent by colleagues are often the principal way in which scientists keep in touch with recent research. Industrial scientists were sometimes sent copies of patents, but generally speaking, scientists were not exchanging patent offprints in the same way as papers. One scientist involved in collaboration with research groups in several countries said that she had never seen the other groups' patents or requested copies of them, even though they had swapped materials, results, and articles.

Patents are classified and located according to a complicated system with which most researchers are completely unfamiliar. Given the constraints of time and limited resources available to academic scientists in the United Kingdom, they are not likely to have time to invest in learning new skills. Even when scientists do get hold of patents, the information they contain does not always leap off the page. Patents differ from academic publications in a number of ways: they make different types of statements, use a different mode of referencing, and have a different balance between methodological and theoretical content (Myers 1995). The different format and the use of legal terminology may make scientists initially uneasy about reading patents, although this is not a permanent barrier. Scientists adapt to reading patents as they had to adapt to reading journal articles after learning from textbooks. There are, however, more long-lasting barriers to understanding. One scientist, referring to a patent he was currently examining, said that it was "deliberately obscure and I think the reason it was obscure was because the thing doesn't work." Another researcher, when asked whether she used patents as a source of information said, "Having written patents and knowing the way you write them is to make it impossible to reproduce the work, no." Some people referred to the extreme length of some patents, which was designed to discourage people from reading them. These comments contrast sharply with the justification of the patent system as a means of promoting disclosure and dissemination of information that would otherwise be kept secret.

As a result of these sorts of disincentives to use patents, the first time many scientists come face to face with a patent is when they have to write one themselves. Previous patents in the field are used as guides as to what sort of language is expected and how information should be structured. In some cases, an education in the patent system can seem to be rather like a crash course in strategic lying. The language used in a patent application would be taken as a lie in any other context. As one scientist commented:

> You had to take it as a joke really, you had to say this will do this . . . and write it in the present tense, and just be over the top in the way you would never be in a publication.

One scientist in our sample used patents as a teaching aid for undergraduate students. They had to examine a patent in their area of study to see what information it disclosed and, more important, what further information one would require if one wanted actually to reduce the patent to practice (thus fulfilling one of the legal requirements of the patent). In this exercise, students were taught not only to recognize methods and theories but also to discover the deviousness of the scientific and legal professions. This scientist acknowledged that he was trying to teach his students that,

> Just because it's been printed and granted by the U.S. patent office doesn't mean to say that it contains anything that is scientifically sensible.

At the level of the laboratory and experimental research, then, patents are not used to enroll peers but, rather, to keep them away. As the above quotations reveal, patents are occasionally constructed as claims that *resist translation*. But even if a claim made in a patent is not "scientifically sensible," everyone recognizes that it can serve as a means of professional control.

Within the institution or scientific community

Both the law and the industry regard patents as a tool for controlling relations between institutions. However, we found that scientists also used their patents *within* scientific institutions, both at the meso level (within their university or the international research community in their specialized area of research) and at the micro level (in their laboratories). Scientists were asked whether they ever used patent claims to demarcate areas of professional or commercial interest. Researchers who had experience in patenting did think that it could be used to help secure their credibility among members of the university with responsibility for commercialization. Several scientists said quite explicitly that a "credibility factor" was involved in patenting. The importance of this factor depends on how links with industry and their potential to generate income are viewed within particular subdisciplines and institutions. One industrial liaison officer said that,

> Some of the big professors would use that as a bargaining tool in another area, use their patent portfolio, use their contact with the pharmaceutical companies in particular, the fact that BTG are just about to sign a deal with somebody. . . . they would push that out as a kind of "look at how much money I am about to make for the university."

One scientist we interviewed had been involved in an institutional debate about how much significance should be attached to activities such as patenting and academic-industry relations in her university. She had argued that not enough credit was given for this work. The increasing importance of patents in a wider range of institutions suggests that patents will also play a greater role in the internal politics of specialisms and universities than they have before.

The relation to corporate links

At the macroeconomic level, the patent system is supposed to facilitate technology transfer in two ways: by granting monopoly rights as an incentive for companies to innovate and by requiring the disclosure of information. The patenting activity of the scientists that we interviewed did not seem to fit into this pattern. Actual products or processes that could be used by consumers were rarely a major theme in our discussions with scientists. For scientists, patents formed part of their links with industry; they were not primarily vehicles for technology transfer. Although some of the scientists had earned substantial sums of money for further research work, through the licensing of patented inventions, others had not been successful in earning money on patents and had only modest hopes of doing so in the future. For many scientists, patents were part of the payoff for the industrial sponsors of their work. Rather than initiating a process of technology transfer, patents were then the result of such activity. Patents do, however, also play a role in the earlier stages of this process when scientists try to attract industrial funding. Here, patents worked as a marketing tool. They indicated that academic scientists were sensitive to the industry need for proprietary knowledge and that they were competent, not only to produce scientific knowledge but also to secure it for commercial use:

> It really is what one is looking at the patent for. It's advertising, it's window dressing, it's to allow you to go to the next stage, to raise funding, to allow you to indicate seriousness in what you are doing.

Companies often use patents as a source of strategic information about their competitors, and they employ information scientists to review and process new patent literature and to carry out specific searches. Most academics in our study did not use patents for these purposes, although some scientists were involved in the monitoring of patents on behalf of their industrial sponsors. For example, they were reviewing competitors' patents in terms of their practical feasibility. Only one scientists said that he reviewed the patent literature in the area to get an idea of "who is working on what and what their approaches have been." Another way in which companies use patents is as a bargaining tool in negotiating licensing or other deals. They may keep spare patents or try and patent in areas of interest to competitors to try to block their work or to have something to swap. Some of the scientists most involved in commercial activities did recognize this practice and tried to use their patents to negotiate the best possible deal with sponsoring companies or to protect their freedom to follow their own research agenda:

> the only reason I would be interested was to make sure that what I do, nobody tells me I can't do something. If you have a patent then that's it covered; nobody can come along later and tell you what you've done belongs to them.

In external relations, then, patents did not appear to serve so much as a channel for the dissemination of information but as a device used to control competitors.

In summary, scientists *do use patents* to maintain and further their position in the credibility cycle. Scientists use patents to initiate and maintain connections with outside bodies such as companies; they use patents to further their position within university academic hierarchies; and they use them in their own laboratories to help in writing their own patents. Scientists do not necessarily see patents as the vehicle for technology transfer, but they use patents with great effect to enable them to carry on with their academic research programs. Within the scientists' credibility cycle, patents do not serve as sources of information or as a means of cementing *academic networks*. They are not a way to *enroll* other academic actors.

Discussion

Our analysis of the emergent patenting practices within academic research science shows that although both in the world of academic research and in the world of patenting, scientists are rewarded for "novel" contribution, the meaning of novelty and the form of rewards are different in these two social worlds. As scientists cross boundaries and move from the localized network of their specific research world to the virtual universe of "state-of-the-art" knowledge in the patenting world, they engage in translations that are initially experienced as problematic and that require a different form of production and articulation work than is the norm in academia.

ANT has sought to explore the construction and the exchange of scientific knowledge in networks of heterogeneous actors, who enroll each other to stabilize or black box particular understandings of nature. The human actors in scientific institutions continually have to create and invest credit to sustain and advance their own involvement in the processes of science. This approach implies the existence of a singular world in which actors translate interests and engage in the enrollment of allies. Our work suggests that when research scientists engage in patenting, the processes at work are different from those that operate in their purely academic work.

The patenting arena calls for distinct forms of production and articulation work, and it is not simply part of the scientists' heterogeneous network as ANT might declare were it to consider patenting. Scientists—even those most experienced with patenting—are not *members* of the patenting world in the same way in which they are members of their research world. In this sense, they are not "enrolled" within the world of patenting. How, then, can we summarize some of the key differences in the way novelty, reward, and credibility operate in the distinct worlds of patenting and research?

Institutional context

There is an important difference in the institutional contexts within which academic and patent production work takes place. Academic status and credibility depend on the production and articulation of knowledge claims that are negotiated with other scientists in local and broader networks. According to Lynch (1985, 264), "what counts as a notable finding, a definitive anatomical entity, a thing's attributes, a procedure of measurement, an adequate display of data, and a plan of methodic action" is secured through interaction with others. Publications and their "inscriptions" are circulated and evaluated among participants and serve as a "social glue" (Knorr-Cetina

1995, 155). The institutional relations within the university facilitate this process and require scientists to engage in academic production to secure research status for themselves and the university, thereby to assure themselves of research funding.

Patenting production cannot be easily mapped onto this set of institutional relations. Scientists find that they must engage with a distinct group of individuals—such as patent agents and industrial liaison officers—and organizations—such as spin-off firms, large corporations and the patent office—with whom they have legal and contractual, rather than collegial, relations. Nevertheless, the demands these actors can place on the scientists can be as great as those of scientific peers. For example, patent production requires scientists to expend intellectual energy and material resources on searching existing patent literature and on experiments that provide a demonstrable "embodiment" of the patent claim. The institutional relations here are distinct from those characteristic of the academic network, not only because different work has to be done but also because different actors judge its merit and set the terms of how and when it is to be delivered. The "social glue" that secures the research network among academics *may* be in jeopardy here because a key feature of patent production is to preempt sharing one's ideas freely with other people. We have a few examples of scientists reporting that the need to secure patents has led them or research groups they have collaborated with to abandon or put on hold such collaborations until the patents are issued. Balancing the distinct institutional demands of the university and the patent system can be problematic. Credibility in one may undermine the credibility in the other.

Enrollment

Engaging with the patent system means engaging with a different set of actors, whose enrollment is open to question. The notion of "enrollment" in ANT assumes that significant others can be enrolled not merely by translating one's claims into those acceptable to them but crucially also by translating *their* ideas into one's own. According to Knorr-Cetina (1995, 159),

> How do I persuade someone to accept my proposal, my method, my invention? By convincing them that it is in their interest to adopt my offer, by redefining ("translating") their interests such that they converge with mine, and thereby by "enrolling" a heterogeneous crowd of "actors" into a network of associations that stabilise the technical object.

As we have argued, scientists do translate their scientific claims into patents. And they do rely on securing the support and expertise of specific actors who are part of a (qualitatively) different network. However, the stabilization of the technical object—here, the patent—depends on the judgment of some actors who are not enrollable in the sense assumed by ANT. The most important in this regard is the virtual actor—the person skilled in the art—with whom one cannot negotiate but whose "voice" is articulated through judgments of patent examiners (and, possibly, the courts). Scientists construct various mental images of this person—from a postdoc with two or three years research experience to a senior scientist one might expect to employ as a defense witness in a patent dispute. These constructs did not necessarily conform to those of the patent examiners. The collective expertise of the person skilled in the art accumulates over time, but this may only be "discovered" during a patent dispute in which the judge might declare that an unpublished patent application constituted prior art for a new claim under consideration by the court. Moreover, a scientist's control over his or her *own* status as *scientific* actor—clearly assumed by ANT's notion of strategic interaction with other scientists—can suddenly be removed by a court's declaration that previous academic production by the same scientist must be regarded as prior art, *even if that scientist insists that this is not the case*.

Moreover, the world of patenting within which scientists operate is a network of actors linked by distinct "social glue." Social ties in this world are based on distinct modes of interaction through which patents, facts, and their novelty are constructed. Although ANT declares that networks are heterogeneous, we believe that it understates the sense of heterogeneity. It is not enough to recognize the variety of social actors that comprise complex networks; one must also recognize that this variety means something: actors do have locales, senses of belonging, and reference group memberships that cannot easily be translated from one part of the network to another.

Time scale

Time scales over which patent claims are developed and evaluated are different from those associated with academic production. How is the speed at which the credit cycle turns affected by the competing demands of patenting and academic publication? If patenting sometimes causes delays in publishing, does patenting have a drag-effect on securing professional credibility within local research communities? Such problems may be most acute in cases of patent dispute that have to be resolved through the courts and that can be subject to especially long delays (we have come across a case that has lasted twenty-nine years; Swinbanks 1994).

Furthermore, scientists need to develop a sense of timing needed to decide when to file for a patent to ensure that it will attract sufficient industrial interest without preempting the possibility of a stronger patent claim filed later when further work had been conducted. A similar dilemma confronts scientists who must decide on the timing of their academic publications to ensure that their ideas are disseminated early enough to claim priority and to attract support. However, in contrast to patents, academic articles do not preempt later versions of the same research whose publication can result in similar (or even higher) returns—hence the phenomenon of the so-called self-plagiarized papers.

Textual differences

Patents and articles also differ as texts (see Myers 1995; Packer and Webster 1995). Articles aim to build on existing work and to disseminate new research findings that can be explicitly linked to the contribution of other members of a scientist's network. Patents must declare novelty and nonobviousness with regard to earlier work. Do these different practices affect what gets cited as part of the academic and the patent claims? Are the citation practices similar or different, and when patent examiners cite earlier "art," is this convergent with the invisible college that the scientist would regard as his or her reference group?

There is evidence that the preparation of texts as part of academic production is influenced by the preparation of texts for patent production. That is, scientists in our study report being more guarded about the speculative claims they might be tempted to make in the conclusions to their academic articles lest these compromise the potential scope and novelty of patents they are drafting at the same time. Thus entering the patent arena may influence the process and content of the dissemination of results within the academic arena. Eisenberg's (1987) belief that patenting can demand as much—if not more—disclosure than articles only makes sense if scientists would actually use the two arenas as equally available sources of information. The evidence presented earlier suggests that they do not.

Scientists have an ambivalent attitude toward patents as sources of information. Some regard them as useful devices for teaching: students might obtain from them knowledge about methodologies for experimental work. Others believe that patents hide as much as they reveal and that even their methodological precision that is ostensibly required by the law can be suspect: some have found it impossible to reproduce the detailed experimental procedures inscribed in the patent text. But this occurs also in academic research (Collins 1985), in which without tacit knowledge, no amount of text-based instruction will enable successful experimental replication of earlier work.

Boundaries and their implications for reward

Our argument that scientists can translate academic production into patent production through rethinking notions of novelty and inventiveness has implications for our understanding of how scientists are rewarded. The credit-cycle model developed by Latour and Woolgar (1986) assumes that scientists work within one academic—even if heterogeneous—network. We have sketched some important differences between the social worlds of academia and patenting.[2] Given that scientists translate their work across the boundaries of these worlds, are the rewards of one social world also transferable to the other?

Income from patents can obviously be used to sustain academic research through the royalties it generates, although typically the amounts are small compared with endowments, contract research grants, or governmental funding. Through a process of translation, the less tangible rewards of one world are also convertible into rewards of the other: we saw this, for example, in the translation of texts and in the defensive use of patents to demarcate academic areas. Moreover, some scientists were prepared to put patents on their curricula vitae and were favorably disposed toward new recruits who had patents, along with traditional scientific publications. In its recent advice to universities relating to the research assessment exercise in March 1996, the British funding council, the Higher Education Funding Council, took basically this position on patenting. Panels responsible for assessing submissions will be asked to regard patents as indicators of the utility to and exploitation of research by industry. Against this, however, the primary task of the panels is to review submissions on the basis of four best "quality" outputs of researchers in different fields. Whether a researcher would risk submitting a patent as his or her best output remains to be seen, but given our earlier argument about the distinct networks and audiences for patents, we suspect that such submissions would be unlikely.

These examples illustrate the translation of credit between the two social worlds, but the qualitative difference between the two social worlds reemerges when we examine how *failure* in one world is kept relatively insulated from the other. For example, if a scientist's patent is ultimately refused by a patent examiner or a court, this is *not* regarded by academic parties as questioning the quality of the research lying behind it. Unlike the rejection of a publication, the rejection of a patent claim does not then damage the status of a scientist in his or her invisible college. Nevertheless, academic scientists acting as witnesses in patent disputes that go to court *can* feel professionally threatened by barristers for the opposition who challenge their judgment: such disputes, orchestrated by the gladiatorial dynamics of the courtroom, can pose difficulties for scientists who enter them as though they were operating in an academic context, in which, as Lynch (1985) has shown, interaction is based on a mutual desire to secure agreement.

The transfer of rewards (or failures) from the social world of patenting to the social world of academic research has been summarized in Figure 3.

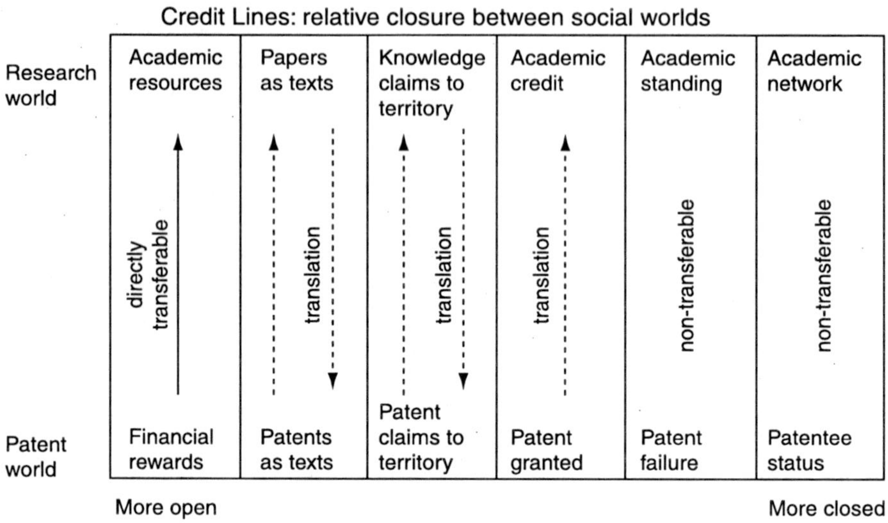

Figure 3 Translating credibility between social worlds.

Conclusion

We have shown how the social worlds of academic and patent production are distinct and how scientists translate novelty between the two, and we have suggested a number of key differences between them in terms of context, actors, timing, texts, and rewards. We suggest that ANT needs to acknowledge the qualitative difference between these social worlds and networks. Without such a distinction, we fail to capture the dynamics of the interaction and the boundaries between these arenas. Although scientists can move between the two worlds and cross boundaries, doing so is often problematic and requires a range of sociotechnical competencies that a tradition of engaging solely in academic production does not provide. As scientists develop new competencies for patenting, they also confront the uncertainties of engaging with a world that evaluates novelty differently than they are accustomed, is held together by a distinct "social glue," and affects the academic credit cycle by bringing new rewards—such as stronger likelihood of securing new industrial research—and new costs—such as holding back on the dissemination of new knowledge claims and delaying or forgoing the reputation this brings. Some might suggest that there are two credit cycles: one for academic and one for patent production. Indeed, inasmuch as industrial scientists also produce patents, one could argue that they also engage with the discrete credit cycle of patenting. However, while acknowledging that there *are* discrete features of the patent system of reward, we would prefer to incorporate these specific dynamics of the patent system within the broader credit cycle of scientists precisely because *they have to map it onto their more central activity as professional scientists*. In doing so, they mediate the effects of patenting by, for example, using a patent to fend off potential competitors, especially private companies, in order to carry on *as before*. In this example, scientists used the novelty of science to translate into a novelty of patenting that could then be used, in turn, to secure additional scientific novelty. The story of academic patenting is both richer and more complicated than the conventional ANT perspective allows.

Acknowledgements

The work reported here was supported by the U.K. Economic and Social Research Council (ESRC) and is based on research project R000 23 3302—Patenting in Academia and Industry: Determining Novelty and Scope in Two Distinct R&D Sectors. We would like to thank a number of colleagues who have commented on the results of this research at various meetings, including Jon Agar, Brian Balmer, Julia Garritt, Simon Shackley, David Skinner, Juliet Webster, and Steve Woolgar. We would also like to thank the editor and referees for their very valuable comments on an earlier draft.

Notes

1 British Technology Group (BTG) is a technology transfer company that used to be publicly owned and used to have right of first refusal on the commercialization of all government-funded research in universities. It lost these monopoly rights in 1985 and was privatized in 1992.

2 The notion of distinct social worlds within which scientists seek distinct forms of recognition and reward has also been explored recently by Rip (1994), who has looked at the social world of research councils and the relationship this has to the social world of academic science. Like us, he sees them as separate reward structures that impinge on scientists' credit cycles, although his analysis does not go on to examine the implications this has for ANT.

References

Bourdieu, P. 1974. The specificity of the scientific field and the social conditions of the progress of reason. *Social Science Information* 6:110–35.

Bozeman, B., and M. Crow. 1991. Technology transfer from U.S. government and university R&D. *Science and Public Policy* 11:231–46.

Brooks, H. 1993. Research universities and the social contract for science. In *Empowering technology*, edited by L. M. Branscomb, 202–34. Cambridge, MA: MIT Press.

Cambrosio, A. 1994. Standards, instruments and techniques. Paper presented at EASST conference, Science, Technology and Change: New Theories, Realities. Institutions, 28–31 August, Budapest, Hungary.

Collins, H. 1985. *Changing order*. New York: Russell Sage.

Eisenberg, R. 1987. Proprietary rights and the norms of science in biotechnology research. *Yale Law Journal* 97:177–231.

Etzkowitz, H. 1993. Entrepreneurs from science: The origins of science-based regional economic development. *Minerva* 31:326–60.

Fujimura, J. H. 1987. Constructing "do-able" problems in cancer research: Articulating alignment. *Social Studies of Science* 17:257–93.

Hagstrom, W. O. 1966. *The scientific community*. New York: Basic Books.

Knorr-Cetina, K. 1995. Laboratory studies: The cultural approach to the study of science. In *Handbook of sociology of science and technology studies*, edited by S. Jasanoff, G. Markle, J. C. Petersen, and T. Pinch, 140–66. London: Sage.

Krimsky, S. 1991. Academic-corporate ties in biotechnology: A quantitative study. *Science, Technology, & Human Values* 16:275–87.

Latour, B., and S. Woolgar. 1986. *Laboratory life: The construction of scientific facts*. Princeton, NJ: Princeton University Press.

Lynch, M. 1985. *Art and artefact in laboratory science*. London: Routledge and Kegan Paul.

Mackenzie, M., P. Keating, and A. Cambrosio. 1990. Patents and free scientific information on biotechnology: Making monoclonal antibodies property. *Science, Technology & Human Values* 15:65–83.

Mullis, K. 1990. The unusual origin of the polymerase chain reaction. *Scientific American* 263:36–43.

Myers, G. 1995. From discovery to invention: The writing and rewriting of two patents. *Social Studies of Science* 25:57–105.

Narin, F. 1994. Patent bibliometrics. *Scientometrics* 31:147–55.

Packer, K., and A. Webster. 1995. Inventing boundaries: The prior art of the social world. *Social Studies of Science* 25:107–18.

Rappert, B. 1995. Shifting notions of accountability in public and private sector research in the UK: Some central concerns. *Science and Public Policy* 22:383–90.

Rip, A. 1994. The republic of science in the 1990s. *Higher Education* 28:3–23.

Seashore Louis, K., M. Anderson, and L. Rosenberg. Forthcoming. The changing context of science and university-industry relations: Academic misconduct and academic values. In *Capitalizing knowledge*, edited by A. Webster and H. Etzkowitz. New York: State University of New York Press.

Star, S. L. 1986. Triangulating clinical and basic research. *History of Science* 26:29–48.

Swinbanks, D. 1994. TI loses patent battle to Fujitsu. *Nature* 371:92.

Webster, A. 1994. University-corporate ties and the construction of research agendas. *Sociology* 28:123–42.

Williams, K. M. 1994. When is a "private" conversation "public" disclosure? *Bio/technology* 12:523–24.

52

CULTURING BIOLOGY

Cell lines for the second millennium

Sarah Franklin

Source: *Health* 5(3) (2001): 321–340.

Abstract

Public concerns about innovative biomedical health technologies, such as human therapeutic cloning, have been the subject of a rapidly expanding social scientific literature. A prominent argument within much of this literature is that 'the social' is itself undergoing fundamental transformation in the context of what some have called 'the age of biological control'. This article interrogates the question of how social relationality is being transformed, or relocated, within the cell line itself by examining the recent merger between Geron Corporation and Roslin Biomed. Arguing certain social concerns are being built in to the cell line, the question of 'biological control' is refigured as one of social relationality.

In the first full-length account of the making of Dolly the sheep, *Clone: The road to Dolly and the path ahead, New York Times* science journalist, Gina Kolata (1997) describes the cloning of a higher mammal from an adult cell as one of the most important scientific accomplishments for which the previous century will come to be known, comparable to the splitting of the atom, the discovery of the double-helix and the elimination of smallpox. In the more recent book about Dolly, *The second creation: The age of biological control by the scientists who cloned Dolly*, Ian Wilmut offers a more specific set of claims:

> As decades and centuries pass, the science of cloning and the technologies that flow from it will affect all aspects of human life – the things that people can do, the way we live, and even, if we choose,

> the kinds of people we are. Those future technologies will offer our successors a degree of control over life's processes that will come effectively to seem absolute. Until the birth of Dolly scientists were apt to declare that this or that procedure would be 'biologically impossible' – but now that expression seems to have lost all meaning. In the 21st century and beyond, human ambition will be bound only by the laws of physics, the rules of logic, and our descendants' own sense of right and wrong. Truly, Dolly has taken us into the age of biological control.
>
> (Wilmut et al., 2000: 17)

Wilmut's argument that Dolly 'might reasonably claim to be the most extraordinary creature ever born' (Wilmut et al., 2000: 17), might refer as easily to her celebrity as to her anatomy, for she is undoubtedly the world's most famous sheep – a millennial animal whose birth is widely interpreted as a 'sign of the times'. Dolly is an animal who reminds us that a defining feature of turn-of-the-millennium culture is the dramatic rise of the new genetics, which has moved very visibly out of the laboratory and the clinic into the spotlight of public attention, in particular through the project to map the human genome. There came a time in the 1990s when it was no longer possible to open a newspaper without encountering some new biomedical or biotechnological discovery directly related to genetic research. Then, there came a time, a bit more recently, when it was not possible to open any particular *section* of the newspaper without encountering a story about a new technique, a new life form, a new company or some other new dimension of the new genetics. In the year 2001, it has become clearer than ever that these changes affect the human condition in its every aspect, from the food we eat, to the way we define health, to our national economies, to our understandings of the human, the future and ourselves. What is therefore particularly striking about the impact of the new genetics on society is *its simultaneous enormity and intimacy*: from our models of the world economy to our understandings of what it is to be a parent, genetics is reshaping the basic concepts through which knowledge about ourselves and our world is produced.

It is for the same reason, and equally striking, that these changes, which are at once remote and deeply personal, engender conflicting feelings of awe and of anxiety. On the one hand, scientific and technical accomplishments are welcomed, especially in the area of medicine, as possible cures for disease or sources of relief from suffering. On the other hand, techniques such as cloning by nuclear transfer or gene therapy are widely viewed with suspicion and distrust by the general public. In areas such as food production, genetic modification of plants, seeds and micro-organisms has proven highly controversial, especially in Europe. In sum, it is fair to say public responses to the new genetics are very mixed. Some science journalists, such as Gina Kolata, argue that public suspicion towards science is higher today in the USA than at any time in post-war history (1997). At the same time, the media is full of enthusiastic reports about progress being made to map the human genome, to extend the human life-span, and to develop new cures for genetic disease.

In the past, one of the main responses by government and industry to public concern over new genetic technologies has been to create ethical advisory committees to address sensitive or unfamiliar issues such as cloning. However, it has become apparent that while this continues to be necessary, it is no longer an adequate response to public anxiety, and is itself often viewed with suspicion and distrust. An important literature has begun to identify the complexity of public responses to the new genetics, and part of the picture is considerable public scepticism not only about the potential benefits of projects such as the human genome initiative, but also suspicion towards government advisory and regulatory intitiatives (Wellcome, 1998).

In some countries, and particularly in Britain, a recent change has been the realization by government and industry that more empirically based social scientific research into public understandings of science is an urgent priority. What has also begun to become more apparent is that it is not adequate simply to enquire about the 'impact on society' of the new genetics, but it is necessary to investigate what might instead be described as their cultural dimensions, that is, their *meanings*. This position assumes that what we mean by society itself, what we understand the social to be, *is itself one of the things that is changing in the context of the new genetics*. In thinking about the new genetics as forms of global culture, public culture, popular culture or national culture we need to be thinking in terms of how they generate effects, how they are productive of new meanings and how we inhabit the new forms of knowledge they provide. The relationship between society and the new genetics is highly mediated, and to understand this dimension of the new genetics requires moving beyond established bioethical approaches to address changing cultural categories. Whereas bioethics often asks directly if something is right or wrong, and how we know what we know, interpretive social scientific disciplines such as anthropology ask instead how and why things mean what they do, how knowledge is constructed and how understandings are produced. This way of asking questions about the impact of the new genetics relocates the question of their 'effects' at the level of daily life, professional life, family life or personal identity (see especially Rabinow, 1999; Rapp, 1999; Edwards, 2000; Finkler, 2000).

A growing literature on social dimensions of the new genetics has complicated the 'impact' model of change, and the 'deficient' model of public understandings by providing nuanced analyses not only of public responses to genetic screening,[1] but of media representations of the new genetics (Nelkin and Lindee, 1995) and how popular images such as the Frankenstein myth shape public understandings (Turney, 1998). While studies of 'lay' or 'public' understandings of the new genetics by people who have no direct relationship to them are only just beginning to emerge (Edwards et al.,

1993, 1998), ethnographies such as Jeanette Edwards' *Born and bred* clearly indicate that genetic knowledges are not simply received, but are woven into complex knowledge formations about personhood, conception, kinship and parenting (Edwards, 2000). Such findings are amplified by the impressive series of studies by Anne Kerr, Sarah Cunningham-Burley and Amanda Amos (Kerr et al., 1998a, 1998b; and see below), who similarly emphasize the two-way traffic between authoritative forms of genetic knowledge and public responses to them. As they describe their findings from a recent study:

> People not only had expert knowledge about cultural attitudes and social relations related to the new genetics, but could also deal competently with technical and methodological issues . . . Views were challenged and moderated; and shared and unique knowledge was revealed.
>
> (Cunningham-Burley and Kerr, 1999: 196)

Building on such findings, Cunningham-Burley and Kerr go so far as to suggest that a particular model of the social, embedded in the discourses of clinicians and scientists, is an impediment to the very social recognition they so frequently and urgently call for. The authors conclude that: 'scientists' and clinicians' emphasis on the separate nature of the social realm, in which their knowledges and technologies are applied, does not square with their [stated] social concerns about this area' (Cunningham-Burley and Kerr, 1999: 162). The problem is not only one of recognition, but of definition: it is not only necessary to recognize social concerns, it is evident we need to rethink how the social is defined.

This article supports such claims by arguing that new definitions of the social are very much at stake in the context of the new genetics and biomedicines, as well as new definitions of ethics, politics and economics. Public responses to the cloning of Dolly the sheep, and to proposed clinical applications of cloning in the form of stem cell therapies, are good places to look for indicative evidence of changes in how the social is being defined in the context of biomedicine – sometimes in unexpected ways. Dolly became a kind of totemic animal, a sign of the times, a millennial sheep, not only because her birth was considered to be biologically impossible, but because her birth became a symbol of the transgressive potential of the new genetics more generally – a potential we could describe as anti-foundational in its capacity to disrupt taken-for-granted limits formerly assumed to be incontrovertible and all but self-evident. What Ian Wilmut describes as the 'age of biological control' is also about the disappearance of the boundary between the social and the biological: biological control, as Wilmut describes it, means socially engineered biology.[2] Life was different after Dolly in both its social and biological senses, in part because she showed us we cannot make that separation anymore, between the social and the biological, the cultural and the genetic or the natural and the man-made. And life was also different after Dolly because she was herself a new life-form, a higher vertebrate cloned from an adult cell. Her breeding was not only selective: it was revolutionary. It was this sense of difference that catalysed such dramatic reactions to Dolly all over the world.

And ever since Dolly was born, events have unfolded rapidly. Initial public reactions to Dolly were particularly focused on the possibility of cloning humans, not sheep. President Clinton announced a ban on human cloning in 1997, and there were similar calls for a worldwide moratorium on cloning a human being from all over the globe. Ian Wilmut has repeatedly stated his opposition to human reproductive cloning, as have many other research scientists active in this field, based on a mixture of practical and ethical criteria. But last May, the patents for cloning by somatic cell nuclear transfer were relicensed to the California pharmaceutical company, Geron, who have formed a merger with the Roslin BioMed[3] in Scotland, to produce a new consortium, Geron Biomed, to develop ways of using cloning to produce human replacement tissue. At present, this proposal depends on a strict separation between cloning *parts* of human beings for therapeutic purposes, such as organ replacement, and cloning of *a whole human* for reproductive purposes, which is considered unacceptable. But it is clear why such a distinction is highly volatile, and potentially controversial, especially in Britain, where a number of controversies have weakened public confidence in scientific expertise.

Britain has been the site of increasing public anxiety about scientific innovation and expertise ever since the outbreak of 'mad-cow disease', or BSE, resulted in a crisis of public confidence in scientific authority.[4] The causes of the human form of encephalopathy. Creutzfeldt-Jakob disease, are still poorly understood, and it remains untreatable. So far 86 people have died, and the rate of new cases has doubled annually, raising grave fears about an epidemic which could claim as many as 100,000 lives, even by conservative estimates. To the dismay of Tony Blair's government, a successor public 'panic' emerged in early 1999, concerning the fact that large quantities of ordinary foodstuffs, from tea biscuits to breakfast cereals, contained genetically modified soy products, which were not labelled as such, and the revelation in the press that some scientific uncertainty surrounded the safety of this new ingredient.[5] Public lack of confidence in both GM products and in government scientific advisers' abilities to determine their safety was played out over the course of 1999 in dramatic tabloid headlines about 'Frankenstein Foods'. As a result of both the BSE and the GM controversies, it is now widely believed by government ministers, policy makers, industry regulators, the scientific community and many representatives of private corporations that there is an intensity of public distrust of science in Britain which now comprises something like an epidemic in its own right. This crisis of 'bad faith' is seen to be a major threat to Britain's future competitiveness in the fast-paced biotech sector, in which it is a world leader, second only to the USA in terms of scientific expertise and innovation.[6]

This sense of crisis in Britain has reached the highest levels. The Royal Society wrote a letter to the *Times* expressing the view that it is not the public who is ignorant of science, but scientists who are ignorant about the public in their dismissal of public concerns. Britain's most elite branch of government, the House of Lords, commissioned for the first time an inquiry into public perceptions of science. In their Report, entitled 'Science and Society', the Select Committee on Science and Technology notes that:

> Society's relationship with science is in a critical phase. Science today is exciting, and full of opportunities. Yet public confidence in scientific advice to Government has been rocked by BSE; and many people are uneasy about the rapid advance of areas such as biotechnology and IT. . . . This crisis of confidence is of great importance both to British society and to British science.
>
> (House of Lords, 2000:5)

Two opposing strategies are proposed in the Report concerning public anxiety about new scientific developments, especially in the realm of biotechnology and biomedicine. One is based on providing the public with *better scientific information*, by improving communication through a range of strategies from encouraging more science journalism and more science education in schools, to institutionalizing more openness and transparency in regulatory mechanisms controlling scientific innovation and research. To a degree, this strategy reflects the traditional assumption that the more the public understands about science (that is, the more scientifically 'literate' as opposed to 'ignorant' they are), the greater their approval of scientific innovation and technological change.[7]

However, the strategy of improving public education about science, has been widely challenged by social scientists.[8] In a series of studies, for example, sociologist Brian Wynne has argued that members of the 'lay' public in a sense understand science all too well, and can generalize on the basis of their own experience to reach the conclusion that *science always involves a certain amount of unpredictability and uncertainty*. As Wynne points out, this view is widely shared within science itself. It suggests that no amount of scientific literacy will ever compensate for what is, rightly, seen by both the general public and scientists themselves as *an inherent unpredictability and risk in any scientific application*. Practically, the consequences of this view are for public spokespersons to be more open about risk, rather than to attempt to offer reassurance by minimizing it. As both the BSE and the GM foods controversies demonstrated, the public is likely to view overconfident or premature reassurances about the safety of new scientific techniques as patronizing and disingenuous.[9] It is exactly this danger of *increasing* public anxiety through reassurances unsuccessfully mobilized to *diminish* it that led the House of Lords Select Committee to propose that 'one of the major factors engendering mistrust is the failure of institutional science at the frontiers of knowledge to admit publicly its own uncertainties' (2000: 23).[10] Rehearsing the take-home lesson from the BSE crisis, the Committee cautions that 'suppressing uncertainty is bound to diminish public trust and respect' (2000: 7).

A similar concern about the limits of the 'public ignorance' paradigm has been raised by the Wellcome Trust Medicine and Society programme in London, which commissioned a study into 'Public perceptions of cloning' shortly after Dolly's birth was announced in February 1997.[11] Using extended focus groups combined with scientific 'briefing' sessions about the technological details of cloning, the Wellcome study found a directly *inverse* relation between improved scientific literacy and public confidence or trust in cloning as a form of scientific research. While, on the one hand, participants in the study proved very adept at assimilating technical scientific details, this ability served *to widen* their appreciation of 'what could go wrong', *rather than to lessen it*. The Wellcome study revealed deep public distrust not only towards the scientific community itself, which was seen to be high-handed, secretive and unreliable, but equally in the ability of government or regulators to do anything about this problem. The resulting picture emerging from the Wellcome study was thus, again, of *increasing* public suspicion towards scientists, who were seen to be an uncontrolled elite wielding considerable power, which might or might not be used in the interest of the public good, or 'society', and could not realistically be prevented from going 'out of control'.

The Wellcome findings illustrate why the House of Lords Select Committee on Science and Technology recommendations about how to 'contain' public anxiety about science, and to diminish public distrust, required the pursuit of slightly contradictory strategies. On the one hand they encourage more transparent communication of unknown risks, as a way of respecting the intelligent lay person who uses his or her own everyday common sense to conclude correctly that there is, in fact, no way to be absolutely certain about scientific and technological risks. The widespread public enthusiasm for mobile phones, despite evidence suggesting possible risks of brain damage associated with their use, is an example that is often used by the former Chief Scientific Adviser to the British Government, Sir Robert May, to point to public acceptance of risk levels if they correspond to obvious consumer benefits and involve activities over which people exercise a degree of choice (as is also the case with driving). However, where there is less of an immediate consumer benefit, as in the case of GM foods, the public has no countervailing incentive to be risk taking, and is consequently less tolerant of uncertainty. The emergence of a hidden, or latent, risk – especially one that is fatal, as in the case of BSE – more closely resembles the view of opponents to nuclear power, that no matter what benefits are on offer, the risks may be too great.

In the case of nuclear power, a rather different strategy of 'containment' must be employed, which is to convince the public of the efficacy of regulatory mechanisms in which they have, at present, apparently very little confidence. This second strategy, which is based on the precautionary principle, is time consuming: government recommendations about public attitudes to controversial techniques such as gene therapy in Britain conclude that it may take up to 20 years for the public to develop confidence. However, it is all very well to recommend that 'where scientific advice is uncertain this should be admitted from the start', but the reasons governments are reluctant to publicize their doubts about scientific expertise are readily understandable, not to mention that governments are primarily concerned with politics, not science. Moreover, the market pressures to produce the shortest possible route from innovation to sales are inhospitable to precaution.

At the same time, it is clear that both governments and industries increasingly fear the costs of miscalculating public distrust in science, and these fears are, if anything, especially predominant in the fields of biomedicine and biotechnology. Perhaps more than any other case, the demise of the former life sciences giant Monsanto, which was downsized in December 1999 to a shadow of its former self, losing most of its operations outside agriculture to Upjohn-Pharmacia because of how badly it misjudged consumer reactions to GM foods in Europe (especially Britain), has sent a chill through the boardrooms of companies around the world. Novartis, AstraZeneca, DuPont and a host of highly competitive multinationals watched the Monsanto saga unfold with grave concern.[12] Both in the large agribusiness industry, where innovative plant breeding is at a premium, and in the realms of new pharmaceutical manufacturing and genetically engineered biomedicine, the issue of correctly managing public distrust is not only an urgent concern: it is critical to successful capitalization.

In sum, the effort to contain public mistrust of scientists in both publicly funded laboratories such as the Roslin Institute, and the private laboratories of multinational corporations such as Monsanto, has become a much more visible dilemma and a more prominent economic and political priority worldwide. Stakes are high for the brand capital of both governments and industries in the preservation of public goodwill. Sensitivity to the costs of making a wrong move is increasing *at the same time competitive market pressures are intensifying*. Correct management of public opinion is now a market factor that can itself determine a company's success or failure.

A new strategy being pioneered by the company that has licensed the technology to make Dolly the sheep is to build a degree of management of public opinion into its product. In turning to the ways that ethics can be 'built in' to new life-forms, we are returning not only to the cloning of Dolly the sheep, but to a whole new possibility of 'containing' public anxiety about the new genetics, which is to re-engineer an ethically sensitive biotechnology. In the second half of this article, then, I suggest that in addition to the concerns raised by Cunningham-Burley et al. that the social is being too narrowly defined by some professional experts, it is, in contrast, being redefined and indeed relocated in other ways as well. If understanding the social dimensions of the new genetics means not only recognizing the 'impact' of the new genetics on society, or the need for a wider definition of the social that is more interactive. I suggest it is also necessary to consider how the social is literally being reinstalled *within* the biological in ways that will continue to ask for imaginative responses from social scientists.

Smart sheep

As is by now well known, Dolly was the first viable offspring derived from nuclear transfer using fully differentiated adult cells from a higher vertebrate. She was produced from a mammary cell, amplified from a frozen tissue sample taken from a pregnant Finn Dorset ewe who had been dead for six years. The entire adult mammary cell from the Finn Dorset was inserted into a donor egg cell (ovum) from another (Scottish Black-face) sheep. The ovum, which is the largest and most powerful of all mammalian cells, had its nucleus removed, but remained full of the powerful cytoplasm which co-ordinates cell metabolism. It was by harnessing the cytoplasm of the donor Blackface ovum to work, in effect, as a kind of super-computer to reprogram the nucleus of the much smaller Finn Dorset mammary cell, that the Roslin team managed to 'reset its clock', to go back in time, and to recommence development as an 'embryo'. This hybrid, rejuvenated embryo was gestated by a Scottish Blackface surrogate, who eventually gave birth to Dolly.[13]

The new 'smart' embryo was not only the product of dedicated scientific innovation and skill: Dolly was also the product of intensive capital investment. In addition to confirming the success of a new scientific technique, and rewriting the biology rule-book, Dolly was also the viable offspring of a successful merger between the private Scottish pharmaceutical company, PPL therapeutics, and Ian Wilmut's team at the Roslin Institute. PPL therapeutics is the world leader in transposing human genes into animals, mostly sheep, in order to harvest peptides from their milk, which are used to supply human enzymes missing from sufferers of genetic diseases such as cystic fibrosis. Cloning by nuclear transfer is useful to reproduce transgenic animals which have had human genes introduced into their genome because, unlike conventional breeding, cloning enables exact reproduction of an animal's complete nuclear genetic blueprint.

Cloning by nuclear transfer eliminates the genetic risk of sexual (recombinant) reproduction by producing an exact replica in the offspring of the complete genome of a single parent.[14] Through this means, the precise genetic composition of prized individual animals can be both preserved in perpetuity, and more efficiently reproduced in other animals. The possibilities opened up by nuclear transfer are indeed for *any* animal, male or female, wild or domesticated, or even extinct, to become a perpetual germ-line repository, a pure gene bank, because it is no longer only the gametes, the eggs and the sperm, which are necessary for mammalian reproduction to be viable. In sum, a single animal can be cloned to produce an entire herd of animals with identical genetic blueprints.

Nuclear transfer technology thus offers a radical redefinition of selective breeding, or live-stock. Dolly's birth was significant as pure research, but her viability also introduced new recombinant models of genealogy, species and reproduction which were flexible in ways that offered great commercial potential. Nuclear transfer is thus also a device for seeding a corporate plan for the production of biocapital in the form of what Roslin and PPL describe as 'bioreactors'. 'Bioreactors' are the transgenic sheep containing human genes who function as living pharmaceutical producers, by excreting up to 65 grams per litre of valuable human proteins in their milk. PPL pharmaceuticals have been breeding flocks of transgenic sheep who carry Alpha-1-Antitrypsin (AAT) in their milk for several years, as have other pharmaceutical companies, such as Genzyme. The enzyme AAT, which helps to control elastase in the lungs and is used in the treatment of cystic fibrosis, came onto the market in Britain in April 2000 after many years in development. This drug has an enormous market as 1 in 25 Britons carry the gene for cystic fibrosis, and 1 in 2500 births are affected by it. Enzymes such as AAT were the initial motivation to develop cloning by nuclear transfer at Roslin, as a means of securing more efficient means of reproducing transgenic animals.[15] The bioreactor model of selective breeding thus offers profitability through the redesign of genealogy: while it is hard to get missing human genes into people, it turns out it is not so difficult to get them into sheep, and if the system of lactation is used as a promoter, to switch the gene on and off, the benefits are two-fold: gene expression can be regulated, and the proteins the desired gene codes to manufacture can be readily extracted from the animal in her milk.

Even smarter cell lines

PPL and Roslin's bioreactor paradigm is somewhat different from the more recently proposed use of cloning by nuclear transfer, which is human therapeutic cloning. In May 1999, the California-based pharmaceutical corporation Geron purchased Ian Wilmut's Dolly division at Roslin for $25.7 million (in stock options), and announced a $20 million research programme over the next six years to combine the use of nuclear transfer with Geron's human cell line technology to create a new market in human tissue replacement therapy, or human therapeutic cloning.

The new company, Geron BioMed, is a wholly owned subsidiary of Geron Corporation, based at Roslin, and is described by its parent company as the world's 'premiere research and development consortium in transplantation biology' (Geron press release, 4 May 1999). The goal of this consortium is to develop novel forms of human cell therapy that combine Geron's expertise in overcoming the cellular ageing process with Roslin's ability to reprogram adult cells to redifferentiate. In other words, the goal of the new consortium is to provide a kind of bespoke tissue therapy for a wide range of human diseases from heart failure to cancer to Parkinson's disease. This will be accomplished by taking body cells *from the afflicted patient*, turning them into pluripotent[16] cell lines that *act like* early embryonic tissue, which can be redirected to produce numerous forms of specialized human tissue suitable for reimplantation back into the patient.

Founded in 1990, and a publicly traded company on NASDAQ since 1996, Geron specializes in producing human cell lines that do not age. The company's name, Geron, is derived from the same root as gerontology, the study of the ageing process, and reflects their marketing strategy, which is to develop means of selling immortality. The company's most well-publicized research is the ability to clone telomerase, the enzyme which controls the mammalian cellular ageing process. In their own words, Geron's greatest breakthrough has been to 'discover the key to cellular mortality'. This 'key' is Geron's patented telomerase reactivation technology. Telomeres are described by Geron as the finger-like protrusions at the ends of each chromosome (Geron uses the analogy of shoelaces, with their protected tips). Common to all mammalian cells, telomeres act as the 'mitotic clock' of cellular replication by shortening with each cell division. Telomere shortening eventually causes the cessation of cell division, in other words, cell death.[17] However, telomere shortening can be reversed by means of reactivating the enzyme telomerase, which adds DNA repeats to the telomeres as cells divide, thus 'resetting' the cellular clock, by making cells more youthful, or, indeed, immortal.[18]

From Geron's point of view, what is very useful about this technology is that it enables significant improvements in human cell line technology.[19] Geron's ability to reactivate telomerase within specially cultured cell lines made of human embryonic stem cells offers, in their words, the capacity 'to produce large quantities of perpetually-young and healthy differentiated cells for use in the repair of degenerating organs' (Geron *Annual report*, 1997). Geron, in a sense, offers greater purity, and in perpetuity, for cell lines producing replacement human tissue. Geron reprogrames cell temporality, by devising means of making normal human body cells immortal (see further in Cha, 1999).

Roslin's expertise is well matched. The version of cloning by nuclear transfer used for Dolly succeeded by reprogramming the nuclear DNA of an adult, differentiated, body cell to make it behave like an embryo: a cell that could produce every kind of tissue – that is, a cell that could become a viable offspring. It was the ability to *reprogram* the adult DNA to go back in time, which was the astounding accomplishment of Wilmut's team, confirmed by

Dolly's viability. In sum, both Geron and Roslin have devised means of reversing cellular processes by resetting the cellular clock. The promise of this technology is not only to make new immortal cell lines that produce human tissue, but to extend human life-spans by offering genetic immortality as their product.

The combined forces of Roslin and Geron thus offer a number of intriguing possibilities. The 'blue skies' set of applications promises the option of using a *body cell* from someone suffering some kind of degenerative illness to make *an embryo* which can be reprogramed as *an immortal cell line* to produce whatever tissue that person needs to survive. Because of their source, such cells would be ideally histocompatible, that is, perfectly matched, thus eliminating the need for immuno-suppressants. These bespoke cells could be produced in large quantities, and would last the lifetime of the patient, indeed beyond. Given their diverse range of applications, it is difficult to imagine a ceiling on the market for such products.

However, Geron Bio-Med promises more than smart cell lines and smart venture capital. They are also offering new methods of avoiding many of the ethical objections to biotechnology by designing life forms that are 'alive' to potential opposition from an anxious or wary public.[20] Until the recent merger, Geron's cell lines came from two sources: human embryonic stem cells (hESs) derived from embryos donated from IVF programmes, and human embryonic germ cells (hEGs), which are derived from foetal material obtained from medically terminated pregnancies. These sources are not only limited, but morally controversial, especially in the United States, where Geron is based. The merger with Roslin is thus not only desirable in terms of making longer-lasting cell lines: it is also desirable ethically, which means strategically, because it can eliminate certain controversial aspects of human cell line manufacture.

This aspect of the Roslin-Geron merger adds an interesting dimension to the story. For, in addition to its vast potential clinical and commercial viability, the proposed technique now known as human therapeutic cloning offers the added value of providing a kind of moral high road around the worrisome possibility of public opposition that often plagues the biotechnology sector. This is a solution to the 'containment' of public anxiety about the new genetics the British House of Lords did not envisage. Objections to the use of human embryos for research, or the production of replacement body parts from aborted foetal tissue, are eliminated if a normal body cell (say, a liver cell) can be reprogrammed to perform *as if it were an embryonic cell.* Whereas eggs, sperm, embryos and aborted foetal tissue all carry very considerable moral and symbolic weight, and must be donated, it is difficult to imagine organized political or religious objection to culturing a patient's own liver cells to prevent him or her dying of liver failure.[21]

The reason public anxiety about the cloning of Dolly the sheep was primarily focused on the possibility that the technology could be used to clone human beings was that humans are *biologically* very similar to sheep. Interestingly, Geron has capitalized on precisely this identity between animals and humans – but in this case in order to *promote* human cloning. The corporation is attempting to reposition a technology many opposed precisely because it threatened to result in the cloning of a human being into an attractive possibility for cloning replacement human tissue. The connection to animals immediately recognized around the world when Dolly was cloned, that we could also be cloned, is in the process of an anxious and delicate transformation – in sum, from a threat into an opportunity. By rebuilding embryonic cell lines *without embryos*, Geron is launching a very practical way of 'containing' one of the prevailing US sources of public anxiety about science by 'building in' ethical concerns to both its research programme *and the life-forms it is creating*.[22] This redefinition of 'embryo' research is paradoxical in the sense that the public's concern about embryos, as opposed to liver cells and pancreatic cells, is based on a recognition of their unique developmental potential to become a human being. What is seen to be 'special' about the embryo is its potential to become one of us – and this is exactly the same message Geron wants to promote, only with a more utilitarian kind of becoming in mind, and a different kind of embryo on offer.[23]

The status of Geron's post-natural, biotechnologically assembled 'embryos' raises a host of questions that are ethical, theological and cultural and at the same time ontological, social, technological and commercial. In presenting a new version of the embryo question, Geron is offering a 'smart' embryo in much the same way Roslin and PPL designed 'smart' sheep to bypass the technical difficulties of getting human genes into human beings for diseases such as cystic fibrosis (and thus avoiding the costly public disapproval accompanying failed human gene therapy). Like all venture-capital funded biotechnology companies, Geron is striving to avoid circumstances which will compromise its potential future profitability. By selecting a route forward which rids the company of one of the most potentially compromising sources of public opposition in the USA – Pro-Life sensitivity to the use of embryos for research – Geron is charting a strategic course between what is practical and viable experimentally, and what is commercially feasible as a means of 'short-cutting' public opposition to research. Time is not on Geron's side: they cannot reset their corporate clock as they can their own patented cell lines. The market is uncompromising, and unlike immortalized embryonic stem cells, shares can only be split so many times. Like most start-up biotech companies, Geron has never made a profit, and does not expect to do so in the near future. Their stock price fluctuated in 1999 from \$3.50 to \$24.50, and in their November 1999 *10-Q* submission to the US Securities and Exchange Commission Geron listed numerous potentially compromising circumstances that may affect their future viability as a company, including the ethically sensitive nature of their research, as well as its newly international organization.

Conclusion

We need, of course, to think both carefully and critically about what Geron is offering. There are many reasons to assume their technology will not work the way it is supposed to do, and so far they are over a billion dollars in debt to a complicated technological promise that may or may not succeed in either medical or market terms. Like other venture-capital funded biotechnology companies, Geron is building a patent estate that has both broad and specific purposes: specifically they are aiming for a particular product line in replacement tissue, and more broadly they are positioning themselves to be able to take advantage of unforeseen developments.[24] Perhaps the Roslin technology that produced Dolly will be useful to Geron for the purposes they now intend, but it may just as well prove instrumental to a completely different kind of future technology that has not yet even been envisaged.[25] If the Geron strategy of patenting the cell functions necessary to produce cellular recapacitation is viable, they will become one of the major stake-holders in the emerging biomedicine sector, and human health will be substantially redefined as a result.

Geron's strategy of producing less controversial human cell lines may also not succeed as well socially or politically as they intend. It is, after all, a strategy likely to eliminate only one kind of opposition, which is particularly prominent in the USA (namely the Pro-Life movement which believes that life begins at conception). A completely different ethical concern might be raised in other countries such as Britain, for example, in relation to questions of health care distribution, and the issue of whose cells will be eligible for high-tech enhancements of the kind Geron is proposing. In the face of the kinds of major medical challenges an ageing population presents in a highly stratified society such as the USA, how helpful or progressive is it to suggest that a normal human life-span is really something more like 120 or 130 years? Perhaps this is a particularly American promise, or maybe even one that is particularly Californian? While there is no reason to doubt Geron's claim that they are 'pursuing treatments that will allow people to remain healthy as they age', it is clear that Geron is suggesting other promises too: promises to take control of the ageing process at a fundamental level in order to harness genetic immortality on behalf of improved human futures, for example, and to use this promise to create a successfully competitive pharmaceutical market.

Geron's promise, and the hope that it is mobilizing, is in this respect very much a part of the same language of scientific progress and technological potency which so strongly characterizes representations of the human genome project, and the new genetics more generally. What is significant about this language of promise in the contemporary era is the extent to which it sits beside a great deal of uncertainty, among the public at large and also within science itself, about the risks of these much-celebrated potential advances. And these risks are not just biological or medical.

Among the many significant aspects of the promises Geron is offering about the future is also the way they are promising to engineer solutions to social and ethical problems *into the cell line itself*. In their corporate publicity, such as their annual reports, Geron tells a very ordinary story through pictures, as if from a family album, of ordinary people living ordinary lives, which are affected by very common diseases, such as cancer, macular degeneration and athero-schlerosis. But a subtler meaning, I suggest, is that we are all, in a sense, ageing cell lines, for which Geron is devising new means of repair. And, importantly, they are suggesting that our social values too can be encompassed by the kinds of cell lines we make, for example by designing new means of producing embryonic human tissue. This is not an ordinary promise – it is a very extraordinary one. If we listen carefully to Geron's promise it suggests that Geron is not only attempting to redefine the relationship between health and disease, life and death, capital and the new biologies, but also the relationship between the new genetics and society *at a conceptual level as well as a technological one*. We must listen carefully to how Geron is offering to provide biological solutions to social concerns because by doing so they are suggesting a redefinition of the human, the social and the future. In other words, we need to look at these proposals very seriously *simply because of their scale*. We need to remember, in short, that one of the major questions we should be asking about the new genetics is not only the kind of genetic repair we might desire to have available to extend our lives, but why that question is inseparable from the issue of what kind of society we want to live in, on the basis of what kinds of cultural values, now and in the future.

We also need to examine Geron's proposals for another reason. Geron is a private company that has to be very careful about public opinion, and in this respect Geron has not made some of the mistakes many governments have in their responses to public unease about scientific innovation. Two common reactions to 'public opinion' are to overestimate it and to underestimate it. Public acceptance of new technologies is easy to over-estimate just because certain techniques are seen to promise an improved future. Similarly, public opinion is easy to underestimate in the sense that opposition to new developments can be wrongly portrayed as the result of sheer ignorance or irrational prejudice. One of the important lessons the Dolly-Geron story tells us is why it is a mistake to think that we can somehow factor out the hype, the media or the work of the imagination to exaggerate either the promises or the risks of new technology. *This is not going to be possible, now or in the future, because it is precisely the importance of imagining a future yet-to-be that fundamentally defines the whole issue of the new genetics and society.* Geron takes the work of the imagination very seriously, and quite rightly. Successful capitalization for Geron is all about recognizing the links between the hype, the product, the market, the science, the Government and the goodwill of the general public. At the level of capitalization, none of these issues can be 'factored out'. There is no bottom-line, pure reason or neutral ground on which to stand other than hypothetically in the fast-paced world of the remaking of life that so visibly asks for our attention at this time.

Increasingly, the recognition that the social contexts in which the new genetics take cultural shape are highly complex has begun to reshape opinion in some unexpected places, such as the British House of Lords. In achieving such a recognition, it is not surprising governments are turning to social science for evidence-based approaches to the analysis of 'public opinion'. Innovative qualitative studies, such as the Wellcome report on cloning, have opened the door for more interpretive approaches both to be recognized as legitimately scientific, and to become a more explicit funding priority.[26] In the face of the dramatic changes in health policy augured by the new genetics, such developments are both strategic and timely. They will also require innovation from within the social sciences themselves, which, like everything else, are also in the midst of transformation in the context of the new genetics and the very fundamental forms of change associated with them.

Notes

1 See, for example, Rayna Rapp's exhaustive ethnography of amniocentesis in New York City which demonstrates the complexity of different clients' relationship to 'genetic information' (Rapp, 1999). See also Finkler's study of the experience of acquiring genetic information, about parentage as well as genetic pathology (Finkler, 2000).

2 This shift is interestingly compared with Rabinow's notion of 'biosociality' (1996), in which the relations of the social and the biological are reversed – society authoring biology in an inversion of the sociobiological model of nature as first author.

3 It should be noted that although Roslin BioMed is part of the publicly funded Roslin Institute, it was already a semi-privatized unit before the Geron merger.

4 The BSE crisis has never been fully 'resolved', and was itself the subject of a major public inquiry early on in the new government of Tony Blair, which found many faults with the manner in which scientific advice was sought by the (previous) Conservative government, as well as the incomplete response to this advice. The BSE crisis is too extensive a subject to explore in depth here, as it knit together the contested politics of European unity with the contamination of the nation's food supply, the uncertain future of the farming industry (and the emergence of a 'politics of the countryside') and the quality and efficacy of scientific advice to the Government – all of which centred on the highly symbolic substance of British beef, a meat in many ways isomorphic with the nation itself.

5 The 'official' beginning of the GM food wars in Britain is often dated to a letter published in the *Guardian* newspaper on Thursday 12 February from a group of 22 scientists supporting the (then unpublished) work of Dr Arpad Pusztai alleging harmful effects of GM potatoes fed to rats. Controversially, Pusztai had been dismissed from his post for irresponsible actions when he made the results of his study available to the media.

6 This crisis of bad faith also corresponds to the sociological model of the 'risk society' thesis, most closely associated with the work of Ulrich Beck (1992). According to this thesis, people have lost faith in many traditional forms of

authority associated with modernity, such as scientific authority, because they now see science, industry and medicine as 'risk producing'. As Wynne notes, the risk society 'is a new logic superseding that of class conflict [according to which] modern science and technology now produce truly global risks from which not even the rich and powerful can escape' (Wynne, 1996: 55). Epitomizing the 'risk society' model is the threat of nuclear contamination.

7 As the committee notes: 'It is widely assumed that one of the roots of public mistrust of science is ignorance, and in particular the public's apparent insistence on zero risk and absolute certainty' (House of Lords, 2000: 14).

8 Interestingly, the Select Committee notes that their Inquiry:

> has taken the Committee further than it usually goes into the realms of social science. We have been told on all sides, and we believe it to be true, that 'hard science' is paying increased attention to insights from social science. The particular issues of this report, public understanding of and engagement with science, have been the subject of increasing study by social scientists over the last twenty years.
>
> (House of Lords, 2000: 14)

9 As the Select Committee noted in its report: 'Some current research suggests that the public in fact understands uncertainty and risk well, on the basis of everyday experience. People use common sense to interpret and evaluate what they hear about technological advances . . .' (House of Lords, 2000: 22).

10 This is similar to the main theme informing the guidelines issued by the Chief Scientific Adviser to the Government, Sir Robert May, in response to the BSE inquiry in 1997 ('The use of scientific advice in policy making'), which called for greater openness, and argued that when scientific advice is uncertain this should be publicly acknowledged from the start.

11 This provocative study of public perceptions of the life sciences also raises interesting methodological issues about the acquisition and use of qualitative data in the assessment of public policy towards the life sciences (see www.wellcome.ac.uk).

12 Monsanto CEO Robert Shapiro has publicly regretted the company's 'arrogance' in misjudging public opinion on numerous occasions since his company brand capital sank somewhere in the mid-Atlantic, irretrievably tarnished by its demonization in Europe.

13 Cloning by nuclear transfer is often represented as combining the nucleus from the donor cell of the animal to be cloned and the enucleated ovum from another sheep. This description, however, contradicts the account of cloning in *The second creation: The age of biological control by the scientists who created Dolly* by Ian Wilmut, Keith Campbell and Colin Tudge (2000), which offers by far the most detailed account of this technique to be published, and illustrates the entire (though much smaller) adult body cell being combined with the enucleated ovum.

14 The exact genetic traits sought by PPL pharmaceuticals are transgenic. The first cloned transgenic sheep was announced in July 1997, named Polly. Polly was created by 'a version' of the technique used to create Dolly, namely the technique used to produce Megan and Morag, the sheep born at Roslin in 1996, using foetal rather than adult cells. The important point about Polly is not only that she carries the targeted human gene, but the marker for it. The Roslin web pages explain that 'earlier techniques have been hit-or-miss for mixing animal DNA but cloning should make that process more precise'.

15 Cloning by nuclear transfer is not the only way to reproduce transgenic animals, nor is it necessarily the most efficient. PPL pharmaceuticals, who recently announced the birth of five cloned transgenic piglets in Virginia, plan to mate these animals in the conventional fashion to make other transgenics suitable for producing organs for human transplant.

16 The useful feature of embryonic cells, or stem cells, is their ability to retain the capacity to differentiate into any kind of tissue: so-called 'toti-potentiality'. Geron distinguishes between their 'pluripotential' cells lines, which can become most, but not all, forms of tissue (for example, they cannot, at present, be directed to become blood cells) and the 'true' toti-potential capacities of embryos.

17 In their 1997 *Annual report*, Geron also uses the visual analogy of the protected ends of shoelaces to represent telomeres.

18 Like early embryonic cells, cancer cells do not lose their telomeres, which is why cancer cells are commonly used to establish immortal cell lines. Initially, Geron aimed for a market in diagnostic kits that would test for telomerase to identify oncogenesis. This is still an important emerging market for Geron, though somewhat displaced by the new avenues made possible through using nuclear transfer to make better immortal ('more immortal') cell lines.

19 A normal cell line made from body tissue has to be mixed with several other ingredients, many of unknown function, in order to activate and become productive. Making a cell line of, for example, the mammary tissue of a Finn Dorset ewe, requires a complicated cocktail of ingredients to achieve a small fibroblast, as the resulting scab-like collection of mammary cells in culture has many needs that are not well understood.

20 As *San José Mercury News* journalist Ariana Cha notes, 'at this precarious moment. Geron finds itself at the center of a growing national debate on human embryo research, with some legislators and religious leaders calling for new limits on the experiments that are key to the company's stem cell program' (1999).

21 In a characteristically pithy summation of the new Geron BioMed cell lines, one of Britain's leading tabloid newspapers, *The Express*, reported that:

> Scientists in Scotland claim to have removed the main ethical obstacle to so-called therapeutic cloning – which would lead to the growing of spare parts and tissues for transplant in the lab. A breakthrough means dozens of embryos will no longer have to be sacrificed to generate genetically compatible tissues for transplant. It could mean a lifting of the ban on therapeutic cloning, which is far more acceptable to scientists than reproductive cloning – the creation of carbon-copy babies.
>
> (*The Express*, 27 January 2000)

22 In Geron's *Form 10-Q* report to the Security and Exchange commission of 15 November 1999 (which they are required to file as a publicly traded company as a form of 'full disclosure' to potential investors), it is stated that:

> Our pluripotent stem cell therapies program may involve the use of human pluripotent stem cells that would be derived from human embryonic or fetal tissue. The use of human pluripotent stem cells gives rise to ethical, legal and social issues regarding the appropriate use of these cells. In the event that our research related to human pluripotent stem cell therapies becomes the subject of adverse commentary or publicity, our name and goodwill could be very adversely affected.
>
> (Geron, 1999: 21)

23 In the Christian tradition, this is what is 'divine' about the conceptus: it has the ability to become a human being, a being made in the image of God, by him, and thus embodying, if in a fallen form, His divine gift, the gift of life. This is why the institution of marriage, the Holy Sacrament, must protect the creation of human life, in which Man enacts his only small portion of divinity through the miracle of procreation, or bringing forth, and this is why procreation outside of marriage is considered blasphemy – a disrespect for the gift of life.

24 In this respect, the venture biotech sector is quite similar to the dot.com sector where the pace of innovation is such that product development and market making are very unpredictable. Consequently, something patented for one purpose may well prove critical to some other, as yet barely imagined, future. Consequently, the significance of creating the future in a form that is profitable becomes the real object – which is yet another way of underscoring how important the work of the imagination is to the production of the new life-forms which will populate it.

25 After all, this is already true of the technique of cloning by somatic cell nuclear transfer, which was developed as a means of producing transgenic sheep, and is now being used for a completely different purpose by Geron.

26 The recent joint funding initiative between two of Britain's national funding councils – the Medical Research Council (MRC) and the Economic and Social Research Council (ESRC) – to fund 21 social science research projects investigating 'innovative health technologies' at a cost of £3.13 million is an example of this distinctively British shift in funding priorities, as is the development within the Wellcome Foundation of a £5 million Medicine and Society programme.

References

Beck, Ulrich. (1992). *Risk society: Towards a new modernity*. Trans. Mark Ritter. London: Sage.

Cha, Ariana Eunjung. (1999). Geron's trials biotech pioneer has medical vision in place but must start achieving results. *San José Mercury News*, 4 July.

Cunningham-Burley, Sarah and Kerr, Anne. (1999). Defining the 'social': Towards an understanding of scientific and medical discourses on the social aspects of the new genetics. In Peter Conrad and Jonathan Gabe (Eds.), *Sociological perspectives on the new genetics*. Oxford: Blackwell.

Cunningham-Burley, Sarah, Kerr, Anne and Pavis, Stephen. (1999). Theorizing subjects and subject matter in focus groups. In Rosalind Barbour and Jenny Kitzinger (Eds.), *Developing focus group research: Politics, theory, practice*. London: Sage.

Edwards, Jeanette. (2000). *Born and bred: Idioms of kinship and new reproductive technologies in England*. Oxford: Oxford University Press.

Edwards, Jeanette, Franklin, Sarah, Price, Frances, Hirsch, Eric and Strathern, Marilyn. (1993). *Technologies of procreation: Kinship in the age of assisted conception*. Manchester: Manchester University Press.

Edwards, Jeanette, Franklin, Sarah, Price, Frances, Hirsch, Eric and Strathern, Marilyn. (1998). *Technologies of procreation: Kinship in the age of assisted conception*, 2nd edn. London: Routledge.

Finkler, Kaja. (2000). *Experiencing the new genetics: Family and kinship on the medical frontier*. Philadelphia, PA: University of Pennsylvania Press.

Geron Corporation. (1997). *Annual report*. Menlo Park, CA: Geron Corporation.

Geron Corporation. (1999). *Form 10-Q*. United States Securities and Exchange Commission, 15 November.

House of Lords Select Committee on Science and Technology. (2000). *Science and society*. London: HMSO.

Kerr, Anne, Cunningham-Burley, Sarah and Amos, Amanda (1997). The new genetics: Professionals' discursive boundaries. *The Sociological Review, 45(2)*, 279–303.

Kerr, Anne, Cunningham-Burley, Sarah and Amos, Amanda (1998a). The new genetics and health: Mobilising lay expertise. *Public Understanding of Science, 7*, 41–60.

Kerr, Anne, Cunningham-Burley, Sarah and Amos, Amanda (1998b). Drawing the line: An analysis of lay people's discussion about the new genetics. *Public Understanding of Science, 7*, 113–33.

Kolata, Gina. (1997). *Clone: The road to Dolly and the path ahead*. London: The Penguin Press.

Nelkin, Dorothy and Lindee, M. Susan. (1995). *The DNA mystique: The gene as cultural icon*. New York: W.H. Freeman.

Rabinow, Paul. (1996). *Essays on the anthropology of reason*. Princeton, NJ: Princeton University Press.

Rabinow, Paul. (1999). *French DNA: Trouble in purgatory*. Princeton, NJ: Princeton University Press.

Rapp, Rayna. (1999). *Testing women, testing the fetus: The social impact of amniocentesis in America*. New York: Routledge.

Turney, Jon. (1998). *Frankenstein's footsteps: Science, genetics and popular culture*. New Haven, CT: Yale University Press.

Wellcome Trust, Medicine and Society Programme. (1998). *Public perceptions of cloning*. London: Wellcome (www.wellcome.ac.uk).

Wilmut, Ian, Campbell, Keith and Tudge, Colin. (2000). *The second creation: The age of biological control by the scientists who created Dolly*. London: Headline.

Wynne, Brian. (1996). May the sheep safely graze? A reflexive view of the expert-lay knowledge divide. In Scott Lash, Bronislaw Szerszynski and Brian Wynne (Eds.), *Risk, environment and modernity: Towards a new ecology*. London: Sage.

Part 4

BIOPROSPECTING AND BIOPIRACY

53

BIOPROSPECTING

Lessons from benefit-sharing experiences

Katy Moran

Source: *International Journal of Biotechnology* 2(12) (2000): 132–144.

Abstract

Biodiversity prospecting, or bioprospecting, is the exploration of biodiversity for commercially valuable genetic and biochemical resources. The Convention on Biological Diversity recognizes the sovereignty of nations over their bioresources and their rights to share in the benefits that accrue from commercialized bioproducts. Since the Convention was introduced in 1992, many biodiversity prospecting arrangements throughout the world have progressed with varying degrees of success. This paper looks at how benefits from bioprospecting have been shared among the stakeholders in the arrangements. It briefly describes lessons learned from three case studies on different continents and among various stakeholders and commercial interests.

1 Introduction

Those concerned with the conservation of bioresources recognize that when biodiversity-rich nations benefit from their sustainable use, conservation opportunities increase. The Convention on Biological Diversity (CBD) codifies this principle in its interrelated goals of:

1 the conservation of biological diversity;
2 the sustainable use of its components, and
3 the fair and equitable sharing of the benefits arising out of its use.

No longer are bioresources considered the 'common property of mankind', as the CBD recognizes the sovereignty of nations over their bioresources and their rights to share in the benefits that accrue from commercialized bioproducts[1].

Since the CBD was introduced in Rio during the 1992 Earth Summit, 178 nations and the European Union have ratified the Convention, agreeing to enact legislation to implement provisions that represent national interests, yet do not impose restrictions that conflict with the objectives of the CBD. At the same time, commercial users of bioresources voluntarily attempt to comply with convention provisions in order to gain access to the raw bioresources that can be developed into bioproducts. Besides the risk of lawsuits and despoiling a company's commercial reputation, the legal search for biological resources protects future business strategies for companies to access raw bioresources. These arrangements have come to be called bioprospecting, or exploring biological diversity for commercially valuable genetic and biochemical resources[2].

It is useful to examine how these practices, by both provider countries and user companies, are actually working today and offer examples to nations in the process of establishing legal frameworks to comply with the CBD. As of the beginning of 2000, some 42 countries have introduced access and benefit sharing regulations or have laws under development. Regional framework, which respond to the occurrence of biodiversity regionally, are also under consideration in the five States of the Andean Commission, the Association of South East Asian Nations (ASEAN), the Organization of African Unity (OAU), and the South Pacific Regional Environment Programme (SPREP).

But there are more stakeholders than provider countries, which include local, state, regional and national government levels, and user industries, including both small and multinational companies. Many other interested parties are involved in bioprospecting relationships, often with competing agendas and conflicting interests. Major stakeholders include both the domestic and international research communities, including botanical gardens and universities. Their participation reflects the increasing commercial value of bioresources and spans a continuum from pure to applied research[3].

Also recognized as valuable to bioprospecting is the ethnobotanical knowledge of the forest-dwelling (or native, tribal, indigenous) peoples who live in or near tropical forests and possess information on the use of plants for medicinal purposes. Of the 120 active compounds isolated from higher plants and used today in Western medicine, 74% have the same therapeutic use as in native societies. This traditional knowledge is embedded in forest peoples' cultural systems and passed down generationally. Accumulated over millennia, traditional knowledge is as rich and diverse as tropical forests' biological resources, and as threatened[4].

Non-governmental organizations (NGOs) such as conservation, development, human rights, health and other non-profit organizations, foundations, *pro bono* law firms and private consultants play growing roles as intermediaries in fostering, facilitating and evaluating bioprospecting partnerships. To level the playing field in negotiating expertise, many countries and culture groups have successfully sought outside expertise at different stages of bioprospecting negotiations from intermediaries that have some vested interest in promoting equitable arrangements. Many intermediaries speak the local language, have worked for many years and often lived in the community and earned its trust. They play many roles in access and benefit-sharing arrangements with source countries and other entities, often becoming partners in the relationship[3].

This paper will review three existing case studies on bioprospecting in Asia, Latin America and Africa, focusing on lessons regarding benefit-sharing, the third goal of the CBD. Although there is a broad array of industrial uses for biodiversity, including biotechnology, cosmetics, crop protection, horticulture and personal care industries, this paper confines case studies and discussions to benefit-sharing from pharmaceuticals and botanical medicines[5]. There are no 'model' bioprospecting arrangements, as the case studies demonstrate. Rather, in one study the impetus came from a government research institute, another from an international environmental organization and another was primarily initiated by a small company. Since one of the most contentious questions of bioprospecting regards the use of indigenous knowledge in the discovery of new medicines, each case study includes indigenous participants, so as best to illustrate and compare such arrangements[6].

2 Comparative experiences in benefit-sharing

2.1 India

Rich in biological and cultural diversity, India houses over 45,000 species of plants, 30% that are higher plants and 400 unique ethnic groups, 75% of which are tribally organized. India's proposed Biodiversity Act strictly regulates international access to bioresources for both research and commercial use with heavy fines for breach of the Act.

The Tropical Botanic Garden and Research Institute (TBGRI) was established by the government of Kerala, India, in 1979. In 1987, the TBGRI held an ethnobotanical field study in the forests of the Western Ghats in southwest India. These forests are home to the Kani tribe, nomadic traditional collectors of non-timber forest products, who now lead a primarily settled life in tribal hamlets. The Kanis use a wild plant for energy that they called 'arogyapacha', and was identified as *Trichopus zeylanicus* by the TBGRI. It provided a lead to the development of the drug 'Jeevani' (giver of life) after the TBGRI transferred the manufacturing licence to a major Ayurvedic drug company in India. The Aryavaidya Pharmacy Coimbatore Ltd. licensed Jeevani as a tonic to bolster the immune system and provide energy for a fee Rs 10 lakhs (approximately US$25,000). The TGBRI agreed to share 50% of the licence fee and the 2% royalty on profits with the Kani[7].

Traditional systems of the Kanis have eroded and been replaced by those of non-tribal local communities. Rather than governance by a tribal chief (Mootha Kani), today's system, referred to as the Panchayati Raj, is based on the principle of devolution of administrative powers to the local village level under the Constitution of India. The predominantly non-tribal Panchayat's decision-making body is elected by members.

Kanis, with a population of under 17,000, live in different areas and have differing opinions on the arrangement with the TBGRI, which has interacted primarily with the Kanis from one area. This group of Kanis supports and appreciates the role of the TBGRI. However, Kanis in other Panchayat areas are offended that the TBGRI only works with one group. Some feel the benefit-sharing arrangement is superficial, since most Kanis are not involved in the process. In September of 1995, a group of nine Kani medicine men wrote to the Chief Minister of Kerala, objecting to the sale of their knowledge to a private company.

The TBGRI acknowledges it has not reached out or communicated to all members of the Kani tribe, so, with the help of NGOs a Trust was established and registered. The TBGRI reported in 1998 that the Trust has over 500 members, but some critics feel this number falls far short of adequate representation of the Kani. In March, 1999, *Science* reported that the first payment of $21,000 would be made and shared by the tribal community and the institute[8].

Dr. P. Pushpangadan, former director of the TBGRI, reports that arogyapacha, a perennial undergrowth, was cultivated to ensure a regular supply of the raw drug to the manufacturer[9]. But the TBGRI scientists learned that the medicinal qualities of the plant are lost, unless grown in natural forest settings. So the TBGRI organized 50 Kani families who live inside the forest to cultivate and pre-process the plant under the supervision of the TBGRI scientists. This plan, writes Pushpangadan, generates employment, as leaves of the plant are bought from the Kani who manage the semi-wild crop. Each family has one or two acres of arogyapacha under cultivation in the first year of the project and earns about Rs. 30,000 per acre. This income is expected to increase in subsequent years, as it is anticipated that the production of leaves will increase and last for 20–30 years. Since the plant can be grown only in the natural forest habitat to maintain its medicinal qualities, it seems the Kani are in an effective bargaining position to regulate and control its harvest. Pushpangadan describes arogyapacha as 'forest friendly' because it is grown under the shade of the natural forest canopy, but he does not refer to any sustainability studies on its management. Although it contributed no men, material or finances to the effort, Pushpangadan recently writes, that the State Forest Department now demands a share of the licence fee and royalties on the grounds that the plant material is collected within a forest area.

In all fairness, it must be remembered that the TBGRI process for benefit-sharing with the Kanis evolved in a policy vacuum, well before the CBD was introduced. The lessons learned from mistakes made in this case study did not have the luxury of precedent, or of guiding legislation by the national government, but developed procedures in an *ad hoc* manner gradually over a dozen years. This occurred despite the fact that all project participants were nationals, and did not involve any international trade, companies or institutions. When governments enact restrictive regulations, it inhibits scientific research, whereas simple, flexible access laws discourage 'biopiracy', encourage innovations and minimize the transaction costs of bioprospecting for all.

Likewise, the absence of an intact traditional system of governance by the Kani crippled the decision-making processes that are enjoyed by culture groups with a lesser degree of acculturation. The trust fund that was initiated after complaints from the Kanis about equity cannot be considered legitimate unless the people who contributed to the process from which revenues are generated, are the same ones who control the funds to some degree. Since wealth and technology are as concentrated in the North as poverty and biodiversity are in the South, what is 'equitable' in the bioprospecting process is difficult to define. Equity is a relative term and can be determined only by the participants in the process. It means not only equitable compensation, but equal standing among participants in making decisions about what form benefits should take.

Even though the Kani are quite powerless in relation to a large Indian government research institution, their uncontested right to deny access to bioresources or traditional knowledge would have supplied strong leverage during access negotiations, had the bioprospecting arrangement been negotiated today. The right to deny access is widely recognized and these processes are now undertaken well before any bioprospecting activities begin. No collecting permits are issued until such contractual agreements are negotiated, after the prior informed consent process. There must be a written understanding among all relevant stakeholders of rights, obligations, roles, and ownership, of who shares how much in what kinds of benefits, when. It defines exactly who the state and/or national authority is, whether it be the State Forest Department, the TBGRI, or some other entity that can make a claim as a stakeholder[10].

In its wording, the CBD lumps together 'indigenous and local communities', even though there are several anthropological criteria that clearly differentiate indigenous peoples from being defined simply as local communities. This case study demonstrates the dangers of assuming that the indigenous Kani and local communities share the same values, traditions and methods of governance.

Another question concerns the sustainability of the present system of sourcing arogyapacha, since there is no information on sustainability studies connected to methods of managing and harvesting the plant. Clearly, environmental impacts on the forest, as well as sustainability studies, are required if it is anticipated that the need for arogyapacha will increase and last for 20–30 years[11].

2.2 *Suriname*

The International Cooperative Biodiversity Group (ICBG) was launched in 1991 with funding from the US government, sponsored by the National Institutes of Health (NIH), the National Science Foundation (NSF) and the US Agency for International Development (USAID). Three interrelated goals of the ICBG include health improvement through the discovery of new drugs from natural sources, conservation of biological diversity and sustainable economic development. By 1994 ICBG projects were held in Suriname, Costa Rica, Argentina, Chile, Mexico, Peru, Cameroon and Nigeria[12].

Suriname is located in the centre of a two billion-year-old Pre-Cambrian formation known as the Guayana Shield. This region of northern South America is the most extensive and least disturbed major tropical wilderness, with more than 90% of its original forests still intact. The interior of Suriname is covered with undisturbed Neotropical Amazonian forest and largely uninhabited. With a land area of 166,000 square kilometers and a human population of 412,000, Suriname is one of the least densely populated countries in the world.

The five year old Suriname ICBG's bioprospecting program was proposed by the Suriname office of Conservation International (CI), an international non-governmental organization (NGO) and the Saramaka Maroons. Maroons are descendants of escaped African slaves who have lived in the Suriname forest for nearly three centuries. Although they are not indigenous to the area, the Saramaka Maroons meet several anthropological criteria in definitions of 'indigenous' including forest dwelling, use of a local language and tribal social organization[13]. The cooperative partnership also includes the Virginia Polytech Institute and State University, Missouri Botanical Garden, Bristol Myers-Squibb Pharmaceutical Research Institute and Bedrijf Geneesmiddelen Voorziening Suriname (BGVS), a pharmaceutical company owned by the Surinamese government.

Through a Statement of Understanding, the Suriname ICGB established a benefit-sharing plan with a US$60,000 total advance payment from Bristol Myers Squibb Pharmaceutical Research Institute into the Forest Peoples' Fund (FPF), with additional contributions of $20,000 a year as the ICBG is renewed. The FPF is a mechanism, which CI Suriname helped broker, through which 'up front' benefits and future royalties from new drugs developed can be returned to the Saramaka people. Along with compensating the Saramaka Maroons for their ethnobotanical contributions, the FPF creates conservation incentives, finances sustainable management projects, provides research and technology exchanges and supports other socially and environmentally sound projects. A board of directors consisting of two representatives from the Indian community (one Saramaka and one Amerindian), two from CI Suriname, and one from Suriname's Department of Interior was created to review proposals for ways to spend the advance payment and any future royalties from drug sales. It was agreed that funds would be used for projects involving community development, biodiversity conservation and health care. If any products are commercialized from ethnobotanical collections, 50% of Suriname's share of any future royalties will go to the FPF and the other 50% will go to various ICGB partners in Suriname. If a drug is derived from random collections, the FPF's share is reduced to 30%, while 70% goes to other ICGB partners.

Green[14], a consultant who speaks both Saramakan and Sranan Tongo, evaluated the Suriname project during a site visit in 1998. He learned that before the project began, Saramaka leaders held numerous palavers (kutus), after which their paramount chief and local chiefs (Kapitenis) voted for Saramakas to participate in the project. They considered ways to use the donated funds that supplied ongoing income generation to benefit whole communities. CI (Suriname and Washington) played a crucial guiding role in this, although the by-laws of the fund were written by Surinamese participants and are governed by the laws of Suriname. About half the available $60,000 funds was spent on local income generating projects. The FPF purchased outboard motors and developed a canoe taxi service to transport people and goods up-river to Saramaka territory. Other funds were used to train local people to cater food for ecotourism; to purchase sewing machines and train women to sew school uniforms for sale to parents; and to supply tools and training for a commercial gardening project. Another project trained people in woodcarving, carpentry and furniture-making. A community-based organization, called Afinga, helped to coordinate these microenterprise initiatives. All of these small scale projects have the potential to generate much-needed cash for the Saramakas of the upper Suriname river area where the ICBG is implemented.

Originally, according to Green, the ICBG agreement held no special provision for royalty sharing with traditional healers, an important consideration of ethnobotanical drug discovery. In Suriname, virtually all ethnobotanical knowledge made available to the IGBC project was supplied by healers, rather than from general or popular plant knowledge. While fully supporting local development initiatives, traditional healers expressed a preference for being paid a small amount for their participation as ethnobotanical collecting developed. Since it was locally recognized that the ICBG and the FPF came to Saramaka territory because of the wisdom of local healers, the presence of the ICBG project and especially the funds made available through the FPF, increased the healers' prestige in the community.

Bioprospecting is complicated. It involves the development of drugs by megacompanies in faraway countries, use of alien technologies, complicated and time consuming legal and governmental approval processes and highly complex relationships between a bewildering array of public and private organizations, both foreign and domestic. Some communities need significant help from intermediaries to participate in the process[14]. The lesson from the Suriname case study is how smoothly the bioprospecting process can move along when time and resources have been spent beforehand and when the stated goals of all partners are not in conflict, but interrelated. In Suriname, for example, the drug discovery process creates economic value for biologically diverse organisms, which in turn creates development funds and incentives for conserving biodiversity, satisfying the goals of all stakeholders[15].

Conservation International[16] helped pioneer the use of trust funds to target specific revenues for communities or projects in many countries. In Suriname, the Saramaka have a social and cultural identity distinct from the dominant society. They are geographically and linguistically isolated, which makes them particularly vulnerable to outsiders. Management of the FPF, however, is independent of outsiders, allowing the communities to determine how to use their benefits through their own cultural institutions without outside interference.

The question of individual compensation arises and is dealt with here. Typically, acknowledgment of the contribution of healers from indigenous societies in the discovery of new medicines that have historically benefited humankind is missing. This recognition is important because it enhances the prestige of individual healers within indigenous communities, countries and internationally, just as it does for Western scientists. And formal recognition of the healers' contributing role generates pride in the Saramaka culture and motivates its conservation.

2.3 Nigeria

Today, Africa has one of the highest rates of deforestation in the world. The continent is estimated to have about 216,634,000 ha. of closed forest areas and an annual deforestation rate calculated at about 1%. Located in West Africa, the Federal Republic of Nigeria is the continent's most populous nation. It suffers an annual deforestation rate of 5.0%, compared to a global rate of 0.6%. To contend with such statistics, Nigeria modified its national parks law to attract and encourage conservation funding by establishing biodiversity prospecting requirements.

The Nigerian population, estimated as over 100 million in 1995, includes over 250 distinct ethnic groups, some numbering fewer than 10,000 people. Most are concentrated in specific geographic areas. The Hausa and Fulani (or Peul) in the north, the Yoruba in the southwest, and the Igbo in the southeast comprise about 65% of the population. Traditional leaders, such as chiefs, healers, clan and lineage heads, are a major social force in communities, where numerous community development associations provide fora for citizens to identify, discuss and prioritize problems and to seek means to solve them[17].

The oldest component of the Nigerian health sector consists of traditional healers and birth attendants, who are the *de facto* providers of primary health care in Nigeria, estimated to serve about 80% of the population. Healers provide client-centred, personalized health care that is culturally appropriate, holistic and tailored to meet the needs and expectations of the patient. Since healers share the cognitive understandings and cultural values of those they treat, they serve a function broader and more complex than their medical counterparts in the modern sector.

Nigerian healers are not strictly 'traditional', however. The term 'traditional', in this case, refers to healing practices that have endured over time because of their continuing usefulness. Traditional healers in Nigeria are adaptive, incorporating into their healing repertoire modern components that adhere to their cultural values. Most healers want to learn how to treat illness more effectively, so they are highly motivated to cooperate with the modern health sector[18].

The Bioresources Development and Conservation Programme (BDCP) is a multiethnic international NGO based in Nigeria that builds technical skills so that bioresources are a viable vehicle for improved health care and sustainable development[19]. The BDCP views science and technology as useful tools to be adapted to the local cultural framework, rather than as a modern alternative to the contributions of community members. One of the objectives of the BDCP is to use local bioresources and knowledge to target therapeutic categories for tropical diseases suffered in Nigeria such as malaria, leishmaniasis, and trypanosomiasis.

In 1990, Shaman Pharmaceuticals, Inc. established a research relationship with Nigerian scientific institutions, and the BDCP became the focal point for collaborative research. Shaman is a small San Francisco company that uses ethnobotany as well as isolation and natural products chemistry to discover and develop novel pharmaceuticals[20]. Nigerian scientists proposed initial discussions with healers and traditional leaders to talk about a collaborative relationship with Shaman for bioprospecting. The BDCP visited communities where members had worked for several years. Typically, each community or village state is autonomous, with its own chief and government. Their community decision-making process includes the village chief, his advisors, traditional healers and the elders. Before bioprospecting began, discussions covered topics such as the intentions and goals of the project; how and where the plants would be analysed; and their potential for commercialization and benefit-sharing. After lengthy discussions over a year, the groups felt that Shaman shared a common purpose with them consistent with their cultural values concerning human health and agreed to collaborate through an Agreement of Principles[21].

Four ethnobotanical field expeditions were conducted. By choice of Nigerian collaborators, immediate and medium-term benefits from the expeditions took the form of workshops and training programs on public health, botany, conservation and ethnobotany; support for a medicinal plant reserve; supplies for village schools; botanical collection supplies for a herbarium; laboratory equipment for scientific research on plants that treat parasitic diseases prevalent in West Africa and support for Nigerian scientists to apply modern analytical techniques. Fulfilling company policy, immediate and medium-term benefits, such as those above, totalling over US$200,000 (two hundred thousand dollars) have been distributed through programs to the various stakeholders in the collaboration as the expeditions occur. The company regularly reports laboratory results to participating communities, and general literature on the project is published with both Nigerian and US authors[22].

At the time of its incorporation in 1990 as a for-profit corporation, Shaman founded the Healing Forest Conservancy (the Conservancy), an independent non-profit foundation. The Conservancy was established specifically to develop and implement a process to return long-term benefits to Shaman's collaborating countries and cultures, after a product is commercialized. Through the Conservancy, Shaman will donate a percentage of profits from commercial products to all company collaborators (some 30 other countries and culture groups) for as long as Shaman has a profit. The Conservancy will distribute these long-term benefits, equally, to all the countries and cultures that are Shaman collaborators, regardless of where the plant sample or traditional knowledge that was commercialized originated[23].

On October 20, 1997, in Abuja, Nigeria, the BDCP launched the Fund for Integrated Rural Development and Traditional Medicine (FIRD-TM) during an international workshop on medicinal plants. The FIRD-TM is the vehicle to receive and channel benefits in an equitable and consistent manner from many contributors, including the IGBC, of which the BDCP is a member. Funds are directed to source communities from which commercially useful bioresources and ethnobotanical knowledge is derived. The BDCP facilitated the establishment of the Fund and is its sponsoring entity. The Fund, however, has an independent board composed of leaders of traditional healers' associations, senior government officials, multiethnic representatives of village councils and technical experts from scientific institutions. Diverse culture groups in Nigeria will receive resources from the Fund through traditional healers' organizations and villages consistent with their governing customs. Town associations, village heads and professional guilds of healers are empowered to make decisions regarding use of the funds for projects in their localities. Those funded will follow the criteria of promoting conservation of biodiversity and drug development, as well as the socio-economic development of rural cultures.

At the Fund's inauguration, the Healing Forest Conservancy donated $40,000 to the FIRD-TM for a pilot project, to test the efficiency and efficacy of the trust fund process, using a step-by-step process to assess the feasibility of future trust funds for benefit-sharing. During the inauguration ceremony, the Association of Indigenous Pharmaceutical Manufacturers and the Orange Drug Company of Nigeria pledged additional monies to complement the Conservancy's donation[18].

The pilot Conservancy project generated a 21 page constitution, which supplies a legally enforceable mechanism, under domestic law, for a trust fund, written *pro bono*, by Morrison & Foerster, Ltd. in Washington DC. Under the Constitution, trustees of the fund, as a collective entity, may institute legal proceedings in their capacity to achieve trust fund objectives. The Conservancy Constitution is a template to use for the benefit-sharing actions of the Conservancy in many different countries that is flexible enough to respond to unique conditions in countries where Shaman collaborates. It supplies a legal mechanism to disburse widely financial resources, over a long time frame, and within varied sectors of society.

The Constitution provides for the distribution of benefits to all stakeholders in accordance with the following guidelines:

1 At least 50%, but not more than 70%, of available funds shall be distributed to traditional healers' organizations and community development funds.
2 At least 10%, but not more than 15%, of available funds shall be distributed to national universities and other national institutions that share a commitment to the aims and objectives of the Fund.
3 At least 10%, but not more than 15%, of available funds shall be distributed to the sponsoring entity for its furtherance of conservation and development activities.

After ten years and $170 million, in early 1999 Shaman Pharmaceuticals abandoned attempts to take any of its discoveries through the Food and Drug Administration regulatory process, as future time and costs for additional clinical trials proved prohibitive[24]. ShamanBotanicals.com was incorporated as a private, wholly-owned subsidiary of Shaman Pharmaceuticals to leverage the company's research and development by launching its first botanical dietary supplement, SB Normal Stool Formula on August 1, 1999[25]. The product delivers a standardized extract from the sap of Sangre de Drago, the Croton lechleri tree, to prevent fluid loss and promote normal stool formation in the intestine to those suffering from side effects of HIV therapies, cancer therapies and irritable bowel syndrome[26]. ShamanBotanicals.com is selling SB Normal Stool Formula through existing retail channels, including mass market (food, drug and mass stores), multi-level marketing and the Internet. This and future products will be differentiated by their origins, their quality and Shaman's continuing commitment to benefit-sharing, sustainability and conservation.

For many years, Shaman invested in studies on managing and harvesting Sangre de Drago in Latin America, as the source of a future product. Ecological and distribution studies, as well as studies on yield and sustainability totaling US$1 million have been carried out at sites from Mexico to Paraguay[27].

A major lesson of this case study is the time, costs and risks associated with drug discovery, a burden shouldered primarily by the company, but with critical implications for benefit-sharing to source countries and culture groups. Spreading the risks and benefits among all stakeholders increases opportunities for benefits and lessens risk. Royalties may never materialize due to the tremendous costs, long time frame, unpredictability and volatility of the market and the many other potential pitfalls of drug discovery. Some sort of up-front benefits, monetary or non-monetary, as well as 'milestone' payments such as those that went to Nigeria, are essential. Shaman Pharmaceuticals, Inc., was restructured due to costs of future clinical trials for the company's first drug, which consequently was launched as a dietary supplement, not a pharmaceutical. In the case of Shaman, however, the company's benefit-sharing principles are still intact and considerable revenues can be generated from the botanical income stream, while pharmaceutical research continues. In 1997, for example, the U.S. dietary supplement market for herbals or botanicals was nearly $4 billion, with a $5 billion 1998 projection and compounded yearly growth rate of 15–25%.

The BDCP goals of discovering and marketing bioproducts for diseases important to Nigeria is itself a lesson in opportunities brought about through bioprospecting. Too often, when the importance of biodiversity conservation is discussed for its value to human health, it refers to the health of residents of industrialized nations. The attitude is that biodiversity must be preserved to enlarge the pharmacopoeia of Western medicine, which provides therapeutics primarily for Western societies. Less discussed is the vitality of biodiversity to the health of 80% of the world, populations that depend solely on medicinal plants for their primary health care. Preserving biodiversity for the benefit of human health means preserving it for those in the tropics already using it, as well as for distant populations that may know it only in some refined or synthetic form at some unspecified future date.

This case study also demonstrates how biodiversity-rich yet financially-poor tropical nations, such as Nigeria, can increase conservation and research funding and gain valuable training and technology. Since the economic values of biodiversity are seldom accountable in market valuations, strategies like bioprospecting bring added financial support for conservation. Countries can choose whether to supply natural products in the form of extracts, rather than raw unprocessed material, to foreign investors or to establish their own medicinal plant, phytochemical or pharmaceutical industry.

3 Lessons and conclusions

3.1 Monetary and non-monetary benefits

Benefit-sharing can take financial, conservation, social and scientific forms and must be decided during the prior informed consent process, before any bioprospecting permits are issued. As case studies demonstrate, opportunities for financial compensation include up-front payments and medium term benefit-sharing as research progresses. Many companies offer stakes in equity, profit sharing and joint venture opportunities. Royalties occur only if and when a drug is marketed, but it is risky to rely only on resources that may never materialize.

Cases that supplied support for conservation activities illustrated how cultural, as well as environmental, conservation can be accomplished. Acknowledging the contribution of healers supplies proof of the continuation of a traditional profession. It promises an economic future from preservation of the cultural patrimony by which healers train and pass on their heritage to future generations of practitioners.

In Nigeria, significant social benefits include development of low cost phytomedicines and conserving the primary source of health care in the country. Participation of nationals in research on tropical diseases through technology transfer, training, and joint research publications builds the technical capacity of nationals during the discovery process.

3.2 Training and technology transfer

The Kani benefit from training and sourcing agreements to cultivate high-value raw material for processing into phytomedicines. Such training supplies income-producing opportunities to rural communities that are best located to manage, collect, and protect species. Inventories that describe local use can be undertaken in the local language, which protects ethnobotanical information from outsiders and leaves a documentation of cultural knowledge within the community. Training also generates employment by outsiders who are interested in bioprospecting and increases the capability of gene-rich countries to perform these services.

National governments benefit from training programs by gaining a technological infrastructure for science and commerce, yielding jobs and taxes. Fees can be charged to outsiders with commercial or research interests, allowing debt-ridden tropical nations to forgo short-term profits from logging, cattle-grazing, and monoculture development projects that destroy forests. There are several entry levels, including: collecting procedures, inventories, bioassays, re-collection, harvesting, herbarium specimen storage, taxonomic identification, cataloguing and management of collections. Training increases the capacity of biodiversity-rich countries to assess and evaluate their resources, to generate biological databases and to enter into the natural products industry if they choose to do so.

3.3 Trust funds

While trust funds are now widely recognized as the method of choice to distribute benefits, it is important to acknowledge the consent and support of each host country government for the activities of the fund. This minimizes the risk that trust funds are perceived as a threat to the sovereign right of the host country government to exploit its own natural resources. However, this acknowledgment should not be viewed as undermining the independence of trust funds and their autonomous operating authority as a non-governmental organization. Their precise legal status will depend on laws of the host country where the fund is established. It is important that the fund be able to obtain the benefits normally associated with non-profit and charitable entities, such as exemption from taxation, so it can offer the added value of attracting other sources of funds. Its NGO status reinforces the independence of the fund from the host government and permits it to participate in other NGO fora.

Acknowledgements

I gratefully acknowledge the critical input to this article from Marianne Guerin-McManus of Conservation International; Ted Green, private consultant; Maurice Iwu, the BDCP and its collaborators in Nigeria; Lisa Conte and Steve King from Shaman Pharmaceuticals, Inc.; and Tom Mays of Morrison and Foerster, LLP., Stas Burgiel of Bionet, and all other members of the Nigerian Trust Fund Pilot Project Committee. Generous funding from ShamanBotanicals.com, the Jocarno Fund, The Nelson Talbott Foundation and numerous individual contributors to the Healing Forest Conservancy made this publication possible.

Notes

1 United Nations International Convention on Biological Diversity. (1992) [Online] Available: http://www.biodiv.org

2 Reid, W. Laird, S., Meyer, C., Gamez, R., Sittenfeld, A., Janzen, D., Gollin, M. and Juma, C. (1993) *Biodiversity Prospecting*, World Resources Institute: Washington, DC.

3 ten Kate, K. and Laird, S. (1999) *The Commercial Use of Biodiversity: Access to Genetic Resources and Benefit-Sharing*. Earthscan Pubs. Ltd.: London.

4 Moran, K. (1977) 'Returning benefits from ethnobiological drug discovery to native communities', in *Biodiversity and Human Health*, pp.243–263. Island Press: Washington, DC.

5 Biotechnology in the Global Economy (1999) [Online] Available: http://www.cid.harvard.edu/cidbiotech/homepage.html

6 Case Studies on Access and Benefit Sharing. [Online] Available: http://www.biodiv.org/chm/techno/gen-res.html
7 Anuradha, R.V. (1998) 'Sharing with the Kanis: a case study from Kerala, India', *Secretariat to the Convention on Biological Diversity*, Fourth Meeting of the COP, Bratislava, Slovakia, May.
8 *Science* (1999) March 12, Vol. 283, pp.1614–1615, News of the Week.
9 Personal communication from Dr. P. Pushpangadan. (1998).
10 Moran, K. (1977) 'Returning benefits from ethnobiological drug discovery to native communities', in *Biodiversity and Human Health*, pp.243–263. Island Press: Washington, DC.
11 King, S.R., Meza, E., Carlson, T., Chinnock, J., Moran, K. and Borges, J. (1999) 'Issues in the commercialization of medicinal plants', *Herbalgram*, No. 47, pp.46–51.
12 Guerin-McManus, M. *et al.* (1998) 'Bioprospecting in practice: A case study of the Suriname ICBG project and benefits sharing under the Convention on Biological Diversity', *Secretariat to the Convention on Biological Diversity*, Fourth Meeting of the COP, Bratislava, Slovakia, May.
13 Price, R. (1976) *The Guiana Maroons*, Baltimore. The Johns Hopkins University Press.
14 Green, E.C. (1998) 'Ethnobotany, intellectual property rights and benefit sharing: The ICBG project in Suriname', *Paper presented at SfAA Symposium on Coming to terms with IPR*, San Juan, Puerto Rico, April 23.
15 Mays, T., Duffy-Mazan, K., Cragg, G. and Boyd, M. (1997) 'A paradigm for the equitable sharing of benefits resulting from biodiversity research and development', in *Biodiversity and Human Health*, pp.267–280. Island Press: Washington, DC. Earth Negotiations Bulletin (1999) *Summary report of the experts' panel on access and benefit-sharing*. International Institute for Sustainable Development: Washington, DC.
16 Conservation International (CI) [Online] Available: http://www.conservation.org
17 Iwu, M.M. (1996) 'Biodiversity prospecting in Nigeria: seeking equity and reciprocity in intellectual property rights through partnership arrangements and capacity building', *Journal of Ethnopharmacology*, Vol. 5, No. 1, pp.209–219.
18 Moran, K. (1988) 'Mechanisms for benefit-sharing: Nigerian case study for the Convention on Biological Diversity', *Secretariat to the Convention on Biological Diversity*, Fourth Meeting of the COP, Bratislava, Slovakia, May.
19 Bioresources Development and Conservation Programme (BDCP). [Online] Available: http://www.Bioresources.org
20 Shaman Pharmaceuticals, Inc. [Online] Available: http://www.shaman.com
21 Carlson, T.J., Iwu, M.M., King, S., Obialor, C. and Ozioko A. (1997) 'Medicinal plant research in Nigeria: an approach for compliance with the Convention on Biological Diversity', *Diversity*. Vol.13, No.1, pp.29–33.
22 Richter, K.R., Carlson, T.S.J. (1998) 'Reporting biological assay results on tropical medicinal plants to host country collaborators', *Journal of Ethnopharmacology*, Vol. 62, pp.85–88.
23 The Healing Forest Conservancy [Online] Available: http://www.shaman.com/Healing_Forest.html
24 *The Economist* (1999) 'Shaman Loses its Magic', 21 February.
25 ShamanBotanicals.com [Online] Available: http://www.ShamanBotanicals.com
26 Holodniy, M., Koch, J., Mistal, M., Schmidt, J.M., Khandwala, A., Pennington, J.E. and Porter, S.B. (1999) 'A double-blind, randomized, placebo-controlled, phase II study to assess the safety and efficacy of orally administered SP-303 for the symptomatic treatment of diarrhea in patients with AIDS', MS accepted for publication in the *American Journal of Gastroenterology*.

27 King, S., Menza, E., Ayala, F., Forero, L.E., Pena, M., Zak, V. and Bastein, H. (1997) '*Croton lechleri* and the sustainable harvest and management of plants in pharmaceuticals, phytomedicines and cosmetic industries', in *International Symposium on Herbal Medicines*, (Eds.) Wozzniak, D., Yuen, S.S., Garrett, M. and Shuman, T.K. *International Institute for Human Resources Development*, College of Health and Human Services, San Diego State University: San Diego, CA.

54

INDIGENOUS PEOPLE INCORPORATED?

Culture as politics, culture as property in pharmaceutical bioprospecting

Shane Greene

Source: *Current Anthropology* 45(2) (2004): 211–237.

Abstract

The ongoing debate over indigenous claims to intellectual and cultural property reveals a series of indigenous strategies of mobilization that both appropriate from and work against the logic of the market. Of particular significance in this regard are the various indigenous strategies used in contemporary pharmaceutical bioprospecting activities to address claims to traditional medical knowledge as cultural property. This article presents field data on a controversial ethnopharmaceutical project among the Aguaruna of Peru's high forest and offers a comparative analysis of the outcomes with attention to several other cases in and beyond South America. In particular, questions are raised about the forms of legitimating authority in the burgeoning international indigenous movement, the role of NGOs, researchers, bureaucracies, and corporations in this process, and the dilemmas that emerge from the politicization and privatization of indigenous culture and identity.

The long-standing Euro-American tradition that imagines indigenous peoples as forever teetering on the brink of cultural collapse and demographic destruction, whether in colonial ideologies of the civilizing process, anthropological scholarship on acculturation, or development policies that project an inevitable triumph of modernization over tradition, has proven profoundly misleading. The native subjects of Europe's imperial expansion, the ethnic and tribal pockets still existent within modern nation-states, and the tradition-laden "obstacles" to capitalist development have demonstrated themselves to be actors of world-historical proportions. Being indigenous in our era has become much more than a question of tenuous cultural survival or merely the discursive product of colonial and modern imaginations. It is a historically validated subject position that is accompanied by a conscious strategy of effective, if still highly contested and multiply configured, political, cultural, and historical action for significant populations across the world.

One of the more remarkable phenomena of the past few decades in this regard has been the increasingly visible efforts by indigenous peoples from different world regions to formalize their own forms of political struggle and representation at local, national, and global levels (see Chirif, García, and Smith 1991, Brysk 2000, Brown 1993, Muehlebach 2001, Kearney and Varese 1995, Smith 1996, Friedman 1999, Warren and Jackson 2002, Van Cott, 1994, Warren 1998, Smith and Ward 2000, Ramos 1998, Greene 2002, Ballón Aguirre 1988, Albó 1991, Urban and Sherzer 1991, Niezen 2003, Montejo 2002, Ewen 1994). In the process they have formed an institutional network of indigenous activists, organizations, and advocates that is global in scope. The most significant and internationally visible forum in which indigenous spokespeople come together is the United Nations Work Group on Indigenous Populations, which conducts debates on the definition of "indigenous" and drafts legislation for rights to territory, self-determination, the environment, and culture (Muehlebach 2001, Niezen 2003). The activities of the UN Work Group and the increasing visibility of indigenous movements are often cited as evidence of indigenous peoples' having gained a "voice" in national and international politics (Muehlebach 2001, Montejo 2002, Ewen 1994). The idea of a "voice" immediately raises the question of who is speaking for whom and to whom, and therefore of central importance to these emergent forms of indigenous mobilization are the issues of indigenous self-representation and the role of persons and organizations that act as mediators between external interests and local constituencies.

Political and cultural mediation is nothing new to the encounter between indigenous peoples and the projects of European colonizers or modern nation-states. Colonial administrations often relied heavily on native "chiefs" of one kind or another to enforce colonial policies on local subjects. The role of native *kurakas* in Spanish America and the system of indirect rule employing tribal chiefs in Africa are classic examples (see Rasnake 1988; Stern 1982; Comaroff and Comaroff 1991:255). Nor is the study of indigenous political mediation new to the anthropology of the modern nation-state context. Eric Wolf's (1956) article on native leaders in Mexico as political, economic, and cultural "brokers" between their home communities and the national society is a classic in this respect.

Some of the more notable controversies in contemporary anthropology suggest that indigenous self-representation, global indigenous mobilization, and the mediation involved in such intercultural politics are still very much at the heart of recent anthropology. David Stoll's (1999) contentious rereading and revision of the Nobel prizewinner Rigoberta Menchú's account of life as a Maya during Guatemala's internal war landed him in a heated debate about his authority to question representations of her as an internationally revered indigenous icon. More controversial still is Patrick Tierney's (2000) critical account of several researchers' work with the Venezuelan Yanomami. Tierney's book includes details about the anthropologist Napoleon Chagnon's attempt to call into question the act of political representation of one of the group's most outspoken leaders, Davi Yanomami, whom he considered a "pawn" of environmental NGOs and his own anthropological rivals (see Graham 2002).

Ultimately these recent controversies reveal that contemporary indigenous spokespeople and representative organizations are negotiating their role as mediators and representatives of indigenous collectivities with a whole host of other nonindigenous peoples and organizations specific to this historical moment. While anthropologists and missionaries have long been key interlocutors with indigenous peoples, they are hardly alone in this emergent debate—productive at times of somewhat uncomfortable encounters—over the nature of indigenous activism and what constitutes legitimate representation. Environmental and human rights nongovernmental organizations (NGOs), state and international bureaucracies, development and funding agencies, corporations, researchers, and activists all now have their own stakes in negotiating the politics of indigenous representation. Indeed, global environmental activism, the objectives of transnational NGOs, sustainable-development initiatives, and even the "green capitalism" practiced by ostensibly eco-conscious corporations have emerged as powerful agendas with respect to indigenous peoples and their forms of representation. Through such environmentally friendly contacts some indigenous leaders and organizations have successfully communicated their plight to global audiences and powerful international funding and development institutions (Conklin and Graham 1995, Smith 1996, Turner 1993, Carneiro da Cunha and Almeida 2000). However, many point to the pitfalls of seeing indigenous peoples as inherently environmentally conservative and thus stereotyped as "ecologically noble savages" (see Redford 1990; Orlove and Brush 1996:334).

Two related phenomena are part and parcel of this greater and more visible indigenous mobilization and its mediation by nonindigenous interlocutors: the politicization of culture and its treatment as property. In an age in which anthropologists and other scholars are increasingly critical of the arbitrary boundaries implied by the "culture" concept (see Gupta and Ferguson 1992), indigenous peoples publicly embrace such boundaries, citing the virtues of indigenous cultures in ways that portray them as relatively fixed and indentifiable wholes associated with delimited ethnic collectivities/territories and specific histories (see Fischer 1999). Indeed, claims to distinct, delimited cultures give validity to the historically specific subnational ethnic identities on which much global indigenous politics is based. Indigenous representatives increasingly speak about their distinct cultures as objects around which political struggle is organized in their efforts to negotiate the legacies of colonial occupation, incorporation into modern national projects, and the pressures of globalization.

The politicization of indigenous culture is accompanied by a tendency to define it as property. Indigenous representatives across the world now commonly speak about themselves not merely as representatives of distinct cultures but also as part-owners of collectively propertied cultures (Brown 2003).[1] Contemporary indigenous cultural property claims range from the material (human remains, artifacts, significant sites) to the intangible (sacred symbols, music, cultural heroes, traditional plant knowledge). Struggles to make culture property represent one of the most active spheres of indigenous political and, I will argue, economic action. Indeed, in some instances cultural property claims have surpassed in importance the concerns over territory and land rights that have historically been central to indigenous mobilization (Muehlebach 2001). Passage of the Native American Graves Protection and Repatriation Act in 1990 gave Native Americans unprecedented rights to reclaim human remains and material cultural artifacts in the possession of U.S. museums (see Haas 1996, Brown 1998). Significant struggles over intangible cultural property are occurring as well. In an extraordinary legal settlement the administrators of the Crazy Horse Estate and the Rosebud Sioux tribe of South Dakota successfully challenged a brewing company's use of the name "Crazy Horse" in the marketing of a malt liquor product. SBC Holdings issued a formal apology and settled with the Native American plaintiffs in 2001 (Gale 2001). Halfway across the world in Southern Africa, the San peoples, now enveloped by a number of ethnotourism enterprises, presented a very similar complaint about the misuse of their cultural property. A group of !Xûn San denounced a South African photographer and a local tourist lodge for the unauthorized use of a photograph that featured a now-deceased leader of the San in tourist brochures (Sylvain 2002:1080).

These examples suggest that at one level indigenous people are struggling to monitor and assert more direct control over their cultural property by challenging and politicizing its use by nonindigenous actors. Their own economic motivations seem to be a secondary concern. The Crazy Horse case, for example, involved not monetary damages but culturally appropriate compensation including racehorses, braids of tobacco and sweetgrass, and Pendleton blankets (Gale 2001). These strategies to reclaim tangible and intangible elements of their culture from nonindigenous private parties (e.g., corporations, museums, researchers, tourist agencies) and effectively remove them from the public domain revolve primarily around the apparent sacrilege and defamation that the misuse, possession, display, and/or commercial exploitation of these elements entails.

The contemporary search for scientific-commercial utility in the world's biological resources known as bioprospecting confronts a similar set of issues with respect to indigenous claims to traditional medicinal knowledge as cultural or intellectual property. In this sphere of practices and politics a set of dynamics is emerging that requires as much attention to its economic aspects as to ever-present concerns about sacrilege and defamation. In this paper I hope to raise a series of provocative questions about what is at stake when indigenous representatives take their own economic aspirations seriously in negotiations over the potential market value of their people's traditional medical knowledge and the kinds of new problems that emerge as a result. In this sense, I depart from a very old Durkheimian structural-functionalist tradition in anthropology. Instead of focusing on the principles of lineage, marriage, descent, and so on, that structure the corporate kin group, I am interested in the contemporary political-economics that give rise to entrepreneurial and identity-based strategies of indigenous incorporation in cultural property claims. What political and economic dilemmas arise from the treatment of indigenous medical knowledge as property in the current era of drug bioprospecting and the pursuit of sustainable development? And what kinds of dilemmas are encountered when exclusive cultural property owners must be legally and contractually defined in a terrain that involves the multiple agendas and differing expectations, of indigenous representatives, drug corporations, state institutions, academics, NGOs, and activists of all sorts?

Drug bioprospecting and indigenous property claims to traditional knowledge

The 1990s witnessed a resurgence of public and private interest in the potential of plant-based sources for new pharmaceutical products after a steady decline in such research since the mid-twentieth century. The United Nations Convention on Biological Diversity, ratified in 1992, is the primary international regulatory agreement concerned with bioprospecting. In an effort to address North/South, developed/developing inequities in the international biotechnology industry, the convention recognizes each state's sovereign control over access to biological resources found in its national territory. This provides poorer nation-states with considerably more leverage in negotiating the terms for the extraction of biological resources from their territories—resources that were previously considered part of a universal public domain or common heritage of humankind. Furthermore, article 8j of the convention suggests that bioprospecting efforts that draw on traditional uses of biological resources and result in successful commercial ventures should entail protection for and benefit sharing with the indigenous or local populations whose knowledge contributes to biologically engineered products.

The possibility of future drugs based on traditional plant remedies is in large part rooted in historical examples—aspirin, antibiotics, and quinine being only a few of the more common. The idea of medicinally valuable plants is often invoked with images of indigenous groups, their traditional medical practices, and the miracle drugs that presumably lie undiscovered in the depths of the tropical forest (see Goering 1995, Dye 2000). However, it is unclear whether the future of pharmaceuticals will depend significantly on bioprospecting, much less on the input of the medical knowledge of indigenous peoples. Recent prognoses (see Artuso 1997, Aylward 1995, Moran, King, and Carlson 2001, *Economist* 1999, Albers-Schönberg 1995, O'Conner 2000, Macilwain 1998) are not nearly as optimistic as those of the early nineties (e.g., Pollack 1992).

In response to current pharmaceutical bioprospecting activities and the UN Convention, many now conceptualize traditional knowledge as a novel form of collective intellectual property (see Rosenthal 1997; Cox and Balick 1994; Elisabetsky 1991; Plotkin 1993; Lewis et al. 1999; King and Tempesta 1994; Boyle 1996; Tobin 1999b, 1994). Rosemary Coombe (1999, 1998) is critical of this move because intellectual property is historically associated with an ideology of possessive individualism and romanticized individual authorship, a peculiar feature of and for capitalist societies. She is nevertheless cognizant of the political-rhetorical potential of intellectual property claims when they are deployed by historically subjugated indigenous groups eager to gain some control over the market forces transforming their societies (1998: chap. 5). Indigenous activists offer differing opinions: some enthusiastically promote the idea of traditional knowledge as intellectual property while others reject it as an imposition of Western categories on their own cultural forms (see Brush 1999, Coombe 1999). Environmental NGOs, also crucial actors in the debate over bio-prospecting, often position themselves as a counter-balance to the exploitative tendencies of transnational capital interests and as allies of indigenous, local, and Third World peoples more generally.

Darrell Posey was one of the first anthropologists to engage and promote the idea of traditional knowledge as indigenous groups' collective intellectual property (see Posey 1990).[2] Other anthropologists have since cited problems with the arbitrary exclusivity and enclosure implied in intellectual property, its history as a legal mechanism of capitalist societies, the incompatibility of collective and culturally fluid noncapitalist knowledge systems with individualized, exclusive Western property laws, and the threat that such a move represents to the public domain of knowledge (see Brush 1994, 1996, 1999; Posey and Dutfield 1996; Brown 1998, 2003; Strathern 1996). Dove (1995), Coombe (1999), and Greene (2002) make the related points that many indigenous peoples are plagued more by problems of territorial invasion, loss, dislocation, and marginalization by their own state elites than by the appropriation of their traditional knowledge by bio-pirates from the North.

Much of the early literature on bioprospecting and traditional knowledge as intellectual property is rife with polemics and ideological positioning. Further debate should be tempered with more extensive evaluation of actual cases, the complicated negotiations involved, and, as Cori Hayden (2003, n.d.) shows, the increasing avoidance by bioprospectors of such negotiations. On the basis of my own field studies and related documentation, I examine a controversial example of pharmaceutical bioprospecting negotiations involving the Aguaruna people of the Peruvian Amazon. The Amazon is a particularly good context in which to begin such an evaluation, given its central position in much of the contemporary debate over bioprospecting, indigenous and environmental activism, and cultural property protection (see, e.g., Conklin 2002). The Aguaruna case will be compared with related cases in and beyond the Amazon in an attempt to formulate some preliminary conclusions about the dilemmas of indigenous representation, indigenous incorporation, and contemporary claims to culture as property.

ICBG–Aguaruna negotiations, round 1

One of the most commonly cited examples of renewed interest in bioprospecting is the public grants program entitled the International Cooperative Biodiversity Group (ICBG).[3] The ICBG program is the product of a workshop in 1991 sponsored by the National Institutes of Health (NIH), the National Science Foundation (NSF), and the United States Agency for International Development (USAID). The NIH emerged as the administrator of the program and announced a competition for large grants for research into the pharmaceutical potential of international biodiversity. According to Rosenthal (1997: 281–82), the ICBG program is essentially "experimental," since biodiversity's pharmaceutical utility is far from certain, and is designed to operate with goals that are clearly responsive to the principles outlined in the UN Convention.[4]

Two rounds of ICBG grants, one in 1993 and one in 1998, were awarded to a total of eight research teams with principal investigators from U.S. universities or health institutes (see the ICBG web site: www.nih.gov/fic/opportunities/icbg.html). ICBG grants are based on a collaborative funding, research, and mutual-benefits relationship between U.S. and developing-country institutions, commercial partners, and, in a few cases, specific indigenous/local communities. The ICBG Request for Applications (1997:II) promotes "active participation" of corporate-sector sponsors, effectively obliging potential grant recipients to secure private partners.

In 1993 a team from Washington University headed by Walter Lewis, a career ethnobotanist, won an ICBG award for the collection of plant samples in collaboration with the Aguaruna, a sizable Amazonian group (population 45,137 in Peru's 1993 census) that inhabits titled lands in Peru's high jungle region near the Ecuadorian border. Long reputed to be dedicated warriors, the Aguaruna have been at the forefront of Amazonian indigenous political organizing in Peru since the 1970s. The vast majority of the 187 individually titled Aguaruna communities are affiliated (with varying degrees of loyalty and participation) with local Aguaruna-run organizations or in some cases organizations comanaged with other Amazonian ethnic groups (the Huambisa and Chayahuita peoples). As of this writing I can identify 13 distinct organizations of this type, the most recent having surfaced in 2002. The local organizations typically coordinate with or are claimed as local affiliates by one of two national indigenous confederations based in Lima, the Peruvian capital. All of this makes for a complicated set of political alliances and divisions among the larger Aguaruna collective, as the ICBG researchers soon discovered.

Under the ICBG protocols, Washington University's research was carried out as a partnership with Peruvian and indigenous institutions. Washington University, the Universidad Peruana Cayetano Heredia (UPCH), the Museo de Historia Natural de la Universidad San Marcos, and a category termed "Aguaruna People" constituted the four partners. All three institutions participated in plant collections with the Aguaruna, with Washington University and UPCH collecting for use in biological screening and the museum for an inventory of Peru's biodiversity. On the basis of prior contacts with Monsanto Corporation in the St. Louis area and in order to comply with the ICBG program's "encouragement" of private-sector collaboration, Washington University secured the participation of G. D. Searle & Co. (then Monsanto's pharmaceutical division).

According to Lewis (interview, 2000), the Organización Central de Comunidades Aguarunas del Alto Marañon (OCCAAM) had been considered as a potential indigenous partner in the original ICBG grant application, but having received the award the ICBG team entered into negotiations with the Consejo Aguaruna Huambisa, historically a much more influential indigenous organization in the region. Lewis (interview, 2000) says that the NIH referred him to the Consejo and its internationally famous founder and president, Evaristo Nugkuag, but the current director of the ICBG program, Josh Rosenthal (personal communication), disputes this. In any case, the Consejo was clearly the more prominent institutional actor among the Aguaruna. Until recently it was commonly presumed to represent a large proportion of the Aguaruna population, and without question it has a more illustrious history than the smaller, less visible Aguaruna organizations. Furthermore, Nugkuag's involvement in local, national, and international indigenous organizing had equipped him with substantial experience in international networking and won him several prestigious prizes. His fame was, of course, not without substantial controversy in the heated politics of indigenous representation.[5]

In April 1994 the ICBG team signed a very simple letter of intent to collaborate with the Consejo Aguaruna Huambisa in which the indigenous organization was promised an up-front annual payment for plant collections and, in relatively ill-defined terms, a share of future royalties. Having secured a preliminary agreement with the Consejo, Washington University researchers returned to the United States and formalized the terms of Searle's participation with a license option agreement that made Washington University responsible for delivering the annual payments to the Consejo and any future royalties to be shared among the four partners. Under this agreement and the letter of intent signed with the Consejo, Washington University essentially became the legal intermediary between the U.S. pharmaceutical interest and all the Peruvian partners (fig. 1).

When the Consejo Aguaruna Huambisa heard of the separate agreement between Washington University and Searle, it began stalling the research and demanding more information about the content of the agreement. As Nugkuag reasons, "They [the ICBG team] were not very clear. The earnings [i.e., the future royalty scheme], logically, are established between Monsanto and Washington University. And we were not included in that agreement, thus, our claim [to share in the earnings] is in vain" (interview, 2000). With these reservations about the potential benefits and the separate agreement, the Consejo consulted with a North American NGO, Rural Advancement Foundation International (RAFI), which acts as a watchdog with regard to bioprospecting activities.[6] It was through contacts at RAFI that the Consejo obtained a copy of the license agreement (Evaristo Nugkuag, interview, 2000). RAFI published the royalty rates on the World Wide Web and accused Searle and the ICBG team of being "bio-pirates" with the intention of exploiting Aguaruna knowledge in the interest of corporate gain (RAFI 1994). During the summer of 1994 Brendan Tobin, an Irish-born Peruvian-naturalized

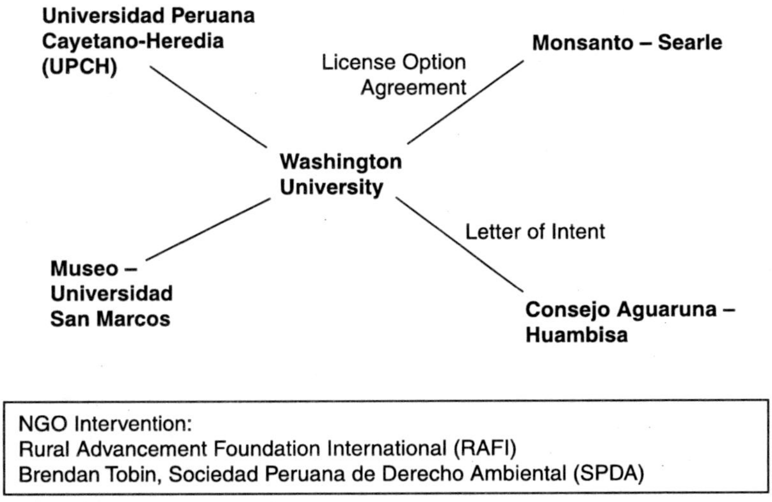

Figure 1 Agreement structure of the first round of ICBG-Aguaruna negotiations.

lawyer working in an NGO called the Sociedad Peruana de Derecho Ambiental (SPDA), received a copy of the agreement and then requested a copy of the letter of intent from the Consejo (Brendan Tobin, personal communication). Tobin objected to the legal inconsistencies between the letter of intent and the agreement, criticized the royalty rate as too low, and judged the ICBG-Searle research "directly linked to the furtherance of the economic and development interests of U.S. industry" (1994:11).

The idea that the Consejo was being offered a skimpy benefit package thus arose from consultation with the two NGOs. It had in fact already signed the letter of intent and accepted $10,000 as part of an annual payment specified in the letter as compensation for collections (see Lambrecht 1998). The two NGO opinions advocated more direct and transparent negotiations between Searle and the Aguaruna (see Tobin 1994, RAFI 1995). Both RAFI's polemical attack on the ICBG program and Tobin's legal analysis assumed that bioprospecting with Aguaruna traditional knowledge was of great importance to Searle and therefore the flaws of the agreement structure represented an attempt to cheat the Aguaruna out of highly valued traditional knowledge.

With the Consejo's concerns unresolved, the ICBG team entered the field in late 1994. It is impossible to verify exactly how and where it collected plants during this conflictive period. Lewis (interview, 2000) maintains that the collections were made in conjunction with local Ministry of Agriculture officials and largely without the use of native informants in the hills around a non-indigenous settlement called Imazita. Nugkuag (interview, 2000) says that the ICBG researchers made a critical mistake in choosing to work with the Ministry of Agriculture and that "without having authorization to enter into communities with the community chiefs they went astray in order to collect orchids. They collected other species of medicinal plants in what could be called a discrete fashion." Lewis (personal communication) adamantly denies this latter claim. He says that of the roughly 300 samples collected during the disputed period, only about 10 samples were collected with Aguaruna informants and that these came from the home community of the OCCAAM, from which the researchers had received consent.

In early 1995, as its distrust of the ICBG team's field activities grew, the Consejo withdrew from the ICBG project. The ICBG team returned to the Aguaruna area to talk to the leaders of OCCAAM in the hope of finding an indigenous partner that would be receptive to the research. OCCAAM leaders showed support for the project and soon signed onto the ICBG program with a separate and much more elaborate agreement than the original letter of intent. On learning of the new agreement, the Consejo sent a strongly worded letter, signed by various Aguaruna community chiefs and Consejo leaders, to officials at the NIH and Washington University protesting the ICBG activities. It made the letter public through its contacts at RAFI, who immediately translated it and posted parts of it on its web site (see RAFI 1995). The letter alleged, among other things, that Washington University had denied the Consejo adequate information about the license agreement and entered Aguaruna territory without authorization and vigorously demanded that the project be terminated. Significantly, it also implied that OCCAAM did not exist (referring to it as "the phantom group"). The organizational efforts of OCCAAM (initiated around 1975) had begun prior to the Consejo's (initiated around 1977), however, and the relation between the two organizations had been conflictive from the beginning.

In its web-site publications RAFI portrays the Consejo Aguaruna Huambisa as the sole organizational representative of Aguaruna people. Its report on the ICBG conflict with the Consejo states that "without consultation with or approval from indigenous people, Washington University researchers unilaterally decided to initiate collection of samples and ethnographic material (to be provided to chemical giant Monsanto) in remote native communities in northeastern Peru" (RAFI 1995:5). They make no mention of the ICBG's relations and subsequent written agreement with OCCAAM, instead presenting "the Aguaruna" as a single homogeneous mass unproblematically represented by the Consejo—a representation that does not at all reflect contemporary or customary Aguaruna reality. OCCAAM clearly suffered from international invisibility and a lack of transnational NGO contacts while the Consejo did not. But the tide was about to change.

Although the desire to have the project removed from Aguaruna territory did not materialize, the Consejo's protest letter and its translation/publication by RAFI for an English-speaking activist audience proved to be a useful weapon. ICBG program officials at the NIH took the protest seriously, launched their own investigation, and insisted that the ICBG researchers return all the samples collected during the period in question to Peru. The NIH also insisted that the ICBG researchers clarify the terms of consent to the project with legitimate representatives of Aguaruna communities if they wished to continue with the grant.[7] A new round of negotiations was in order.

ICBG–Aguaruna negotiations, round 2

In response to the Consejo's protest letter and in the hope of salvaging the project, an Aguaruna named Ricardo Apanú involved in OCCAAM (who later became the ICBG field coordinator) began to organize a defense of the ICBG team's research activities with the support of two other Aguaruna organizations, the Federación de Comunidades Nativas Aguarunas del Río Nieva (FECONARIN) and the Federación Aguaruna del Río Domingusa (FAD). In April 1995 these three organizations signed, stamped, and sent a written rebuttal to the recipients of the Consejo protest letter. The "Memorial" clarified that the ICBG researchers had worked appropriately with the leaders of OCCAAM and its affiliated communities. It also harshly criticized Evaristo Nugkuag, forcefully rejected his "representativeness" of the Aguaruna people ("desconocer su representatividad"), and even questioned the authenticity of the protest letter.

In a further effort to turn the political tides in their favor, the three local Aguaruna organizations put the ICBG researchers in contact with a national-level actor: the Confederación de Nacionalidades Amazónicas del Perú (CONAP). Based in Lima, CONAP is a multiethnic Amazonian indigenous confederation to which OCCAAM, FECONARIN, and FAD claim affiliation, and its president is of Aguaruna origin. Initial discussions culminated in a large meeting in December 1995 in Santa María de Nieva, a small river port in Aguaruna territory. In the absence of the Consejo Aguaruna Huambisa and several other Aguaruna organizations, CONAP convoked a meeting of some community and organizational leaders, among them leaders and delegates of the three organizations already interested, two other Aguaruna organizations, the Organización de Desarrollo de las Comunidades Fronterizas del Cenepa (ODECOFROC) and the Organización Aguaruna del Alto Mayo (OAAM), and a Huambisa organization, the Federación de Comunidades Nativas del Río Santiago (FECONARSA).

Also present at the meeting were the ICBG researchers, Brendan Tobin from the SPDA, and a Searle representative, establishing the first direct contact between the drug company and Aguaruna representatives. After a long debate about the ICBG project and much discussion about internal issues, OCCAAM, FECONARIN, and FAD accepted the ICBG research in their affiliated communities, confided in their national-level associate, CONAP, and unofficially incorporated a consortium of Aguaruna organizations which hereafter I will refer to as CONAP and Affiliates. While OAAM communities initially fell outside of the research area, they were later included in the ICBG agreement and constituted a fourth local Aguaruna partner to CONAP and Affiliates.[8] An important issue discussed during the 1995 meeting was the possible inclusion of ODECOFROC and FECONARSA. CONAP leaders explained that while the other organizations had registered as nonprofits (*asociaciones civiles*), these two lacked formal recognition within the Peruvian legal system and, further, had previously maintained close affiliations with the Consejo Aguaruna Huambisa. Both the president of ODECOFROC and a delegate from FECONARSA used the occasion to criticize the Consejo. The seemingly eternal association between the Consejo and its once-celebrated founder and visionary leader converted the event into an opportunity to pass judgment on Evaristo Nugkuag, now a common target in the Aguaruna's internal blame game. In fact, it was during this meeting that ODECOFROC's founder and first president announced that his organization would soon formally disassociate itself from the Consejo. Although both organizations eventually gained legal status as nonprofits and disassociated themselves from the Consejo, neither ever became a member of CONAP and Affiliates or received benefits from the research.[9]

As a result of this historic meeting, three persons were designated to negotiate the contractual details of CONAP and Affiliates' acceptance with the ICBG researchers and Searle: César Sarasara, CONAP's Aguaruna president, Mercedes Manríquez, a lawyer from Lima working with CONAP, and Brendan Tobin of the SPDA. After various discussions and drafts of potential agreements, in May 1996 a team including Sarasara and the two lawyers traveled to St. Louis to negotiate the final details with Searle and its lawyers face to face. Figure 2 diagrams the resulting agreement structure.

In this second round CONAP and Affiliates, represented by Sarasara and the two lawyers, were in direct contractual negotiations with Searle, the rela-

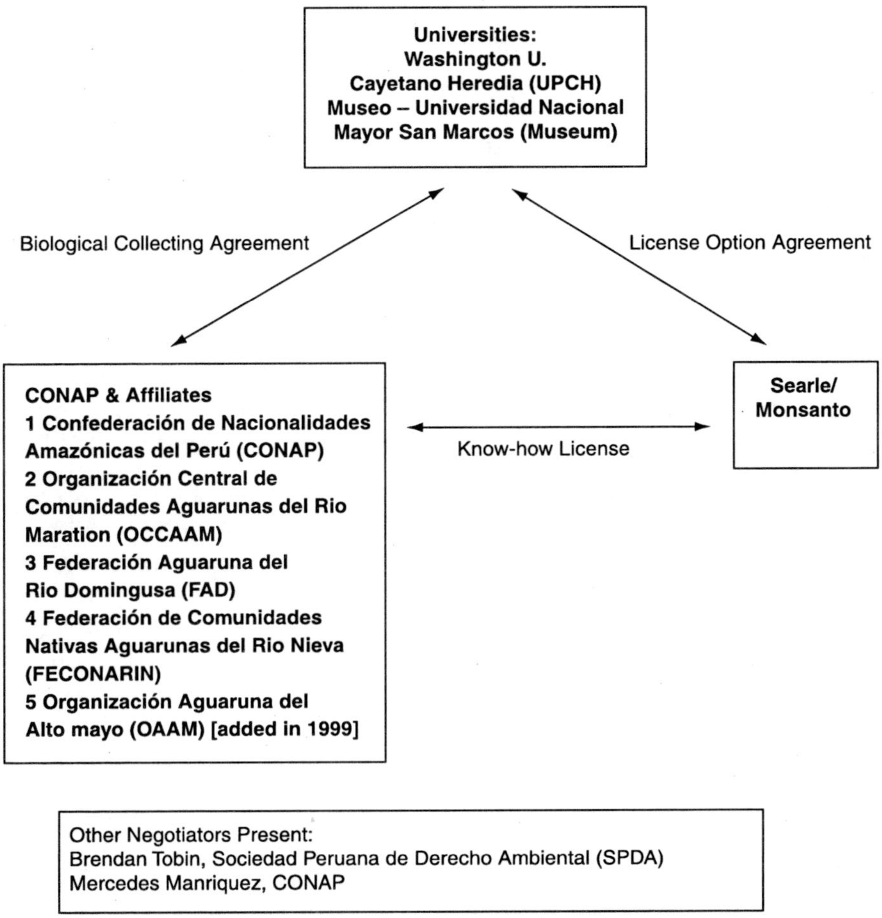

Figure 2 Agreement structure of the second round of ICBG-Aguaruna negotiations.

tions no longer being mediated by Washington University. This is a significant achievement for the indigenous peoples involved, and they recognize it as such. It is certainly one of the first instances of an indigenous group's representing its own interests—albeit as mediated by non-indigenous legal experts—with a large pharmaceutical entity in negotiations over the potential commercialization of traditional knowledge. This round of negotiations resulted in a different royalty and annual collection payment scheme and, most significant, a know-how license agreement. Tobin (personal communication) believes that the know-how license is a truly novel step in contract law, for the first time giving a group of indigenous peoples control and full ownership of its traditional knowledge. In legal terms the representatives of CONAP and Affiliates, who constituted the only indigenous signing partners, licensed directly to Searle the traditional medicinal knowledge, conceived as collective know-how, of all the Aguaruna for an initial four years (see Tobin 1999*b*, n.d.). The annual payments distribut able by Searle were reformulated and divided between a flat "collection payment" for the plant samples and "license fees" for the Aguaruna medicinal know-how. The know-how license also included a stipulation of two early, nonreturnable royalty payments (termed "milestones") to CONAP and Affiliates contingent on Searle's reaching two very advanced stages of new-drug approval based on a lead from the ICBG research.

In addition to the know-how license, CONAP and Affiliates signed a biological collecting agreement with the ICBG researchers specifying the terms and conditions of the field collections. This agreement included a stipulation "to ensure the fair and equitable sharing of benefits among the Aguaruna People" (article 3.01). The agreement thus remains open to other Aguaruna communities, provided that they apply for inclusion by affiliating themselves with an existing Aguaruna organization, and to other Aguaruna organizations, provided that they are approved by CONAP and Affiliates in a traditional assembly and dialogue called the Ipaamamu that has become central to CONAP's strategy for dealing with the local constituency (article 7). This measure of inclusion and the promotion of equitable sharing are significant because the Aguaruna consortium represented by CONAP and Affiliates in fact accounts for less than half the Aguaruna population (see Tobin 1999*a*). Nonexclusivity in the contracts reflects the fact that the property rights claim being made to indigenous medicinal know-how is a collective one in the name of the "Aguaruna People," the category that has been consistently used in the ICBG agreements.

The definition of "Aguaruna People" used in the biological collecting agreement identifies them as those "who live in the collection area," while the "collection area" is defined as the areas "inhabited by members of the Collaborating Organizations" (article I). "Collaborating Organizations" (article 1.06) refers to what I have chosen to call "CONAP and Affiliates" as a way of reflecting the contractual nature of this attempt to incorporate the specific subset of Aguaruna organizations that signed the agreement. This definition excludes Aguaruna in other organizations or other nonorganized areas who choose not to associate themselves with CONAP and Affiliates and may not even know that such a possibility exists. Ironically, these non-participants, including the Consejo Aguaruna Huambisa, constitute a clear Aguaruna majority. Further, it is quite clear from the second-round agreements that any and all monies are and will continue to be funneled directly through CONAP and Affiliates (article 3.01). In short, the resulting ICBG agreements, while they may contain pretensions to no-nexclusivity among the indigenous partners, effectively produce the opposite. The legal arrangement clearly moves in the direction of contractual and financial legitimation of CONAP and Affiliates as representatives of the "Aguaruna People."[10]

In essence, therefore, the relatively arbitrary exclusivity and oversimplification represented by RAFI's move to legitimate the Consejo Aguaruna Huambisa in protest of a bioprospecting project in round I was undone and then effectively duplicated at another level by ICBG and Searle's countermove to legitimate CONAP and Affiliates in support of the project in round 2. The major difference was in their opposed agendas and the indigenous organizations they chose to back as the legitimate "voice" of the "Aguaruna People."

Calculating the results: bad odds and big expectations

With the multiple agreements in place, the ICBG-Aguaruna project finally got under way with renewed collections in 1996 and worked in communities affiliated with OCCAAM, FAD, FECONARIN, and OAAM over the next four years. The researchers visited 22 Aguaruna communities and collected approximately 3,500 medicinal extracts.[11] In terms of global biodiversity equity and the goals of the Convention on Biological Diversity, the host-country benefits were not insignificant. According to Lewis (interview, 2000), the value of the grant funds, resources, and technology brought to Peru by the project totaled nearly $1 million. The majority of these benefits were made possible by public grant funds and were channeled into the two Peruvian universities, providing research and training opportunities for Peruvian students and faculty (Lewis et al. 1999:81). The ICBG project also brought several environmental training and higher-education opportunities to the Aguaruna involved (Lewis, interview, 2000). Two Aguaruna in particular were trained so thoroughly in botanical methods and scientific classification of plants that they eventually went on to direct their own field collections (Lewis et al. 1999:81). CONAP and Affiliates used a portion of the annual know-how license and collection fees contributed by Searle for their own organizational needs and distributed the rest to their affiliated communities in the form of small loans, scholarships for Aguaruna students, and individual reimbursement to field informants who worked with the ICBG researchers in identifying medicinal plants.

One of the most significant effects of the ICBG is the amount of political and symbolic capital it has bestowed on CONAP and its Aguaruna president, César Sarasara. When I first visited in early 1997, CONAP consisted of little more than César Sarasara in a small Lima office with a couple of desks, a phone, and a typewriter and was constantly borrowing money to pay the rent and utility bills. A couple of years later it was staffed by a number of salaried workers, sponsored events, and had the visible material items found in any functioning NGO (computers, fax machine, active web site, etc.). Sarasara, whose business degree and Aguaruna origin make him an appealing leader and a savvy multicultural negotiator, has become a frequent guest on the global indigenous-representative circuit. He is much sought-after by the press, has been awarded a prize by the International Society of Ethnobiology, and has become a coauthor of a scientific paper—all events directly related to the project.[12]

The rise of CONAP and Affiliates has its counterpart in the fall of the Consejo Aguaruna Huambisa, whose momentary success in questioning the legitimacy of the ICBG research via RAFI was soon overshadowed by CONAP and Affiliates' successful renegotiation of the project. Questioning the legitimacy of the Consejo, in part because of its strong association with its former controversial leader, has become commonplace in Peru for both external actors and the indigenous peoples themselves. One of the founders of the Consejo, Pancho Juwau (interview, 2000), remarked cynically that when people mention the Consejo now they think, "Eso apesta desde lejos" (That stinks from far away).[13]

There is another important and sobering result of the ICBG-Aguaruna experience: a clear clash between the odds and the expectations generated in the course of the negotiations (see Greene 2002). CONAP and Affiliates' attempt to privatize Aguaruna traditional medicinal knowledge has resulted in local expectations of large sums of money. Indeed, raising expectations of a substantial economic gain over the long term became an essential part of their strategy to persuade local communities to accept the project. A past president of FAD reported, for example, that the Aguaruna on the Domingusa River—in good millenarian fashion—were expecting millions of dollars in the year 2000, merely four years after the initiation of field collections. In meetings of CONAP with OAAM's affiliated communities I witnessed leaders several times creating high expectations by speaking about the arrival of millions of dollars almost as if it were guaranteed. The realization that these expectations were exaggerated has already begun to manifest itself in local resentment toward the leaders of CONAP, who are deemed responsible.

From the outset Searle used a screening process employing high-throughput machinery that tested plant samples for bioactivity against certain diseases that the company had identified beforehand in the hope of randomly producing interesting leads. This approach effectively ignored the specific ethnomedicinal information collected by the ICBG researchers that had been the object of so much political controversy in the first place.[14] Further, the tests were limited to diabetes, cardiovascular problems, and inflammatory diseases, health conditions that are much more common in Searle's consumer base in wealthy countries than among the Aguaruna.[15] In September 1999, at the final ICBG annual meeting with CONAP and Affiliates, Walter Lewis of Washington University announced on behalf of Searle (which had sent no representative) that the pharmaceutical company had found no leads worthy of pursuit and did not intend to extend the know-how license. From Searle's economic point of view, the deal was closed.

The fact that Searle never really explored the Aguaruna know-how that it had gone to so much trouble to acquire does not mean that the company did not gain from participating in the ICBG program. Monsanto featured the ICBG-Aguaruna project in an annual report to its consumers and shareholders as an important effort to conserve the tropical forest and promote benefit sharing in accordance with the Convention on Biological Diversity, praising the Aguaruna's traditional medical knowledge (Monsanto 1998:28). This suggests complicated motives that involve a concern with marketing and company image in an era in which sustainable development is the discourse of the day as much as or perhaps more than a concern with successful cooperation with indigenous peoples on an environmentalist agenda. Searle's cancellation of the know-how license was, of course, entirely legitimate from a financial and contractual point of view. After all, business is business.[16]

Bioprospecting and indigenous property claims in the broader view

The case studies that are now emerging complicate the clichéd accounts of bioprospecting negotiations, indigenous cultural property claims, and the politics of indigenous representation. Two of the best-known cases involve patent challenges, a successful one against a transnational corporation and a less successful one against a U.S. researcher. After several years of public protest and concerted legal effort by a coalition of Indian and U.S. research institutes, NGOs, and farm organizations, the European Patent Office revoked a patent on a fungicidal compound derived from the neem tree held by the agro-industrial company W. R. Grace (see Raghavan 2000). Vandana Shiva (2000), one of the most visible Indian activists and scientists involved in the protest, has shown that the medicinal and pesticidal properties of neem have been known to farmers in India for millennia and that the tree figures as a sacred symbol in Sanskrit and Islamic traditions. In arguing that neem has "become a symbol of Indian indigenous knowledge," she reveals an apparent conflation of the modern Indian nation with the many ancient and contemporary peoples native to the subcontinent that have recognized the tree's multiple uses.

In 1999 a coalition formed between North American environmental law NGOs and the Coordinadora de las Organizaciones Indígenas de la Cuenca Amazónica (COICA), based in Quito, Ecuador, filed a similar petition with the U.S. Patent and Trademark Office (PTO) requesting the cancellation of a U.S. researcher's patent on an ostensibly "new" variety of ayahuasca. Indigenous peoples throughout the Amazon have used this vine for centuries in the preparation of hallucinogenic remedies that are central to local cosmologies and traditional forms of healing. The PTO initially recognized the validity of the petition but later, on very shaky legal grounds, declined to revoke the patent. Representatives of COICA considered this decision a "profoundly disturbing sacrilege and assault on their traditional values" (Wiser 2001:13).

These patent disputes show that some indigenous peoples and even some spokespersons and activists who identify their entire Third World nations with indigenous peoples (as in the neem case) feel deeply affronted by some cases of politicized bioprospecting. The apparent sacrilege committed and the global historical inequities reenacted when foreign companies and researchers claim culturally sacred remedies as patented property clearly play a crucial role in the way in which indigenous and Third World indigenous nationalists conceive of their cultural property claims. In the name of all Amazonian peoples, for example, the president of COICA, Antonio Jacanamijoy, defended his organization's position by stating that ayahuasca is "a sacred plant used to cure our illnesses, clean our spirits, and predict our future. Ayahuasca belongs to all our communities that use it and therefore it is impossible that it could be the property of just one man" (COICA 2002:2). In the neem case Shiva (2000) makes it clear that the patenting of a method of neem seed emulsion by W. R. Grace is sacrilege. Furthermore, she argues that this case of international "bio-piracy" not only was legally unfounded (a claim that was upheld by the European Patent Office) but would have resulted in a price hike that would have adversely affected traditional farmers who relied on neem seeds.

The ICBG-Aguaruna negotiations raise a set of issues that have received less attention in the bioprospecting and indigenous rights literature. In particular, I have emphasized the politics involved in legitimating (and delegitimating) indigenous forms of representation, the ramifications of the incorporation of indigenous groups in claiming culture as property, and the possible clash of expectations associated with the pharmaceutical sector's tenuous commitment to bioprospecting. These problems are not unique to the ICBG-Aguaruna case. The events leading to the collapse of an ICBG project with the Maya of Chiapas, Mexico, in 2000 are strikingly similar (see Brown 2003, Nigh 2002). Using a similarly aggressive Internet campaign, RAFI again inserted a biopiracy-watchdog agenda into the negotiations in conjunction with a faction of Maya representatives opposed to the research and this time helped bring the project to a complete halt (RAFI 2000). Similarly labyrinthine politics involving rival indigenous representatives, NGO intervention, and lack of consensus about legitimate representation of the Maya people played a prominent part. The corporate partner in this case was a start-up drug company with capital resources so limited as to make the likelihood of a major drug discovery roughly equivalent to that of winning the Mexican lottery, according to Brent and Elois Ann Berlin, the principal investigators of the Maya project (2002:467). Controversy over indigenous representation and distrust of the motives of foreign bioprospectors were matched again by a heightened and somewhat distorted perception of the potential economic ramifications of such research into traditional plant knowledge.

Of the eight ICBG grants awarded, the two that were clearly the most controversial, ICBG-Aguaruna and ICBG-Maya, entailed direct negotiations between indigenous peoples and bioprospectors and the intervention of interested third-party NGOs claiming to act on behalf of or in conjunction with those indigenous peoples. Meanwhile, other ICBG projects have seemingly been carried off without the slightest scandal in collaboration with host-country universities, research institutes, and governmental agencies. Cori Hayden (n.d.) has been tracking another ICBG project that is collecting plants in Chile, Argentina, and Mexico, working with public and private universities and research institutes. ICBG-Latin America has received two rounds of ICBG funding and established apparently cordial relations with its host-country partners (see Timmerman et al. 1999). Hayden attributes this apparent success to a strategic avoidance on the part of the researchers of the indigenous-knowledge question and of identifiably "indigenous" communities that might claim their own cultural property rights. The ICBG researchers avoid this issue by buying plant samples in urban herbal markets from local vendors. They consider this "de-indigenized" plant knowledge circulating in urban markets in an already commodified form as part of the national public domain. Thus, while they purchase plants at a "fair" price from local vendors, the only long-term benefit-sharing obligation they feel compelled to recognize is to the nation, represented by their host-country research partners' institutions.

The University of Illinois at Chicago (UIC) ICBG program in Laos and Vietnam seems to share some of these dynamics (see Soejarto et al. 1999). The UIC researchers make collections primarily in national parks (e.g., areas demarcated as "public" national space) and in contractual collaboration with state and private research institutions in the host countries. For the ethnobotanical component of the project, contact between the foreign researchers and local communities is mediated by state institutions such as the National Committee of Ethnic Communities in Vietnam and the Ministry of Health in Laos. As beneficiaries of the project, the host-country institutions essentially speak for and must represent the interests of the local peoples in the ICBG research (p. 107). Bioprospectors of this sort approach benefit-sharing obligations primarily as a national problem to be resolved with state or other host-country institutions that do not present uncertainty about institutional legitimacy and representation.

Contrary to the tendency toward hyperbole in accusations of global biopiracy and rampant exploitation of indigenous knowledge, the current state of affairs in pharamaceutical research suggests that we need a more subtle understanding of the dynamics at work. Several observers have noted that from an investment standpoint drug companies allocate very little to bioprospecting of any sort compared with their investment in the synthetic, computer-modeled and engineering-based research which constitutes the preferred method of drug development (see Aylward 1995; Albers-Schönberg 1995; Goering 1995:10). Discourses of sustainable development and environmental conservation will no doubt continue to influence drug companies to consider demands in the public sphere for an environmentally responsible capitalism. A greener, moral politics requires a greener, more moral capitalism. Yet, there are inherent contradictions in the idea of "green capitalism" when it seems more profitable to maintain the capitalist status quo (see Gersh 1999). Interests in capital accumulation and survival among corporate competitors easily outweigh interests in a renewed moral and environmental order. A recent report reveals that despite the industry emphasis on pioneering research and development, drug manufacturers are in fact remarkably uninventive entrepreneurs that dedicate more financial resources to advertising and corporate-image campaigns that boost sales than to discovering new medicines (Public Citizen 2001). The study also finds that a considerable proportion (more than half) of new drug products released between 1982 and 1991 were not miraculous new medicines but "me-too" drugs, slightly modified imitations of already existing products.

The tenuous commitment to bioprospecting by the private pharmaceutical sector is especially pronounced with regard to the kind of bioprospecting that directly relies on indigenous knowledge. The failure of Shaman Pharmaceuticals, a start-up company founded in 1990, to discover a profitable drug based on traditional plant knowledge only adds to the evidence provided by the ICBG-Aguaruna and ICBG-Maya projects. Although Shaman maintains a stance consistent with the Convention on Biological Diversity, ready to implement profit-sharing mechanisms with local peoples who contribute traditional plant knowledge, such a scenario has yet to materialize. Despite some significant advances with an experimental drug named Provir, based on the traditional Amazonian remedy *sangre de grado*, Shaman was forced to close down its pharmaceutical research division and is currently marketing an unprofitable herbal supplement, still operating with the same environmental and moral ideals under a mountain of corporate debt (see Brown 2003, *Economist* 1999). This is reason enough for pause and reflection.

Quandaries of indigenous representation and incorporation

Pharmaceutical companies' relative uncertainty with respect to bioprospecting and concerns over corporate image in an environmental era and indigenous peoples' increasingly common efforts to treat knowledge as property produce widely differing expectations, assumptions, and strategies for negotiation. A panel of external experts identified the creation of unrealistic expectations surrounding the potential economic rewards of bioprospecting in developing regions as one of the major dangers of the ICBG program's design (see Report of a Special Panel of Experts on the International Cooperative Biodiversity Group 1997). Others have pointed to the raising of expectations in economically poor but biodiversity-rich nations as a general problem created by the rhetoric of bioprospecting (see Macilwain 1998). I would add that while the creation of inflated expectations may be a problem within bioprospecting research as a whole, current cases suggest that it is most exaggerated and politicized when negotiations involve the question of indigenous cultural property and indigenous representation. Indeed, scholars such as Cori Hayden suggest that in evading the indigenous-knowledge question and designing bioprospecting activities in terms of obligations to the nation-state some bioprospectors consciously attempt to avoid controversy, misunderstanding, and allegations of bio-piracy. Why do contemporary bioprospecting arrangements with indigenous peoples produce such contention?

One answer, as some of these cases reveal, is that the emergent global politics of representation and mobilization in which many indigenous peoples are involved has yet to be fully recognized, legitimated, and routinized by national and international societies and indigenous communities. The debate over who speaks for whom in indigenous affairs continues to create crises of representation in which indigenous leaders, organizations, and spokespeople are forced to re/negotiate their representativeness both among themselves and in relation to the agendas of foreign actors who typically hold positions of greater power, resources, influence, and prestige.

Coming to terms with the influence of NGOs in this process is an important step, since relatively little has been written about them from an anthropological point of view. While critics see the more powerful international NGOs as part of a neoliberal project to perpetuate a false consciousness of marginalized peoples' "empowerment" (see Petras 1997; Hardt and Negri 2000:36), others argue that NGOs represent the primary force of democratization, social service, and development in the age of the shrinking state (see Meyer 1999, Bebbington et al. 1993, Clark 1990). In the case of indigenous peoples, external NGO allies and indigenous institutions, often of the NGO sort themselves, can and do provide important leverage against private and state interests. But as bureaucratic institutions with their own political and social agendas they also often operate at considerable geographic, cultural, and linguistic distance from local constituencies, which can result in a tendency to oversimplify and romanticize indigenous realities. For example, the currently popular image of Amazonian Indians as guardians of the tropical forest is in significant part due to the alliance created between environmental NGOs and indigenous peoples in the past two decades (Conklin and Graham 1995). Similarly, examination of bioprospecting negotiations indicates that NGOs operating in conjunction with indigenous representatives play a determinative role in supporting indigenous claims while also creating exaggerated expectations about the economic potential of indigenous intellectual property claims and playing into the fear associated with the widespread discourse of pharmaceutical bio-piracy. Such organizations sometimes provide key resources and scientific and legal expertise, as in the ayahuasca and neem patent cases. In other cases their activities, however well intended, effectively result in legitimating the indigenous groups that most clearly conform to their own institutional, political, and social objectives and forms of anticorporate, antibioprospecting self-representation without ever confronting the complex political realities of local populations.

NGOs, however important, are in fact just one subset of actors among many that influence the negotiation and de/legitimation of indigenous representativeness. Researchers of all sorts, religious organizations, international funding agencies, corporations, and state bureaucracies all influence and complicate the politics of indigenous representation in accordance with their own agendas. The tendency of NGOs to distort claims of indigenous representation in protest of bioprospecting are matched by the tendency of researchers, agents of public institutions, and pharmaceutical corporations to do the same in their own support. Both sides seek out and attempt to legitimate indigenous allies in accordance with their institutional, political, or economic goals. Furthermore, as we have seen here, dissent and internal discussion within indigenous groups over the issue of representation often gives potential credence to both sides of the argument, just as it reveals the problems inherent in negotiating collectively owned cultures with only a selection of indigenous brokers. Because of the collective nature of claims to culture as property, there is a common assumption on all sides of the debate that indigenous collectives must possess a centralized structure of representative authority comparable to that of consolidated nation-states with which external actors can negotiate. Establishing who are the legitimate representatives of indigenous collectives is, however, often a matter of internal and external debate. In circumstances where indigenous institution building is still in process, it is no surprise that the legitimacy of any particular indigenous organization varies with the prevailing political winds and the degree of access to more powerful external actors.

This does not, I suggest, make indigenous peoples innocent pawns of more powerful players but rather demonstrates their active engagement in wider networks of power as political strategists whose decisions have direct repercussions on their legitimacy as indigenous representatives. Key legal victories and the successful insertion of indigenous agendas into the decisions of more powerful economic and political institutions and actors indicate that indigenous strategies can under the right circumstances prove extremely effective, but "victory" is usually relative. The costs of politicizing culture and treating it as property must also be taken into account. While global indigenous mobilization is a sign of significant strength for some of the world's most marginalized peoples, the politics of indigenous authorities and spokespeople remains a highly contested sphere. Emergent indigenous political representation is simultaneously supported and constrained by national and international alliances that are continually identifying legitimate mediators between external interests and local communities in ways that are as much dependent on their own agendas as on those of indigenous collectivities.

While some of the cases examined here highlight primarily concerns about sacrilege, misrepresentation, and exploitation, others force us to consider the ramifications of the economic expectations and aspirations born from indigenous cultural property claims. Some have emphasized that cultural property claims, based on Western intellectual property laws, imply a privatization derived directly from market models and ideals (see Brush 1996, 1999; Coombe 1998, 1999). The cases analyzed here demonstrate that cultural property claims are in fact marshaled both against outside market interests and in accordance with indigenous peoples' own internal interests in putting the market to use for their own mobilization. The neem and ayahuasca cases, for example, revolve around the perceived moral, cultural, and economic attack on a nation's and an indigenous people's right to prohibit collectively propertied knowledge from becoming the exclusive patented property of a corporation or a private individual. The material presented here on the ICBG-Aguaruna case suggests that we must consider the ways in which indigenous claims to cultural property also lead to strategies of local entrepreneurship and examine the potential ramifications of these strategies. Treating culture as property is a strategy indigenous groups use to adopt and transform the logic of their Western counterparts—to market themselves as Indigenous People Incorporated and create a form of politically informed economic activity that in another context has been labeled "tribal capitalism" (see Rata, quoted in Friedman 1999:9).

The ICBG-Aguaruna project demonstrates that indigenous entrepreneurship is also very much a part of claims to cultural property. The know-how license signed with Searle is one clear instance of it, and another is the application for a utility patent that Washington University researchers filed in January 2003 (Lewis and Ramani n.d.). The lack of corporate support since Searle's withdrawal has not hindered Walter Lewis's ethnobotanical research on Aguaruna plant samples, an endeavor that has produced leads in the area of antimalarials for which he and his colleagues are seeking patent protection. In the patent application the inventors assign equal ownership

(25% each) to members of the original four ICBG partners. One of those four potential patent owners is CONAP, which represents the local Aguaruna federations that are partners to the agreements. Should the patent be granted, CONAP would become probably one of the first indigenous institutions anywhere to own patented knowledge. The likelihood of a patent's being converted into profits is, of course, small, but one cannot fail to acknowledge the apparently economically interested nature of this particular brand of indigenous mobilization.[17]

What exactly is at the core of the contemporary indigenous strategies of mobilization witnessed not only in bioprospecting negotiations but in traditional crafts and folk music marketing and eco- and ethno-tourism? Ultimately what is being represented, discussed, negotiated, and in some instances sold is indigenous identity itself (see Brysk 2000: chap. 4; Sylvain 2002; Conklin and Graham 1995; Friedman 1999). It is not merely Amazonian plant knowledge, Andean music, or tours of a San village that are the object of privatization. At a more fundamental level, indigenous group identities themselves are subject to these strategies, since it is often the connection to an indigenous culture and identity that makes these commodities potentially attractive to global consumers and producers.

Privatizing and marketing parts of their traditional cultures or resources is another strategy used in a broader politics of self-determination that emerges in connection with the economic demands of a global market. It is precisely through this kind of indigenous entrepreneurship that incorporated groups such as CONAP and Affiliates promote a specifically *indigenous* mobilization. They argue that indigenous mobilization needs more than a solid historical and cultural foundation in today's multicultural identity politics. It also requires economic viability, for survival as an organized social movement is directly tied to the financial solvency of the institutions indigenous groups have founded and the communities they represent in a world dominated by market relations. This complicated state of affairs makes untenable any explanation of indigenous incorporation as a simple sell-out to corporate ideals. It also calls into question the simplified images of the indigenous masses living in unity and clamoring against the alleged evils of corporate America. Both strategies are at work, and they present opportunities for advancing an indigenous agenda in some circumstances and harbor potential dangers to it in others. The adoption of corporate ideals is part of indigenous mobilization, but it cannot be separated from the broader issues of historical inequity, cultural sacrilege, and misrepresentation around which indigenous peoples also organize struggle. Strategies of indigenous incorporation mix with local cultural frameworks and are deployed as part of a broader indigenous politics of identity and self-determination (see Coombe 1998: chap. 5).

Two principal dangers emerge, however, with regard to such moves to privatize indigenous knowledge, culture, and identity. One of these, as the history of capitalist expansion has repeatedly made clear, is that the incorporation of indigenous groups could just as likely lead to the monopolization of cultural identity and property by certain indigenous classes and the marginalization of others as to collective political-economic mobilization. The other is that the creation of unrealistic expectations about the market value of traditional knowledge, cultural property, and incorporated identities will continue to sow discontent not only between indigenous groups and the various agents of development and modernity but between the indigenous constituency and its brokering leadership. In either case, there is still space for ample debate in the search for a way forward.

Comments

SUBHABRATA BOBBY BANERJEE
International Graduate School of Management, University of South Australia, City West Campus, North Terrace, Adelaide, SA 5000, Australia (apache-@unisa.edu.au). 14 XI 03.

Greene's paper addresses an important and controversial issue. Rather than talking about its strengths and weaknesses, however, I want to comment on what it leaves unsaid.

Greene rightly points out that indigeneity is a highly contested process, but I do not think that we can dismiss the neocolonial influences on this process as easily as he does. If indigenous self-representation is "at the heart of recent anthropology," then we cannot dismiss the discursive effects of the colonial project (Banerjee 2000). In fact, we must place the discussion of colonialism and the role of anthropology in producing indigenous subjects in the center of the debate. Greene raises the question of "representative legitimacy." What is not made part of the discussion is why indigenous communities have to prove that they are legitimate and why the grounds for establishing legitimacy are in most cases a product of the colonial system. To claim representative legitimacy they are forced to conform to the categories of what Wolfe (1999) calls "repressive authenticity." "Indigenous ecology" thus comes to be represented as a pristine source of knowledge separate from indigenous economy and society. Like Spivak's subaltern, the indigenes can have only a *quoted* existence in a larger statement that belongs to the anthropologist alone. We talk about them and occasionally quote them, but in the end we are their authors.

As Osuri (2003) argues, these discourses do not explain the conflations of indigeneity as identity and indigeneity as a series of events that make visible the relationships between non-indigenous and indigenous communities. Rather, they obscure the notion that the production and consumption of indigenous identity occur within Western (post)modern modes of theorizing identities, a process that disavows the colonial context within which "indigenous policies" are developed and indigenous identities are regulated. Indigeneity can also be represented as a set of relationships between indigenous communities and their colonizers that produces empowering or disempowering representations in the context of colonial relations of power. Despite acknowledging the domination effects of colonialism, Greene does not seem to accept that indigenous peoples continue to be regulated by structures that are a legacy of the colonial era and must conform to what for many of them are still alien categories.

In fact, the politicization and treatment as property of culture that Greene talks about are precisely the result of applying Western categories to non-Western sites. "Value" as defined by Western capitalism can be created only if indigenous culture and knowledge are converted to property, and indigenous communities all over the world are forced to employ these categories if they want to protect their knowledge. What is missing from Greene's analysis is the consequences of this process for the community. While he discusses the range of stake-holders involved in the bioprospecting arrangements, one voice is missing, and that is the voice of the people. In revealing the complexities of interactions between First World corporations, Third World governments, indigenous organizations, and NGOs, he is silent about the disparities in power among them—disparities that ensure particular outcomes while precluding others.

Greene rightly identifies the vagaries of the market and the "tenuous commitment" of the pharmaceutical sector as factors that can influence the future of indigenous involvement in bioprospecting. For instance, on November 11, 2003, the president of the Aveda Corporation announced that it was discontinuing its indigenous product line to "demonstrate our ongoing support and respect for indigenous peoples in their efforts to protect their traditional knowledge and resources" (Conseil 2003). He also declared that the company would begin the formalities necessary to abandon any rights it might have in the "indigenous" trademark. Leaving aside the problematic question of why a multinational corporation should possess rights in an "indigenous" trademark in the first place, we can only speculate as to the motives behind the company's decision.

If "Indigenous People Incorporated" is to be a genuine agent for promoting indigenous self-determination, significant structural barriers in the political economy will need to be overcome. The high failure rate of indigenous enterprises has as much to do with lack of access to infrastructure as with profound incompatibility with many elements of the market system. The business success of indigenous enterprises is always predicated on a separation of commerce and culture that is alien to most indigenous communities. It also denies the historical legacy of indigenous dispossession and assumes the existence of a level playing field on which indigenous communities can compete in the mining, tourism, forestry, and arts sectors of a global economy.

The indigenous economy is a hybrid economy with market, state, and customary components, all of which need to be recognized and integrated in assessing indigenous economic development (Altman 2001). Developing a sustainable indigenous enterprise implies understanding the relationships among these components. It is crucial that such a configuration be informed by an explicit social-justice agenda. The challenge for indigenous communities is to determine the best form of political and cultural mediation for delivering the economic and social outcomes they desire, and this will depend on the political, economic, social, cultural, and environmental contexts of each community.

KELLY BANNISTER
School of Environmental Studies, University of Victoria, P.O. Box 3060, Victoria, B.C., Canada 8W 3R4 (kel@uvic.ca). 14 XI 03

Greene's article provides a much-needed case-based approach to analyzing bioprospecting arrangements. The broader global context in which he situates his detailed analysis is effective in drawing attention to some fundamental assumptions about these kinds of projects that merit further consideration.

Having also traveled to Santa María de Nieva and Lima in 2002 to interview some of the same actors about their involvement in the ICBG-Aguaruna bioprospecting project, I find that Greene has presented a fair and reasonable account. As he notes, some details cannot be reconciled because of conflicting versions of the same incident by different actors or the unavailability of documents. However, the case still offers a rich opportunity to understand some of the political, cultural, and economic complexities that underlie international, multi-institutional, and cross-cultural bioprospecting arrangements. Such an opportunity is also rare—more typically, those involved in contentious projects tend to protect themselves from further scrutiny and criticism. In this respect, I think that the individuals who agreed to be interviewed and share redacted copies of contracts and other agreements for this case study should be commended.

Greene identifies several important contradictions inherent in indigenous self-representation, political mediation, and the legitimacy of who is speaking, for whom, and to whom. An irony exists whether it is state governments or indigenous political leaders who lay claim to cultural knowledge when they are not the traditional heirs of that knowledge. Furthermore, from a practical point of view, it seems that the more the knowledge is distanced from its traditional sources, the more likely that the scientific application of that knowledge will be assessed out of context and the less likely that it will provide drug leads—at least, leads related to the traditional use. Rather than which of the national Aguaruna organizations or regional federations is the legitimate authority to contract with Searle on Aguaruna medicinal plant knowledge, the question—on both moral and practical grounds—becomes whether these are the appropriate signatories for such a contract. The "indigenous entrepreneurship" that Greene discusses may, in fact, be based on intracultural appropriation.

Also worth noting is a fundamental challenge that is not raised by Greene but has plagued the ICBG-Aguaruna and other bioprospecting projects—communication. Not only did actors in the ICBG-Aguaruna have to overcome the geographical distances between U.S. and Peruvian institutions and Aguaruna communities (where there are no roads, let alone an Internet) but they faced challenges in translating lengthy and complicated legal documents and conducting business in three languages (English, Spanish, and the local Jívaroan language). While political agendas to some degree influenced the events that transpired, including the generation of unrealistic financial expectations (i.e., "millions of dollars"), it cannot be ruled out that communication problems leading to a genuine *lack of understanding* also contributed to some of the conflict.

Given Greene's insightful discussion of roles, motivations, and strategies on the indigenous side of bioprospecting agreements, I was somewhat disappointed in his shallower treatment of NGOs and corporations. Since one NGO was responsible for the majority of the controversy around both the ICBG-Aguaruna and the ICBG-Maya projects mentioned and no other NGO is cited in the context of "distort[ing] claims of indigenous representation in protest of bioprospecting," the broader generalization of this practice across NGOs seems unwarranted. With respect to corporations, the portrayal of Searle is slightly misleading in this instance, and it is unfortunate that the company's perspective was unavailable. For example, regarding the expiration of the know-how license, Greene concludes that "after all, business is business," which does not adequately account for the fact that Searle was, coincidentally, eliminating its natural-products program as the license term came up. Likewise, the biological screens that Searle used to test the Aguaruna plant extracts (i.e., "diabetes, cardiovascular problems, and inflammatory diseases") were, indeed, largely irrelevant to the types of diseases of primary importance to Aguaruna peoples. However, at the time of the ICBG grant application, Searle did have an infectious-diseases division, which had closed down by the time the grant was awarded.

I do not raise these points in defense of any NGO or corporation. Rather, I offer them as a reminder that conclusions may differ depending on the availability of facts under consideration and as an encouragement to probe beneath the generalizations that have, thus far, largely impeded our understanding of the complicated dynamics at work among the various actors in today's bioprospecting arrangements. Overall, Greene makes an important contribution toward a deeper understanding in this regard.

STEPHEN BRUSH
Department of Human and Community Development, University of California, Davis, Calif. 95616, U.S.A. (sbbrush@ucdavis.edu). II XI 03

A decade and more has passed since the United Nations Conference on the Environment and Development in Rio de Janeiro that launched the Convention on Biological Diversity. A key objective of that convention was to create a formal regime of access to biological resources and benefit sharing from their use. This regime was envisioned as a solution to the "bio-piracy" that was perceived to result from the previous system of common heritage. The direct beneficiaries of the new regime were nation-states, which the convention recognized as sovereign owners of genetic resources. Indigenous people and other stewards of biological resources were implied beneficiaries, but their participation in the imagined benefit stream was to be mediated by the state. The decade since the convention was launched has been a disappointment. Very few countries have negotiated or implemented access and benefit-sharing regimes, the flow of genetic resources has slowed, no new drugs have resulted from bioprospecting, large pharmaceutical companies have withdrawn from the screening of plants, and the loss of biological diversity has proceeded apace. While the imagery of the "tragedy of the commons" dominated the formulation of the convention, an increasingly popular image after a decade of convention-inspired politics is the "tragedy of the anticommons" (Heller and Eisenberg 1998). Perhaps the most direct result of the convention has been to raise consciousness in governments and some social groups about the importance of biological resources, but this has been achieved by creating unrealistic expectations about the value of these resources and confusion about "ownership" of genetic resources and intellectual property.

Greene's exploration of the ICBG-Aguaruna is an insightful and important discussion of the way the idea of access and benefit sharing played out in the Peruvian Amazon. The ICBG program is an innovative effort to bring together different sectors engaged in creating, using, conserving, and studying genetic resources. Like other bioprospecting programs. ICBG projects have been most successful when community participation is minimized, but the ideal of the Convention on Biological Diversity and the ICBG has been to create reciprocal flows of genetic resources and benefits that would reach into communities to support traditional knowledge and foster biological stewardship. Greene's article expands our knowledge of the experience of these more participatory projects, and it confirms the suspicion that the difficulties experienced in the ICBG-Maya are not unique.

The logic of the convention and the framework for bioprospecting that follows it are based on the premise that the public-domain or common-heritage approach offered fewer benefits than could be realized by commoditizing genetic resources through contracts between "suppliers" and "users." The benefit stream, however, is powered by granting intellectual property rights to discoveries based on biological resources. The scope of exclusion relating to intellectual property is expanded through the contracting process that is the keystone of bioprospecting. The benefit stream from bioprospecting is defined by the contracting parties—a principle that is relatively unproblematic at the nation-state level but fraught with difficulties at the community level, where political strife within and between communities ensues from the exclusion inherent in contracting.

Greene's description of the power struggles among different Aguaruna groups confirms the observation that development projects have very different meanings for their nominal beneficiaries than for national and international participants, sponsors, and advocates who depend on local clients. It also shows us that actors at the national and international level have multiple and complicated motives and that political influence and opportunities for publicity are as important as genetic resources in the approach to local communities. The history of the Maya and Aguaruna ICBG projects would seem to suggest that future bioprospecting will eschew direct community participation. Nevertheless, the World Intellectual Property Organization is actively developing a model for the protection of "traditional knowledge" (Wendland 2002). Greene's paper is an important cautionary tale of how good international intentions can go awry at the local level. Similar political asymmetries, local conflicts, ambiguous social boundaries, broad sharing of poorly documented knowledge, and complicated agendas are numerous. If traditional knowledge and biological resources are precious but diminishing resources, we might wonder whether to pursue policies that will invite conflicts within groups like the Aguaruna. Although the architects of the ICBG believed in nonexclusivity, the program's reliance on intellectual property produced the opposite. A policy that begins with exclusivity is bound to produce similar results or worse.

NOEL CASTREE
School of Geography, Manchester University, Manchester M13 9PL, U.K. (mfvssnc2@man.ac.uk). 31 x 03

One of the peculiarities of bioprospecting is that its champions and critics—despite their differences—both misunderstand the object of their fervour and censure. Greene's measured essay capsizes some of the verities commonly uttered in debates about the three main forms of biological prospecting (chemical, mechanical, and genetic). More specifically, he criticizes the way in which indigenous peoples are typically figured in what Hayden (2003:36) evocatively calls "the bioprospecting imaginary." His illuminating analysis of the ICBG-Aguaruna relationship in Peru shows how programmatic beliefs about bioprospecting founder on the complexities of real-life bioaccords. Embracing the sort of "situational pragmatism" that Michael Brown (1998:205) has recommended, Greene demonstrates the ambivalent potentials of attempts to reward indigenous "stewards" of material and informational biodiversity. Bioprospecting, he shows, fails to deliver on the ideal-typical aspirations of its cheerleaders while managing, simultaneously, to offer certain indigenous peoples the hope of something better. Greene thus wisely resists easy denunciations. His undogmatic approach is to see bioprospecting as a doublesided phenomenon which requires contextual evaluation.

It seems to me that Greene's principal contribution to the debate about bioprospecting and indigenous peoples is twofold. First, he urges us to resist theoretical essentialism. Bioprospecting has become an abstract signifier that is understood either through a pejorative reading of the market (among the left) or through a positive assessment of the privatizing and pricing of hitherto "open-access" goods (among liberals and conservatives). The problem here, it seems to me, is that preconceived theoretical-normative ideas become the rigid lenses through which examples of bioprospecting are viewed (Castree 2003). Determined to show that their preferred world-view is the right one, critics and fans of bioprospecting selectively assess cases in such a way that the conclusions, are, in effect, known from the start. Greene's essay shows why measured empirical analysis defeats theoretical simplicities. What we have been calling bioprospecting in the singular is, he implies, really a cluster of ideas and practices whose effects are *contingent*. Bio-prospecting's successes/failures on the ground cannot, as he shows, be predicted from the supposed essential character of the phenomenon.

Secondly, Greene takes us beyond the now tired debate over whether indigenous peoples stand to gain or lose from bioprospecting. His alternative is to focus on the complexities *within* the indigenous movement. This dehomogenizing move not only points up the impossibility of a unified indigenous voice (even at the regional scale) but also highlights the precarious and shifting webs of connection that allow some indigenous peoples to be recognized as legitimate stakeholders in any and all bioaccords. The "problem of representation" is not simply that of the identification by Western biotech companies, botanical gardens, and the like of the "right" indigenous partners. It is also a problem of how different indigenous peoples step onto the political stage locally and regionally while others are marginalized or silenced. This is complicated by the role played by NGOs, national governments, and other nonindigenous institutions. As Greene shows, over time the "Aguaruna" in the ICBG-Aguaruna relationship were, variously, the Consejo Aguaruna Huambisa, OCCAAM, FECONARIN, FAD, ODECOFROC, OAAM, and CONAP. The perceived legitimacy and representativity of these organizations were, he demonstrates, contingent on their changing relationships with ICBG partners, national indigenous federations, and a host of other players.

Ultimately, as Greene notes, prospectors for novel biological resources will increasingly try to avoid the need to identify indigenous stakeholders, but his wider arguments have a longer-term purchase. The politicization and treatment as property of "culture" will continue apace, whether in the realm of the biological or elsewhere. Those of us who care about the fate of subaltern communities cannot afford to see the commodification of culture as an unqualified menace, but neither can we see it as an unproblematic means for the subaltern, at long last, to speak in their "true voice."

SHIVCHARN S. DHILLION
Centre for Development and the Environment, Box 1116 Blindern, 0317 Oslo, Norway (shivcharn. dhillion@ina.nlh.no). 24 XI 03

The privatization of indigenous knowledge, culture, and identity is indeed an emerging problem in both the developing and the developed world. Indigenous peoples in India, Malaysia, Thailand, and many countries in South America are forming consortia working towards such privatization at the expense of other, less conspicuous or "glamorous" indigenous peoples. As in the case of CONAP and Affiliates, individual communities often challenge the authority of heads of consortia to speak on their behalf. This has occurred recently in Sabah, South and North-East India, Vietnam, the Lao P.D.R., Thailand, and China (Tibet). In Thailand indigenous organizations at the national or regional level are made up of a few dominant groups that represent all local knowledge, culture, identity, and diversity. Their representation is further enhanced and championed by NGOs and researchers who focus on select indigenous communities, thus globalizing the indigenous culture and identity across Thailand and beyond its borders to include Burmese, Laotian, and Vietnamese indigenous people. In the Sichuan District of China, for instance, nomads who are now sedentary and owners of land resources allocated by the Chinese government now speak for all nomadic pastoralists in the region. Among the Sami of the northernmost part of Europe, including Norway, Sweden, Finland, and Russia, local Sami communities have repeatedly contested the structure of the central Sami organization as not articulating their differential needs. Regional indigenous consortia can easily lead to the monopolization and structured dilution of cultural identities.

Although the Convention on Biological Diversity explicitly champions recognition of national sovereignty and provides measures for securing "environmentally sound use," access and benefit-sharing regimes, and acknowledgment of indigenous "groups" and their knowledge and cultural systems, few parties to it in the North have implemented detailed regulations for compliance with it (Dhillion et al. 2002). Broad-scale plans exist in many nations but are far from any clear definition. Then there are nations like the U.S.A., which have not ratified the convention but have schemes for biodiversity transfer from the South through programs such as the ICBG without regulatory regimes in place. The question arises how indigenous communities can protect themselves when the bioprospector's state has no regulatory regime. A number of nations in the South are beginning to implement the convention's provisions, especially in relation to access and benefit sharing, although in some parts of the South progress is very slow (see Aguilar-Støen and Dhillion 2003). Thus, given the current rather loosely woven regulatory system, there is room for flexible negotiations between consortia of indigenous groups and those interested in genetic resources, fostering corporate-like ideals with consequences for the marginalization of selected indigenous communities, distortion of cultural identity, and equity in benefit-sharing regimes. Thus in most cases there are no clear regulatory regimes or models that begin to take into account the complexities of traditional knowledge, ownership, equity, and sustainable management of the resource itself that communities can draw upon—such regimes have to be outlined to provide templates that can be built upon by all stakeholders, including communities within organizations. The growing role of the convention in constructing and influencing regulatory regimes for biological diversity and its access cannot be ignored (see www.biodiv.org and Svarstad and Dhillion 2000). There is also a growing body of good and bad cases that lessons can be learnt from.

Of relevance here, as pointed out by Greene, is the market value of the biological resource related to traditional knowledge. The current political climate with regard to the need to implement the provisions of the convention, coupled with the cropping up of traditional knowledge institutes, healers' associations, and biodiversity resource and inventory centers, is leading bio-prospectors to move away from using traditional knowledge as potential leads to new discoveries. The high processing costs associated with post-collection of the biological resource from the field are far from recognized and understood by biodiversity-rich states, let alone indigenous group organizations. Furthermore, the future of bioprospecting, like that of most economic activities, is unpredictable and dependent on biotechnological developments that appear to facilitate moving from the use of traditional knowledge to random collection. The more cost-effective the technological advances, the less the value of bioprospecting and the less will be allocated to R&D budgets on bioprospecting (Rausser and Small 2000). Although the arguments and observations in the case presented by Greene are all relevant, overestimates of expected income can play a even bigger role in influencing the continual interest of bioprospectors in indigenous cultural properties than is assumed. The monopolization of indigeneous culture through organization of indigeneous communities can also lead to the misrepresentation of the actual value of the biological resource, because traditional knowledge may vary widely among communities within an organization. In addition, if such organizations are to be responsible for ensuring the sustainable management of biological resources through their negotiations, then biodiversity conservation may be jeopardized.

CORI HAYDEN
Girton College, University of Cambridge, Cambridge CB3 0JG, U.K. (cph28@cam.ac.uk). 13 XI 03

At the core of Greene's detailed, thoughtful account are the questions of property and politics and the way in which the treatment of culture as property itself becomes an idiom of (political) representation. His discussion graphically illustrates some of the paradoxical effects of bioprospecting: in the harrowing twists and turns of different Aguaruna groups' struggles to define the terms of their participation in the ICBG project, we see yet again that intellectual property and benefit sharing are chronically double-edged avenues to "inclusion" and enfranchisement. For all of its resonance with other prospecting experiments, the Aguaruna story is also a singularly important one, with CONAP's unprecedented negotiation of a "know-how" license with Monsanto-Searle and now the possibility of a jointly owned patent with Washington University. It is telling, perhaps, that these "victories" have come not only with consequential political realignments and marginalizations in Aguaruna communities but with corporate retreat as well.

Greene is right to argue that the eventual profitability of these arrangements is only part of the picture (an undeniably important one for those who have high expectations). He also asks what these mobilizations are *about*, more broadly. In answer, he turns to wider trends toward privatizing or "incorporating" indigenous identity. While he has many useful things to say about how efforts to treat culture as property work both with and against the market, I worry that boiling it all down to the incorporation of "identity" runs the risk of resedimenting or flattening out what is clearly a dynamic, molten set of political and social processes—a series of ongoing experiments with different languages and mechanisms of recognition in which indigenous peoples, NGOs, researchers, and companies are all implicated. As we have seen here and in related work, these efforts cluster strategically around a wide but not arbitrary range of idioms: "property" and "identity" join community, territorial sovereignty, and self-determination as languages of representation that sometimes stand for one another. I would thus be more inclined to keep the account *open* in the terms that Greene, in dialogue with one of his Peruvian interlocutors, productively sets out: as a question about the forms that "representativity" can take.

I cannot help thinking here of an ongoing discussion in the field of bioethics. Some ethicists have recently wondered whether the ethical protocols developed to protect indigenous and aboriginal communities involved in epidemiological or clinical research could be applicable to other kinds of collectivities, such as people with HIV, women with breast cancer, or Ashkenazi Jews. Noting the hauntingly familiar qualities associated with "aboriginal" communities in these discussions (territorial boundedness, shared customs, a common language), one critical review noted that the borrowing seemed doomed—*unless*, the authors argued, we were to jettison those definitions of community and take into account the most important feature that any group, indigenous or not, must have in order for informed consent to work: a system of legitimate political representation. This move—a reminder that in these contexts, community *is* as community *represents*—condenses much for us here.

The requirement of representativity—suffesed with notions of territoriality and identity but readily stripped down to an essential political functionality (the capacity to speak for)—is certainly one of the most striking aspects of benefit-sharing agreements and the intellectual property logics to which they are pegged. One of the biggest problems with intellectual and cultural property, Greene argues, is that they assume and thus must produce "a centralized structure of representative authority . . . with which external actors can negotiate." The effects are often jarring, as in the poetic manner in which the ICBG project defined its contract-signing "community": the "Aguaruna People" are those "who live in the collection area," and the "collection area" is the area "inhabited by members of the Collaborating Organizations." This reads quite smoothly alongside the above-mentioned definition of community as a collective that "works" in political terms. There is something worth attending to here. The problem Greene identifies with the idiom of cultural or intellectual property is not a problem *limited to* or driven by property in any simple sense. The requirement for an authorized quasi-sovereign has become central to the bioscience research enterprise more generally; here, property claims and commercialization arrangements absolutely make their mark, but they also work with and alongside other idioms and mechanisms for "representing interests"—most notably, informed consent—in researchers' and funding bodies' (such as the NIH's) efforts to identify and define their legitimate interlocutors. Greene's piece gives us much to think about regarding bioscience research as a crucial site of struggle over what will count as a collective, as representation, and as a "legitimate" form of political sociality. Property is but one inflection here, and an intriguingly volatile one at that: it seems important to note that, as indigenous peoples increasingly gain the floor with the language of intellectual property, many commentators argue that global intellectual property regimes are scrambling to hold onto some political legitimacy themselves.

WALTER H. LEWIS, GERARDO LAMAS, ABRAHAM VAISBERG, N. ROGERIO CASTRO, AND MEMORY ELVIN-LEWIS

Department of Biology, Washington University, One Brookings Dr., St. Louis, MO 63130, U.S.A. (lewis@biology.wustl.edu). 18 XI 03

Greene has provided a relatively balanced view of parts of the ICBG-Peru project during its 1994-to-2000 tenure. This is true even though he became involved in the project only in September 1999, during the last annual meeting and field collecting in the Department of San Martín, when the majority of the field research had been conducted in the Department of Amazonas. Clearly, for the initial and more controversial years of the project he has had to rely on the opinions, recollections, and hearsay of others, and this has resulted in factual inaccuracies that call for comment.

The ICBG-Peru project grant was awarded not in 1993 (in the first round, three ICBG projects were funded by NIH in 1993 and two in 1994) but on July 1, 1994. The award followed the signing of a letter of intent on April 7, 1994, by the Consejo Aguaruna y Huambisa's chairman, Evaristo Nugkuag, Wraiz Pérez Ramírez, Jorge Fernández, director of the Ministry of Agriculture at Jaén in collaboration with the vice minister of agriculture in Lima, and the four original applicants (Searle, the UPCH, the USM, and WU). Greene refers to this five-page document as a "very simple" letter, but we disagree in the strongest way. We believe that at that time, and even now, it was a forceful and meaningful outline of a collaborative effort to conduct ethnobotanically oriented research with indigenous people. Greene points to "relatively ill-defined terms" with regard to the sharing of future royalties, but he forgets that at this point there was no license option agreement. This agreement, which established the royalty rate, was signed on June 29, 1994. Speed was critical, for without the agreement there would be no grant, and the grant had to be awarded by July 1 or be lost. What was included in the letter of intent was a clear outline of how royalties and other benefits would be shared, the one adopted from the original grant application (submitted in 1992) and the one that we adhere to today, namely, that the four parties would share benefits equally, with a total of 75% of any income returning to Peru. This remains one of the most liberal and progressive arrangements for the sharing of profits of any agreement on bioprospecting known and recognizes the major contribution of the host country in a substantial way. Other minor corrections are that two, not three, institutions participated in collecting, and in 1994 only 4 regional Aguaruna organizations were recognized officially in Lima (of which 3 were our collaborators), far from the current 13 that Greene notes.

With the signing of the license option agreement, royalty rates were essentially established by Searle. The university researchers unfortunately had little guidance on this except for the NIH's assurance that, while the proposed rates were low, they were within the range of other proposals. Clearly this was something that could be reinvestigated later, and under time pressure the agreement was signed on June 29 despite its imperfections. New and more beneficial royalty rates along with other considerations were incorporated into the license option amendment agreement signed in September 1996, prior to any royalties' having been paid at the original rates.

Immediately after the approval of the grant, all three university parties met with the CAH's chairman and staff in Lima on July 7, 1994. At this time a $10,000 partial (annual) up-front payment was provided the CAH for collecting privileges as agreed, along with a copy of the confidential license option agreement. No request was made for a copy in Spanish, but this was provided later, along with a copy of the original grant application of 300 single-spaced pages. It is not the case that the CAH never received a copy of the agreement. A copy was presumably provided by the CAH to RAFI, which posted it on the web without the permission of the parties to the agreement.

Following the July 7 meeting, and carrying a letter of permission from Nugkuag to interact with CAH leaders and communities, we traveled to Imazita in Amazonas, where Aguaruna representatives took us to visit a number of CAH communities along the Río Marañón. In violation of the letter of intent, however, we were not allowed to collect plant material. On our return from CAH-controlled territory, and with the full permission of the Ministry of Agriculture in Lima and Jaén and collecting permits from the ministry's National Institute of Natural Resources (INRENA), we spent two to three weeks collecting material among the mestizo communities (with their permission) and in their privately owned forests as well as on lands owned by the Peruvian government. When people were involved, the prior-informed-consent document was signed before collecting began.

Even with the adverse publicity generated by RAFI regarding the project, we learned that the CAH assembly of communities, meeting in November 1994, had at least tentatively approved it.

In January 1995, with CAH concurrence, the WU team returned to Peru to discuss a formal agreement to extend the content of and replace the letter of intent and to engage in a long-term collecting expedition in collaboration with the CAH. All aspects of the letter of intent and the license option agreement were discussed by the researchers representing the three universities and the CAH chairman and staff between January 5 and January 18, 1995. A number of issues were agreed to, such as the maximum amount of collecting per year, collecting areas, assistance in the field, and security, but the single most important concern was the royalty rate. There was consensus that the rate was too low, and for the sake of all parties WU agreed to challenge Searle and negotiate a more favorable value. Unfortunately, there was no representative of Searle at the meeting, and therefore this negotiation had to be done in St. Louis, with counsels present. We requested permission to collect in the CAH controlled territory as planned and then, on our return to St. Louis, to open a dialogue with Searle on the question of royalties. To our surprise, on January 18 the meeting was abruptly terminated by the CAH, and although Nugkuag was to call Lewis that afternoon to arrange the next meeting, no further contact was made despite repeated attempts by the latter. Another CAH agenda had obviously intervened.

The multiuniversity team left for the Imazita region on January 24, 1995 (not late 1994) to collect again in mestizo and government regions and also to make contact with members of OCCAAM, whose headquarters at

Yamayakat is immediately across the Río Marañón from Imazita. This Aguaruna federation is often described as small and isolated, but its major community is the most readily accessible of all federation communities in the entire Upper Marañón, and it had 27 registered communities and annexes and a population at that time of about 2,150. Renewed contacts were made with its leaders, and on February 26, 1995, an initial agreement was signed in Lima.

During this month-long period many collections were made strictly in the mestizo/*colones*/government areas of the Departments of Cajamarca and Amazonas with full permission, plus a few collections around Yamayakat with OCCAAM Aguaruna informants. No plant collections were made in the CAH communities, and our field collection notes, with full data on precise locality, habitat, informants if applicable, and dates clearly attest to this fact. Greene's statement that during this time from January 26 (first collection in Cajamarca en route to Imazita) to February 23 (last collection well beyond the Aguaruna at 25.6 km north-northeast of Bagua Chica in Amazonas) of 1995 "It is impossible to verify exactly how and where the ICBG team collected" is incorrect.

A few months later (April 1995), RAFI wrote the NIH a letter replete with errors, and later a similar letter was sent to WU's chancellor, which after a thorough examination was rejected by the university. The non-CAH federations in the Río Marañón rallied against this attack on the project, and after an NIH investigation of the "charges" full funding of the project was restored following the new round of discussions (December 1995 and May 1996 in St. Louis with all parties and counsels present) and agreements (September 1996) outlined by Greene. Thereafter, with the assistance of CONAP (the umbrella organization for the four collaborating Aguaruna federations and many other indigenous groups), headed by César Sarasara, and with the full collaboration of the four federations, field research was completed successfully in 1999, to be followed by continued research at the universities. As noted by Greene, Searle withdrew from the consortium that year (although it never cancelled the know-how license agreement as he reports). Regarding the intellectual property shared by the four parties equally as part of a utility patent submitted to the U.S. Patent and Trademark Office in 2003, the portion (25%) assigned to CONAP was accepted solely on behalf of the four collaborating federations—OCCAAM, FAD, FECONARIN, and OAAM.

KATHLEEN MC AFEE
School of Forestry and Environmental Studies, Yale University, Sage Hall, 205 Prospect St., New Haven, CT 06511, U.S.A. (kathleen.mcafee@yale.edu).
24 XI 03

As Greene shows, the notion of intellectual property is an unwieldy framework for conceptualizing and trading cultural knowledge, genetic information, or indigenous symbols. He asks what is at stake when indigenous authorities negotiate the market value of their peoples' knowledge and rightly emphasizes the pitfalls of inflated expectations and the problematic politics of indigenous representation. Similar dissection of the representational politics of the ICBG and other intellectual-property brokers would reveal that even more is at stake than Greene's account indicates. Contracts to trade genetic resources and related knowledge were first promoted as the model for managing "global" natural resources by means of market exchange. This strategy of selling nature to save it has since been extended to markets in carbon sequestration and other ecosystem services.

In the negotiations over biodiversity "benefit sharing" between countries and communities in the global South and nations and corporations in the North, such bilateral contracts are counterposed to alternative approaches that aim more directly at redressing past plunder and present inequalities in resource control. In UN fora in the 1980s, the mirage of windfalls from exports of genetic gold persuaded many Southern governments to drop their opposition to the patenting of medicines and crop varieties. Some are only now reviving their earlier claims that staple crops and vital drugs, many of which are derived from Southern source materials, ought to be available to everybody. In the 1990s, hopes for income from bioprospecting deals helped persuade delegates from diversity-rich states to accede to U.S. insistence that the Convention on Biological Diversity recognize intellectual property rights and that the World Trade Organization include provisions for the worldwide applicability of intellectual property to plants and microorganisms. Similar illusions of gains from genetic-information trade underlie current maneuvers by China and Brazil to exclude important food crops from the new International Treaty on Plant Genetic Resources for Food and Agriculture. This accord is significant precisely because it is a multilateral alternative to the market-contract model, one that would keep vital crop germplasm in the public domain and prevent beggar-thy-neighbor competition among communities and countries to undersell their genetic resources.

The policy discourse of biodiversity benefit sharing and bioprospecting focuses on marketable genetic information as if organisms and genes acquired value only when privatized, removed from their unique ecological and cultural contexts, and transformed by high technology. This construction devalues the benefits that communities everywhere *already* derive from the ecosystems that sustain them materially and culturally.

In their efforts to aid their communities, legitimate indigenous representatives may be as entrepreneurial as anyone else, but claiming and selling property rights is a weak basis for economic bargaining. Contracts for trade of genetic resources and associated knowledge are dressed up as if they were economically rational exchanges based upon values (prices) objectively determined by supply and demand. What indigenous participants in biopropecting arrangements are actually offering is legitimacy—their own in the eyes of outsiders—based on some mix of outsiders' guilt, goodwill, and perceptions of special indigenous relationships with nature. The commodities purchased are rarely related to any local knowledge; organic samples are assayed for bioactivity of interest in entirely different technological and economic contexts. Now, supply exceeds demand in what has become a biobuyers' market. Economic and political power asymmetries ensure that, even where local knowledge leads prospectors to useful materials, substantial local returns are unlikely. Unrealistic hopes for royalties from genetic-information sales have led some to advise communities to keep knowledge secret, but hiding and hoarding are likely to speed the loss of that knowledge.

If trading cultural "property" is understood as a political negotiation rather than a market transaction, the risks and potential gains for knowledge providers can be analyzed more realistically. Indigenous and other communities can rescue, preserve, and increase useful local knowledge by documenting it in community registries and publications of resources, inventions, practices, and traditions, thus helping to establish it as "prior art" that cannot legally be appropriated or patented. They can claim collective rights to documented knowledge and negotiate know-how licenses or other terms for allowing access to it. Backed by the requirements of the Convention on Biodiversity that collectors obtain "prior informed consent," they can bargain for remuneration, with total payments specified in advance. Classification, analysis, and processing of plant samples and botanical information can add value to local resources for local purposes and for sales to others. Such arrangements can highlight continuing processes of collective invention, bring status to knowledge contributors, and strengthen group identity and organization. But selling genetic information and cultural capital has in common with selling handicrafts, forest products, crops, or ecotourism packages that export markets are rarely adequate, reliable livelihood sources. The extent to which they are helpful depends on how much they contribute to broader strategies for applying local resources and knowledge to local development objectives.

The same criterion applies to damage caused by biopiracy. Unauthorized use of genetic resources and information does not prevent local people from continuing to use such resources. However, it can prevent them from commercializing crop varieties, medicinal plants, etc., if these are legally appropriated by competitors, as in the case of the yellow beans from Mexico patented by a U.S. bean dealer. Whether biopiracy causes material harm depends on how the resources involved fit into local strategies for cultural survival and economic development.

HANNE VEBER

Department of Language and Culture, Roskilde University, P.O. Box 260, DK-4000 Roskilde, Denmark 14 XI 03

Greene's paper highlights a feature of contemporary politics that invites our consideration of not only the politics of culture but also the political economy of culture-as-politics. Revolving around a sequence of bioprospecting arrangements among the Aguaruna in Peru's high forest, the paper exposes the absurdities of strategies that claim traditional local knowledge as cultural property. Obviously, Greene is not the first to question the sobriety of efforts toward copyrighting local traditional culture, whether on the part of indigenous groups or corporate interests, but his case represents a timely exposé of the implications of such efforts in terms of evoking unrealistic expectations of easy economic windfalls, creating new cleavages or aggravating existing divisions, and accelerating rates of leadership turnover within indigenous groups. Greene's presentation concedes that temporary advantages flow from the economic conjunctures implicated in the politics of culture as property, including opportunities for leadership training or funding for intellectual training related to bioprospecting and environmental conservation. However, the Aguaruna experiences also suggest a real risk of exhausting indigenous leadership resources among their own constituency or subverting the credibility and legitimacy of indigenous leaders in the eyes of the nonindigenous world.

Looking into the role of NGOs, researchers, bureaucrats, and private corporations in the process of raising indigenous claims to intellectual and cultural property in bioprospecting contexts, Greene calls attention to the tendency for naiveté on the part of prospectors as to the complexities of indigenous politics to be matched by an equally naïve idealism on the part of NGOs, legal experts, and others advocating indigenous rights to culture as property. He demonstrates how these agencies are exploited through the cunning of indigenous leaders aiming to reap whatever profits—symbolic, economic, organizational, or political, large or small—may be extracted from the situation. In this way indigenous strategies emerge as modes of appropriating sources of power from the dominant society—or from the logic of the market, as Greene has it—and working against it. This is a welcome perspective that could become revelatory of much that remains to be explored in the fissures between culture-as-historical-isolate ethnography and activism-informed globalizing orientations in contemporary Amazonian research.

The paper offers the contours of a political-economic perspective on the dynamics of indigenous mobilization centered on bioprospecting arrangements and their varied effects on the shaping of indigenous political culture; with some analytical elaboration and specification the approach may lead research toward a truly productive interpretation of the interconnections and contingencies that give rise to indigenous and other strategies of converting culture into capital through a politics of culture and identity. Insofar as such strategies remain highly taxing for leadership credibility and indeed for the authenticity of indigenous identity, they are potentially subversive of indigenous culture as a resource for engagement. The phenomenon reflects its origin as a product of the localization aspect of a globalization process in which the meaning of the local is established in a global framework.

Greene rightly points out that being indigenous in our era has become a historically validated subject position tied to a conscious strategy of multiply configured identity-based political action that appears to take on world-historical proportions. Obviously, indigenous leaders are in the forefront of emergent indigenous activist institutions. Far from being innocent pawns in others' games, indigenous peoples actively engage in and influence regional, national, and transnational networks of power. If they are making history, they are making history under conditions not of their own making. The million-dollar question remains what exactly is the status of the action aspect of indigenous strategies of mobilization for or against the logic of the market. Indeed, deciphering the contradictory elements inherent in the cultural politics of propertied culture not separately as economics *and* cultural symbolics but *in their unity* calls for a perspective beyond a one-sided logic of the market. Yet the cultural politics of Aguaruna bioprospecting arrangements are certainly tied into market contexts; culture and the symbolics involved here reflect these links in ways that may hardly be conceived exhaustively if these contexts are not made central to the analysis. Greene brings this point home nicely.

Reply

SHANE GREENE
Chicago, Ill., U.S.A. 9 XII 03

The author's reply in CA tends to be either a sharp countermove aimed at intellectual self-defense, a tactical circumvention of well-founded criticisms, or some combination of the two. This response will probably be interpreted no differently. The comments are in my view best grouped into four broad categories: congratulatory (Brush, Castree, Veber), complementary (Dhillion, McAfee), substantively critical and engaging (Hayden, Banister, Lewis et al.), and absurd (Banerjee).

Brush, Castree, and Veber congratulate me on an effort that exposes the complexities of contemporary bioprospecting and indigenous cultural politics by linking them to questions related to the Convention of Biological Diversity, indigenous representation, and global political economy more generally. Dhillion and McAfee appear to be convinced by the argument as well and seek only to expand its ramifications. The former draws parallels to cases in Southeast Asia, and the latter places pharmaceutical bioprospecting in the wider context of bio-genetic engineering as a whole and the need to reconsider the local dimension of development schemes.

Hayden's and Bannister's appreciative comments are qualified by requests for clarification on key points of the argument. Hayden's main concern is that I "flatten out" the issue by "boiling it down" to the "incorporation of identity." She suggests that I should have left the argument more "open" by aligning property and identity with community, territorial sovereignty, and self-determination. I don't know what she means by "open," but I wonder if she isn't flattening me out a bit. I explicitly relate the indigenous knowledge, identity, and property issue to a social reality of shifting corporate and leadership alliances among titled, administratively autonomous communities. Likewise, I expose the mysterious reliance on the category "Aguaruna People" in the ICBG's contract negotiations, which implies an abstract ethno-political, corporate actor roughly equivalent to an Aguaruna "imagined community." In fact, as I show, these two facts explain much about why the ICBG caused such controversy and worked to legitimate exclusionary legal fictions in the first place (part of a broader problem, as the Maya ICBG and other cases suggest). Further, the introductory and concluding sections clearly articulate cultural property claims with a broader politics of indigenous recognition, self-determination, and territorial claims that have now found a reception not only in local social movements but also in international legislative forums like the UN. I even suggest that indigenous territorial claims effectively trump cultural property claims despite the increased attention to the latter in global indigenous debates.

Bannister believes that I generalize too much about the role of corporations and NGOs on the basis of the Aguaruna case. She adds some important details about Searle's participation and withdrawal but nothing that detracts from the generally accepted observation that pharmaceutical bioprospecting creates many more expectations than it does realistic possibilities for addressing, much less fulfilling, them. She is probably right that I extrapolate too easily from RAFI's role in two ICBG projects. There are other examples. The NGO Third World Network provided much of the public forum for Vandana Shiva and others' discussion of the neem seed patent case. The common conflations deployed in that case between indigenous people, Third World identity, and the Indian nation-state rely on some of the same simplistic representations I discuss in the ICBG-Aguaruna case. Further, it is worth reiterating that I do not merely criticize the actions of NGOs. I explicitly recognize their importance in providing support, counsel, and leverage in other instances. Indeed, the indigenous organizations in question are most often NGOs as well. The point is that all the actors mentioned, including NGOs, indigenous leaders, corporations, bureaucracies, and researchers (among them myself), are implicated in the process of distorting representations to fit exclusionary legal fictions that become legal realities through the practices and politics of indigenous incorporation and legitimation.

I interpret Lewis et al.'s "just the facts" response as an attempt to set the record straight. They respond directly and only to the factual basis of specific events related to early ICBG-Aguaruna negotiations. Leaving aside a few "factual inaccuracies," they believe, as does Bannister (who came to know many of the same actors), that I present a "relatively balanced view." Therefore, I must assume they do not strongly disagree with the overall argument. Lewis read an earlier version of the text before it was submitted for publication, and unfortunately many of the points of clarification are surfacing only now as the paper goes to press. This is, however, the second time that he has had to instruct me to say that Searle "withdrew" from the know-how license instead of "canceled" or "terminated" it in order to be legally correct (that is my semantic oversight entirely). They are justified in linking the original chronology of the Consejo letter of intent and the license option agreement to the difficult position the ICBG researchers were placed in as a result in round one. As they state, Searle initially unilaterally dictated the royalty rates (a point I failed to make) and made no effort to negotiate directly with Aguaruna representatives until forced to do so by the need to renegotiate in round two. They draw attention to their early intention to pursue higher royalties even while still negotiating with the Consejo Aguaruna Huambisa and clarify that the agreements signed in the second round included a royalty rate increase. Too late I realized that in one of the many attempts to shorten this piece I cut a sentence that remarked on that royalty increase (which I acknowledge only as "different").

They seem surprised that I mention 13 local Aguaruna-Huambisa organizations, 2 of which (ODECOFROC and FECONARSA) existed but were not legally recognized in the Department of Amazonas at the time of the negotiations. They would be less surprised if they related this to my broader argument about the ongoing historical shifts in Aguaruna authority, institutional legitimacy, and attempts at indigenous incorporation. It would also help if they took a clearer position on the contractual use of ethnonyms like "Aguaruna People," which is limited to a specific subset of legally recognized "collaborating" Aguaruna institutions and exclusive unless other indigenous organizations apply to CONAP and Affiliates for inclusion (thus legitimating CONAP and Affiliates' intent to incorporate the "Aguaruna People"). Even if we indulge in the use of such legal fictions to define contractual indigenous partners, my data (from Peruvian public records) show that at least 9 local organizations with Aguaruna members were legally recognized at the time of the negotiations (CAH, FAD, FECONADIC, FECONARIN, Chapi Shiwag, OAAM, OCCAAM, ONAPAA, and ORIAM). Some of these organizations are from other provincial departments (Loreto and San Martín) that were initially excluded from the "collection area" in the contractual agreements, but that is my point. "Aguaruna People" does not exclusively correspond to any consortium of these organizations in this or that provincial department outside these kinds of business-model social relations and contractual negotiations.

Lewis et al. offer a clear response to the statements of Evaristo Nugkuag that I cite as evidence about the early negotiations. I commend them for clarifying their position and citing their field notes as evidence for what happened during the controversial field collections in late 1994 and early 1995. They argue that my dependence on "opinions, recollections, and hearsay" is responsible for the factual inaccuracies they identify in my account. However, I relied not just on oral accounts but also on texts from indigenous institutions and authorities. The danger here is in the implied contrast between untrustworthy local sources and scientific/legal statements portrayed as "factual." At the very least, the dialogue between history, anthropology, and politics now forces us to acknowledge that "opinions, recollections, and hearsay" are not as far from "factual evidence" as the great positivist scientists once believed. Indeed, the very impulse to separate things out into two mutually exclusive categories of discourse—"facts" versus "opinions, recollections, and hearsay"—reveals a reliance on a cultural framework that still places "science" (as factual knowledge) at the apex of a hierarchy of human knowledge types. Such an understanding fails to recognize that "opinions, recollections, and hearsay" may have their own value as legitimate knowledge in a given cultural, political, or historical context (not necessarily excluding ours) and may even be as valued there as the scientific "fact" is by the intelligentsia and lawmakers of our society. The controversial aspects of the ICBG-Aguaruna case were as much about a tremendous clash of culturally, politically, and socially defined expectations as anything else. The only possible solution to resolving such intercultural conflict entails trying to understand the role of differing expectations, how they were produced in the first place, and how history is retold (and "recollected") by different actors. This is in essence the long-term project to which anthropology is committed, but it is also of major relevance for the other actors now engaged in such highly politicized engotiations.

Ironically, Lewis et al.'s comment stands in stark opposition to Banerjee's. While Lewis et al. criticize me for letting too many indigenous sources speak, Banerjee says that "one voice is missing, and that is the voice of the people." If this is so, what are all my direct quotations from indigenous leaders (Nugkuag, Juwau, Jacanamijoy) and indigenous-produced texts that claim to speak for indigenous people? What about the citations of the work of indigenous scholars and Third World activists (Montejo, Shiva, Ewen) addressing global audiences and gaining their own "voices" at the UN rather than simply being "quoted" by the anthropologist? We live in an age when indigenous actors help bring multimillion-dollar international pharmaceutical projects to a halt, successfully revoke patents, win Nobel Prizes, prevail in cultural property law suits, and routinely meet to discuss international legislation. And yet Banerjee remains convinced that "Spivak's subaltern" (his phrase not mine; notice the possession) cannot speak and that "we are their authors." Far from being "silent about the disparities in power" and failing to "accept . . . the legacy of the colonial era," the entire article is about "disparities in power" and premised on the idea that indigenous identity is configured within a colonial legacy. Banerjee's comment stands out from the others in failing to address the actual argument of the paper or the indigenous struggles it chronicles. Indeed, I wonder if in fact his recourse to Spivak (who is more in dialogue with French poststructuralists than with any subalterns) as the relevant authority does anything more than reduce anthropology to the "bourgeoisie scratching its head," to employ one of Edmund Leach's famous phrases.

Acknowledgements

At different stages various institutions made possible the field-work on which this article is based: the Social Science Research Council, the Wenner-Gren Foundation, the Fulbright program, the Tinker Foundation, and the University of Chicago's Center for Latin American Studies. Manuela Carneiro da Cunha and the participants in her intellectual property rights seminar at the University of Chicago read and commented on a very early draft of the paper in the winter of 2000. Since then several other readers have pushed me to improve the argument: Jean Comaroff, Michael Brown, Josh Rosenthal, Steve King, three anonymous reviewers, and probably others along the way. Most of my debt is due to Brendan Tobin, the members of the ICBG team, and several Aguaruna leaders and community members. Particularly deserving of mention are Walter and Memory Lewis, Rogerio Castro, Steve Caspers, Ricardo Apanu, César and Jorge Sarasara, Evaristo Nugkuag, José Catip, Adolfo Juep, and all the other federation leaders. I retain full responsibility for any flaws in the ideas expressed here.

Notes

1 Richard Handler's (1985) essay on the politics of cultural property legislation and the expansion of the concept of *patrimoine* in Quebec demonstrates that such politics is also central to certain forms of nationalism and subnationalism. National and subnational polities, he suggests, especially emerging ones, claim to "have" culture and construct policies of legal exclusion, rights, and ownership around it in an effort to bolster their political legitimacy. Here I concentrate more on issues of representation and political-economics than on the forms of nationalism and subnationalism of which such claims are part.

2 Posey's work brought anthropologists into the debate about biodiversity and the ethics of ethnobotanical appropriations of traditional medical knowledge for pharmaceutical research just before the Second International Congress of Ethnobiology in 1990, where the issue was more thoroughly discussed. In a later publication with Dutfield, Posey places the indigenous struggle over traditional knowledge and intellectual property in the context of the larger struggle over indigenous and local groups' traditional resource rights (Posey and Dutfield 1996).

3 Other forms of bioprospecting, agro-industrial and human-genetic, are certainly tied to the larger debate about biodiversity and access to and ownership of biogenetic resources. While the acquisition of new seed varieties apparently makes up

a significant part of agricultural industry practices (see Brush 1999), in every case this also involves access to the human innovation (i.e., cultivation practices) attached to it. In pharmaceutical bioprospecting, in contrast, many companies and researchers prefer a random or taxonomic methodology for collecting biological material to ethnobotanical methods that involve traditional knowledge and indigenous rights (see Aylward 1995; Moran, King, and Carlson 2001:506).

4 The United States did not sign the convention, and most interpretations of its refusal assume that the measures favoring benefit sharing and national sovereignty over biodiversity resources represented a potential threat to U.S. corporate interests in foreign biological resources. However, this does not neatly fit with the virtually simultaneous appearance of the ICBG program, in which the basic principles of the convention are clearly taken into account.

5 He was awarded the Right Livelihood (Alternative Nobel) Prize in 1986 and the Goldman Environmental Prize in 1991 in recognition of his leadership in the Asociación de Desarrollo Interetnica de la Selva Peruana (AIDESEP) and the Coordinadora de Organizaciones Indígenas de la Cuenca Amazónica (COICA) and for work in the areas of indigenous rights and environmental conservation. While he is clearly an effective and charismatic leader, the degree to which he is representative of indigenous Amazonian peoples has been questioned by Amazonian indigenous organizations that compete for member support (see Smith 1996).

6 RAFI has since changed its name to ETC Group (see www.etcgroup.org).

7 Notably, the 300 samples collected during the disputed time period were returned not to the Aguaruna but to the Peruvian Ministry of Agriculture (Tobin 1999*a*). Regardless of the truth or falsity of the various accusations leveled at him by the Consejo (which I cannot judge, not having been present), Lewis now earnestly acknowledges his "polítical naiveté" (see Lambrecht 1998) as regards the complexity of Aguaruna politics and organization structure.

8 Acceptance by CONAP and Affiliates of course did not automatically mean acceptance by all Aguaruna communities formally affiliated with those organizations. In many instances, individual communities challenged CONAP and Affiliates' authority to accept the project on their behalf and refused to permit the ICBG researchers to work in their communal territory despite their affiliation with one of the participating organizations. While there is not enough space to document all this local dissent, it is important that it be mentioned, since even the apparent incorporation of these organizations provoked substantial internal debate, discussion, and disagreement among Aguaruna community leaders.

9 The information presented about this meeting of Aguaruna organizations with the ICBG researchers, SPDA, and Searle is taken from notes that form an appendix to the ICBG biological collecting agreement signed in 1996. Ironically, despite the absence of then-active leaders of the Consejo Aguaruna Huambisa, the secretary, Santiago Manuig, was a former president of the Consejo (1989–93). Having developed his own conflictive relationship with Nugkuag, he used the meeting to call into question the activities of his successor.

10 The Peruvian state, despite its signing of the Convention on Biological Diversity and the participation of two major Peruvian universities (one public and one private), was notably absent (except for the requirement of licenses from the Ministry of Agriculture) in both rounds of the ICBG-Aguaruna negotiations (Tobin n.d.:6–7). Interestingly, however, on the basis of its experience with the case the state has begun promoting the opposite approach. The agreements between the Aguaruna and the ICBG and Searle have been used as a model for proposed

legislation for access to genetic resources and traditional knowledge in which the state will take more of an interventionist role (see Tobin n.d.:31–32).

11 The number 3,500 is slightly misleading in that, according to Lewis (personal communication), many of these extracts are duplicates of the same plant. It is also important to note that other nonmedicinal plants (some of them unidentifiable by the Aguaruna) were collected for the purposes of a biodiversity inventory (samples to be sent both to the Museo de San Marcos in Lima and to the Missouri Botanical Gardens) that was not made available for screening. Therefore the total number of plants collected was in the neighbourhood of 4,800 (Lewis, interview, 2000).

12 Sarasara's interactions with the press are various (for examples, see Lambrecht 1998, Chatterjee 1997). He was awarded the first José Guallart Prize by the International Society of Ethnobiology (Lewis, personal communication) and is a coauthor of Lewis et al. (1999).

13 The detailed history of indigenous organizations is best left for another paper. While the Consejo was originally intended to represent several river areas (Marañon, Chiriaco, Nieva, Santiago, Cenepa, Domingusa) and managed many projects, OCCAAM was largely isolated. FECONARIN and FAD were founded around 1990 by Aguaruna disillusioned with the Consejo, and by the mid-1990s ODECOFROC and FECONARSA had appeared in the same area with familiar criticisms of the Consejo. The current leaders of the Consejo are extremely self-conscious about the organization's fragmentation and loss of prestige. The past two or three presidents have all begun taking measures to address what they call the Consejo's "crisis." So far the results are fairly meager.

14 Some would argue, of course, that the mere fact that the plants have already been identified by the Aguaruna as medicinally useful is in itself an intellectual contribution (a sort of prescreening). My own opinion is that the contribution is restricted at that point to a purely abstract utility without any contextual, practical, medical, cultural, or social content. Generally ethnobotanists believe that the chances for a plant identified by local informants to become useful in research and development are much greater if detailed and specific information about its "usefulness" is used as a guide in researching how the plant might be developed (see Farnsworth 1990).

15 For an account of the division of research interests between Searle, UPCH, and Washington University and their results thus far, see Lewis et al. (1999).

16 Gaining a fuller impression of Searle's interests and motivations has proved difficult, especially given corporate turnovers and takeovers. I spoke with an ex-Searle representative who had previously worked closely with the ICBG researchers and had contact with CONAP and Affiliates. He chose not to go on record with any information about the project, given that Searle was no longer his employer. Press releases from Searle related to the project were unavailable because of the corporate restructuring. When Monsanto merged with Pharmacia/Upjohn, Searle was dissolved and restructured into Pharmacia Corporation. A media representative from Pharmacia generously provided me with some articles that mentioned the ICBG project from the company archive, but he was unable to locate any press releases distributed by Searle itself and had no information on current contacts within the company with knowledge about the ICBG project.

17 The patent makes specific mention of the individual informants who identified the plants from which the patented compounds have been isolated and the Aguaruna communities in which they live (Walter Lewis, personal communication). The decision to include only CONAP in representation of the other four Aguaruna federations, rather than all of them together as in the biological collection

agreement, in the patent application is curious. As a national-level actor CONAP is obviously more stable and convenient for coordinating such matters. Unlike the other local federations involved, however, it is not an exclusively Aguaruna organization but an interethnic confederation, and other indigenous Amazonians constitute the majority of CONAP's local affiliates (particularly important are the Shipibo and Yanesha peoples). None of them had anything to do with the ICBG collections, but as participating members of a democratically run organization they presumably have as much claim to the patented property as the Aguaruna organizations that CONAP represents. The complexities of identifying stakeholders, contractual partners, and property owners of indigenous knowledge seem only to multiply.

References

AGUILAR-STØEN, M., AND S. S. DHILLION. 2003. Implementation of the Convention on Biodiversity in Mesoamerica: Environment and development aspects. *Environmental Conservation* 30:131–38. [SSD]

ALBERS-SCHÖNBERG, GEORG. 1995. "The pharmaceutical discovery process," in *Intellectual property rights and biodiversity conservation.* Edited by T. Swanson, pp. 67–92. Cambridge: Cambridge University Press.

ALBÓ, XAVIER. 1991. El retorno del indio. *Revista Andina* 9: 299–345.

ALTMAN, JON. 2001. *Sustainable development options on aboriginal land: The hybrid economy in the twenty-first century.* Australian National University Centre for Aboriginal Economic Policy Research Discussion Paper 226. [SBB]

ARTUSO, ANTHONY. 1997. "Capturing the chemical value of biodiversity: Economic perspectives and policy prescriptions," in *Biodiversity and human health.* Edited by F. Grifo and J. Rosenthal, pp. 184–204. Washington, D.C.: Island Press.

AYLWARD, BRUCE. 1995. "The role of plant screening and plant supply in biodiversity conservation, drug development, and health care," in *Intellectual property rights and biodiversity conservation.* Edited by T. Swanson, pp. 93–126. Cambridge: Cambridge University Press.

BALLÓN AGUIRRE, FRANCISCO. 1988. *Política de la supervivencia.* Centro de Investigatigación y Promoción Amazónica, Temas Amazónicas 4.

BANERJEE, S. B. 2000. Whose land is it anyway? National interest, indigenous stakeholders, and colonial discourses: The case of the Jabiluka uranium mine. *Organization and Environment* 13 (1):3–38. [SBB]

BEBBINGTON, ANTHONY, GRAHAM THIELE, PENELOPE DAVIES, MARTIN PRAGER, AND HERNANDO RIVEROS. 1993. *Non-governmental organizations and the state in Latin America.* London: Routledge.

BERLIN, BERLIN, AND ELOIS ANN BERLIN. 2002. Comment on: Maya medicine in the biological gaze: Bioprospecting research as herbal fetishism, by Ronald Nigh. CURRENT ANTHROPOLOGY 43:466–67.

BOYLE, JAMES. 1996. *Shamans, software, and spleens.* Cambridge: Harvard University Press.

BROWN, MICHAEL. 1993. Facing the state, facing the world: Amazonia's native leaders and the new politics of identity. *L'Homme* 33:307–26.

——. 1998. Can culture be copyrighted? CURRENT ANTHROPOLOGY 39:193–222.

——. 2003. *Who owns native culture?* Cambridge: Harvard University Press.

BRUSH, STEVEN. 1994. "A non-market approach to protecting biological resources," in *Intellectual property rights for indigenous peoples*. Edited by T. Greaves, pp. 133–43. Norman, Okla.: Society for Applied Anthropology.

——. 1996. "Whose knowledge, whose genes, whose rights?" in *Valuing local knowledge*. Edited by S. Brush and D. Stabinsky, pp. 1–21. Washington, D.C.: Island Press.

——. 1999. Bioprospecting the public domain. *Cultural Anthropology* 14:535–55.

BRYSK, ALISON. 2000. *From tribal village to global village*. Stanford: Stanford University Press.

CARNEIRO DA CUNHA, MANUELA, AND MAURO W. B. DE ALMEIDA. 2000. Indigenous people, traditional people, and conservation in the Amazon. *Daedalus* 129:315–38.

CASTREE, N. 2003. Bioprospecting: From theory to practice (and back again). *Transactions of the Institute of British Geographers* 28:35–55. [NC]

CHATTERJEE, PRATAP. 1997. Environment: Peru native peoples' stake in medicinal plant deal. *Inter Press Service, Global Information Network*, March 6.

CHIRIF, ALBERTO, PEDRO GARCÍA, AND RICHARD SMITH. 1991. *El indígena y su territorio son uno solo*. Lima: OXFAM and COICA.

CLARK, JOHN. 1990. *Democratizing development: The role of voluntary organizations*. West Hartford: Kumarian Press.

COICA (COORDINADORA DE LAS ORGANIZACIONES INDÍGENAS DE LA CUENCA AMAZÓNICA). 2002. Patente de la planta sagrada ayahuasca. Web site accessed 1/30/2003: www.coica.org/interna.asp?s = 5&r=10.

COMAROFF, JEAN, AND JOHN COMAROFF. 1991. *Of revelation and revolution*. Vol. 1. Chicago: University of Chicago Press.

CONKLIN, BETH. 2002. Shamans versus pirates in the Amazonian treasure chest. *American Anthropologist* 104:1050–61.

CONKLIN, BETH, AND LAURA GRAHAM. 1995. The shifting middle ground; Amazonian Indians and eco-politics. *American Anthropologist* 97:695–710.

CONSEIL, DOMINIQUE. 2003. Aveda announces discontinuation of indigenous product CollectionAveda. Accessed on November 14, 2003 (http://www.aveda.com/about/press/indigenous.asp). [SBB]

COOMBE, ROSEMARY. 1998. *The cultural life of intellectual properties*. Durham: Duke University Press.

——. 1999. Intellectual property, human rights, and sovereignty; New dilemmas in international law posed by the recognition of indigenous knowledge and the conservation of biodiversity. *Indiana Journal of Global Legal Studies* 6:59–116.

COX, PAUL, AND MICHAEL BALICK. 1994. The ethnobotanical approach to drug discovery. *Scientific American*, June.

DHILLION, S. S., H. SVARSTAD, C. AMUNDSEN, AND H. C. BUGGE. 2002. Bioprospecting: Effects on the environment and development. *AMBIO* 31:491–93. [SSD]

DOVE, MICHAEL. 1995. "Center, periphery, and biodiversity: A paradox of governance and a developmental challenge," in *Valuing local knowledge*. Edited by S. Brush and D. Stabinsky, pp. 41–67. Washington, D.C.: Island Press.

DYE, LEE. 2000. Cures from the forest. August 24, Web site accessed 8/15/03: abcnews.go.com/sections/science/DyeHard/dyehard000824.html.

Economist. 1999. Shaman loses its magic. February 20, p. 77.

ELISABETSKY, ELIZABETH. 1991. Folklore, tradition, or know-how? *Cultural Survival Quarterly*, Summer, pp. 9–13.

EWEN, ALEXANDER. 1994. *Voice of indigenous peoples: Native people address the United Nations*. Santa Fe: Clear Light Publishers.

FARNSWORTH, NORMAN. 1990. "The role of ethnopharmacology in drug development," in *Bioactive compounds from plants*, pp. 2–11. Ciba Foundation Symposium 154. Chichester: Wiley.

FISCHER, EDWARD. 1999. Cultural logic and Maya identity: Rethinking constructivism and essentialism. CURRENT ANTHROPOLOGY 40:1–61.

FRIEDMAN, JONATHAN. 1999. Indigenous struggles and the discreet charm of the bourgeoisie. *Australian Journal of Anthropology* 10(1):1–14.

GALE, DENNIS. 2001. Company settles suit over Crazy Horse Malt Liquor. *Montana Forum*, Web site accessed 1/20/03: http://www.montanaforum.com/rednews/2001/04/27/build/tribal/27liquorsuit.php?nnn = 3.

GERSH, JEFF. 1999. Capitalism goes green? *San Diego Earth Times*, May. Web site accessed 3/10/02: www.sdearthtimes.com/eto599/eto599s3.html.

GOERING, LAURIE. 1995. Rain forests may offer new miracle drugs. *Chicago Tribune*, September 12.

GRAHAM, LAURA. 2002. "How should an Indian speak?" in *Indigenous movements, self-representation, and the state in Latin America*. Edited by K. Warren and J. Jackson, pp. 181–228. Austin: University of Texas.

GREENE, SHANE. 2002. "Intellectual property, resources or territory? Reframing the debate over indigenous rights, traditional knowledge, and pharmaceutical bioprospection," in *Truth claims: Representation and human rights*. Edited by M. Bradley and P. Petro, pp. 229–49. New Brunswick: Rutgers University Press.

GUPTA, AKHIL, AND JAMES FERGUSON. 1992. Beyond "culture": Space, identity, and the politics of difference. *Cultural Anthropology* 7:6–23.

HAAS, JONATHAN. 1996. Power, objects, and a voice for anthropology. CURRENT ANTHROPOLOGY 37:S1–22.

HANDLER, RICHARD. 1985. "On having a culture," in *Objects and others*. Historical Anthropology 3. Edited by George Stocking, pp. 192–217. Madison: University of Wisconsin Press.

HARDT, MICHAEL, AND ANTONIO NEGRI. 2000. *Empire*. Cambridge: Harvard University Press.

HAYDEN, CORI. 2003. From market to market: Bioprospecting's idioms of inclusion. *American Ethnologist* 30(3):1–13.

——. n.d. "Prospecting's publics," in *Property in question*. Edited by K. Verdery and C. Humphrey. Oxford: Berg. In press.

HELLER, M. A., AND R. S. EISENBERG. 1998. Can patents deter innovation? The anticommons in biomedical research. *Science* 280:698–701. [SB]

KEARNEY, MICHAEL, AND STEFANO VARESE. 1995. "Latin America's indigenous peoples: Changing identities and forms of resistance," in *Capital, power, and inequality in Latin America*. Edited by S. Halebsky and R. Harris. Boulder: Westview Press.

KING, STEVEN, AND MICHAEL TEMPESTA. 1994. "From shaman to human clinical trials," in *Ethnobotany and the search for new drugs*. Ciba Foundation Symposium 185. Edited by D. Chadwick and J. Marsh, pp. 197–206. Chichester: Wiley.

LAMBRECHT, BILL. 1998. Bioprospectors find they must be deal-makers as well as scientists. *St. Louis Post-Dispatch*, October 11.

LEWIS, WALTER, GERARDO LAMAS, ABRAHAM VAISBERG, DAVID CORLEY, AND CESAR SARASARA. 1999. Peruvian medicinal plant sources of new pharmaceuticals

(International Cooperative Biodiversity Group-Peru). *Pharmaceutical Biology*, suppl., 37:69–83.

LEWIS, WALTER, AND VEENA RAMANI. n.d. Ethics and practice in ethnobiology: Analysis of the International Cooperative Biodiversity Group Project in Peru. MS.

MACILWAIN, COLIN. 1998. When rhetoric hits reality in debate on bioprospecting. *Nature* 392:535–40.

MEYER, CARRIE. 1999. *The economics and politics of NGOs in Latin America*. Westport: Praeger.

MONSANTO. 1998. *1997 report on sustainable development*. St. Louis.

MONTEJO, VICTOR. 2002. "The multiplicity of Mayan voices: Mayan leadership and the politics of self-representation," in *Indigenous movements, self-representation, and the state in Latin America*. Edited by K. Warren and J. Jackson, pp. 123–48. Austin: University of Texas.

MORAN, KATY, STEVEN KING, AND THOMAS CARLSON. 2001. Biodiversity prospecting: Lessons and prospects. *Annual Review of Anthropology* 30:505–26.

MUEHLEBACH, ANDREA. 2001. "Making place" at the United Nations: Indigenous cultural politics at the U.N. Working Group on Indigenous Populations. *Cultural Anthropology* 16: 415–48.

NIEZEN, RONALD. 2003. *The origins of indigenism*. Berkeley: University of California Press.

NIGH, RONALD. 2002. Maya medicine in the biological gaze: Bioprospecting research as herbal fetishism. CURRENT ANTHROPOLOGY 43:451–77.

O'CONNER, TANYA. 2000. Interest drops in rainforest remedies. *Health24News* 1, July 7.

ORLOVE, BENJAMIN, AND STEPHEN BRUSH. 1996. Anthropology and the conservation of biodiversity. *Annual Review of Anthropology* 25:329–52.

OSURI, GOLDIE. 2003. Discourses of aboriginality and governmentality in Australian television news. MS. [SBB]

PETRAS, JAMES. 1997. Imperialism and NGOs in Latin America. *Monthly Review* 49(7):10–27.

PLOTKIN, MARK. 1993. *Tales of a shaman's apprentice*. New York: Viking.

POLLACK, ANDREW. 1992. Drug industry going back to nature. *New York Times*, March 5.

POSEY, DARRELL. 1990. Intellectual property rights: What is the position of ethnobiology? *Journal of Ethnobiology* 10: 93–98.

POSEY, DARRELL, AND GRAHAM DUTFIELD. 1996. *Beyond intellectual property*. Ottawa: International Development Research Centre.

PUBLIC CITIZEN. 2001. Rx R&D myths: The case against the drug industry's R&D "scare card." Web site accessed 8/1/01: www.citizen.org.

RAFI (RURAL ADVANCEMENT FOUNDATION INTERNATIONAL). 1994. Bioprospecting/biopiracy and indigenous peoples. *RAFI Communiqué*, November. Web site accessed on 1/9/03: http://www.etcgroup.org/article.asp?newsid = 212.

——. 1995. Biopiracy update: A global pandemic. *RAFI Communiqué*, September/October. Web site accessed on 1/9/03: http://www.etcgroup.org/article.asp?newsid = 206.

——. 2000. Stop biopiracy in Mexico! *RAFI News*, October 23. http://www.etcgroup.org/article.asp?newsid = 18 Accessed 2/9/03.

RAGHAVAN, CHAKRAVARTHI. 2000. Neem patent revoked by European Patent Office. Web site accessed 1/28/03: www.twnside.org.sg/title/revoked.htm.

RAMOS, ALCIDA. 1998. *Indigenism: Ethnic politics in Brazil*. Madison: University of Wisconsin Press.

RASNAKE, ROGER. 1988. *Domination and cultural resistance*. Durham: Duke University Press.

RAUSSER, G. C., AND A. A. SMALL. 2000. Valuing research leads: Bioprospecting and the conservation of genetic resources. *Journal of Political Economy* 108:173–206. [SSD]

REDFORD, KENT. 1990. The ecologically noble savage. *Orion Nature Quarterly* 9(3):25–29.

REPORT OF A SPECIAL PANEL OF EXPERTS ON THE INTERNATIONAL COOPERATIVE BIODIVERSITY GROUP. 1997. Web site accessed on 2/12/01: www.nih.gov/fic/opportunities/finalreport.html.

REQUEST FOR APPLICATIONS ICBG. 1997. International Cooperative Biodiversity Group RFA TW-98-001, August 15. National Institutes of Health, National Science Foundation, Foreign Agricultural Service.

ROSENTHAL, JOSHUA. 1997. "Integrating drug discovery, biodiversity conservation, and economic development: Early lessons from the International Cooperative Biodiversity Groups," in *Biodiversity and human health*. Edited by F. Grifo and J. Rosenthal, pp. 281–301. Washington, D.C.: Island Press.

SHIVA, VANDANA. 2000. The neem tree: A case history of biopiracy. Web site accessed on 1/28/03: www.twnside.org.sg/ title/pir-ch.htm.

SMITH, CLAIRE, AND GRAEME WARD. 2000. *Indigenous cultures in an interconnected world*. St. Leonards, Australia: Allen and Unwin.

SMITH, RICHARD CHASE. 1996. "Las políticas de la diversidad: COICA y las fedraciones étnicas de la Amazonia," in *Pueblos indios, soberania y globalismo*. Edited by S. Varese, pp. 81–125. Quito: Abya-Yala.

SOEJARTO, DJAJA, CHARLOTTE GYLLENHAAL, JACINTO REGALADO, JOHN PEZZUTO, HARRY FONG, GHEE TENG TAN, NGUYEN TIEN HIEP, LE THI XUAN, DO QUANG BINH, NGUYEN VAN HUNG, TRUONG QUANG BICH, NGUYEN NGHIA THIN, PHAN KE LOC, BUI MINH VU, BOUN HOONG SOUTHAVONG, KONGMANY SYDARA, SOMSANITH BOUAMANIVONG, MELANIE O'NEILL, JANE LEWIS, XINMIN XIE, AND GREGG DIETZMAN. 1999. Studies on biodiversity of Vietnam and Laos: The UIC-Based ICBG Program. *Pharmaceutical Biology*, suppl., 37:100–13.

STERN, STEVE. 1982. *Peru's Indian peoples and the challenge of Spanish conquest*. Madison: University of Wisconsin Press.

STOLL, DAVID. 1999. *Rigoberta Menchú and the story of all poor Guatemalans*. Boulder: Westview Press.

STRATHERN, MARILYN. 1996. Potential property: Intellectual rights and property in persons. *Social Anthropology* 4(1):17–32.

SVARSTAD, H., AND S. S. DHILLION. Editors. 2000. *Responding to bioprospecting: From biodiversity in the South to medicines in the North*. Oslo: Spartacus Press. [SSD]

SYLVAIN, RENÉE. 2002. "Land, water, and truth": San identity and global indigenism. *American Anthropologist* 104:1074–85.

TIERNEY, PATRICK. 2000. *Darkness in El Dorado*. New York: Norton.

TIMMERMAN, BARBARA, GERALD WACHTER, SUSANNE VALCIC, BARBARA HUTCHINSON, CARLA CASLER, JERRY HENZEL, SUDHA RAM, FAIZ CURRIM, RITA MANAK, SCOTT FRANZBLAU, WILLIAM MAIESE, DEBORAH GALINIS, ENRIQUE SUAREZ, RENEE FORTUNATO, EDGARDO SAAVEDRA, ROBERT BYE, RACHEL MATA, AND GLORIA MONTENEGRO. 1999.

The Latin American ICBG: The first five years. *Pharmaceutical Biology*, suppl., 37:35–54.

TOBIN, BRENDAN. 1994. Putting the commercial cart before the cultural horse. Paper written for the Sociedad Peruana de Derecho Ambiental (SPDA). MS.

——. 1999*a*. Know-how licenses: The Peruvian ICBG project and protection of indigenous collective property rights. Paper presented at the Seminario Internacional de Conocimientos Tradicionales y Recursos Genéticos: Una Oportunidad en el Mercado Global, Lima, Peru, May 19–21.

——. 1999*b*. "Protegiendo los derechos de propiedad colectiva: La busqueda de una solución provisional," in *Comunidades y conflictos socioambientales*. Edited by Pablo Ortiz, pp. 383–95. Quito: Abya-Yala.

——. n.d. Putting the commercial cart before the cultural horse: Part 2. MS.

TURNER, TERENCE. 1993. The role of indigenous peoples in the environmental crisis: The example of the Kayapó of the Brazilian Amazon. *Perspectives in Biology and Medicine* 36: 526–45.

URBAN, GREG, AND JOEL SHERZER. 1991. *Nation-states and Indians in Latin America*. Austin: University of Texas Press.

VAN COTT, DONNA LEE. 1994. *Indigenous peoples and democracy in Latin America*. New York: St. Martin's Press.

WARREN, KAY. 1998. *Indigenous movements and their critics*. Princeton: Princeton University Press.

WARREN, KAY, AND JEAN JACKSON. 2002. *Indigenous movements, self-representation, and the state in Latin America*. Austin: University of Texas Press.

WENDLAND, W. B.. 2002. Intellectual property, traditional knowledge, and folklore: WIPO's exploratory program. *International Review of Industrial Property and Copyright Law* 33: 485–504, 606–21. [SB]

WISER, GLENN. 2001. U.S. Patent and Trademark Office reinstates ayahuasca patent. Web site accessed on 12/15/02: www.ciel.org/publications/ptodecisionanalysis.pdf.

WOLF, ERIC. 1956. Aspects of group relations in a complex society: Mexico. *American Anthropologist* 58:1065–77.

WOLFE, PATRICK. 1999. *Settler colonialism and the transformation of anthropology: The politics and poetics of an ethnographic event*. London: Cassell. [SBB]

55

FROM MARKET TO MARKET

Bioprospecting's idioms of inclusion

Cori Hayden

Source: *American Ethnologist* 30(3) (2003): 359–371.

Bioprospecting is the new name for an old, though arguably ambivalent, legacy: pharmaceutical companies' use of plants and "traditional knowledge" as leads for developing new drugs. Following roughly fifty years of qualified reticence on this front (despite a few high-profile successes based on such leads, industry priorities largely rested in synthetic chemistry), the late 1980s saw many life sciences firms in the United States and Europe renew an interest in nature—and, in some cases, traditional knowledge—as a valuable shortcut for finding new leads. The drug industry's much-hyped "return to nature" took visible form in a number of high-profile prospecting arrangements, in which multinational companies joined partnerships with indigenous organizations and newly formed research institutes in biodiversity-rich climes.[1]

The novelty of these partnerships did not lie only in corporations' (renewed) interest in nature. In line with the 1992 UN Convention on Biological Diversity (CBD)—and a growing body of supporting legislation in signatory nations—these Northern "takings" now come with some form of "giving" built in.[2] In the name of sustainable development as well as redistributive justice, the CBD mandated that drug and biotechnology companies share economic benefits with source nations and communities if they desire continued access to Southern resources. It is in many ways a fragile mandate, running up against, for example, contravening WTO and other trade accords and concerted opposition from the U.S. biotechnology and pharmaceutical lobbies. Nonetheless, the CBD has produced both an idiom and an institutional framework that together have had noteworthy effects on the collection of and traffic in biological resources. Chief among these are the benefit-sharing arrangements that have proliferated across the world, in which resource collection—whether undertaken by corporate entities or academic scientists—is paired up with measures to return benefits to indigenous peoples, rural communities, or national institutes in exchange for their stewardship and provision of potentially valuable raw material.

The generativity of these agreements does not, I argue, lie in their ability to deliver on their promises: To date, no major prospecting initiative has produced any patented products.[3] Yet these promises have indeed set new relationships in motion. In this article, I draw on ethnographic research in Mexico to explore some of the paradoxical effects of bioprospecting's fragile incitement to share.

I depart from a particular ethnographic problem and the questions it points up both for prospecting politics and for anthropological engagements with globalization, science, indigenous rights, and intellectual property more broadly. The "problem" comes to me, and to the Mexican scientists with whom I work, in the form of the foundational question that defines and haunts most contemporary bioprospecting endeavors: not just *how much* should corporations pay, but also *who* shall "come with" collected plants these days? In other words, who will be included as benefit-recipients, and on what grounds? In the U.S.–Mexican prospecting collaboration in which my colleagues at Mexico's National Autonomous University (UNAM) are participating, the sponsoring agency, the U.S. National Institutes of Health (NIH), has a fairly specific idea of how to answer this question. Simply stated, participating developing country researchers are supposed to sign contracts with each individual who provides them with plants and information. As the presumptive authors and stewards of resources, these providers—and, by implication, the communities of which they are a part—are the people who are supposed to be future benefit-recipients. As with many 1990s ventures in participatory development, the crucial lubricant here is an appeal to the idea of local or community knowledge, resources, and people—all of which, it is assumed, are to be found together in discrete, identifiable, and thus rewardable packages (see Cooke and Kothari 2001; Zerner 2000).

The Mexican ethnobotanists, for their part, offer an alternative strategy to this view of seamless transactions and neatly bundled community resources. In the first two years of the implementation of this contract, the lead UNAM ethnobotanist, Harvard-trained Robert Bye, chose some different starting points altogether for identifying his plants, his "local participants," and the relations between them. Rather than starting in places marked as "communities," Bye, drawing on a long-standing and well-established research method, culled most of his initial plant samples from urban marketplaces. The UNAM researchers' strategy of prospecting in markets has created a vivid breach in the bioprospecting imaginary, both disrupting and reinscribing some of the fundamental assumptions shaping this kind of enterprise—most notably, the idea that plants and knowledge "come with" identifiable local authors, claimants, or stewards attached.

This conversation between prospecting strategies invokes and provokes some big questions about the idioms of inclusion and exclusion on offer in a new era of biodiversity entrepreneurialism. "Market" and "community," as Stephen Gudeman (2001) has written recently, are indeed worth thinking about as competing idioms for the redistribution of corporate returns.[4] In fact, the relationship between market and community is, on a grand scale, one of *the* defining questions for many anthropologists writing about globalization, economy, and identity (not to mention for Aristotle and Marx). In these discussions, market often stands for capitalism itself, and community has been its endangered or, alternately, intimate other (see Joseph 2002). Thus, many anthropologists have argued that global capitalism is running roughshod over indigenous communities, thereby endangering ways of life that exist outside of capitalist logics (see Nash 2001); others have argued that the very intensity of contemporary invocations of community or indigeneity as a site of resistance suggests that these concepts actually come built into "capitalist modernity" in the first place (Joseph 2002; Lowe and Lloyd 1997).

Inevitably running through these discussions and their many permutations are the lines that (readings of) Marx and Mauss have so enduringly carved into the world: market as the site of abstraction, commodity transactions, rational actors, and disembedded and disentangled relations; community as bearer of the gift, home to barter, shared values, and embedded relations. A great deal of worthy anthropological work has, to be sure, gone into unraveling these stark binaries and the assumptions that underpin them (see Carrier 1997; Gregory 1982; Gudeman 2001; Strathern 1988; Thomas 1991). At the same time, it must be said that some of this work has arguably resedimented those categories precisely in the attempt to mix them up (I would place Gudeman's *The Anthropology of Economy* [2001] in this category) or fully naturalized some of the fundamental characteristics of market-mediated man—most notably, the calculating, rational actor—even as they attempt to track, sociologically, how market and nonmarket entanglements get made (I think here especially of Michel Callon's *The Laws of the Markets* [1998]). In this article, I aim to embark on a different kind of project, an ethnographically grounded one in a particular sense. The schema of market and community will, in fact, permeate this account; but, I do not have any particular commitments to adhering to them myself, nor do I endeavor explicitly to debunk them. My analytical point of departure is slightly off to the side: I am interested precisely in how these well-rehearsed binaries, and, indeed, even the challenges that anthropologists and sociologists have leveled at them, permeate practices of bioprospecting and, in the process, lead us to different questions altogether.

With these issues in mind, I set out to do two things. First, I explore *how* the notion of community and its classic foil, the market, fare in some slightly unexpected ways as idioms for structuring bioprospecting's modes of taking as giving. What are the presumed efficacies, contents, and boundaries of these concrete "places" and abstract "forces?" What, in fact, *do* market and community stand for? This entails asking not just about the redistribution of corporate rewards but also, by implication, about idioms for "placing" traditional knowledge and medicinal plants in particular kinds of domains—compensable or not.

My second agenda, somewhat submerged, but crucially present in the framing of my analysis, is to think about these highly charged articulations through the lens of the anthropology of science.[5] For the notions of market and community that I explore arise directly through the encounter between fascinating, long-standing scientific collecting methods and the new mandates and accountabilities that prospecting contracts bring to scientific research itself. Thus, my intention is not to deconstruct "the market" and its foils. Rather, I explore how these ideas are explicitly animated, short-circuited, and redeployed in the world of bioprospecting, particularly in the hands of one team of scientists who, through their collecting strategies, propose provocative models of their own about prospecting as a mode of inclusion and exclusion.

Market mediations and the contours of community

The prospecting initiative on which I focus in this essay is part of an international collaboration based at the University of Arizona and is funded by the U.S. government's International Cooperative Biodiversity Groups (ICBG) program. The Arizona-based "Latin America ICBG" involves researchers not just from Mexico but also from Chile and Argentina; it counts the U.S. conglomerate American Home Products (which owns the drug company Wyeth-Ayerst and, until recently, the agrochemical company American Cyanamid as well) as its "industrial outlet." In return for sending extracts of medicinal plants to the University of Arizona and American Home Products, the source country researchers receive minimal research funds each year, and promises of a percentage of royalties ten to 20 years in the future, should those companies develop a drug or pesticide based on these collections.[6] The project is also designed to collect "ethnobotanical knowledge" about plant uses and to direct royalties back to the people or communities from which these resources have been culled.

As this combination of agendas might suggest, the ICBG program is in many ways calibrated directly to the terms set by the UN Convention on Biological Diversity. (Notably, despite the fact that the U.S. Congress has not ratified the convention and trade representatives continue to militate strongly against it [because of industry objections to its benefit-sharing measures], this government initiative has been one of the world's primary engines of benefit-sharing contracts.) Funded by the National Institutes of Health, the National Science Foundation, and, initially, the U.S. Agency for International Development, the ICBG program took off in 1993 and has funded eight projects, all linking U.S. academic researchers to, on the one hand, developing country collaborators, primarily in Latin America and Africa, and on the other hand, drug or biotechnology companies.[7]

In many ways, this initiative is an amplified version of the NIH's long-standing support for plant and microbe screening carried out under the auspices of the National Cancer Institute (see Goodman and Walsh 2001; laird et al. 2000). With this new program, the NIH infuses a long-standing tradition of plant-based drug discovery with the language and mechanisms of sustainable development, hoping to link profits from pharmaceutical development to conservation and rural economic development (Schweitzer et al. 1991). The ICBG, then, like the CBD itself, belongs to the well-populated annals of "market-mediated" conservation and development initiatives that took root in the late 1980s and early 1990s. Here, biodiversity-derived drugs promise not only corporate wealth, but, if harnessed and redirected through the proper redistribution mechanisms, they might well generate more biodiversity *and* economic prosperity in the South to boot (Eisner and Beiring 1994; World Resources Institute et al. 1993; World Resources Institute et al. 1992; see also Orlove and Brush 1996).[8]

Central to prospecting's ostensible transformation from simply a drug discovery initiative to a new form of market-mediated sustainable development is the redistribution of biodiversity-derived profits. I noted above that the ICBG's preferred destination for these benefits is something called "community." But how do they arrive at this vivid abstraction? In many prospecting initiatives forged in the early 1990s, including the ICBG, benefits are often described as both reward and incentive. The slippage, I would argue, matters a great deal. To invoke reward or even compensation is, in the context of the history of colonial resource extraction that so powerfully marks biodiversity politics, to invite the calibration of future exchanges to some egregious histories of violence and extraction. It is also, of course, to imply that indigenous peoples and other benefit-sharers have a legitimate property right in both knowledge and plants. This proposition is a highly contentious one in the world of prospecting.

Securing recognition of intellectual property rights (IPR) (or various proxies) for indigenous peoples is a battle that many engaged anthropologists, ethnobiologists, legal scholars, and indigenous activists have been fighting since the mid-1980s. Through professional codes of conduct, proposals for developing copyright for indigenous knowledge, the development of "traditional resource rights," and other ever-evolving mechanisms, researchers and the indigenous coalitions with which they work have made the question of compensating indigenous peoples for their resources a powerful aspect of contemporary ethnoscientific research and of indigenous and biodiversity politics more widely (see Brown 1998; Brush 1994; Brush and Stabinsksy 1996; Coombe 1998; Greaves 1994; Posey 1996). Here, too, the question of competing market and community principles looms large. In discussions leading up to the 2000 Conference of Parties to the UN Convention on Biological Diversity, members of the International Indigenous Forum on Biodiversity wrote that

> indigenous peoples have a legitimate right . . . to deny access to our knowledge and to say NO to bioprospecting, exploration, or the application of intellectual property rights when these procedures are contrary to the principles and collective rights of our peoples.
>
> (In Muelas Hurtado 2000:3, emphasis added)

Recognizing that rights to intellectual property can be a highly problematic form of empowerment, researcher-activists such as Darrell Posey have been working for years to develop alternate mechanisms, such as traditional resource rights, that would forge new bases for rights without reproducing the view of the world required by IPR (Posey 1996). These movements and efforts have indisputably gone a long way toward shaping the contemporary politics of corporate resource extraction, academic research, and various combinations thereof. Nonetheless, in most prospecting contracts, benefits-shared usually stop far short of actually granting intellectual property rights or similar alternatives for "local people."[9]

Wary of entering into the kinds of negotiations that property claims and their proxies would entail, corporate and academic participants often prefer other idioms of return altogether: "donation," technology transfer, or the future-looking neoliberal tropes of incentive-building and stakeholding. In this vein, many prospecting initiatives (most notoriously, perhaps, those emanating from Costa Rica's National Biodiversity Institute [INBio]), describe their ideal participants as people who might be enticed, through the promise of future returns, to be stewards of endangered biodiversity (see Reid et al. 1993). In such a view, incentive building rather than intellectual property looms large as the idiom through which participants are to be brought into the prospecting fold.

And yet for all this, "property" is not so easily excised from the mix. In the ICBG, for example, the definition of *benefit-recipients* takes root in the decidedly Lockean idea that proper benefit-recipients are those people who can be rewarded for their identifiable input of labor and innovation into, and stewardship of, nature. It is in this spirit that the U.S. National Institutes of Health, the primary funding body of the ICBG program, enjoins participating Latin American researchers to sign contracts with each individual who provides them with plants and information. The idea is that granting people financial stakes in their resources will encourage them to conserve these resources; to borrow an apt phrase from Marilyn Strathern's (2000) analysis of British "audit culture," local participants are being encouraged here very explicitly to "value their values." And in turn, this logic of incentives and valuation comes with a fairly powerful set of assumptions about the relation among plants, people, and knowledge: They must be found together—localized and localizable—in one discrete package. (That is, it does not make sense to send benefits back to people who aren't managing the resources from whence the lead came). An emphasis on incentive, filtered through the idiom (but not the fact) of intellectual property, thus requires the territorialized local community as the bundled-together source of (semi-) raw material and destination for future benefits.

I would follow Miranda Joseph's lead here, not to mention that of a host of anthropological critics of sustainable development in particular, and suggest that *this* notion of (modified, green) capitalism requires—or at least strongly requests—community (Joseph 2002).[10] And community, in this instance, is a curious and uneasy mix, animated by a host of sometimes competing notions: incentive, locality, IPR, reward, territoriality. But there is another element that might be added to this list of the projected efficacies of (green) capitalism's intimate other: legitimacy, both political and juridical.

In Mexico, as in many places, prospecting has become embroiled in historically charged struggles over sovereignties and rights. The Convention on Biological Diversity has redrawn jurisdiction over "genetic resources" and "traditional knowledge" by taking them out of the global commons and placing them squarely within the sovereignty of nation-states—a victory, of sorts, for Southern nation-states. It also recognizes community rights over traditional knowledge. This dual move has, in effect, provided fuel for reinvigorating competing sovereignty claims by indigenous peoples and nation-states. With Mexico's entry into NAFTA in 1994, these long-sensitive questions flared up dramatically. The day the accord was to take effect, the Zapatista army declared war on the Mexican government and its neoliberal "reforms"—including the end of land reform, withdrawal of rural subsidies, and privatization of various forms of "national patrimony" (Nash 2001; see also Ferry 2002; Gledhill 2001; Stephen 1998).[11] Alongisde and intertwined with these concerns, the question of indigenous sovereignty and rights to resources nationwide has been a powerful and contentious rallying cry.

Prospecting has, not surprisingly, pushed some fairly sensitive buttons on these fronts (Hayden 2003). Indeed, in part because of the political sensitivity of matters of indigenous rights to land and natural resources. the Mexican government, both under former President Ernesto Zedillo and now under Vicente Fox, has pointedly shied away from drafting any binding regulatory measures for prospecting activities, even though Mexico is a signatory nation of the CBD and draft legislation on "access to genetic resources" has been "in process" since 1997. Thus, although the secretariat of the environment (now, Semarnat) has been approving prospecting projects and granting permits under previous protocols for "scientific" collections, there is no clear-cut juridical framework established for prospecting activities in Mexico—a point to which many activists and prospecting critics repeatedly turn in denouncing these contracts as illegitimate by definition (see Nadal 2000).

In these circumstances, NIH project managers have placed a premium on the benefit-sharing contract as a mode of informed consent (an arguably overloaded and underqualified one at that): a hoped-for guarantee that collections are proceeding with the proper authorizations from both individuals and communities. And here, too, or perhaps especially, plants, (sovereign, authorized) people, and knowledge *must* cohere. It has proven a difficult and contentious request to fulfil. Several ICBG projects in which researchers have sought to put this notion of community-based prospecting in motion have run into serious difficulties (see Greene 2002; Hayden 2003). Perhaps the most notorious example is that of the Maya ICBG project, sited in Chiapas and run by University of Georgia researchers Brent Berlin and Elois Ann Berlin. This program ran aground quite spectacularly shortly after it was funded in 1998, in large part because of powerful opposition from a coalition of Chiapan healers and midwives (along with some very vocal North American nongovernmental organizations, or NGOs), who argued that the collecting and negotiating processes undertaken in the area were illegitimate.[12]

As I have noted, prospecting "in communities" is not the only model that has been mobilized in Mexico under the (sometimes unapproving) mantle of the ICBG initiative. In fact, in the Latin America ICBG project, based at the University of Arizona and funded five years before the Maya ICBG, none of the researchers from the participating source countries are going directly to "community" as their first (or only) access point for finding plants and the people that come with them. In Argentina, researchers have been collecting plants on large, privately held ranches; in Chile, participating researchers have been working on government lands and cultivating some of their compounds in petri dishes in their labs (Gloria Montenegro, personal communication June 1997). And, as I noted above, the participating Mexican researchers in this project have mobilized their long-standing interest in urban markets to craft an alternate strategy of their own for linking plants-collected to benefits-returned. This is not, as we might surmise, the kind of "market-mediated participation" that ICBG architects had in mind.

Market research

For the ethnobotanists with whom I work, as for many of their colleagues in Mexico and elsewhere, the iconic twin images of plant-based drug discovery—the intrepid explorer crashing through virgin jungle and the ethnobotanist as shaman's apprentice—fall far from the realities of everyday research. So does the vision of "traditional knowledge" as an objectlike thing, fixed in place, isomorphic with easily identified and bounded communities. Since the early 1980s, Mexican researchers such as anthropologist Carlos Zolla, chemist Xavier Lozoya, and ethnobotanists Robert Bye and Edelmira Linares (among many others) have demonstrated through their research and writing that medicinal plants are also well traveled, cosmopolitan, dynamic, even urban phenomena (Bye and Linares 1983; Lozoya and Zolla 1984). Anthropologist Valentina Napolitano reminds us that "traditional" Mexican herbal medicine "has always been strongly linked to popular medicine as it is practiced in low-income [urban] neighborhoods" (Napolitano and Mora in press:4). Urban plant markets are a crucial locus for the study (not to mention the practice) of herbal therapeutics. Robert Bye has been studying the plants circulating through these markets for well over 15 years. Market plants are particularly rich research objects, full of some intriguingly specific material, biochemical, and, it would seem, *social* properties.

Whether dedicated entirely to medicinal plants or offering dried herbs alongside piñatas and fresh vegetables, bustling markets such as La Lagunilla in Mexico City or the central city market in Chihuahua were considered promising venues by researchers long before the advent of prospecting contracts. From an ethnobotanical standpoint, there is a great deal to learn from even a quick pass through these sites: Popular plant names and uses, for example, are on vivid display, alongside other key dimensions of ethnobotanical inquiry such as which part of a given plant people consider most useful. (Since vendors sell plants in the forms in which people use them, a cursory look around any given market will indicate, for example, that *matarique's* efficacy rests in its roots, *gordolobo's* healing properties lie in its flowers, and *copalquín's* curative powers are found in its bark.)

Although this much information is on ready display, a great deal of additional work is required if markets are to become points of departure for more detailed studies. Most pointedly, buying plants in these sites becomes an entry point for more extended conversations with vendors about how a plant-based remedy is prepared, where the plant is collected, and, with luck, how to locate the collectors who have brought particular specimens in from afar. For the most part, the people who sell plants in urban markets are, indeed, not the people who collect them. The vendors I know in mid-to-large cities buy their merchandise from a heterogeneous array of suppliers. Sometimes these are wholesalers; but, in the northern state of Chihuahua, where Bye has long conducted research, suppliers are just as likely to be Tarahumara men, women, and girls who come in from the mountains several hours' bus ride away; local miners who pick up plants in their spare time, hoping to make some extra money (prospecting in double time, we might say); or farmers who, every two weeks, gather the "weeds" that grow on the edges of their corn and bean fields and bring them to the city.

A number of Mexican researchers have placed such collectors at the very center of their studies of plants and their uses. Thus, anthropologist and medical doctor Paul Hersch Martínez has investigated the fragile livelihoods of plant collectors in the southeastern state of Puebla (Hersch Martínez 1996), and ethnobotanist Myrna Mendoza Cruz (personal communication 1997) tracks the creative substitutions (which often become the basis of new popular classifications) that rural collectors finesse when wholesalers push them to bring in specific plants that are simply not in season. Bye, for his part, has been interested in the relationship between commercial demand and biological distributions: Do popular plants become more scarce in the field, and if so, does this scarcity prompt shifts in therapeutic practices and plant uses (Robert Bye, personal communication, 1996)? These are consequential questions that require an understanding of plants (and "biodiversity") as always already shot through with entanglements: with commerce, human traffic, management, and, indeed, creativity.

But market plants prove interesting in many other terms as well. For ethnobotanists who are also interested in economic botany—a field that focuses not on "local" or "popular" uses but, rather, on the industrial potential of plant life—such specimens certainly hold distinctive appeal. Bye works in and across these two subfields of botany, and market research is indeed an important trading zone between them. Market plants, he seeks to demonstrate in his research, are promising leads in the search for "bioactivity" (the potential of plant chemical compounds to show some kind of biological efficacy). Bye argues that these plants have been used and tested by generations of consumers and, thus, can be counted on to work—for something, and, perhaps, even for the purposes commonly attributed them. In a preliminary progress report for the ICBG prospecting program, the UNAM team wrote,

> [Market] plants have been tested empirically by generations of Mexicans, and, as a consequence, their demand is constant, there are established routes of supply and processing, and it is possible that they have been subject to processes of selection by man, improving their therapeutic activity.
>
> (Bye et al. n.d.:3)

Bye and his partners in the ICBG prospecting agreement, UNAM chemists Rachel Mata and Rogelio Pereda, have worked together for years on precisely this kind of corroborative–collaborative project. The ICBG project is in many ways just another source of funding for these scientists to continue their work together, analyzing the chemical activities of the compounds hiding in plants both picked and purchased.

If markets serve as excellent first-tier bioassays (see Bye and Linares 1986), they also hold other kinds of appeal for plant researchers. In logistical terms, of course, market studies are a notably quick and efficient way to acquire plants. In the context of the ICBG project, with its quota of 100 extracts per year that the Mexican researchers must meet to continue receiving funds, such efficiency has been especially attractive—even more so, given that the Mexican team began two years behind schedule because of an extra round of institutional negotiations and thus spent the first year and a half catching up on their missed collections. Of equal significance for the viability of the team's collections, the plants purchased in markets are, most often, already dried and so are relatively resistant to mold in the VW bus *(combi)* during a several-weeks-long collecting trip. Mold is often a danger with freshly collected plants, and it ruins them as candidates for future chemical analysis.

Thus, the materiality of market plants—the ease with which they may be collected and stored, their heightened (one might say, selected-for) biochemical efficacy, the ways that they "come with" a host of useful information embodied in their form—holds multiple appeal for ethnobotanical studies. For precisely these reasons, market work might seem ideally suited for a project in which collections are directly destined for an "industrial outlet" with drug discovery in mind. But in the view of some NIH program administrators, markets are an unsettling interruption. Market plants may well be shot through with human relations, but, as government officials and concerned observers on both sides of the border have asked, do market-based collections contain the *right* mixtures? Do they produce the right kinds of relationships between communities and biodiversity-derived benefits? When the market meets the NIH's "local community," both end up significantly and interestingly transformed.

Market relations

I have already alluded to what might be found in a market from an ethnobotanical point of view. But when mobilized as a point of entry for prospecting, market research takes on other striking features. First and foremost, Bye and his coworkers very pointedly do not sign benefit-sharing contracts with plant vendors. As one UNAM ethnobotanist who works in markets told me in 1996, "Plant vendors are merely vectors of transmission of information—they're not sources, so they don't merit part of the royalty benefits." Rosemary Coombe (1998) reminds us that such attributions of authorship are significant indeed: The distinction between those who make knowledge and those who merely distribute it is *the* difference that matters where the idiom of intellectual property is concerned (see also Boyle 1996). But it is not just a recourse to authorship that serves as a crucial idiom for determining who shall be "inside" and "outside" the reach of potential benefits-shared. The commodity transaction itself looms large in this context; the very act of a purchase, strictly speaking, sets market collections free of long-term obligations. Bye explained to me that it doesn't make sense to treat vendors as sources of knowledge that deserve compensation: When you buy the plants, you buy the information, and no further obligations are involved.

Like the NIH's invocation of the local community, this invocation of the market—as a site of short-term obligations, of limited entanglements, of the distribution but not the generation of "knowledge"—is a strategic essentialism, one that will sound provocatively familiar to an anthropological or sociological audience (Carrier 1997; Granovetter 1985; Gudeman 2001). And, indeed, Bye himself does not dwell for very long in this idealized world of truncated transactions. Markets, as I suggested above and as I will demonstrate further below, prove to be such rich ethnobotanical sites precisely because of the webs of relations and entanglements that can be traced through them. This emphasis on the "disembedded-ness" of market transactions (see Callon 1998) serves a very particular and, indeed, temporary purpose: It is the first step in Bye's efforts to construct an alternative strategy for relating plants-collected to benefits-redistributed.

If not through market transactions, where, then, are benefit-recipients to be found? In considering this question, I introduce another powerful boundary-making device: not just authorship and the commodity but the nation itself. Medicinal plants, specifically those that travel though markets, hold a privileged place in a powerful ethnobotanical and popular story about Mexico as a mestizo nation. Building on the national origin story promulgated in the aftermath of the Mexican revolution (1910–1917), many people both inside and outside of Mexican public intellectual circuits name medicinal plants a national resource, both a reflection and a result of the hybridization of indigenous cultures and plants with those coming from Europe and Africa. Though having identifiable roots in the indigenous "past," these plants now, this story tells us, belong to a diffuse national culture. (As anthropologist Carlos Zolla phrased the common sentiment to me in a June 1997 interview, "With medicinal plants, as with many other things in Mexico, we are a country of mixtures.")

These stories of mixture have, over the years, been granted extraordinary tangible institutional form. Certainly looming large here is the legacy of *indigenismo*, an official policy in which the assimilation of indigenous peoples into the nation has been a powerful modernizing trope.[13] And where medicinal plants in particular are concerned, I would note efforts in the late 1970s by nationalist-populist President Luís Echeverria to resurrect a 19th-century project to develop a domestic drug industry based on "distinctively Mexican" resources—plants and traditional knowledge (see Sherwood 1991:168–169; Soto Laveaga 2001). The resulting institute, IMEPLAM (the Mexican Institute for the Study of Medicinal Plants), became home to many prominent Mexican chemists, anthropologists, and ethnobotanists (including Xavier Lozoya and Carlos Zolla), and the research generated out of this effort and its subsequent incarnations has helped constitute the Mexican *herbolaria*, or storehouse of medicinal plants, as a distinctly national, hybrid resource (see Hayden 2003; Lozoya 1976).[14] Significantly, many renderings of these centuries of mixture begin precisely with descriptions of the extraordinary market systems through which medicinal plants (among many other things) have traveled since the height of the Aztec empire, and certainly well before.

These market-mediated notions of the nation (or, more specifically, of a Mexican national herbolaria) constitute the second step in the Mexican researchers' efforts to refashion the relation between plants-collected and benefits-redistributed. In a report on the project's progress in Mexico, Bye and his colleagues in the Faculty of Chemistry explain their strategy of working in markets thus:

> Medicinal plants were central to the quotidian life of precolombian Mesoamerica. As such, medicinal plants and knowledge thereof are a national legacy that have been shared by many generations and cultures over several centuries. Given the antiquity and broad diffusion of this Mexican medicinal flora, the ICBG program is concentrating on studies of plants from different regions of the country. To understand the evolution of a "national medicinal flora" that has developed over centuries, this broad sample has also incorporated the contemporary system of commercialization, which conserves many prehispanic characteristics.
>
> [Bye et al. n.d.:7]

Arguing that it is impossible to trace this prehispanic-national legacy to particular individuals and communities, they have made a bid for redefining the relationship between resource acquisition and benefit-sharing:

> Market work gives you a bit of flexibility (in terms of sharing benefits) . . . some of the plants have a very wide distribution in terms of their use and to a certain degree, one could say, this is Mexican, this is not from this community . . . remember, the main objective here is to promote conservation and maybe there's a greater need for conservation and social-economic development in another community. And if we stick to this straight definition that the person who gave the information gets all the credit, then there's a problem if we want to help somebody else.[15]

With these broad distributions in mind, the UNAM team has made an explicit case for not going directly to people such as local healers and the communities of which they are a part as their first points of contact for gathering both plants and benefit-recipients. Rather, in the early years of the project, they paired up market collections with a bid for considering non-governmental organizations, indigenous collectives, and productive organizations as their interlocutors (Bye et al. n.d.:2). In this way, Bye does not seek out authors or owners as his participants but, rather, a diverse array of community enterprises whose goals, or projects, line up with those that might meet the criteria of the program. Among these are small *artesania*-production enterprises in Sonora, indigenous collectives in Chihuahua dedicated to cultivating and commercializing medicinal plants for local and regional markets, and a group of traditional healers in Oaxaca, organized through the government's Instituto Nacional Indigenista (INI), with whom Bye has worked to establish an ethnobotanical garden. These forms of alliance follow the traces, then, of neopopulist government development projects formulated to "cushion the blow" of the withdrawal of government subsidies for rural production under NAFTA, or of government-sponsored indigenist initiatives. They are decidedly *not* arrangements forged as acknowledgement of the provision of information and plants. The complex material and ideological legacy of national(ized) medicinal plants is the crucial condition of possibility for this rerouting of benefit-sharing relations away from a direct exchange of resources for (promises of) benefits and toward more diffuse but, in a certain sense, also more targeted relations.

Market interruptions

The Mexican scientists working on this project have been quite explicit about their rationales for collecting in markets. In fact, many of the explanations I have been quoting have been articulated precisely as answers to the objections and unease of ICBG project directors in Washington, D.C. The sponsors of this agreement at the NIH have never been particularly content with market research as a mode of prospecting. Why? Market collections, as one high-level administrator told me in 1998, "break the link" among people, plants, and benefits. And it is precisely this link that projects such as the ICBG are supposed to be both encouraging and recognizing in the first place. Bye's program, one might say, injects *too much market* into the NIH's market-mediated strategy.

One of the many things at stake in this excess is the work that the paired-up idioms of intellectual property and community are meant to do within the ICBG as a "green" drug discovery initiative. This powerful pair is meant precisely to recognize and thus create connections, or links. Many anthropological discussions of intellectual property dwell precisely on the question of connections, though they lean in quite the opposite direction. Marilyn Strathern has shown how intellectual property claims interject in the kinds of connections, flows, and travels that we might identify in processes of knowledge production: "Ownership," she argues, "re-embeds ideas and products in an organism (whether a corporation, culture, or individual author). Ownership gathers things momentarily to a point. . . . [It halts] endless dissemination". (1999:177). Certainly such truncations are precisely what intellectual property claims effect—even as they set new processes in motion.[16]

Both Strathern and Georgina Born (among others) have made powerful cases for thinking not just about the institution and fact of intellectual property but also about the idiom of IPR—or, in Born's terms, the "informal practices and discourses of intellectual property" (1996:102). And in the present case, just as patents themselves generate a multiplicity of effects, the efficacy of IPR (as a legitimating trope for certain practices) takes some interesting turns.

In the "dilemma" that market-based prospecting poses for the NIH, the idiom of intellectual property—the idea that plants and knowledge come with identifiable, remunerable author–owners—is laid bare as a strategy not for severing but for producing long-term chains of connections, legitimacies, and obligations. These connections are supposed to take the form of an "uninterrupted" chain of transactions—starting and ending with resource providers *as* benefit-recipients, and to be guaranteed, as it were, by the benefit-sharing contract. Certainly, this transaction would, even in its idealized form, be marked by a significant interruption: the period of up to 20 years as "resource providers" wait for that other form of intellectual property—a patented drug—to materialize and thus to generate royalty payments. As Shane Greene (2002) has noted in his discussion of an ICBG collaboration between Monsanto-Searle and Aguaruna communities and coalitions in Peru, this delay looms large in the growing discontent of Aguaruna with their coalition leaders and the ICBG itself.

For his part, Robert Bye has identified some additional breaches in the ICBG's notions of a seamless, contractually mediated, and therefore legitimate connection. The "links" produced by this model of compensation seem, in the context of the kind of widely circulating knowledge that interests Bye, unduly limiting, counterintuitive, or just plain wrong. Medicinal plants (or at least the kinds that he has been studying), he argues, *don't work that way in Mexico*. Displacing the curiously twinned idiom of IPR-community as *the* mode of entitlement allows Bye to channel prospecting-derived benefits where he argues they are most appropriate and needed. It is precisely the truncations produced by "the market" that facilitate his efforts to extend the network of ICBG beneficiaries in creative and positive ways.

Bye's benefit-sharing counterproposal to the NIH does not, I hope it is clear, simply dig up the familiar disjuncture between "market" and "community." Nor does it plant us solely in a predictable debate about the difference between talking about *a* (concrete) marketplace and *the* (abstract) market. Rather, at stake in this counterproposal and the NIH's response thereto is an intriguingly vivid clash of abstractions. The NIH, in its attempts to wed drug discovery to the social goods promised by sustainable development, stands as the guardian of a somewhat romanticized local that cannot tolerate the presence of market transactions. The Mexican scientists, the very people in charge of identifying and enrolling these local participants, counter with a few powerful idealized models of their own—nationalized mixtures and obligation-free, pure market transactions.

Yet as I hinted above, market research has more complex temporal dimensions to it, which make it difficult to dwell for long in this highly charged if not slightly odd topography of long-term, encumbered, IPR-requested community and short-term, truncated, mestizo, market transactions. For the Mexican ethnobotanists, market research is also an ideal prospecting technology precisely because it does, indeed, contain social relationships that may be good candidates for future, long-term cultivation.

Market connections

For ethnobotanists working in urban markets, prospecting contracts obviously do not represent the first instance in which science and commerce have mixed. Nor do the collaborations underlying such contracts mark the first time that "reciprocity" figures centrally in the work they do and the collections they undertake. Like many of their colleagues across Mexico and elsewhere, the UNAM ethnobotanists have long made efforts to "give back" to the communities with whom they work. Among their modes of reciprocity have been community education campaigns, in which researchers offer (and often are asked) to teach people about the plants found near to hand. Indeed, many Mexican ethnobotanists have, since the 1970s, considered the project of "returning knowledge" to communities as a political and activist act (Toledo 1995)—one that takes its force and meaning against the backdrop of extended histories of territorial displacement and linguistic assimilation fomented by decades of government *indigenista* policies. These histories—alongside the fact that academic knowledge production is one of the many forces that have helped place "traditional knowledge" into the public domain—conjoin to make ethnobotanists themselves, often, the de facto "guardians" of traditional knowledge (see Brush 1999).

In this prospecting project, the idea of teaching "local people" about their plants makes its way into the UNAM scientists' view of potential benefits they can offer, in several senses. For instance, they have offered to join forces with local schools in areas where they have set up potential benefit-sharing relations—such as Mayo and mixed mestizo and Tarahumara communities—to produce illustrated dictionaries of medicinal plants. But so too do markets themselves play an important role in identifying and producing these kinds of potential relationships. The process of starting in markets actually helps researchers root their traveling plants and knowledges "back" in identifiable (community?) places. Consider this, then, step three in Bye's market-mediated efforts to refashion the relationship between plants acquired and participants-enrolled.

Yerba del conejo is one of these relation-laden market plants.[17] The leaves of this plant were of interest to the Mexican research team as a possible pharmaceutical "hit" that also, Bye hoped, could serve as the basis for a community-level cultivation and commercialization project. After numerous visits to the northern city market where they had spotted this plant in 1997, the researchers managed to track down one of the collectors who brings yerba del conejo to the city. He consented to meet the researchers at his brother's ranch, where the plant grows along the weedy borders of the property. The collector's brother had little use for the weed, much less a name for it, and so was a bit bemused that these *cientificos* from Mexico City were curious enough about it to ask his permission to unearth and take with them two or three live plants, roots and all, in plastic bags.

For the researchers, these plants were not, I was told, "community" resources yet. How might they *become* community resources? The plants were taken for replanting in the nursery at UNAM's Botanical Garden; depending on how well they weathered Mexico City's rather distinctive atmospheric conditions, they would then be subject to cultivation and germination studies. If these studies yielded enough information about the optimal conditions for cultivating this plant, then Bye might venture back to the ranch or other sites nearby, to replace yerba del conejo in hospitable soil and, through ICBG-funded workshops and public education projects, to teach people that it is "a valuable plant." Rather than "mining" the knowledge and plants of a particular community, then, ICBG community development and conservation activities are more likely headed in this direction, cultivating people's interest in the plants nearby and creating opportunities for cultivation and commercialization projects.

Optimally, this is how markets function in the service of bioprospecting. Commercial circuits lead researchers first to plants that can be sent off to the participating companies in short order, and then, in the longer term, those plants might lead them to vendors, interlocutors, and even communities who may in turn elect to become local experts—and benefit-recipients—themselves. Market work is, in this sense, not an undoing but, rather, a postponement of prospecting's modes of community gathering.

The NIH's notions of compensation and Bye's are not. in this sense, as distinct as they might have seemed. Both aim to find ways to enroll the proper kinds of subjects into longer-term projects of stewardship and sustainable resource management. Both work with an idea of community as a proper site of these relationships. And for both parties, the market transaction or the commodity relation does not itself a proper participant make. For, in the above example, too, the collector who gathers and sells yerba del conejo does not, by virtue of this activity alone, find himself within reach of benefit-sharing status.

The point where these two models of benefit-sharing diverge is not, then, where they end up but, rather, where they begin.

Research strategies

What is one to make of these two uneasily articulated modes of inclusion and exclusion? Answering this question, I would suggest, entails a point of departure anchored neither in market nor in community. Rather, I would look to the question that has animated both idioms throughout this account: the issue of legitimacy. Borrowing a page from the annals of science studies, I would argue that this is not simply a "political" question. Rather, I think it is worth attending to the ways in which these matters of governance, of political and social relations, are written both large and small in the material worlds of scientific research practice itself. As sociologists and historians of science have shown, the viability of particular research strategies or research objects is inseparable from the social and institutional relations in which they are embedded (see Clarke 1995; Fujimura 1996)—and I might add here, the liabilities that they (ostensibly) help avoid.

Certainly, for my colleagues in Mexico, plant work has always been social work. But the dimensions and shape of this sociality, its renderings of accounts, look different now. For the UNAM ethnobotanists, prospecting contracts are both symptoms of and potential solutions for a relatively novel imperative: to inject the right kind of publics into their collections.[18] This imperative, they worry, comes with its risks as well: accusations of biopiracy, the risk of sullying long-term relations with the highly charged yet elusive promises of benefit-sharing. It is in anticipation of this kind of fallout and those kinds of accusations—and the activist mobilizations that can and will accompany them—that the articulations of market and community outlined in this essay have taken shape.

For the NIH, the viability of resource collection for plant-derived drug discovery starts and stops with community—an idealized and necessary source or site of quasi-sovereignty, collective authorship, and ethical capacity. But for UNAM ethnobotanist Robert Bye, this model is perhaps the *least* promising way to fashion benefit-sharing partnerships in a country in which the politics of indigenous resource rights are extraordinarily sensitive, and in which popular and ethnobotanical notions of a diffuse herbolaria—at once, I argue, "cosmopolitan" and "nationalist"—provide an alternate model of plants and the relations that come with them. Through his use of market research, Bye argues that it is possible to channel socially and biochemically enhanced plants into drug discovery circuits *and* to enroll indigenous and rural communities and organizations as benefit-recipients, without having to link these processes through (slippery or contentious) attributions of authorship and ownership. The "disagreement" over methods here is thus not precisely about *who* to enroll as a benefit-recipient but *how*. In what idiom—territoriality, nationalism, cosmopolitanism, authorship—will biological collections be attached to social relations, interests, and claims? This is a science studies-inflected way of framing *the* major questions haunting both bioprospecting and indigenous rights movements more broadly.

Coda

While researchers, activists, and funding bodies (who often, of course, occupy quite different positions relative to each other and to the prospecting enterprise) continue their attempts to navigate the post-CBD world of taking and giving, the terrain of bioprospecting continues to shift underfoot. Fully consonant with wider trends in the world of prospecting, it seems that Mexican plants (and their people, whoever they may be) will soon be cut out of the Latin American ICBG project altogether, as the ICBG program in general turns its emphasis to ever more liability-free political and natural environments (Robert Bye, personal communication 2003). Safer and more lucrative shores seem to be in the offing: most prominently, microbes, rich in easily manipulable genetic material and, unlike medicinal plants, ostensibly "free" of thorny "cultural" claims (see Helmreich 2003). Indeed, invoking medicinal plants' unruly mix of enhanced political entanglement and "questionable" biochemical activity, major drug companies seem content to see plant and ethnobotanically derived screening return to the margins of corporate drug discovery (see Parry 2003).

Such attributions of viability matter not just for our understandings of the politics of prospecting but also for the analytic project of the anthropology of science in what are, arguably, rapidly transforming circumstances, both North and South. As the idioms of market and community continue to infuse new and familiar taxonomies of inclusion and exclusion, so too will it be imperative to attend to negotiations of governance, legitimacy, and entanglements through the practice of scientific research itself, and all of its ever-shifting accountabilities.

Acknowledgements

Funding for the research on which this article draws was provided by the Wenner Gren Foundation for Anthropological Research and the Fulbright Commission; support for writing was provided by Girton College at the University of Cambridge. I also thank my colleagues in Mexico, particularly Robert Bye, Edelmira Linares, Rachel Mata, Myma Mendoza Cruz, and Gustavo Morales, for their generous cooperation as I conducted my research; I thank as well Barbara Timmermann at the University of Arizona and Joshua Rosenthal at the U.S. National Institutes of Health. Anna Tsing, Lisa Rofel, Donna Haraway, Hugh Raffles, Jonathan Fox, Teresa Rojas Rabiela, Catherine Alexander, Marc Stears, and María Carranza provided comments on previous versions of the arguments that appear here. Finally, I would like to acknowledge with gratitude the detailed attention and constructive engagements of Virginia Dominguez and Linda Forman at *AE*, as well as the very helpful comments of three anonymous reviewers.

Notes

1 See Parry 2000 and Takacs 1996 for accounts of this "return to nature."

2 See Povinelli 2002 for an extended discussion of what she calls "late-liberal" modes of taking-as-giving in the context of the multicultural state.

3 And even this (elusive) promise is a complicated one: Not entirely without merit, many prospecting critics vociferously contest these agreements as dressed-up versions of familiar forms of corporate plunder.

4 Gudeman argues, "Rethinking the distribution of corporate returns lies at the heart of a new political economy" (2001:151).

5 I have written at length elsewhere about the ways in which bioprospecting requires academic plant collecting practices, with their own long-standing complexities, rationales, and socialities, to do new kinds of representational work (Hayden 2003). Scientists involved in bioprospecting contracts must in a sense *do* science studies, and do them one better: Much like classic sociologists of science, my ethnobotanist colleagues are endeavoring to "trace the relations" of people and interests that lie embedded, latent in knowledge and artifacts (see Latour 1987)—and not only that, but to compensate them (Hayden 2003).

6 The exact percentage of royalties to be shared remains confidential, but project administrators in the United States tell me it is somewhere between 2 and 15 percent—this sum would pass through the University of Arizona before heading south. Arizona will then keep 45 percent of that first redistributed sum directing the remaining 55 percent to the country that was the source of the plant material (Timmermann 1997).

7 For information on the ICBG program, see Grifo 1996; Rosenthal 1997; and Timmermann 1997; see also Greene 2002.

8 For critical commentary on these questions, see, among others, Escobar 1995; Foster 1997; Goldman 1997; Hayden 2003; Zerner 2000.

9 See Shane Greene 2002 for an account of an ICBG project in Peru in which, exceptionally, the participating indigenous coalitions were able to negotiate "trade secret" status for their knowledge and, thus, to license its use to the participating company.

10 For critiques of notions of community within development see, among many others, Brosius et al. 1998: Cooke and Kothari 2001; Escobar 1995; Zerner 2000.

11 See, in particular, Ferry 2002 for an insightful analysis of these articulations of national patrimony and neoliberal markets in relation to mining and mineral resources, Stephen 1997 on *ejidal* "reform," and Nash 2001 on the relationship between globalization and neoliberal reforms and indigenous political mobilizations in Chiapas.

12 The demise of the Maya ICBG is a long and complicated story, much of which transpired through Internet exchanges, campaigns, and bulletins. See the websites of the following organizations for an indicative guide to the issues: RAFI (now the ETC group), at www.etcgroup.org; CIEPAC (in Spanish and English), at www.ciepac.org; the University of Georgia, at http://guallart.dac.uga.edu/ICBGreply .-html. For continued updates and extended links, see the very useful site maintained by University of Chicago graduate student Michelle Day, at http://home.u-chicago.edu/-mmday/Mayanmedicine.html. See also Hayden 2003 for a more extended discussion than I have offered here.

13 Whereas ethnobotany and anthropology in Mexico have certainly been crucial participants in official indigenismo, many practitioners in these fields have also used their research to resist or help protect indigenous peoples from the kinds of cultural assimilation and deterritorialization that various incarnations of

indlgenismo—not to mention other forms of development—have set in motion (see Alcorn 1984; Caballero and Mapes 1985; Toledo 1996).

14 Echeverría's efforts on this front were part of his wider move to reinvigorate "indigenous" (i.e., national) scientific capacities in the name of Mexico's self-sufficiency (Schoijet and Worthington 1993). Joining a wave of nationalizations throughout Latin America, Echeverría not only attempted to resurrect a domestic, ethnobotanically based drug industry but also effectively nationalized the extant pharmaceutical industry by expropriating foreign companies (see Lozoya 1976).

15 Interview of Robert Bye by Tamara Dionne Stout, July 1997.

16 That is, when a product is patented and marketed, it is thereby launched into the public domain, continuing, we might say, its dissemination. At the same time, the matter of who shall benefit from such commercialization is highly controlled through the patent.

17 A pseudonym.

18 As suggested earlier, this imperative has emerged as much from the terms written into their research grants as from national and international activist sensibilities and multilateral accords.

References

Alcorn, Janis B.

1984 Huastec Mayan Ethnobotany. Austin: University of Texas Press.

Born, Georgina

1996 (Im)materiality and Sociality: The Dynamics of Intellectual Property in a Computer Software Research Culture. Social Anthropology 4(2):101–116.

Boyle, James

1996 Shamans, Software, and Spleens: Law and the Construction of the Information Society. Cambridge, MA: Harvard University Press.

Brosius, Peter J., Anna Tsing, and Charles Zerner

1998 Representing Communities: Histories and Politics of Community-Based Resource Management. Society and Natural Resources 11(2):157–168.

Brown, Michael

1998 Can Culture Be Copyrighted? Current Anthropology 39(2): 193–222.

Brush, Stephen B.

1994 Indigenous Knowledge of Biological Resources and Intellectual Property Rights: The Role of Anthropology. American Anthropologist 95(3):653–686.

1999 Bioprospecting the Public Domain. Cultural Anthropology 14(4):535–555.

Brush, Stephen B., and Doreen Stabinsky

1996 Valuing Local Knowledge: Indigenous People and Intellectual Property Rights. Washington, DC: Island Press.

Bye, Robert A., Jr., and Edelmira Linares

1983 The Role of Plants Found in the Mexican Markets and Their Importance in Ethnobotanical Studies. Journal of Ethnobiology 3(1):1–13.

1986 Medicinal Plants of the Sierra Madre: A Comparative Study of Tarahumara and Mexican Market Plants. Economic Botany 40(1):103–124.

Bye, Robert, Rachel Mata, and Rogelio Pereda

n.d. Avance en el programa del International Cooperative Biodiversity Group en México. Mexico. DF: National Autonomous University.

Caballero, Javier, and Cristina Mapes
1985 Gathering and Subsistence Patterns among the Piuahepecha Indians of Mexico. Journal of Ethnobiology 5(1):31–47.
Callon, Michel, ed.
1998 The Laws of the Markets. Malden, MA: Blackwell.
Carrier, James G., ed.
1997 Meanings of the Market: The Free Market in Western Culture. Oxford: Berg.
Clarke, Adele
1995 Research Materials and Reproductive Science in the United States, 1910–1940. *In* Ecologies of Knowledge: Work and Politics in Science and Technology. Susan Leigh Star, ed. Pp. 183–225. Albany: State University of New York Press.
Cooke, Bill, and Uma Kothari, eds.
2001 Participation: The New Tyranny? New York: Zed Books.
Coombe, Rosemary J.
1998 The Cultural Life of Intellectual Properties: Authorship, Appropriation and the Law. Durham, NC: Duke University Press.
Eisner, Thomas, and Elizabeth A. Beiring
1994 Biotic Exploration Fund—Protecting Biodiversity through Chemical Prospecting. BioScience 44(2):95–98.
Escobar, Arturo
1995 Encountering Development: The Making and Unmaking of the Third World. Princeton: Princeton University Press.
Ferry, Elizabeth
2002 Inalienable Commodities: The Production and Circulation of Silver and Patrimony in a Mexican Mining Cooperative. Cultural Anthropology 17(3):331–358.
Foster, John, ed.
1997 Valuing Nature? Economics, Ethics, and Environment. New York: Routledge.
Fujimura, Joan
1996 Crafting Science: A Sociohistory of the Quest for the Genetics of Cancer. Cambridge, MA: Harvard University Press.
Gledhill, John
2001 "Disappearing the Poor?": A Critique of the New Wisdoms of Social Democracy in an Age of Globalization. Urban Anthropology 30(2–3):123–156.
Goldman, Michael, ed.
1997 Privatizing Nature: Political Struggles for the Global Commons. New Brunswick, NJ: Rutgers University Press.
Goodman, Jordan, and Vivien Walsh
2001 The Story of Taxol: Nature and Politics in the Pursuit of an Anti-Cancer Drug. Cambridge: Cambridge University Press.
Granovetter, Mark
1985 Economic Action and Social Structure: The Problem of Embeddedness. American Journal of Sociology 91(3):481–510.
Greaves, Tom, ed.
1994 Intellectual Property Rights for Indigenous Peoples, A Sourcebook. Oklahoma City: Society for Applied Anthropology.
Greene, Shane
2002 Intellectual Property, Resources, or Territory? Reframing the Debate over

Indigenous Rights, Traditional Knowledge, and Pharmaceutical Bioprospection. *In* Truth Claims: Representation and Human Rights. Mark Philip Bradley and Patrice Petro, eds. Pp. 229–249. New Brunswick, NJ: Rutgers University Press.

Gregory, Christopher A.
1982 Gifts and Commodities. London: Academic Press.

Grifo, Francesca
1996 Chemical Prospecting: A View from the International Cooperative Biodiversity Groups Program. *In* Biodiversity, Biotechnology, and Sustainable Development in Health and Agriculture: Emerging Connections. Pp. 12–28. Washington, DC: Panamerican Health Organization.

Gudeman, Stephen
2001 The Anthropology of Economy: Community, Market, Culture. Malden, MA: Blackwell.

Hayden, Cori
2003 When Nature Goes Public: The Making and Un-Making of Bioprospecting in Mexico. Princeton: Princeton University Press.

Helmreich, Stefan
2003 Life@Sea: Networking Marine Biodiversity into Biotech Futures. *In* Remaking Life and Death: Towards an Anthropology of the Biosciences. Sarah Franklin and Margaret Lock, eds. Pp. 272–308. Santa Fe: School of American Research Press.

Hersch Martínez, Paul
1996 Destino común: Los recolectores y su flora medicinal. México, DF: Instituto Nacional de Antropología e Historia.

Joseph, Miranda
2002 Against the Romance of Community. Minneapolis: University of Minnesota Press.

Laird, Sarah, A. B. Cunningham, and Estherine Lisinge
2000 One in Ten Thousand? The Cameroon Case of *Ancistrocladus korupensis*. In People, Plants, and Justice: The Politics of Nature Conservation. Charles Zerner, ed. Pp. 345–373. New York: Columbia University Press.

Latour, Bruno
1987 Science in Action: How to Follow Scientists and Engineers through Society. Cambridge, MA: Harvard University Press.

Lowe, Lisa, and David Lloyd, eds.
1997 The Politics of Culture in the Shadow of Capitalism. Durham, NC: Duke University Press.

Lozoya, Xavier
1976 El Instituto Mexicano para el Estudio de las Plantas Medicinales. A.C. (IMEPLAM). *In* Estado actual del conocimiento en plantas medicinales de México. Xavier Lozoya and Carlos Zolla, eds. Pp. 243–255. México, DF: Folios.

Lozoya, Xavier, and Carlos Zolla, eds.
1984 La medicina invisible: Introduction al estudio de la medicina tradicional de México. México, DF: Folios.

Muelas Hurtado, Lorenzo
2000 Appeal to the Indigenous Peoples Representatives at COP5. Unpublished MS, submitted by the author to bio-ipr@cuenet.com, February 7.

Nadal, Alejandro
2000 Biopiratería: El debate político. La Jornada (Mexico City), September 13: 22.
Napolitano, Valentina, and Gerardo Mora
In press—Complementary Medicine: Cosmopolitan, Popular Knowledge and Transcultural Translations—Cases from Urban Mexico. Theory, Culture, and Society 20(4).
Nash, June C.
2001 Mayan Visions: The Quest for Autonomy in an Age of Globalization. New York: Routledge.
Orlove, Benjamin, and Stephen B. Brush
1996 Anthropology and the Convention on Biological Diversity. Annual Review of Anthropology 25:329–352.
Parry, Bronwyn
2000 The Fate of the Collections: Social Justice and the Annexation of Plant Genetic Resources. *In* People, Plants, and Justice: The Politics of Nature Conservation. Charles Zerner, ed. Pp. 374–402. New York: Columbia University Press.
2003 The Fate of the Collections: Exploring the Dynamics of Trade in Bio-Information. New York: Columbia University Press.
Posey, Darrell Addison
1996 Traditional Resource Rights: International Instruments for Protection and Compensation for Indigenous Peoples and Local Communities. Gland, Switzerland: IUCN and the World Conservation Union.
Povinelli, Elizabeth
2002 The Cunning of Recognition: Indigenous Alterities and the Making of Australian Multiculturalism. Durham, NC: Duke University Press.
Rosenthal, Joshua
1997 Integrating Drug Discovery, Biodiversity Conservation, and Economic Development: Early Lessons from the International Cooperative Biodiversity Groups. *In* Biodiversity and Human Health. Francesca Grifo and Joshua Rosenthal, eds. Pp. 281–301. Washington, DC: Island Press.
Schoijet, Mauricio, and Richard Worthington
1993 Globalization of Science and Repression of Scientists in Mexico. Science, Technology, and Human Values 18(2):209–230.
Schweitzer, Jeff, F. G. Handley, J. Edwards, W. F. Harris, M. R. Grever, S. A. Schepartz, G. Cragg, K. Snader. and A. Bhat
1991 Summary of the Workshop on Drug Development, Biological Diversity, and Economic Growth. Journal of the National Cancer Institute 83(18):1294–1298.
Sherwood, Robert
1991 Pharmaceuticals: U.S. Perspective. *In* U.S.-Mexican Industrial Integration: The Road to Free Trade. Sidney Weintraub, ed., with Luis Rubio F. and Alan D. Jones. Pp. 161–179. Boulder: Westview Press.
Soto Laveaga, Gabriela
2001 Root of Discord: Steroid Hormones, a Wild Yam, "Peasants," and State Formation in Mexico (1941–1986). Ph.D. dissertation, Department of History, University of California, San Diego.
Stephen, Lynn
1998 Between NAFTA and Zapata: Responses to Restructuring the Commons in Chiapas and Oaxaca, Mexico. *In* Privatizing Nature: Political Struggles for

the Global Commons. Michael Goldman, ed. Pp. 76–101. New Brunswick, NJ: Rutgers University Press.

Strathern, Marilyn

1988 The Gender of the Gift: Problems with Women and Problems with Society in Melanesia. Berkeley: University of California Press.

1999 Potential Property: Intellectual Rights and Property in Persons. *In* Property, Substance, Effect: Anthropological Essays on Persons and Things. Marilyn Strathern, ed. Pp. 161–178. London: Athlone Press.

2000 Accountability . . . and Ethnography. *In* Audit Cultures: Anthropological Studies in Accountability, Ethics, and the Academy. Marilyn Strathern, ed. Pp. 279–304. London: Routledge.

Takacs, David

1996 The Idea of Biodiversity: Philosophies of Paradise. Baltimore: Johns Hopkins University Press.

Thomas, Nicholas

1991 Entangled Objects: Exchange, Material Culture, and Colonialism in the Pacific. Cambridge, MA: Harvard University Press.

Timmermann, Barbara

1997 Biodiversity Prospecting and Models for Collections Resources: The NIH/NSF/USAID Model. *In* Global Genetic Resources: Access, Ownership, and Intellectual Property Rights. K. Elaine Hoagland and Amy Y. Rossman, eds. Pp. 219–234. Washington, DC: Association of Systematics Collections.

Toledo, Victor M.

1995 New Paradigms for a New Ethnobotany: Reflections on the Case of Mexico. *In* Ethnobotany: Evolution of a Discipline. Richard Evans Schultes and Siri von Reis, eds. Pp. 75–92. Portland, OR: Dioscorides Press.

1996 The Ecological Consequences of the 1992 Agrarian Law of Mexico. *In* Reforming Mexico's Agrarian Reform. Laura Randall, ed. Pp. 247–260. New York: M. E. Sharpe.

World Resources Institute, Instituto Nacional de Biodiversidad, Rainforest Alliance, and African Centre for Technology Studies

1993 Biodiversity Prospecting: Using Genetic Resources for Sustainable Development. Washington, DC: World Resources Institute, Instituto Nacional de Biodiversidad, Rainforest Alliance, and African Centre for Technology Studies.

World Resources Institute, the World Conservation Union, United Nations Environment Programme, Food and Agriculture Organization of the United Nations, and United Nations Educational Scientific and Cultural Organization

1992 Global Biodiversity Strategy: Guidelines for Action to Save, Study, and Use Earth's Biotic Wealth Sustainably and Equitably. Washington, DC: World Resources Institute.

Zerner, Charles

2000 Introduction: Towards a Broader Vision of Justice and Nature Conservation. *In* People, Plants, and Justice: The Politics of Nature Conservation. Charles Zerner, ed. Pp. 3–20. New York: Columbia University Press.

56

GENOME

Donna Haraway

Source: *Modest_Witness@Second_Millennium.FemaleMan©_Meets OncoMouse™*, London: Routledge, 1997, pp. 244–265.

If universal humanity was plastic under the sign of the population at mid-century, then human nature is best described as virtual in present, end-of-the-millennium regimes of biological knowledge and power. Specifically, human nature is embodied, literally, in an odd thing called a genetic database, held in a few international locations such as the three large public databases for genetic map and sequence data: the U.S. GenBank©, the European Molecular Biological Laboratory, and the DNA Data Bank of Japan. The Genome Data Base at Johns Hopkins University is a massive central repository of all gene-mapping information. In the world of gene sequencing, intellectual property rights vie with human rights for the attention of lawyers and scientists alike. Criminal as well as corporate lawyers have a stake in the material and metaphoric representation of the genome. Funding and policy strongly support rapid public access to genome databases in the interests of research and development. For example, in 1993 the French researcher Daniel Cohen, of the Centre d'Etude du Polymorphisme Humaine in Paris, made his first complete draft map of the human genome available through the Internet. GenInfo, developed by the U.S. National Center for Biotechnology Information of the National Library of Medicine, is a kind of metadatabase containing both protein and nucleic acid sequence data "to which other databases can add, refer, annotate, interpret, and extrapolate" (Corteau 1991:202).[1] In part because of the tremendous physical computing power and human expertise that resulted from nuclear weapons research, informatics development in the U.S. Human Genome Project began under the auspices of GenBank© at the U.S. National Laboratories at Los Alamos, New Mexico. It was there also that the expertise and machines existed that built the matrix for the flourishing of artificial life research at the nearby Santa Fe Institute.

A database is an information structure. Computer programs are the habitats of information structures, and an organism's genome is a kind of nature park among databases. Just as racial hygiene and eugenics were committed to science and progress, and populationist doctrines of human universals were unambiguously on the side of development and the future, the genome is allied with all that is up-to-the-minute. Yet, something peculiar happened to the stable, family-loving, Mendelian gene when it passed into a database, where it has more in common with LANDSAT photographs, Geographical Information Systems, international seed banks, and the World Bank than with T.H. Morgan's fruitflies at Columbia University in the 1910s or UNESCO's populations in the 1950s. Banking and mapping seem to be the name of the genetic game at an accelerating pace since the 1970s, in the corporatization of biology to make it fit for the New World Order, Inc.[2] If the modern synthesis, ideologically speaking, tended to make everyone his brother's keeper, then, in its versions of kin selection and inclusive fitness-maximization strategies, the sociobiological synthesis runs to making everyone his or her sibling's banker.[3]

Biotechnology in the service of corporate profit is a revolutionary force for remaking the inhabitants of planet Earth, from viruses and bacteria right up the now repudiated chain of being to *Homo sapiens* and beyond. Biological research globally is progressively practiced under the direct auspices of corporations, from the multinational pharmaceutical and agribusiness giants to venture-capital companies that fascinate the writers for the business sections of daily newspapers. Molecular biology and molecular genetics have become nearly synonymous with biotechnology as engineering and redesign disciplines. Beings like Man the Hunter and Woman the Gatherer reappear for their roles on the stage of nature enterprised up as Man™ and Woman™—copyrighted; registered for commerce; and, above all, highly flexible.[4] In a world where the artifactual and the natural have imploded, nature itself, both ideologically and materially, has been patently reconstructed. Structural adjustment demands no less of bacteria and trees as well as of people, businesses, and nations.

The genome is the totality of genetic "information" in an organism, or, more commonly, the totality of genetic information in all the chromosomes in the nucleus of a cell. Conventionally, the genome refers only to the nucleic acid that "codes" for something and not to the dynamic, multipart structures and processes that constitute functional, reproducing cells and organisms. Thus, not even the proteins critical to nuclear chromosomal organization or DNA structures such as mitochondrial chromosomes outside the nucleus are part of the genome, much less the whole living cell. Embodied information with a complex time structure is reduced to a linear code in an archive outside time. This reduction gives rise to the curious, ubiquitous, mixed metaphor of "mapping the code," applied to projects to represent all the information in the genome. DNA in this view is a master molecule, the code of codes, the foundation of unity and diversity. Much of the history of genetics since the 1950s is the history of the consolidation and elaboration of the equation of "gene = information" in the context of master-molecule metaphors. I consider this representational practice for thinking about genetics to constitute a kind of artificial life research itself, where the paradigmatic habitat for life—the program—bears no necessary relationship to messy, thick organisms.

The convergence of genomics and informatics, in technique and personnel as well as in basic theory and shared tropes, is immensely consequential for bioscientific constructions of human nature. The technical ability to manipulate genetic information, in particular to pass it from one kind of organism to another in a regulated manner in the lab, or to synthesize and insert new genes, has grown exponentially since the first successful genetic engineering experiments of the early 1970s. In principle, there is no naturally occurring genome that cannot be experimentally redesigned. This is a very different matter compared to the genetic traffic among populations of a species studied within the midcentury evolutionary synthesis, much less compared to the genetic, natural racial types that inhabited the biological world earlier in the century. Genetic engineering is not eugenics, just as the genome does not give the same kind of account of a species as does organic racial discourse.[5]

From the point of view of the 1990s, the genome is an information structure that can exist in various physical media. The medium might be the DNA sequences organized into natural chromosomes in the whole organism. Or the medium might be various built physical structures, such as yeast artificial chromosomes (YACs) or bacterial plasmids, designed to hold and transfer cloned genes or other interesting stretches of nucleic acid. The entire genome of an organism might be held in a "library" of such artifactual biochemical information structures. The medium of the database might also be the computer programs that manage the structure, error checking, storage, retrieval, and distribution of genetic information for the various international genome projects that are under way for *Homo sapiens* and for other model species critical to genetic, developmental, and immunological research. Those species include mice, dogs, bacteria, yeast, nematodes, rice, and a few more creatures indispensable for international technoscientific research.

The U.S. Human Genome Project officially began in 1988 under the management of the Department of Energy and the National Institutes of Health. As a whole, the global Human Genome Project is a multinational, long-term, competitive and cooperative, multibillion-dollar (yen, franc, mark, etc.) effort to represent exhaustively—in genetic, physical, and DNA sequence maps—the totality of information in the species genome.[6] The data are all entered into computerized databases, from which information is made available around the world on terms still very much being worked out. Computerized database design is at the leading edge of genomics research. Design decisions about these huge databases shape what can be easily compared to what else, and so determine the kinds of uses that can be made of the original data. Such decisions structure the kinds of ideas of the species that can be sustained. National science bodies, tax- and foundation-supported universities, international organizations, private corporations, communities, indigenous peoples, and many configurations of political and scientific activists all play a part in the saga.

Questions about agency—who is an actor—abound in the world of the genome, as in the worlds of technoscience in general. For example, in the discourse of genome informatics, data are exchanged among "agents" and sent to "users" of databases. These entities could as easily be computers or programs as people (Erickson 1992).[7] It does not solve the trouble to say that people are the end users. That turns out to be a contingent, technical, design decision—or a way or representing ongoing flows of information—more than an ontological necessity. People are in the information loop, but their status is a bit iffy in the artificial life world. Compared to the biological humanism of the modern synthesis, technohumanism has had to make a few timely ideological adjustments. Genomics is neither taxidermy nor the reconstruction practices of the new physical anthropology, and the emerging techniques of animation occupy the minds of more than the *Jurassic Park* special-effects programmers at Industrial Light and Magic.

Issues of *agency* permeate practices of *representation* in many senses of both terms: Who, exactly, in the human genome project represents whom? A prior question has to be a little different, however. Who, or what, is the human that *is* to be exhaustively represented? Molecular geneticists are consumed with interest in the variability of DNA sequences. Their databases are built to house information about both stable and variable regions of genes or proteins. Indeed, for actors from drug designers to forensic criminologists, the uniqueness of each *individual's* genome is part of the technical allure of the human genome projects' spinoffs. More fundamentally, however, the genome projects produce entities of a different ontological kind than flesh-and blood organisms, "natural races," or any other sort of "normal" organic being. At the risk of repeating myself, the human genome projects produce ontologically specific things called databases as objects of knowledge and practice. The human to be represented, then, has a particular kind of totality, or species being, as well as a specific kind of individuality. At whatever level of individuality or collectivity, from a single gene region extracted from one sample through the whole species genome, this human is itself an information structure whose program might be written in nucleic acids or in the artificial intelligence programming language called Lisp®.

Therefore, variability has its own syntax in genome discourse as well. There is no illusion in the 1990s about single "wild-type" genes and various mutant deviants.[8] That was the terminology of Mendelian genetics of the early twentieth century, when the languages of the normal and the deviant were much more sanitary. Racial hygiene and its typological syntax are not supported by genome discourse, or by artificial life discourses in general. Genetic investment strategies, in the sense of both evolutionary theory and business practice, *are* supported. The populationist thinking of the modern synthesis blasted an entire toolkit of resources for believing in norms and types. Flexibility, with its specific grammars of human unity and diversity, is the name of the game at the end of this millennium. However, for all of their commitment to variability, most molecular geneticists are not trained in evolutionary population biology, or even in population genetics. This disciplinary fact has given rise to a most interesting project and ensuing controversy for the purposes of this chapter. Let us pick up questions of agency and representation, as well as unity and difference, though the Human Genome Diversity Project (HGDP).

If the human genome databases are exhaustively to represent the species—and to provide information to users who demand that kind of knowledge, in dreams of totality as well as in practical projects—the repositories must contain physical and electronic data about the specific molecular constitution and frequency of genes on a truly global scale. Population geneticists were critical both of molecular biologists' sampling protocols for human genetic material and of their woeful statistical grasp of the structure, distribution, history, and variability of human populations. The population geneticists were also worried that many human populations around the world were becoming extinct—either literally or through interbreeding and swamping of their diversity in larger adjoining populations—with the consequent loss of genetic information forever impoverishing the databases of the species. What it means to be human would have irredeemable informational gaps. There would be a biodiversity information loss in the lifeworld of the genome. Like the vanishing of a rainforest fungus or fern before pharmaceutical companies could survey the species for promising drugs, the vanishing of human gene pools is a blow to technoscience. Prompt and thorough genetic collection and banking procedures as well as preservation of the source of variation, if possible, are the solution.

I am being a bit mordant in my reading of purposes in this account, for the organizers of the Human Genome Diversity Project were largely liberal biological humanists of the old stamp. Also, I remain sympathetic to the desire to produce a human species database that draws from as large a concept of humanity as possible. I want there to be a way to reconfigure this desire and its attendant humanism. However, it was precisely the doctrines of difference, representation, and agency of "universal" humanism that got the project and its well-meaning organizers into well-deserved trouble.[9]

Beginning about 1991, the organizers of the Human Genome Diversity Project proposed to amend the evolutionary population thinking, or lack of thinking, of the mainline Human Genome Project by collecting hair-root, white blood-cell, and cheek-tissue samples, to be held in the American Type Culture Collection, from over 700 groups of indigenous peoples on six continents. Over five years, the cost would be about $23–35 million (compared to more than $3 billion for the Human Genome Project as a whole). Unfortunately, unself-conscious, modernist perspectives distorted the definition of the categories of people from whom samples were to be sought, leading to a vision of dynamic human groups as timeless "isolates of historic interest." Also, other potentially genetically distinct ethnic communities did not appear on the sampling list.

The planning of the project did not involve members of the communities to be studied in any formative way in the science. The people to be sampled might give or withhold permission, to be more or less carefully sought and thoroughly explained, but they were not regarded as partners in knowledge production who might have ends and meanings of their own in such an undertaking. Their versions of the human story, complexly articulated with the genetic science of the visitors, did not shape the research agenda. Permission is not the same thing as collaboration, and the latter could lead to fundamental changes in who and what would count as science and as scientists. All the trappings of universal science notwithstanding, amending a database is a pretty culturally specific thing to want to do. Just why should other people, much less folks called "isolates of historic interest," help out with that project? That is not a rhetorical question, and there can be very strong answers coming from counterintuitive as well as obvious viewpoints for any actor. The question is a fundamental one about the rhetoric of persuasion and the practical processes through which people—including scientists and everybody else—get reconstituted as subjects and objects in encounters. How should the many discourses in play within and between people like the Guaymi of Panama and the Population Geneticists of California be articulated with each other in a power-sensitive way? This is an ethical question, but it is much more than that. It is a question about what may count as modern knowledge and who will count as producers of that knowledge.[10]

Not surprisingly, it turned out that indigenous people were more interested in representing themselves than in being represented in the human story. The encounter was most certainly not between "traditional" and "modern" peoples but between contemporaneous people (and peoples) with richly interlocking and diverging discourses, each with its own agendas and histories. Functioning as boundary objects, "genes" and "genomes" circulated among many of the languages in play.[11] Members of communities to be sampled, as well as other spokespeople, had several concerns. Some were adamant that genes or other products derived from indigenous material not be patented and used for commercial profit. Others were worried that the genetic information about tribal and marginalized peoples could be misused in genocidal ways by national governments. Some argued that medical and social priorities of the communities could be addressed by the money that would go to funding the genetic sampling, and the HGDP did not give benefits back to the people. Some were quite willing to have indigenous genetic material contribute to a medically useful world knowledge fund, but only under United Nations or similar auspices that would prevent exploitation and profit-making. Ethics committee members of the HGDP tried to assure skeptics that the project had no commercial interests and that the HGDP would try to make sure that any commercial benefits that did result from the sampled material flowed back to the communities. But overall, the general issue was the question of the agency of people who did not consider themselves a biodiversity resource. Diversity was about both their *object* status and their *subject* status.

In May 1993, at a nongovernmental conference meeting parallel to the UN Human Rights Conference in Vienna, the Rural Advancement Foundation International (RAFI) and indigenous peoples urged the HGDP to "halt current collection efforts, convene a meeting with Indigenous peoples to address ethical and scientific issues, incorporate Indigenous organizations in every aspect of the HGDP and grant them veto power, and place the HGDP under direct United Nations control, with decision making delegated to a management committee dominated by Indigenous people" (RAFI 1993:13). Leaders of the HGDP tried to address the objections, but by fall of 1993 they had not set up mechanisms acceptable to the critics to include indigenous peoples in project organizing. The World Council of Indigenous Peoples monitored the project skeptically. It is important to me to note, however, that the HGDP was a minority effort in the Human Genome Project (HGP) and not at the center of the prestigious action. To get the research done at all in the face of the nonpopulationist molecular genetic orthodoxy that guided ordinary practice in the HGP would have been no small trick. It has proved easier to slow down or stop the HGDP, a kind of oppositional effort, than to question the powerful HGP itself. That makes the trouble with "difference" built into this potentially positive scientific project all the more disturbing—and important.

Inescapably, independently of the HGDP but fatally glued onto it, the all-too-predictable scandal happened. Like all pathologies, the scandal revealed the structure of what passes for normal in bioscientific regimes of knowledge and power. The Guaymi people carry a unique virus and its antibodies that might be important in leukemia research. Blood taken in 1990 from a 26-year-old Guaymi woman with leukemia, with her "informed oral consent," in the language of the U.S. Center for Disease Control in Atlanta, was used to produce an "immortalized" cell line deposited at the American Type Culture Collection. The U.S. Secretary of Commerce proceeded to file a patent claim on the cell line. Pat Moony of the Rural Advancement Foundation International found out about the claim in August 1993 and informed Isidoro Acosta, the president of the Guaymi General Congress. Considering the patent claim to be straight-forward biopiracy, Acosta and another Guaymi representative went to Geneva to raise the issue with the Biological Diversity Convention, which had been adopted at the 1993 Earth Summit in Brazil.[12]

That convention had been intended to deal with plant and animal material, but the Guaymi made strategic use of its language to address technoscientifically defined human biodiversity. The Guaymi also went to the GATT secretariat to argue against the patentability of material of human origin in the intellectual property provisions of the new GATT treaty then being drafted.

In late 1993, the U.S. Secretary of Commerce withdrew the patent application, although by early 1994 the cell culture had not been returned, as demanded, to the Guaymi. The property and sovereignty battles are far from being resolved; they are at the heart of bioscientific regimes of knowledge and power worldwide. Scientific and commercial stakes are high. The stakes are also the ongoing configuration of subjects and objects, of agency and representation, inside of and by means of these disputes about biopower. The stakes are about what will count as human unity and diversity. The human family is at stake in its databases. I am instructed by the encounter of discourses, where genes are the circulating boundary objects. The Guaymi and the U.S. actors engaged each other in biogenetic terms, and they struggled for shaping those terms in the process. Perhaps the Guaymi did not initiate biotechnological and genetic engineering discourses, including their business and legal branches, but the indigenous Panamanians are far from passive objects in these material and linguistic fields. They are actors who are reconfiguring these powerful discourses, along with others they bring to the encounter. In the process, the Guaymi are changing themselves, the international scientists, and other policy elites.

The organizers of the HGDP continued to try to reorganize the research plan to satisfy both funding agencies and people to be sampled, and in late 1994 the project's International Executive Committee released a document that aimed to establish trust with indigenous peoples' organizations (Kahn 1994). The revised plan promised local control over the survey and protection of the research subjects' patent rights as well as an independent committee established by UNESCO to advise project organizers on ongoing ethical and other controversial matters. A key provision is that in order to develop scientific priorities and ethical guidelines based on local conditions and cultures, the research be done as much as possible in the countries or regions where the sampled populations live. But localism will not solve key problems. International biodiversity property issues will not go away, and the cosmopolitan nature, as well as local cultural dimensions, of science provide both the attraction and the danger in the HGDP. Issues of cultural meaning, as well as technical and financial matters, are at stake in the global-local dialectic of technoscience, and people categorized as "indigenous" might well be more cosmopolitan than those labeled "Westerners" in key respects. Global/local does not translate as western/elsewhere or modern/traditional.[13] The biotechnology involved in the HGDP is of interest to prospective host countries, and several groups have also expressed interest in possible medical benefits as well as in participating in a project that contributes to defining humanity transnationally.

Europeans were among the first indigenous peoples to proceed with HGDP research. In 1994, the European Union provided $1.2 million to set up 25 labs from Barcelona to Budapest to study questions about European genetic diversity and paleoanthropological history. Of course, the "races" of Europe were also central to the scientific constructions of human unity and diversity in the nineteenth century, and people elsewhere in the world have not always been so convinced this is the way to think about the matter. But regional committees to pursue the HGDP have been set up in North America, South and Central America, and Africa as well as Europe, while India, China, and Japan had declined by late 1994 (Kahn 1994:722). Organized Native Americans in the United States predictably have been divided. The Euchees and Apaches of Oklahoma decided to participate in the HGDP, in part because of their interest in research on the genetics of diabetes, a major health problem of Native Americans. At the same time, in the summer of 1994 a broad coalition of consumer, indigenous, environmental and nongovernmental organizations working on development issued a statement calling on all participants "to work with parallel movements led by indigenous nations to eliminate federal funding to the Human Genome Diversity Project" (Bereano 1995). The major reason was the potential for commercialization, especially in the form of patents on human genes and proteins, without benefit to the sampled populations whose body parts would become museum specimens in an updated form. The Europeans have also shown considerable resistance to the patent fever that grips biotechnology in North America, and the European Parliament legislated that publicly funded research should not give rise to privately held patents (Bereano 1995).

A troubling leitmotiv in the Guaymi cell-line dispute returns us to the narratives, images, and myths with which I want to conclude this meditation on the human family. In the midst of the polemics, Pat Mooney of the RAFI was quoted as saying, "When a foreign government comes into a country, takes blood without explaining the real implications to local people, and then tries to patent and profit from the cell line, that's wrong. Life should not be subject to patent monopolies" (RAFI 1994:7). The patent monopoly part is true enough, but penetration by a foreign power to take blood evokes much more than intellectual property issues. Indeed, some of the indigenous organizations critical of the HGDP called it the "vampire project" (Kahn 1994:721). I cannot help but hear Mooney's quote in the context of periodically surfacing stories in Latin America about white North Americans stealing body parts, sucking blood, and kidnapping children to be organ donors. The factual accuracy of the accounts is not the point, even though the dubious standards of evidence to which commentators have been held when the stories appear in U.S. news articles and radio talk shows appall me. What matters in this chapter is the stories themselves, that is, the ready association of technoscience with realms of the undead, tales of vampires, and transgressive traffic in the bloody tissues of life. Sampling blood is never an innocent symbolic act. The red fluid is too potent, and blood debts are too current. Stories lie in wait even for the most carefully literal-minded. Blood's translations into the sticky threads of DNA, even in the aseptic databases of cyberspace, have inherited the precious fluid's double-edged power. The genome lives in the realm of the undead in myriad ways that cannot be contained by rational intentions, explicit explanations, and literal behavior. The stories get at structures of power and fantasy that must be faced in all their displaced, uncanny truth.

Table 1, "Night Births and Vampire Progeny," is a rough guide through a tiny region of the mine-strewn territory. My chart is indebted to three mainline publications within technoscientific professional and popular culture. Pursuing the symptomatic logic of this chapter, my technique is resolute over-reading. I know no better strategy to deal with the vermin-infested normality of rational discourse. Just state the obvious. Say what should not have to be said.

Running several times in *Science* magazine in 1989–1990, Du Pont's wonderful advertisement for OncoMouse™, the first patented animal in the world, provides my first text [Figure 1 Stalking Cancer].[14] OncoMouse™ contains a cancer-causing bit of DNA, called an oncogene, derived from the genome of another creature and implanted by means of genetic engineering techniques. A model for breast cancer research, the redesigned rodent is like a machine tool in the workshops for the production of knowledge.

Table 1 Night Births and Vampire Progeny

IMAGE	OncoMouse™	Gorilla-suited bride	SimEve
SOURCE	*Science* magazine	*American Medical News*	*Time* magazine
KIN CATEGORY	species	family	race
REPRODUCTIVE PRACTICE	genetic engineering	professional investment	cybergenesis by morphing
NARRATIVES AND MYTHS	night births in the laboratory scientific enlightenment Plato's allegory of the cave heroic quest	Bad investments yield polluted off-spring. Reverse alchemy turns gold into base metal. racialized heterosexuality vampire-toothed bride	masculine parthenogenesis mind children Orestian Trilogy Pygmalion and Galatea
SLOGAN	"where better things for better living come to life"	"If you've made an unholy alliance . . ."	"love that will forever remain unrequited."

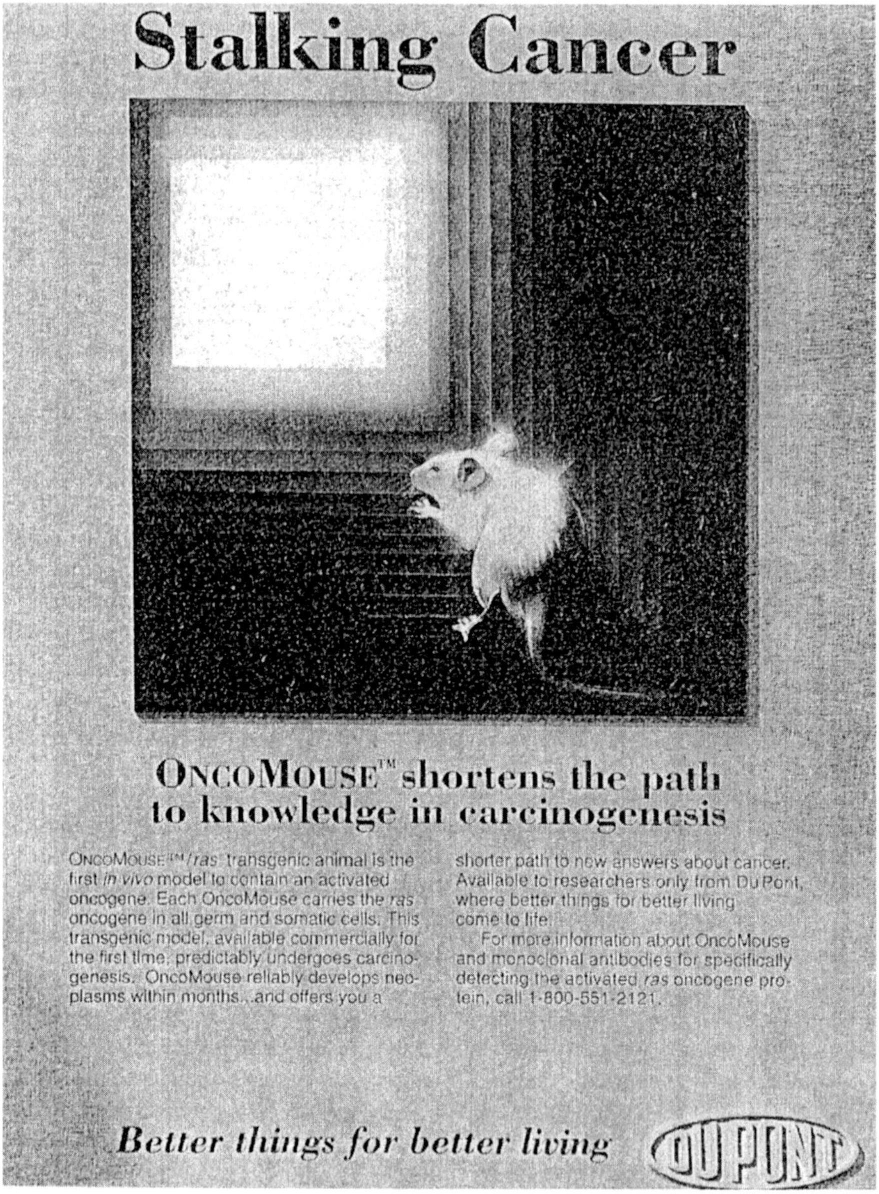

Figure 1 Du Pont advertisement from Science magazine for OncoMouse™, April 27, 1990. Courtesy of Du Pont NEN products. on May 19, 1995 Du Pont announced its intent to divest its medical products businesses. The former Du Pont NEN products busness will become NEN life science products.

OncoMouse™ is a transgenic animal whose scene of evolution is the laboratory. Inhabiting the nature of no nature, OncoMouse™'s natural habitat is the fully artifactual space of technoscience. Symbolically and materially, OncoMouse™ is where the categories of nature and culture implode for members of technoscientific cultures. For that very reason, the mouse has been at the center of controversy since its production. Defined by a spliced genome, identified with a spliced name, patented, and trademarked, OncoMouse™ is paradigmatic of nature enterprised up. What interests me here, however, are the stories that are crusted like barnacles onto the striking advertising image.

Du Pont's white mouse is in the midst of a heroic travel or quest narrative and part of a noble hunt in which the cancer enemy is stalked. Epistemophilia, the lusty search for knowledge of origins, is everywhere. The mouse climbs out of a womblike, geometric cave toward the light of knowledge, evoking the narrative elements of the Western Enlightenment and of Plato's allegory of the cave. OncoMouse™ is "available to researchers only from Du Pont, where better things for better living come to life." Like it or not, we are catapulted into the narrative fields that contain Frankenstein and his monster and all the other alluring scenes of night births in the mythological culture of science. The laboratory repeatedly figures as an uncanny place, where entities that do not fit, do not belong, cannot be normal—that transgress previously important categories—come into being. I am drawn to the laboratory for this essential narrative of epistemological and material power. How could feminists and antiracists in this culture do without the power of the laboratory to make the normal dubious? Raking ambivalence and strong visitations from a culturally specific unconscious, however, are the price of this alliance with the creatures of technoscience. Reproduction is afoot here, with all of its power to reconfigure kinship. In the proliferating zones of the undead, the kin categories of species are undone and redone, all too often by force. Consciously or unconsciously, whoever designed this ad knew all the right stories. Enlightenment has never been more pregnant with consequences—semiological, financial, and technological—for the human family.

Family imagery is much more explicit and far more ominous in my next text, an ad for Prepaid Medical Management, Inc. (PreMed), which was published in American Medical News on August 7, 1987 [Figure 2 If you've made an unholy HMO alliance, perhaps we can help]. PreMed tells physicians that it can help get them out of unprofitable contracts with health maintenance organizations (HMOs) that had promised a financially sound patient base and quality care but instead delivered profits for distant shareholders and high administrative fees for doctors. PreMed claims to have aided physicians in establishing locally controlled and fiscally sound HMOs in which doctors could determine whom they treated and how they practiced medicine. There is little question that these are pressing concerns in the con-

Figure 2 Courtesy of Premed. Advertisement from American Medical News, August 7, 1987.

text of a medicine-for-profit system, in which many patients are uninsured, underinsured, or covered by public plans that pay much less for services than private insurers. Although not referring directly to the larger context, the ad appeared in the midst of an epidemic of national publicity about high Medicare and Medicaid patient loads in urban HMOs, African American crack-addicted and AIDS-infected mothers and babies in inner cities, and astronomical malpractice insurance costs, particularly for urban obstetricians.

The PreMed verbal text makes no reference to race, gender, or class, but I think these codes structure the ad. "Accepting reduced fees and increased risks" is a code for accepting too many poor patients who do not have private insurance. The code, if not a more complicated reality, biases readers to see those high-risk, poor patients as overwhelmingly people of color, especially African American. The visual scene of a wedding and the verbal text about an unholy alliance propel the reader to see the patient as female and black and the doctor as male and white. An unholy alliance is "miscegenation", the bloodsucking monster at the heart of racist and misogynistic terror.[15]

Finally, it is the double disguise, the twice-done veiling of the bride that makes the ad so flagrantly about what it literally covers up with a joke: the class-structured, racialized, sexual politics of U.S. reproductive health and the further withdrawal of medical services from already underserved populations. A white-medical-coat-clad, stethoscope-wearing, prosperous-looking white man with just the right amount of graying hair is putting a gold wedding band on the ring finger of a black gorilla-suited bride in a white wedding gown and veil. The bride is doubly not there. Present are only two disguises: the wedding dress and gorilla suit. The implied infected or addicted pregnant Black woman who is always, in the code, on welfare, is denied in advance.[16] The surface of the ad insists that it is I, not PreMed, who is both making the connection of the gorillasuited bride with African American women and putting the wedding scene into the context of reproductive health care. Can't I take a joke? But my power to be amused is vitiated by the searing memory of just where African American women fit historically into systems of marriage and kinship in white heterosexual patriarchy in the United States. *Miscegenation* is still a national racist synonym for infection, counterfeit issue unfit to carry the name of the father, and a spoiled future. The bitter history of the scientific and medical animalization of people of African descent, especially in the narratives of the great chain of being that associated apes and Black people, further accounts for my poor sense of humor. The gorilla suit cannot be an innocent joke here, and good intentions are no excuse. The lying disguises cannot hide what they deny.

But this bride is less a living—or a reconstructed—gorilla than an undead monster. She is not a creature in an Akeley diorama, whose natural types always glowed with health. The gorilla-suited bride is the type of no type. Her lips are parted just enough to show the gleam of a bright white tooth.

The bride is a vampire, equipped with the tool for sucking the blood of the husband and polluting his lineage. The shining tooth echoes the brilliant gold of the wedding ring. The wedding night bodes ill. The conventional trope of the scientist-husband of nature generating the legitimate, sacred fruit of true knowledge in the womb of the wife's body is engaged here with chilling modifications. A metaphor for the magical power of science, alchemy is about the generative sexual practices of the craft, which are a kind of marriage that yields gold from base metal. Alchemy is about holy alliances, true marriages with gleaming children. In the PreMed advertisement, the narrative is reversed, and an "unholy alliance" threatens to mutate the promised gold of a medical-career investment into the base metal of a nonproductive practice. "If you've made an unholy alliance, perhaps we can help." Call upon PreMed and enjoy the fruit of a productive union. Be flexible; make the required structural adjustments to stimulate the production of wealth—and its flow upward to the deserving professional classes. Leave that unnatural and unprofitable alliance with infected bodies. A healthy family life demands no less.

The PreMed ad almost seems out of its time. It shouldn't still be possible to publish such an image in a scientific medical magazine. But it is possible. The fierce resurgence of explicit racist, sexist, and class-biased discourse of many kinds all over the world, and exuberantly in the United States, give all too much permission for this merely implict and latent joke.

My third text, by contrast, wants to be firmly on the side of the antiracist angels. All the signs of liberal multiculturalism pervade *Time* magazine's cover image for its special fall 1993 issue on immigration [Figure 3 The New Face of America]. These angels, however, turn out to exist in cyberspace. The *Time* cover is a morphed portrait of a being I call SimEve. In the background is a matrix of her mixed cybergenetic kin, all resulting from different "racial" crosses effected by a computer program. "Take a good look at this woman. She was created by a computer from a mix of several races. What you see is a remarkable preview of . . . The New Face of America." Indeed. We are abruptly returned to the ontology of databases and the marriage of genomics and informatics in the artificial life worlds that reconstitute what it means to be human. Here, the category so ethereally and technically reconfigured is race. In an odd computerized updating of the typological categories of the nineteenth and early twentieth centuries, the programmer who gave birth to SimEve and her many siblings generated the ideal racial synthesis, whose only possible existence is in the matrices of cyberspace. Genetic engineering is not yet up to the task, so it falls to the computer sciences alone for now. Full of new information, the First Family reconstructed by Jay Matternes has had a transgenic change of form, to reemerge from *Time*'s computer womb as morphed ideal citizens, fit for the "Rebirthing of America." If the biotechnological genetics laboratory was the natural habitat and evolutionary scene fusing nature and culture for OncoMouse™'s version of

Figure 3 *Time* magazine's morphed simeve, Backed by the racial-ethnic, computer generated matrix for *Time*'s "Rebirthing America" special issue, Fall 1993. Photograph by Ted Thai. Reprinted with permission.

the origin of life, SimEve's primal story takes place in the first morphing program for the personal computer, called Morph 2.0, produced by Gryphon Software Corporation.[17]

This technology has proved irresistible in the United States for 1990s mass cultural racialized kinship discourse on human unity and diversity. Never has there been a better toy for playing out sexualized racial fantasies, anxieties, and dreams. The reverie begins in cross-specific morphing, with the compelling computer-generated composite of human and chimpanzee faces on the cover of the 1992 *Cambridge Encyclopedia of Human Evolution*.[18] Like all portraits, this photograph records and shapes social identity. Soberly looking straight at the reader, the mature face is intelligent and beautiful. Like Carl Akeley's taxidermic reconstructions, this morphed face feeds a deep fantasy of touch across the ethnospecific categories of nature and culture. Unframed by any such specificity, the face seems to bring word about an original transformation in universal natural history.

On the contemporary human register, Gillette's shaving ads on television show the transformation of men's faces into each other across a racial spectrum, producing a utopic multiethnic male bonding. In the September 1994 Great American Fashion issue of the feminism-lite magazine *Mirabella*, the prominent photographer Hiro produced the computer-generated cover image from many photos of exquisitely beautiful multiracial, multiethnic women. Asked by the editors to give them a photo to represent "the diversity of America," Hiro did a simulated (and very light-skinned) woman.[19] A tiny microchip floats through space next to her gorgeous face. I read the chip as a sign of insemination, of the seminal creative power of Hiro, a modern Pygmalion/Henry Higgins creating his Galatea/Eliza Doolittle.[20] But the seminal power is not just. Hiro's; it is the generative power of technology. Pygmalion himself has been morphed; he has become a computer program. Internationally, Benetton's ads, including its morphed racial transforms and its magazine *The United Colours of Benetton*, are the most famous. As Celia Lury put it, eschewing the distinction between cloth and skin, Benetton deals with the color of skin as a fashion palette (Lury 1994). Benetton produces a stunningly beautiful, young, stylish panhumanity composed by mix-and-match techniques. Diversity, like DNA, is the code of codes. Race, in Sarah Franklin's words, becomes a fashion accessory (Franklin 1994).

Pop star Michael Jackson brings this last point to its highest perfection. Spanning the range of chosen and imposed bodily "technologies" from cosmetic surgery, genetic skin disease, erotic performance in "private" and "public" life, clothing, costume, music videos—and mortal aging in spite of it all—Jackson's morphing practices have reshaped him by race, sexuality, gender, species, and generation. In the music video "Black and White," Jackson racially morphed himself by computer. In "real life," while a skin disease blanched his skin, he altered his facial features through cosmetic surgery, which produced race, generation, and gender effects. His childlike persona and his alleged transgressive relations with young boys morphed him into a permanent, if not altogether popular or safe, Peter Pan figure. His performance in Walt Disney's Epcot Center in the 3-D, 15-minute science-fiction production of *Captain E/O* caps the picture. As Ramona Fernandez put it, Jackson "tropes his body constantly. . . . In [*Captain E/O*] Jackson is both Mickey and a postmodern Peter Pan accompanied by bodies created by Lucas. . . . His transmuting body enacts and re-enacts the multiple problematics of race, generation, and gender" (1995:245). Analyzing Jackson's transmogrifications of himself and others through computer video technology into Cleopatra, ghoul, panther, machine, and superhero, Fernandez locates Jackson's socially significant shape-shifting within the traditions of African American tricksterism. The difference between human and machine, as well as the differences among species, are all fair game for Jackson's anti-origin narratives. As biologist Scott Gilbert writes, "If one wanted to find the intermediate morph of race, gender, and class, Michael Jackson may well be it. This makes him a science fiction 'representative' of humanity: and this is exactly how he depicts himself in *Captain E/O*."[21] This is humanity according to Epcot, where a potent trickster slipped into the monument to a clean and healthy America.[22]

Beginning unambiguously as an African American boy with striking talent, Jackson became neither black nor white, male nor female, man nor woman, old nor young, human nor animal, historical person nor mythological figure, homosexual nor heterosexual. These shape changes were effected through his art, the medical and computer technology of his culture, and the quirks of his body. Surely not even his brief marriage, least of all to Elvis Presley's daughter, could save him from the oxymoronically ineradicable stigmata of morphing. Science and fiction implode with special force in Jackson's iconic body, which is a national treasure of the first order. Jackson, however, is a much less safe representative for rebirthing the nation than the smoothly homogenized SimEve of *Time* magazine.

Not limited to specialists working for transnational corporations, weekly news magazines, official encyclopedias, or world-class entertainers, morphing is a participant reproductive sport too. In Las Vegas, in the Luxor, at the entrance to the gambling casino's reconstructed tomb of the eighteenth-dynasty Egyptian king Tutankhamen, there is a morphing machine that looks like the ordinary photomats in which one can get a quick snapshot. For five dollars a picture, one can enter the box, select the "gene machine" option, indicate whether one will be reproducing with a live partner or with a video model (human or animal), and then make further choices to determine the race and sex of the resulting child. The morphing machine is not choosy about the biological sex of the parent material. The racial menu for the child is African American, Hispanic, Asian, and Caucasian. Only if one chooses Caucasian are there any further choices, not an unfamiliar belief, but the choices are limited to hair and eye color. Then the machine photographs the parents-to-be, digitally combines them, and shoots out a child of the desired specifications. The child comes out at various ages, from toddler to adolescent. The gene machine is just another way of playing the combinations in Las Vegas at the end of the millennium.[23]

All this is surely not the naturalized typologies of Teddy Bear Patriarchy's early-twentieth-century racial discourse. Nor, in these popular cultural examples, including *Time*'s SimEve, are we subjected to PreMed's version of racial-sexual crossing. So why do I feel so uncomfortable? Shouldn't I be happy that the patently constructed nature of racial and gender categories is so obvious? In the face of resurgent racial hatred all around, what's wrong with a little obvious ideology for butterbrickle multiculturalism? Do we always have to order rocky road? Am I just having a dyspeptic attack of political correctness inevitably brought on by indulging in the pleasures of high-technology commodification within multinational capitalism? Why shouldn't the United Nations' *Family of Man* be morphed into the New World Order's *United Colours of Benetton*? Certainly the photography has advanced, and the human family seems naturally to be the story of the progress of technology.

To address the discomfort, let us look more closely at the *Time* special issue on immigration. In the note from the managing editor on page 2, we learn that *Time* imaging specialist Kin Wah Lam created the matrix of progeny in Figure 3 out of photographs of seven male and seven female models, each assigned to a racial-ethnic category. The top (female) and side (male) photos were electronically "mated" to produce the cybergenetic offspring. Each figure is a pleasant-faced but undramatic nude bust, a "natural" man or woman, enhanced modestly by the understated makeup and minimal hairstyling. All the figures are young adults, and all the unions are chastely heterosexual, although presumably the computer could do a bit better than the technology of eggs and sperm on that score. In their defense, the editors' purpose was "to dramatize the impact of multiethnic marriage, which has increased dramatically in the U.S. during the latest wave of immigration." Still, the trope of reproductive heterosexual marriage is as firmly ensconced here as in the worlds of *The Fossil Footprint Makers of Laetoli* or *The Family of Man*. The mixing of immigration could be dramatized by many other practices. The sense of utter homogeneity that emanates from *Time*'s matrix of diversity is numbing. The blacks are not very black; the blonds are not very blond; the range of skin color would require the best chromatography to distinguish one promising golden hue from another. These figures of the new humanity look like I imagine a catalog of replicants for sale off-world in *Blade Runner* might look—young, beautiful, talented, diverse, and programmed to fulfill the buyer's wishes and then self-destruct. Unlike the terrible white-supremacist scenes of *Birth of a Nation* in 1915, nothing about race and ethnicity in *Time's* "Rebirthing of a Nation" speaks about racial domination, guilt, and hatred. Nothing here is scary, so why am I trembling?

As Claudia Castañeda put it in her argument about "morphing the global U.S. family," "the racism here does not consist in the establishment of a hierarchy for domination based on biologized or even culturized racial difference. Its violence consists in the evacuation of histories of domination and resistance (and of all those events and ways of living that cannot be captured in those two terms) *through* technological (but still decidedly heterosexual) reproduction" (Castañeda 1994).[24] The denials and evasions in this liberal, antiracist, technophilic exercise are at least as thick as they are in the PreMed ad. All the bloody history caught by the ugly word *miscegenation* is missing in the sanitized term *morphing*. Multiculturalism and racial mixing in *Time* magazine are less achievements against the odds of so much pain than a recipe for being innocently raptured out of mundane into redeemed time. It is the resolute absence of history, of the fleshy body that bleeds, that scares me. It is the reconfirmation of the Sacred Image of the Same, once again under the sign of difference, that threatens national rebirth. I want something much messier, more dangerous, thicker, and more satisfying from the hope for multiculturalism. To get that kind of national reproductive health delivery is going to take addressing past and present sexualized racial power, privilege, exclusion, and exploitation. I suspect the nation will have to swallow the castor oil of sober accountability about such racialized sex before morphing looks like much fun to most of its citizens.[25]

Alongside a photo of the imaging specialist, labeled with a classically orientalist caption, "Lam creates a mysterious image," *Time*'s managing editor tells us still more about the cybergenesis of the woman on the cover: "A combination of the racial and ethnic features of the women used to produce the chart, she is: 15% Anglo-Saxon, 17.5% Middle Eastern, 17.5% African, 7.5% Asian, 35% Southern European and 7.5% Hispanic. Little did we know what we had wrought. As onlookers watched the image of our new Eve begin to appear on the computer screen, several staff members fell in love. Said one: 'It really breaks my heart that she doesn't exist.' We sympathize with our own lovelorn colleagues, but even technology has its limits. This is a love that must forever remain unrequited."

Themes running throughout the essay implode in this unlikely black hole. Early-century racialized ethnic categories reappear as entries in an electronic database for a truly odd statistical population analysis. A virtual woman is the result, fathered like Galatea, Pygmalion's creature, with which he fell in love. The curious erotics of single-parent, masculine, technophilic reproduction cannot be missed. SimEve is like Zeus's Athena, child only of the seminal mind—of man and of a computer program. The law of the nation, like that laid down by Athena for Athens in the Orestian trilogy, will be the Law of the Father. The Furies in cyberspace will not be pleased. In the narrative of romantic love, SimEve forever excites a desire that cannot be fulfilled. This is precisely the myth infusing dreams of technological transcendence of the body. In these odd, but conventional, technoscientific erotics, the actual limits of technology only spur the desire to love that which cannot and does not exist. SimEve is the new universal human, mother of the new race, figure of the nation; and she is a computer-generated composite, like the human genome itself. She is the second-and third-order offspring of the ramifying code of codes. She ensures the difference of no difference in the human family.

PostScript™

Throughout this chapter, racial discourse has persistently pivoted on sexual hygiene, and the therapeutic scene has been the theater of nature in the city of science. I am sick to death of bonding through kinship and "the family," and I long for models of solidarity and human unity and difference rooted in friendship, work, partially shared purposes, intractable collective pain, inescapable mortality, and persistent hope. It is time to theorize an "unfamiliar" unconscious, a different primal scene, where everything does not stem from the dramas of identity and reproduction. Ties through blood—including blood recast in the coin of genes and information—have been bloody enough already. I believe that there will be no racial or sexual peace, no livable nature, until we learn to produce humanity through something more and less than kinship. I think I am on the side of the vampires, or at least some of them. But, then, since when does one get to choose which vampire will trouble one's dreams?

Notes

1 The special pull-out section of this *Science* magazine annual issue on the genome was dedicated to databases. See also Nowak 1993:1967.
2 Making life into a force of production and reorganizing biology for corporate convencience can be followed in Yoxen 1981; Wright 1986; and Shiva 1993.
3 The incisive critique of human sociobiology is Kitcher 1987. On unit-of-selection debates, see Brandon and Burian 1984. Defying classification as technical or popular, Dawkins 1976 and 1982 are the best expositions of the logic of the fierce competitive struggle to stay in the game of life, relying on strategies of flexible accumulation that strangely seem so basic to postmodern capitalism as well. For the theory of flexible accumulation in political economy, see Harvey 1989. For multilevel feminist working of the theme of flexibility in the American biomedical body, see Martin 1994.
4 The idea of nature and culture "enterprised up" is borrowed from Marilyn Strathern 1992, a treatment of assisted conception and English kinship in the period of British Thatcherism.
5 In eugenics thinking, the good of the "race" is the central ideological value. The collective aspect is hard to overstress. In 1990s genetic biomedical discourse, the "race"—either humanity as a whole or a particular racial category such as "white people"—plays little or no role, but individual reproductive investment decisions and individual genetic health are central.
6 A good place to start reading on the subject is Kevles and Hood 1992. Flower and

Heath (1993) show how the semiotic-material definition of the human species in the world's genetic databases works through the multiple and heterogeneous processes that construct a reference sequence, or "consensus DNA sequence," as "the" human genome.

7 On "agency" in Internet habitats, see Waldrop 1994.

8 The supplement to the *Oxford English Dictionary* puts the first uses of the term *genom* (sic) in the 1930s, but the word did not then mean a database structure. That sense emerged from the consolidation of genetics as an information science, and especially since the 1970s.

9 I am indebted to an unpublished manuscript by the UCSC anthropology graduate student Cori Hayden (1994a). See Cavalli-Sforza et al. 1991; RAFI 1993:13; Spiwak 1993; and RAFI 1994.

10 I have been instructed by Giovanna DiChiro (1995a and 1995b) on what and who will count as science and as scientists. I draw also on Tsing 1993a and Cussins 1994. All three analysts trouble inherited categories of body and technology, nature and culture, wilderness and city, center and margin—all of which are part of producing the ideological distinction between modern and traditional that makes it seem odd for indigenous peoples to be savvy users and producers of genome discourse. For excellent analysis of problematic discourses of racial difference in ecofeminism, partly rooted in continuing separation of nature and culture and turning to "native" women as resources against the violations of industrial culture, see Sturgeon forthcoming.

11 See Star and Griesemer 1989 for development of the concept of boundary objects.

12 The scramble for the control of "biodiversity," itself quite a recent discursive object, is complex, global, and fraught with consequences for ways of life. Hayden 1994b discusses the 1991 "biodiversity prospecting" agreement between INBio, a Costa Rican nonprofit environmental institute, and Merck, Sharpe and Dohme, the world's biggest pharmaceutical firm. The agreement is a controversial effort to control biopiracy and turn biodiversity resources in "generich" developing countries to their advantage. Biodiversity prospecting arrangements, the Human Genome Diversity Project, debt-for-nature swaps, the Biodiversity Convention, and GATT are just a few examples of the emerging institutional structure shaping human relations to nature in a world where the relations of technoscience to wealth and well-being have never been tighter. See World Resources Institute et al. 1993; Juma 1989; Shiva 1993.

13 See Tsing 1993b for a subtle ethnographic treatment of the complexities of what counts as marginal/central and local/global in an area of Indonesia that is also at the heart of environmental controversies.

14 Reprinted with permission of Du Pont NEN Products. On May 19, 1995, Du Pont announced its intent to divest its Medical Products businesses. The rmer Du Pont NEN Products business will become NEN Life Science Products.

15 According to the *Oxford English Dictionary*, the term *miscegenation* was coined in the United States in 1864.

16 A $3-million National Pregnancy and Health Survey of 2,613 women who gave birth at 52 hospitals around the nation in 1992 suggests how many and which U.S. pregnant women actually use substances that could harm the fetus (and the bottom line for an HMO). Conducted for the National Institute on Drug Abuse and released in September 1994, the study concludes that more than 5 percent of the four million U.S. women who gave birth in 1992 used illegal drugs, while about 20 percent used cigarettes and/or alcohol. Smokers and drinkers were more likely to use illegal drugs than were ethanol and nicotine abstainers. White women were more likely to drink or smoke during pregnancy than women of color (23 percent

of white women drank, compared to 16 percent African American and 9 percent Hispanic; 24 percent of white women smoked, compared to 20 percent Black and 6 percent Hispanic). The racial categories here are crude and partial, but they still have limited utility. Poor, less-educated, unemployed, and unmarried women were more likely to use illegal drugs than more privileged women. About 11 percent of pregnant African American women used such drugs, compared to 5 percent of white and 4 percent of Hispanic mothers-to-be. That still means that more than half of the 221,000 pregnant women who used illegal drugs were white, 75,000 were Black, and 28,000 Hispanic. Alcohol and tobacco can harm a developing fetus as much or more than illegal drugs but with less social and financial stigma. Overall, about 820,000 babies were born to smokers and 757,000 to imbibers. The same baby can show up in all the user categories. The study showed that most women tried to avoid illegal drugs, alcohol, and smoking during pregnancy, but few who used these powerful substances succeeded entirely. See Connell 1994:A7. The need for supportive, nonpunitive treatment for women trying to have a healthy pregnancy could hardly be clearer. Along with readily available, prowoman, substance-treatment programs for those with any of these addictions, raising the incomes and improving the educations of women would likely be the most successful public health measures. Such measures would far outstrip the benefit to child and maternal health from intensive neonatal care units in high-tech hospitals, not to mention the dubious health results from criminalizing users. There is an unholy alliance between medicine as a system and millions of pregnant women in the United States, and it is reflected in the incomes of physicians compared to the incomes of at-risk mothers-to-be. The direction of flow of precious bodily fluids is the reverse of that suggested by the gleaming tooth and gold wedding band of the PreMed ad.

17 Selling in early 1994 for $239 for Macintoshes and $169 for Windows-using machines, Morph was widely used by scientists, teachers, special-effects designers for Hollywood movies, businesspeople making presentations, and law enforcement personnel, for example, for aging missing children. A competitor in the market, PhotoMorph, came with graphics for practicing—"women turning into men, a girl turning into an English sheepdog, a frog "women turning into men, a girl turning into an English sheepdog, a frog turning into a chicken" (Finley 1994:F1–2). Finley illustrated his article with a series of morphed transformations between the competing personal computer giants, Apple Computer confounder Steve Jobs and Microsoft founder Bill Gates. Mergers in the New World Order can be effected by many means. Needless to say, anyone still believing in the documentary status of photographs had better not get a copy of Morph, go to the movies, or look at the missing children on milk cartons.

18 Morphed photograph by Nancy Burson in Jones, Martin, and Pilbeam 1992. Thanks to Ramona Fernandez of the University of California at Santa Cruz for sending me this example.

19 Thanks to Giovanna DiChiro, University of California at Santa Cruz, for the tip on this image and for Hiro's comments from the *Today Show* of August 17, 1994.

20 The computer chip "impresses" its form on the morphed woman; the chip "informs" its electronic progeny in enduring Aristotelian doctrines of masculine self-reproduction that have "impressed" thinkers in the West for many centuries. The perfecting of the copy of the father in the child could be marred by the lack of transparency in the medium of the mother. Mutations on this theme proliferate in cyberspace, as in many other technoscientific wombs at the end of the Second Christian Millennium. For a discussion, which informs my chapter, of doctrines of impression, reproduction, and sanctity in medieval women saints, see Park 1995.

21 Scott Gilbert, personal e-mail communication, September 26, 1995, in response to a previous version of "Universal Donors." Thanks to Gilbert for insisting that I include "Black and White."

22 Fernandez (1995b) emphasizes the trickster theme in her essay on traveling through Disney's many worlds, reading with the mixed cultural literacies required in the turn-of-the-century United States.

23 Thanks to Rosi Braidotti and Anneke Smelik, new parents of two lovely morphed offspring, for this description of what they found possible in America in 1995. These sober European feminist theorists testified that they bonded instantly with their cyberchildren when they saw the compelling photographs of offspring so like and unlike themselves. The emotions were quite potent, even if the children were a little ethereal. I think there is potential here for population-reducing ways of having one's own children after all, in as great a number as one's willingness to put $5 in the machine will allow.

24 Castañeda's and my interpretations of the figures in this issue of *Time* evolved together in Conversation, her hearing of my talk for a History of Consciousness colloquium Feb. 9, 1994, and my reading of her paper. I also draw on undergraduate students' readings of these images in a final exam in my fall 1993 course Science and Politics.

25 Meanwhile, fitting the analysis found in Emily Martin's *Flexible Bodies*, U.S. corporations attempt to capitalize on a particular version of multiculturalism. For an unembarrassed argument, see J. P. Fernandez 1993. See also Kaufman 1993.

References

Bereano, Philip. 1995. "Broad Coalition Challenges Patents on Life." Press release, June 6.

Brandon, Robert N., and Richard M. Burian, eds. 1984. *Genes, Organisms, Populations: Controversies over the Units of Selection*. Cambridge: MIT Press.

Castañeda, Claudia. 1994. "Transnational Adoption as U.S. Racist Complicity?" Paper read at American Ethnological Society Meeting, April.

Cavalli-Sforza, L. L., A. C. Wilson, C. R. Cantor, R. M. Cook-Deegan, and M. C. King. 1991. "Call for a Worldwide Survey of Human Genetic Diversity: A Vanishing Opportunity for the Human Genome Project." *Genomics* 11:490–91.

Connell, Christopher. 1994. "Pregnancy Study Shows Bad Habits." *The Santa Rosa Press Democrat*, September 13, A7.

Courteau, Jacqueline. 1991. "Genome Databases: Special Pull-out Section." *Science* (Genome Issue, Maps and Databases) 254:201–07.

Cussins, Charis. 1994. "Ontological Choreography: Agency for Women Patients in an Infertility Clinic." Unpublished manuscript. Science Studies Program, University of California at San Diego. Under review for *Social Studies of Science*.

Dawkins, Richard 1976, 1989. *The Self Gene*. New York: Oxford Unviersity Press.

——. 1982. *The Extended Phenotype: The Gene as a Unit of Selection*. London: Oxford University Press.

DiChiro, Giovanna. 1995a. "Local Actions, Global Visions: Women Transforming Science, Environment and Health in the U.S. and India." Ph.D. diss., History of Consciousness Board, University of California at Santa Cruz.

——. 1995b. "Nature as Community: The Convergence of Environment and Social Justice." In *Uncommon Ground: Towards Reinventing Nature*, edited by W. Cronon. New York: Norton, 298–320.

Erickson, Deborah. 1992. "Hacking the Genome." *Scientific American* (April):128–37.

Fernandez, John P. 1993. *The Diversity Advantage: How American Business Can Out-Perform Japanese and European Companies in the Global Marketplace*. New York: Lexington Books, Macmillan.

Fernandez, Ramona, 1995. "Pachuco Mickey." In *From Mouse to Mermaid: The Politics of Film, Gender, and Culture*, edited by E. Bell, L. Haas, and L. Sells. Bloomington: Indiana University Press, 236–53.

Finley, Michael. 1994. "The Electronic Alchemist." *San Jose Mercury News*, March 6, F1–2.

Flower, Michael, and Deborah Heath. 1993. "Anatomo-politics: Mapping the Human Genome Project." *Culture, Medicine and Psychiatry* 17:27–41.

Franklin, Sarah. 1994. "Comments on Paper by Celia Lury." Center for Cultural Studies, University of California at Santa Cruz.

Hayden, Cori. 1994a. "The Genome Goes Multicultural." Unpublished manuscript, Anthropology Board, University of California at Santa Cruz.

——. 1994b."Purity, Property and Preservation." Paper read at American Anthropological Association Meetings, November 27–December 4, at Atlanta, GA.

Jones, Steve, Robert Martin, and David Pilbeam, eds. 1992. *The Cambridge Encyclopedia of Human Evolution*. New York: Cambridge University Press.

Juma, Calestous. 1989. *The Gene Hunters: Biotechnology and the Scramble for Seeds*. Princeton: Princeton University Press.

Kahn, Patricia. 1994. "Genetic Diversity Project Tries Again." *Science* 266 (November 4):720–22.

Kauffman, L. A. 1993. "The Diversity Game: Corporate America Toys with Identity Politics." *Village Voice* (August 31):29–33.

Kevles, Daniel J., and Leroy Hood, eds. 1992. *The Code of Codes: Scientific and Social Issues in the Human Genome Project*. Cambridge: Harvard University Press.

Kitcher, Philip. 1987. *Vaulting Ambition: Sociobiology and the Quest for Human Nature*. Cambridge: MIT Press.

Lury, Celia. 1994. "United Colors of Diversity: Benetton's Advertising Campaign and the New Universalism of Global Culture, a Feminist Analysis." Paper delivered at the Center for Cultural Studies, University of California at Santa Cruz.

Martin, Emily, 1994. *Flexible Bodies: Tracking Immunity in American Culture from the Days of Polio to the Age of AIDS*. Boston: Beacon Press.

Nowak, Rachel. 1993. "Draft Genome Map Debuts on Internet." *Science* 262 (December 24): 1967.

Park, Katherine. 1995. "Impressed Images: Reproducing Wonders." Paper read at Histories of Science/Histories of Art, November 3–5, at Harvard and Boston Universities.

RAFI (Rural Advancement Foundation International). 1993. "Patenting Indigenous Peoples." *Earth Island Journal* (Fall):13.

——. 1994. "Following Protest, Claim Withdrawn on Guaymi Indian Cell Line." *GeneWatch: A Bulletin of the Council for Responsible Genetics* 9 (3–4): 6–7.

Shiva, Vandana. 1993. *Monocultures of the Mind: Perspectives on Biodivesity and Biotechnology*. London: Zed Books.

Spiwak, Daniela. 1993. "Gene, Genie, and Science's Thirst for Information with Indigenous Blood." *Abya Yala News* 7 (3–4): 12–14.

Star, Susan Leigh, and James R. Griesemer. 1989. "Institutional Ecology, 'Translations,' and Boundary Objects: Amateurs and Professionals in Berkeley's Museum of Vertebrate Zoology, 1907–39. *Social Studies of Science* 19:387–420.

Tsing, Anna Lowenhaupt. 1993a. "Forest Collisions: The Construction of Nature in Indonesian Rainforest Politics." Unpublished manuscript.

——. 1993b. *In the Realm of the Diamond Queen: Marginality in an Out-of-the-Way Place*. Princeton: Princeton University Press.

Waldrop, M. Mitchell. 1994. "Software Agents Prepare to Sift the Riches of Cyberspace." *Science* 265 (Aug 12):882–83.

World Resources Institute, INBio (Costa Rica), Rainforest Alliance (U.S), and the African Centre for Technology Studies (Kenya), eds. 1993. *Biodiversity Prospecting: Using Generic Resources for Sustainable Development*. A contribution to the WRI/IUCN/UNEP, *Global Biodiversity Strategy*, WRI/IUCN/UNEP, May 1993.

Wright, Susan. 1986. "Recombinant DNA Technology and Its Social Transformation." *Osiris* 2 (second series):303–60.

Yoxen, Edward. 1981. "Life as a Productive Force: Capitalizing the Science and Technology of Molecular Biology." In *Science, Technology and the Labour Process*, edited by L. Levidow and B. Young. London: CSE Books, 66–122.